Ferroelectric Materials

Ferroelectric Materials

Editor:

Mulmudi Hemant Kumar, Ph.D.

ArclerPress

www.arclerpress.com

Ferroelectric Materials

Editor: Mulmudi Hemant Kumar, Ph.D.

708 3rd Avenue, 6th Floor
New York
NY 10017
United States of America
www.arclerpress.com

ISBN: 978-1-68094-361-0

Library of Congress Control Number: 2016913595

About the Editor

Mulmudi Hemant Kumar, Ph.D.

Hemant obtained his PhD from Nanyang Technological University, Singapore in 2014. His interests span from material synthesis via solution processed methods to it application in energy harvesting devices such as photovoltaics and water splitting. He is currently working as a research fellow at the Australian national university on silicon/perovskite solar cells.

Contents

List of Contributors

E.B. Araújo,
Universidade Federal de São Carlos Departamento de Física C.P. 676, 13565-670 São Carlos - SP, Brazil

J.A. Eiras
Universidade Federal de São Carlos Departamento de Física C.P. 676, 13565-670 São Carlos - SP, Brazil

D. J. Goossens
Research School of Chemistry, Australian National University, Canberra ACT 0200, Australia

José Eduardo García
Department of Physics, Universitat Politècnica de Catalunya—BarcelonaTech., 08034 Barcelona, Spain;

Xiaobing Li
Key Laboratory of Inorganic Functional Material and Device, Shanghai Institute of Ceramics, Chinese Academy of Sciences, 215 Chengbei Road Jiading, Shanghai 201800, China

Chao Chen
Key Laboratory of Inorganic Functional Material and Device, Shanghai Institute of Ceramics, Chinese Academy of Sciences, 215 Chengbei Road Jiading, Shanghai 201800, China

Hao Deng
Key Laboratory of Inorganic Functional Material and Device, Shanghai Institute of Ceramics, Chinese Academy of Sciences, 215 Chengbei Road Jiading, Shanghai 201800, China

Haiwu Zhang
Key Laboratory of Inorganic Functional Material and Device, Shanghai Institute of Ceramics, Chinese Academy of Sciences, 215 Chengbei Road Jiading, Shanghai 201800, China

Di Lin
Key Laboratory of Inorganic Functional Material and Device, Shanghai Institute of Ceramics, Chinese Academy of Sciences, 215 Chengbei Road Jiading, Shanghai 201800, China

Xiangyong Zhao
Key Laboratory of Inorganic Functional Material and Device, Shanghai Institute of Ceramics, Chinese Academy of Sciences, 215 Chengbei Road Jiading, Shanghai 201800, China

Haosu Luo
Key Laboratory of Inorganic Functional Material and Device, Shanghai Institute of Ceramics, Chinese Academy of Sciences, 215 Chengbei Road Jiading, Shanghai 201800, China

Rashmi Gupta
Crystal Growth & Materials Research Laboratory, Department of Physics and Electronics, University of Jammu, Jammu 180006, India

Seema Verma
Crystal Growth & Materials Research Laboratory, Department of Physics and Electronics, University of Jammu, Jammu 180006, India

Vishal Singh
Crystal Growth & Materials Research Laboratory, Department of Physics and Electronics, University of Jammu, Jammu 180006, India

K. K. Bamzai
Crystal Growth & Materials Research Laboratory, Department of Physics and Electronics, University of Jammu, Jammu 180006, India

S. K. Barbar
Materials Science Laboratory, Department of Physics, J.N. Vyas University, Jodhpur, Rajasthan 342001, India

M. Roy
Materials Science Laboratory, Department of Physics, M.L. Sukhadia University, Udaipur, Rajasthan 313002, India

Michael Lorenz
Institut für Experimentelle Physik II, Fakultät für Physik und Geowissenschaften, Universität Leipzig, Linnéstraβe 5, 04103 Leipzig, Germany

Holger Hochmuth
Institut für Experimentelle Physik II, Fakultät für Physik und Geowissenschaften, Universität Leipzig, Linnéstraβe 5, 04103 Leipzig, Germany

Christoph Grüner
Institut für Experimentelle Physik II, Fakultät für Physik und Geowissenschaften, Universität Leipzig, Linnéstraβe 5, 04103 Leipzig, Germany

Helena Hilmer
Institut für Experimentelle Physik II, Fakultät für Physik und Geowissenschaften, Universität Leipzig, Linnéstraβe 5, 04103 Leipzig, Germany

Alexander Lajn
Institut für Experimentelle Physik II, Fakultät für Physik und Geowissenschaften, Universität Leipzig, Linnéstraβe 5, 04103 Leipzig, Germany

Daniel Spemann
Institut für Experimentelle Physik II, Fakultät für Physik und Geowissenschaften, Universität Leipzig, Linnéstraβe 5, 04103 Leipzig, Germany

Matthias Brandt
Institut für Experimentelle Physik II, Fakultät für Physik und Geowissenschaften, Universität Leipzig, Linnéstraβe 5, 04103 Leipzig, Germany

Jan Zippel
Institut für Experimentelle Physik II, Fakultät für Physik und Geowissenschaften, Universität Leipzig, Linnéstraβe 5, 04103 Leipzig, Germany

Rüdiger Schmidt-Grund
Institut für Experimentelle Physik II, Fakultät für Physik und Geowissenschaften, Universität Leipzig, Linnéstraβe 5, 04103 Leipzig, Germany

Holger von Wenckstern
Institut für Experimentelle Physik II, Fakultät für Physik und Geowissenschaften, Universität Leipzig, Linnéstraβe 5, 04103 Leipzig, Germany

Marius Grundmann
Institut für Experimentelle Physik II, Fakultät für Physik und Geowissenschaften, Universität Leipzig, Linnéstraβe 5, 04103 Leipzig, Germany

Fiodoras ANISIMOVAS
Center for Physical Sciences and Technology, A. Goštauto 11, LT-01108 Vilnius, Lithuania

Steponas AŠMONTAS
Center for Physical Sciences and Technology, A. Goštauto 11, LT-01108 Vilnius, Lithuania

Oleg KIPRIJANOVIČ
Center for Physical Sciences and Technology, A. Goštauto 11, LT-01108 Vilnius, Lithuania

Andrius MANEIKIS
Center for Physical Sciences and Technology, A. Goštauto 11, LT-01108 Vilnius, Lithuania

Bonifacas VENGALIS
Center for Physical Sciences and Technology, A. Goštauto 11, LT-01108 Vilnius, Lithuania

Ingo Dierking
School of Physics and Astronomy, University of Manchester, Schuster Building, Oxford Road, Manchester M13 9PL, UK;

Manfang Mai
College of Materials Science and Engineering, Shenzhen University, Shenzhen 518060, China
Key Laboratory of Optoelectronic Devices and Systems of Ministry of Education and Guangdong Province, College of Optoelectronic Engineering, Shenzhen University, Shenzhen 518060, China

Shanming Ke
College of Materials Science and Engineering, Shenzhen University, Shenzhen 518060, China

Peng Lin
College of Materials Science and Engineering, Shenzhen University, Shenzhen 518060, China

Xierong Zeng
College of Materials Science and Engineering, Shenzhen University, Shenzhen 518060, China
Key Laboratory of Optoelectronic Devices and Systems of Ministry of Education and Guangdong Province, College of Optoelectronic Engineering, Shenzhen University, Shenzhen 518060, China

Meng H. Lean and
QEDone LLC, Santa Clara, CA 95054, USA

Wei-Ping L. Chu
QEDone LLC, Santa Clara, CA 95054, USA

Kriti Ranjan Sahu
Physics Department, Egra SSB College, Egra, Purba Medinipur, West Bengal 721429, India

Udayan De
Kendriya Vihar, C-4/60, V.I.P. Road, Kolkata 700052, India

Dhar Liquid Crystal Research Lab, Physics Department, University of Lucknow, Lucknow 226007, India
mendra Pratap Singh

Swadesh Kumar Gupta
Liquid Crystal Research Lab, Physics Department, University of Lucknow, Lucknow 226007, India

Rajiv Manohar
Liquid Crystal Research Lab, Physics Department, University of Lucknow, Lucknow 226007, India

N. Kartheeswari
Department of Physics, Karpagam University, Tamilnadu, Coimbatore 641021, India

K. Viswanathan
Department of Physics, Karpagam University, Tamilnadu, Coimbatore 641021, India

Lan Chen
School of Physics and Engineering, Sun Yat-sen University, Guangzhou 510275, China

Di Wu
School of Physics and Engineering, Sun Yat-sen University, Guangzhou 510275, China

Xiao-Min Xiong
School of Physics and Engineering, Sun Yat-sen University, Guangzhou 510275, China

Hai-Yan Wang
Guangzhou Institute of Measurement and Testing Technology, Guangzhou, Guangdong 510663, China

Jin-Xiu Zhang
School of Physics and Engineering, Sun Yat-sen University, Guangzhou 510275, China

Nail Migranov
Department of General and Theoretical Physics, Bashkir State Pedagogical University, Okt. Revolutsii Street 3A, Ufa 450000, Russia

Aleksey Kudreyko
Department of Physics, Ufa State Petroleum Technological University, Kosmonavtov Street 1, Ufa 450062, Russia

Denis Kondratyev
Academy of Sciences of Bashkortostan, Kirova Street 15, Ufa 450008, Russia

Naïma Aïssa
Centre de Mathématiques Appliquées (CMAP), UMR 7641 CNRS, Ecole Polytechnique, Palaiseau Cedex 91128, France
Faculté de Mathématiques, Université des Sciences et de la Technologie Houari Boumediene (USTHB), BP 32, Bab Ezzouar, Alger 16123, Algeria

Kamel Hamdache
Centre de Mathématiques Appliquées (CMAP), UMR 7641 CNRS, Ecole Polytechnique, Palaiseau Cedex 91128, France

Haribabu Palneedi
Department of Materials Science and Engineering, Korea Advanced Institute of Science and Technology (KAIST), Daejeon 34141, Korea
Functional Ceramics Group, Korea Institute of Materials Science (KIMS), Changwon 51508, Korea

Venkateswarlu Annapureddy
Functional Ceramics Group, Korea Institute of Materials Science (KIMS), Changwon 51508, Korea

Shashank Priya

Jungho Ryu
Bio-inspired Materials and Devices Laboratory (BMDL), Center for Energy Harvesting Materials and Systems (CEHMS), Virginia Tech, Blacksburg, VA 24061, USA

Luis E. Fuentes-Cobas
Centro de Investigación en Materiales Avanzados, Miguel de Cervantes 120, Compejo Industrial Chihuahua, Chihuahua 31136, Mexico

María E. Montero-Cabrera
Centro de Investigación en Materiales Avanzados, Miguel de Cervantes 120, Compejo Industrial Chihuahua, Chihuahua 31136, Mexico

Lorena Pardo
Instituto de Ciencia de Materiales de Madrid, Consejo Superior de Investigaciones Científicas, Cantoblanco, Madrid E-28049, Spain

Luis Fuentes-Montero
Diamond Light Source Ltd., Beamline I24, Diamond House, Harwell Science and Innovation Campus, Didcot, Oxfordshire OX11 0DE, UK

Aditya Chauhan
School of Engineering, Indian Institute of Technology Mandi, Mandi 175 001, India

Satyanarayan Patel
School of Engineering, Indian Institute of Technology Mandi, Mandi 175 001, India

Rahul Vaish
School of Engineering, Indian Institute of Technology Mandi, Mandi 175 001, India

Chris R. Bowen
Department of Mechanical Engineering, University of Bath, Bath BA2 7AY, UK

Preface

Ferroelectric materials have applications ranging from memories, capacitors, sensors and actuators, optical switches and infrared detectors. These materials can exist as single crystals, thin polycrystalline films or organic polymers. This book is a combination of articles/review papers based on oxides, polymers and composite ferroelectric materials.

Chapters 1-7 describe various ways of synthesizing lead containing and lead free ferroelectric oxides and single crystals. Their electrical and optical properties are discussed in detail. Chapters 8-10 focus on the emission properties of ferroelectric organics and their application in liquid crystal displays which is an important display technology in today's world. Chapters 11-15 describe various models by which charge transport in ferroelectric materials can be investigated. Impedance spectroscopy which is a commonly employed technique to understand charge transport in metal oxides is also introduced in chapter 12. Chapters 16-20 include miscellaneous topics of ferroelectric materials such as ferroelectric liquid crystals, multiferroic composite materials, interaction of ferroelectric materials with synchrotron light and application of ferroelectric ceramics in high energy density capacitors.

Editor
Mulmudi Hemant Kumar, Ph.D.

INTRODUCTION

Eerroelectricity, property of certain non-conducting crystals, or dielectrics, that exhibit spontaneous electric polarization (separation of the center of positive and negative electric charge, making one side of the crystal positive and the opposite side negative) that can be reversed in direction by the application of an appropriate electric field. Ferroelectricity is named by analogy with ferromagnetism, which occurs in such materials as iron. Iron atoms, being tiny magnets, spontaneously align themselves in clusters called ferromagnetic domains, which in turn can be oriented predominantly in a given direction by the application of an external magnetic field.

Ferroelectric materials—for example, barium titanate ($BaTiO_3$) and Rochelle salt—are composed of crystals in which the structural units are tiny electric dipoles; that is, in each unit the centers of positive charge and of negative charge are slightly separated. In some crystals these electric dipoles spontaneously line up in clusters called domains, and in ferroelectric crystals the domains can be oriented predominantly in one direction by a strong external electric field. Reversing the external field reverses the predominant orientation of the ferroelectric domains, though the switching to a new direction lags somewhat behind the change in the external electric field. This lag of electric polarization behind the applied electric field is ferroelectric hysteresis, named by analogy with ferromagnetic hysteresis.

Ferroelectricity ceases in a given material above a characteristic temperature, called its Curie temperature, because the heat agitates the dipoles sufficiently to overcome the forces that spontaneously align them.

Ferroelectric Materials

Ferroelectricity is a property of certain materials that have a spontaneous electric polarization that can be reversed by the application of an external electric field. The term is used in analogy to ferromagnetism, in which a material exhibits a permanent magnetic moment. Ferroelectricity is a property of certain materials that have a spontaneous electric polarization that can be reversed by the application of an external electric field. The term is used in analogy to ferromagnetism, in which a material exhibits a permanent magnetic moment. Ferromagnetism was already known when ferroelectricity was discovered in 1920 in Rochelle salt by Valasek. Thus, the prefix ferro, meaning iron, was used to describe the property despite the fact that most ferroelectric materials do not contain iron.

Polarization

When most materials are polarized, the polarization induced, P, is almost exactly proportional to the applied external electric field E; so the polarization is a linear function. This is called dielectric polarization (see figure). Some materials, known as paraelectric materials, show a more enhanced nonlinear polarization (see figure). The electric permittivity, corresponding to the slope of the polarization curve, is not constant as in dielectrics but is a function of the external electric field.

In addition to being nonlinear, ferroelectric materials demonstrate a spontaneous nonzero polarization (after entrainment, see figure) even when the applied field E is zero. The distinguishing feature of ferroelectrics is that the spontaneous polarization can be reversed by a suitably strong applied electric field in the opposite direction; the polarization is therefore dependent not only on the current electric field but also on its history, yielding a hysteresis loop. They are called ferroelectrics by analogy to ferromagnetic materials, which have spontaneous magnetization and exhibit similar hysteresis loops.

Typically, materials demonstrate ferroelectricity only below a certain phase transition temperature, called the Curie temperature, T_c, and are paraelectric above this temperature: the spontaneous polarization vanishes, and the ferroelectric crystal transforms into the paraelectric state. Many ferroelectrics lose their piezoelectric properties above T_c completely, because their paraelectric phase has centrosymmetric crystallographic structure.

Applications

The nonlinear nature of ferroelectric materials can be used to make capacitors with tunable capacitance. Typically, a ferroelectric capacitor simply consists of a pair of electrodes sandwiching a layer of ferroelectric material. The permittivity of ferroelectrics is not only tunable but commonly also very high in absolute value, especially when close to the phase transition temperature. Because of this, ferroelectric capacitors are small in physical size compared to dielectric (non-tunable) capacitors of similar capacitance.

The spontaneous polarization of ferroelectric materials implies a hysteresis effect which can be used as a memory function, and ferroelectric capacitors are indeed used to make ferroelectric RAM for computers and RFID cards. In these applications thin films of ferroelectric materials are typically used, as this allows the field required to switch the polarization to be achieved with a moderate voltage. However, when using thin films a great deal of attention needs to be paid to the interfaces, electrodes and sample quality for devices to work reliably.

Ferroelectric materials are required by symmetry considerations to be also piezoelectric and pyroelectric. The combined properties of memory, piezoelectricity, and pyroelectricity make ferroelectric capacitors very useful, e.g. for sensor applications. Ferroelectric capacitors are used in medical ultrasound machines (the capacitors generate and then listen for the ultrasound ping used to image the internal organs of a body), high quality infrared cameras (the infrared image is projected onto a two dimensional array of ferroelectric capacitors capable of detecting temperature differences as small as millionths of a degree Celsius), fire sensors, sonar, vibration sensors, and even fuel injectors on diesel engines.

Another idea of recent interest is the ferroelectric tunnel junction (FTJ) in which a contact made up by nanometer-thick ferroelectric film placed between metal electrodes. The thickness of the ferroelectric layer is small enough to allow tunneling of electrons. The piezoelectric and interface effects as well as the depolarization field may lead to a giant electro resistance (GER) switching effect. Yet another hot topic is multiferroics, where researchers are looking for ways to couple magnetic and ferroelectric ordering within a material or heterostructure; there are several recent reviews on this topic.

Ferroelectric effect

- The ferroelectric effect is an electrical phenomenon whereby certain crystals may exhibit a spontaneous dipole moment (which is called ferroelectric by analogy with ferromagnetic - exhibiting a permanent magnetic moment).
- A ferroelectric material is a material that exhibits, over some range of temperature, a spontaneous electric polarization that can be reversed or reoriented by application of an electric field [poling].
- A necessary criterion is the requirement of an ever-present spontaneous polarization, with the requirement of reversibility or reorientation of that spontaneous polarization being a sufficient criterion for a ferroelectric phase.
- An exclusion from the definition of ferroelectrics is those materials belonging to non-polar crystal classes at all temperatures, and in which a metastable polarization can be induced by an external electric field.

Applications of Ferromagnetic materials

In recent years, ferroelectric materials and thin films have attracted much attention and exhibited potential in many important applications such as

- dynamic random access memories (DRAMS),
- non-volatile ferroelectric random access memories micro-armours and
- infrared sensors.

At present, the ferroelectric materials suitable for these devices are Pb(Zr,Ti)O3 (PZT) systems, SrBi2Ta2O9 (SBT) systems, Bi4Ti3O12 (BIT) systems and BaTiO3 (BT) systems that are studied with a great deal of interest. In these ferroelectric materials, BaTi0.91 (Hf0.5, Zr0.5)0.09O3 (BTHZ-9), one of the BT systems, which has several advantages such as

- an extremely low coercive field,
- a high remnant polarization,
- better mechanical strength and
- small deviation in composition, could have a strong potential application for ferroelectric thin film devices.

In addition, the BTHZ systems, lead and/or bismuth-free material, present a great interest both for applications in the field of environmental protection and for fundamental studies. Therefore, the BTHZ system is expected to be one of the attractive materials suitable for ferroelectric thin film devices.

Pyroelectric Properties and Spontaneous Polarization

All ferroelectric materials are pyroelectric, however, not all pyroelectric materials are ferroelectric. Below a transition temperature called the Curie temperature ferroelectric and pyroelectric materials are polar and possess a spontaneous polarization or electric dipole moment. However, this polarity can be reoriented or reversed fully or in part through the application of an electric field with ferroelectric materials. Complete reversal of the spontaneous polarization is called "switching".

The non-polar phase encountered above the Curie Temperature is known as the paraelectric phase.

The direction of the spontaneous polarization conforms to the crystal symmetry of the material. While the reorientation of the spontaneous polarization is a result of atomic displacements.

The magnitude of the spontaneous polarization is greatest at temperatures well below the Curie temperature and approaches zero as the Curie temperature is neared.

Piezoelectric Properties

Since all pyroelectric materials are piezoelectric, this means ferroelectric materials are inherently piezoelectric. This means that in response to an applied mechanical load, the material will produce an electric charge proportional to the load. Similarly, the material will produce a mechanical deformation in response to an applied voltage.

Properties including the piezoelectric, dielectric and electrooptic co-efficient may vary by several orders of magnitude in the narrow temperature band around the Curie temperature. Especially when compared to other temperature ranges, the changes to these co-efficient is much more gradual.

The piezoelectric co-efficient is much greater in the region of the Curie temperature. Other properties such as dielectric strength and electrooptic properties also change more markedly in the region of the Curie temperature when compared to other temperature ranges.

Applications for Ferroelectric Materials

- Capacitors
- Non-volatile memory
- Piezoelectrics for ultrasound imaging and actuators
- Electro-optic materials for data storage applications
- Thermistors
- Switches known as transchargers or transpolarizers
- Oscillators and filters
- Light deflectors, modulators and displays

Ferroelectric Materials

Ferroelectricity is the effect of spontaneous electric polarization of a material, which can be reversed by application of an electric field. The term ferroelectric has been chosen in analogy to the permanent magnetic moment exhibited by ferroelectric materials. Valasek discovered the first ferroelectric material, namely Rochelle salt, in 1920. At this time ferromagnetism was already known and therefore the prefix iron relates to ferromagnetism and not to actually iron, thus resolving the contradiction caused by the missing iron atoms in most ferroelectric lattices. Unfortunately, the industrial exploitation is rather limited, since Rochelle salt is only ferroelectric for a certain composition and slight variations already lead to a loss of ferroelectricity. Therefore, for several decades it remained as an interesting physical effect, until 1945, when ferroelectric behavior was found for $BaTiO_3$.

$BaTiO_3$ material belongs to the stable perovskite type, which represents one of the fundamental crystal lattice structures. This discovery initiated investigations of perovskite type materials. Soon other perovskites with ferroelectric properties were discovered, thus opening the path to industrial application. Perovskites are still the most important ferroelectric materials.

Initially Lead Zirconate Titanate (PZT) was the material of choice for Ferroelectric Random Access Memory (FRAM) applications. The properties of PZT depend strongly on the composition of the alloy. Currently Ti/Zr ratios of 60/40 and 70/30 are employed, offering clearly defined switching and high switchable polarization.

Strontium Bismuth Tantalate (SBT) ($SrBi_2Ta_2O_9$) is another ferroelectric material of layered perovskite type. It posseses a smaller remanent and switchabel polarization compared to PZT, but does not show polarization fatigue like PZT due to repeated switching with platin electrodes. Furthermore, it exhibts two more processing disadvantages: due the to biaxial nature of the polarization vector the occuring crystal growth direction can lead to ferroelectric films with unwanted switching directions regarding to the externally applied field. The processing temperatures for high quality SBT films are higher ($100-250°C$) than for PZT.

While for the polarization fatigue feasible solutions exist, the time and temperature depend development of preferred polarization orientations, also known as *imprint*, and concerns all of the thin film ferroelectric materials.

Park et al. presented an alternative material based on $Bi_4Ti_3O_{12}$ in 1999. They substituted lanthanum and neodymium into the *A* site of bismuth titanate, resulting in materials known as Bismuth Lanthanum Titanate (BLT) and Bismuth Neodymium Titanate (BNT). These materials exhibit larger remanent polarization and lower processing temperatures compared to SBT in conjunction with an increased coercive field.

Commonly ferroelectric materials are integrated as polycristalline films, but also ferroelectrics epitaxially grown on silicon show good results.

What Properties do Ferroelectric Materials Possess?

Pyroelectric Properties and Spontaneous Polarisation

All ferroelectric materials are pyroelectric, however, not all pyroelectric materials are ferroelectric. Below a transition temperature called the Curie temperature ferroelectric and pyroelectric materials are polar and possess a spontaneous polarization or electric dipole moment. However, this polarity can be reoriented or reversed fully or in part through the application of an electric field with ferroelectric materials. Complete reversal of the spontaneous polarization is called "switching".

- The non-polar phase encountered above the Curie temperature is known as the paraelectric phase.
- The direction of the spontaneous polarization conforms to the crystal symmetry of the material. While the reorientation of the spontaneous polarization is a result of atomic displacements.
- The magnitude of the spontaneous polarization is greatest at temperatures well below the Curie temperature and approaches zero as the Curie temperature is neared.

Materials

The internal electric dipoles of a ferroelectric material are coupled to the material lattice so anything that changes the lattice will change the strength of the dipoles (in other words, a change in the spontaneous polarization). The change in the spontaneous polarization results in a change in the surface charge. This can cause current flow in the case of a ferroelectric capacitor even without the presence of an external voltage across the capacitor. Two stimuli that will change the lattice dimensions of a material are force and temperature. The generation of a surface charge in response to the application of an external stress to a material is called piezoelectricity. A change in the spontaneous polarization of a material in response to a change in temperature is called pyroelectricity.

Generally, there are 230 space groups among which 32 crystalline classes can be found in crystals. There are 21 non-centrosymmetric classes, within which 20 are piezoelectric. Among the piezoelectric classes, 10 have a spontaneous electric polarization that varies with the temperature, therefore they are pyroelectric. Among pyroelectric materials, some of them are ferroelectric

<table>
<tr><th colspan="4">32 Crystalline classes</th></tr>
<tr><th colspan="3">20 classes piezoelectric</th><td rowspan="3">non piezoelectric</td></tr>
<tr><th colspan="2">10 classes pyroelectric</th><td rowspan="2">non pyroelectric</td></tr>
<tr><th>ferroelectric</th><td>non ferroelectric</td></tr>
<tr><td>e.g. : $BaTiO_3$, $PbTiO_3$</td><td>e.g. : Tourmaline</td><td>e.g. : Quartz</td><td></td></tr>
</table>

Ferroelectric phase transitions are often characterized as either displacive (such as $BaTiO_3$) or order-disorder (such as $NaNO_2$), though often phase transitions will demonstrate elements of both behaviors. In barium titanate, a typical ferroelectric of the displacive type, the transition can be understood in terms of a polarization catastrophe, in which, if an ion is displaced from equilibrium slightly, the force from the local electric fields due to the ions in the crystal increases faster than the elastic-restoring forces. This leads to an asymmetrical shift in the equilibrium ion positions and hence to a permanent dipole moment. The ionic displacement in barium titanate concerns the relative position of the titanium ion within the oxygen octahedral cage. In lead titanate, another key ferroelectric material, although the structure is rather similar to barium titanate the driving force for ferroelectricity is more complex with interactions between the lead and oxygen ions also playing an important role. In an order-disorder ferroelectric, there is a dipole moment in each unit cell, but at high temperatures they are pointing in random directions. Upon lowering the temperature and going through the phase transition, the dipoles order, all pointing in the same direction within a domain.

An important ferroelectric material for applications is lead zirconate titanate (PZT), which is part of the solid solution formed between ferroelectric lead titanate and anti-ferroelectric lead zirconate. Different compositions are used for different applications; for memory applications, PZT closer in composition to lead titanate is preferred, whereas piezoelectric applications make use of the diverging piezoelectric coefficients associated with the morphotropic phase boundary that is found close to the 50/50 composition.

Ferroelectric crystals often show several transition temperatures and domain structure hysteresis, much as do ferromagnetic crystals. The nature of the phase transition in some ferroelectric crystals is still not well understood.

In 1974 R.B. Meyer used symmetry arguments to predict ferroelectric liquid crystals, and the prediction could immediately be verified by several observations of behavior connected to ferroelectricity in smectic liquid-crystal phases that are chiral and tilted. The technology allows the building of flat-screen monitors. Mass production between 1994 and 1999 was carried out by Canon. Ferroelectric liquid crystals are used in production of reflective LCoS.

In 2010 David Field found that prosaic films of chemicals such as nitrous oxide or propane exhibited ferroelectric properties. This new class of ferroelectric materials exhibit "spontelectric" properties, and may have wide ranging applications in device and nano-technology and also influence the electrical nature of dust in the interstellar medium. Other ferroelectric materials used include triglycine sulfate, polyvinylidene fluoride (PVDF) and lithium tantalate.

CHAPTER 1

FERROELECTRIC THIN FILMS USING OXIDES AS RAW MATERIALS

E.B. Araújo, J.A. Eiras

Keywords: ferroelectric films, PZT, bismuth titanate

INTRODUCTION

Several researchers have used solid or solution deposition techniques to produce thin films. Solution depositions are normally less expensive and enable better stoichiometric control of complex mixed oxides than other physical techniques such as sputter deposition[1] and laser ablation[2] or chemical vapor deposition (CVD)[3]. One of the difficulties of the physical methods, RF sputter or laser ablation for example, is that the film composition is not the same as of that the target for complex compositions. The most frequently used solution-preparation approaches may be grouped into three categories: sol-gel process that uses alkoxide compounds as starting precursors[4], hybrid processes that use chelating agents such as acetic acid[5] and metalorganic decomposition (MOD) that uses large water insensitive carboxylate compounds[6]. Metalorganic starting reagents such as alkoxide ($M(OR)_n$ where M is a metal and R is an alkyl group) and carboxylate ($M(OOCR)_n$) are employed to prepare the solution for film deposition. As a limitation of these can be pointed out the difficulty to obtain the precursor reagents and sometimes the low stability of complex solution.

In recent years, considerable attention has been devoted to the development of thin film technology of ferroelectric materials[7,8], with a view toward the increase of application possibilities in many electronic and optical devices. The most studied ceramic compositions for film preparation are the same as those well established for bulk ferroelectric ceramics, which normally have a complex composition. Large-scale processing of high-quality thin films requires low-temperature synthesis, high reproducibility, and simplicity in all processing steps with low cost.

Due to this fact, the search for new routes for film preparation remains as an interesting and open subject in order to improve the stability of complex solutions, the control of the stoichiometry of the film composition or to reduce the cost of the process.

This work describes an alternative solution method for preparation of ferroelectric thin films starting on a pre-calcination of oxides or carbonates. The proposed method is applied in the processing of lead zirconate titanate ($PbZr_{0.53}Ti_{0.47}O_3$ - PZT) and bismuth titanate ($Bi_4Ti_3O_{12}$ - BIT) thin films. The method is based on the fact that not all precursor oxides (PbO, TiO_2, ZrO_2 or Bi_2O_3) are soluble in acid media, but a reacted oxide may be soluble. X-ray Diffraction (XRD), Scanning Electron Microscopy (SEM), Infrared Spectroscopy (IR) and Rutherford Backscattering Spectroscopy (RBS) characterization were used to evaluate the efficiency of the method and to compare the film quality to those of films obtained through other methods.

EXPERIMENTAL

It is well know that some oxides and carbonates may be not soluble in acid or alkali media[9]. However, a complex oxide, formed by the reaction of these oxides or carbonates, could be soluble. For example, while lead oxide (PbO) is soluble in HNO_3, titanium oxide (TiO_2) and zirconium oxide (ZrO_2) are soluble H_2SO_4 [9]. On the other hand, PZT is only soluble in aqueous solution of HNO_3 [10]. This fact was used in this work to produce the ion solution, instead of the conventional metalorganic or alkoxide precursors, as starting materials for production of thin films.

To produce the solution containing the desired ions, PZT and BIT powder were previously prepared by conventional mixed oxide method. For preparation of PZT powder, lead

Citation: E.B. Araújo, J.A. Eiras, Ferroelectric Thin Films Using Oxides as Raw Materials, doi.org/10.1590/S1516-14391999000100004.

oxide (PbO), titanium oxide (TiO_2) and zirconium oxide (ZrO_2) were mixed at Zr/Ti = 53/47 ratio using a ball mill during 4 hours. The mixed powder was calcined at 850 °C for 3.5 h in an electric furnace using oxidant atmosphere, obtaining finally the $PbZr_{0.53}Ti_{0.47}O_3$ powder. For preparation of BIT powder, bismuth oxide (Bi_2O_3) and titanium oxide were mixed in appropriate proportion and mixed in a ball mill during 4 h. The mixture was then calcined at the same conditions as the PZT powder to obtain $Bi_4Ti_3O_{12}$ powder. Under these conditions full reacted PZT and BIT powders were obtained. It is interesting to observe that this procedure is the same used to obtain powders to prepare ceramic bodies or thin films through the proposed method, described below.

PZT powder (2 g) was dissolved in 60 mL of diluted acid solution (10% of HNO_3 and 90% of distilled water) during 1 h at 80 °C. After cooling, the solution was diluted with distilled water at room temperature to complete a stock solution with final concentration of 12.5 g/L. BIT powder (1 g) was dissolved in 4 mL of concentrated chloridric acid during 20 min at 80 °C. At room temperature the clear solution was diluted with ethanol to complete 125 mL of stock solution with final concentration of 8 g/L. Both solutions of ions were clear and transparent, showing that all elements were dissolved. These ions solution was the starting solution of the proposed process.

The general idea to deposit a film on a substrate is to distribute the ions homogeneously throughout the polymeric resin, prepared according to the Pechini method[11,12],. The process calls for forming a chelate between dissolved ions with a hydroxycarboxylic acid (citric acid). Heating of the resin in air causes a breakdown of the polymer. Subsequently, the ions are oxidized to form the desired crystalline phases.

For preparation of the resin by the Pechini method, citric acid and ethylene glycol (citric acid/ethylene glycol = 44/56 in mol%) were mixed with 50 mL from the stock solution. The mixture was heated at 90 °C for 1 h, transparent resins being obtained. Films of the resins (PZT or BIT) were deposited at room temperature on fused quartz and Si substrates by dip coating. The as deposited films were previously heated on a hot plate before they were heat treated for crystallization. Deposited films were crack-free, uniform and well adhered on the substrates.

The structure of the annealed film was examined by X-ray diffraction (XRD), using a Rigaku X-ray diffractometer with CuK_a radiation. Scanning electron microscopy (SEM – JEOL 5800LV) was used to study the surface morphology of the films. Infrared spectra (IR) of coating solutions and films were measured in the wavenumber range 400 to 1400 cm^{-1} using a Fourier transform IR spectrometer (Bomem MB-102). The stoichoimetry and thickness of the film were investigated using Rutherford Backscattering (RBS) at 2 MeV and a normal angle of incidence of the He^+ beam. All characterizations were done at room temperature.

RESULTS AND DISCUSSION

The as deposited amorphous films were annealed at temperatures between 500 °C and 700 °C for crystallization. Figure 1 shows the XRD pattern of PZT film deposited on fused quartz and heat-treated for 2.5 h at 600 °C. In this figure we can identify the co-existence of a tetragonal and a rhombohedral phase. The lattice constants *a* and *c* of PZT were calculated for tetragonal phase using the (101), (111) and (200) peaks and were found to be 4.02 and 4.16 Å (*c*/*a*= 1.03), respectively. For the rhombohedral phase we used the (100), (101) and (111) peaks and obtained *a* = 4.13 Å and 90-a = 0.03°. This result agrees with result for PZT thin films, prepared by sol-gel technique[5], with the same nominal composition.

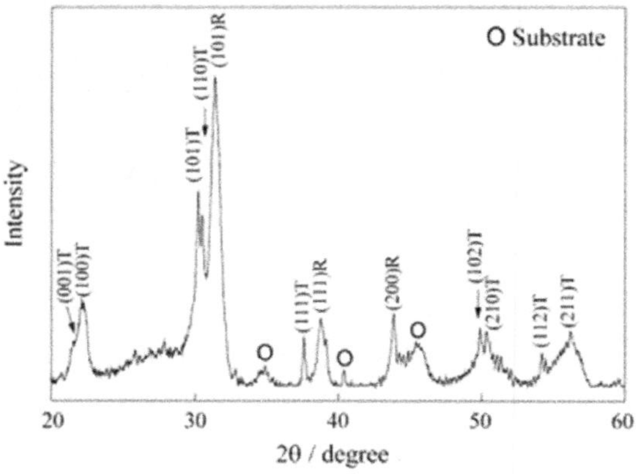

Figure 1: X-ray diffraction pattern at room temperature for $PbZr_{0.53}Ti_{0.47}O_3$ film deposited on fused quartz substrate (O), annealed at 600 °C for 2.5 h.

Figure 2-A, B and C shows the XRD patterns of BIT films on Si substrate after heat treatments for 2.5 h at 500 °C, 600 °C and 700 °C, respectively.

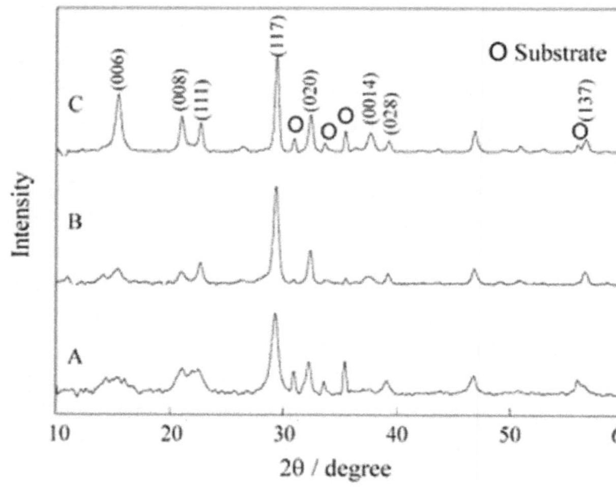

Figure 2: X-ray diffraction patterns of the $Bi_4Ti_3O_{12}$ films deposited on Si substrate (O), annealed at 500°C for 2.5 h (A), at 600°C for 2.5 h (B) and at 700°C for 2.5 h (C).

Some structural fluctuation can be seen in Fig. 2-A for films annealed at 500 °C, when compared with the well defined (111), (008) and (006) peaks, that appear in Figs. 2-B and 2-C. As the annealing temperature was increased, from 500 °C to 700 °C, these peaks in XRD patterns became sharper and the full width at half maximum decreased, indicating better crystallinity. Figure 2-C shows that full crystallized $Bi_4Ti_3O_{12}$ films with no preferential orientation, on the as crystallized surface, were obtained after firing at 700 °C. The calculated lattice constants *a*, *b* and *c* of $Bi_4Ti_3O_{12}$ considering an orthorhombic symmetry and using the (008), (111), (117) and (020) peaks were found to be 5.58 Å, 5.53 Å and 33.67 Å, respectively. These results are very close to $Bi_4Ti_3O_{12}$ thin films obtained by others chemical methods, like metallo-organic solution deposition[13] and sol-gel[14].

Figure 3 shows IR spectra for as deposited (Fig. 3-A) and crystallized at 700 °C PZT film (Fig. 3-B) on Si substrate. The IR spectrum in Fig. 3-A shows several absorption bands. The band centered at 529 cm^{-1}, extending from about 759 cm^{-1} to 414 cm^{-1}, has been associated to M-O bands in organic systems for PZT thin films prepared by sol-gel process[15]. The other bands in Fig. 3-A are attributed to the polymeric resin, formed by citric acid and ethylene glycol[16], which disappear with the heat treatment (Fig. 3-B). The band at about 494 cm^{-1} (Fig. 3-B) has been assigned to the vibrations of TiO_6 octahedra by Last and Spitzer[17,18]. Infrared reference spectra of PZT show that ZrO_6 or TiO_6 octahedra have also a band around the same frequency[15]. In Fig. 3-B the absorption bands around 1277 and 1098 cm^{-1} are associated to traces of carbonate (C_3^{-2}) present in the PZT film[15].

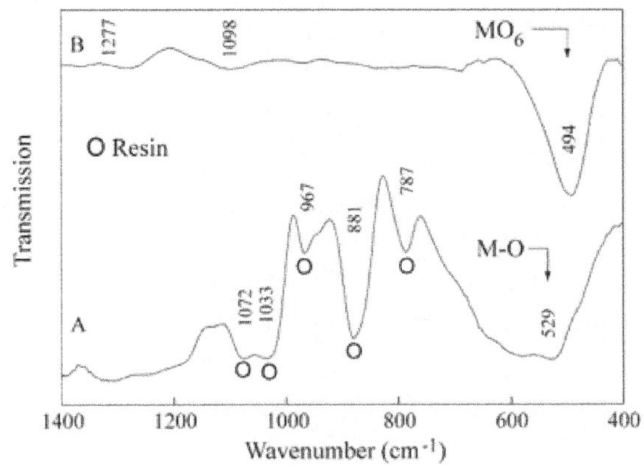

Figure 3. Infrared spectra (IR) for (A) as deposited and (B) crystallized $PbZr_{0.53}Ti_{0.47}O_3$ film (700°C for 2.5 h) on Si substrate.

Figure 4 shows a SEM micrograph for BIT film on Si substrate, heat-treated at 700°C for 2.5 h. The average grain size was estimated to be 76 nm. Figure 4 shows that the film presents good homogeneity, whose morphology is similar with those arising from films prepared by sol-gel technique[14].

Figure 4: SEM micrograph of $Bi_4Ti_3O_{12}$ thin film deposited on Si substrate. The film was annealed in air at 700°C for 2.5 h

Figure 5 shows a typical RBS spectrum for one layer PZT thin film. In this figure, dots represent experimental points while the continuous line represents the simulation for PZT on Si using the RUMP program[19]. The values obtained from the simulation were around 1500 Å for the film thickness and a zirconium to titanium ratio of 53.8/46.2. This Zr/Ti ratio confirms that the deposited film had the same nominal composition of the PZT powder, dissolved in acid solution (see Table 1).

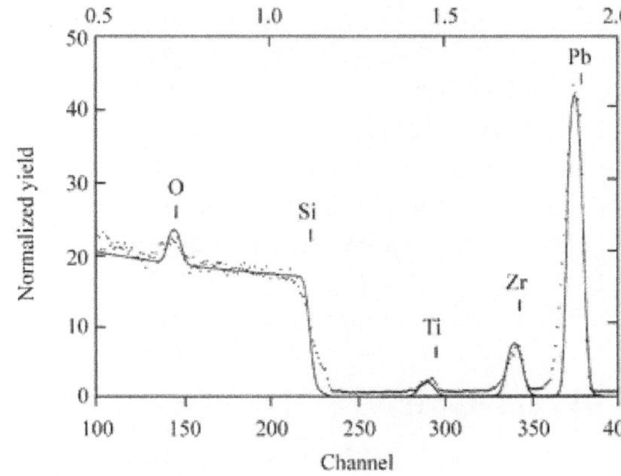

Figure 5: Experimental and simulated Rutherford Backscattering (RBS) spectra for the $PbZr_xTi_{1-x}O_3$ thin film on Si substrate, fired at 600 °C for 2.5 h. Points are experimental while the continuous line represents simulation data of Pb-Zr-Ti-O/Si.

Table 1: Nominal and obtained composition for oxides and thin films of PZT and BIT

Nominal Composition		Powder (as mixed)	Film (RBS simulation)
$Pb(Zr_{0.53}Ti_{0.47})O_3$	Zr/Ti	53/47	53.8/46.2
$Bi_4Ti_3O_{12}$	Ti/Bi	3/4	0.36/0.48

Figure 6 shows a RBS spectra for one layer BIT thin film. Through the simulation an estimated thickness of 1000 Å and titanium to bismuth ratio of 0.36/0.48 were obtained. Also in this case, the Ti/Bi ratio agrees very well with the desired nominal composition (see Table 1).

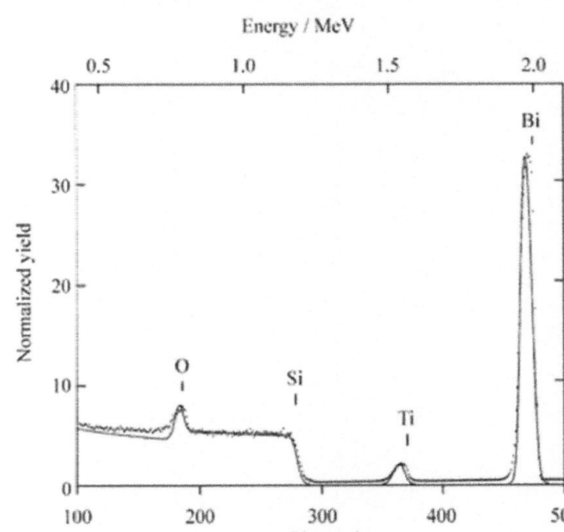

Figure 5: Experimental and simulated Rutherford Backscattering (RBS) spectra for the $PbZr_xTi_{1-x}O_3$ thin film on Si substrate, fired at 600°C for 2.5 h. Points are experimental while the continuous line represents simulation data of Pb-Zr-Ti-O/Si.

The above presented results show that through the proposed method thin films well crystallized have been prepared with the composition of the powders obtained by conventional ceramic method.

CONCLUSION

An alternative chemical method to prepare thin films is presented. The method is based on the dissolution of a ceramic powder in acid media, with the same composition desired for the film, to obtain a solution containing the metallic ions. The method was successfully applied to produce BIT ($Bi_4Ti_3O_{12}$) and PZT ($PbZr_{0.53}Ti_{0.47}O_3$) thin films. The obtained films showed good microstructural homogeneity, the desired composition and a quality similar to those films prepared by sol gel processing. Low cost and high stability of the solution in the preparation process are the main advantages compared with other chemical methods. The presented method can then be particularly useful to produce doped films.

ACKNOWLEDGMENTS

The authors thank Dr. N.I. Morimoto and MSc. A.G. Pedrine, LSI-USP, Dr. M.H. Tabacniks and A.F. dos Reis, LAMFI-USP, for facilities and RBS analysis, Dr. Y.P. Mascarenhas, IFSC-USP, for XRD analysis, Mr. Paulo (DQ-UFSCar) for IR measurements and CNPq (Brazilian agency) for financial support.

REFERENCES

1. Auciello, O.; Kingon, A.I.; Krupanidhi, S.B.; *MRS Bulletin*, p. 25, June 1996.
2. Auciello, O.; Ramesh, R. *MRS Bulletin*, p. 31, June 1996.
3. Keijser. M. de;. Dormans; G.J.M *MRS Bulletin*, p. 37, June 1996.
4. Johnson Jr., D.W. *Am. Ceram. Soc. Bull.* v. 64 n. 12, p. 1597, 1985.
5. 5. Yi, G.; Wu, Z.; Sayer, M. *J. Appl. Phys.* v. 64, n. 5, p. 2717, 1988.
6. Haertling, G.H. *Ferroelectrics* v. 119, n. 51 1991.
7. Sheppard, L.M. *Ceramic Bulletin* v. 71 n. 1, p. 85, 1992.
8. Sayer, M.; Sreenivas, K. *Science* 247, 1056, 1990.
9. Weast, R.C. *Handbook of Chemistry and Physics*, 68th edition, CRC Press, 1988.
10. Araújo, E.B.; Eiras, J.A. *Proceeding of 41° Congresso Brasileiro de Cerâmica* v. 1, p. 413, 1997.
11. Pechini, M. U.S. Pat. N° 3 330 697, July 11, 1967.
12. Lessing, P.A. *Ceramic Bulletin* v. 68, n. 5, p. 1002, 1989.
13. Joshi, P.C.; Krupanidhi, S.B. *J. Appl. Phys.* v. 72, n. 12, p. 5827, 1992.
14. Sedlar, M.; Sayer, M. *Ceramics International* v. 22, n. 241, 1996.
15. Lipeles, R.A. *Ceramic Transactions, Ferroelectric Films* v. 25, p. 147-153, 1992.
16. Ponchert, C.J. *The Aldrich Library of Infrared Spectra*, second edition, 1975.]
17. Last, J.T. *Phys. Rev*. v. 105, p. 1740, 1957.
18. Spitzer, W.G.; Miller, R.C.; Kleinman, D.A.; Howarth, L.E. *Phys. Rev.* v. 126, p. 1710, 1962.
19. Doolittle, L.R. *Phys. Rev.* v. B9, p. 344, 1985.

CHAPTER 2 DIFFUSE SCATTERING FROM LEAD-CONTAINING FERROELECTRIC PEROVSKITE OXIDES

D. J. Goossens

INTRODUCTION

Single crystal diffuse scattering (SCDS) has been the subject of study since the earliest days of crystallography [1] and is seen in many patterns collected using film (e.g., [2]), film being an early variant of "area detector" and therefore very good for surveying large regions of reciprocal space—far better than an electronic point counter. The different forms of disorder present in crystalline materials, and how they are manifested in diffraction patterns have been well explored, both in monographs and papers (these include [3–14]).

Diffuse scattering can be defined for the purposes of this paper as the coherently scattered intensity that is notlocalised on the reciprocal lattice. It is the result of the two-body correlations in a crystalline material (see e.g., [15, 16]). These correlations may exist between atoms, molecules, and, in the case of neutron diffraction, magnetic moments [17–20].

It has long been known that diffuse scattering, an example is shown in Figure 1, contains information about the correlations in atomic and molecular thermal motions [21] as well as static short-range order. Modelling diffuse scattering is not simple, because locally the short-range order, SRO, need not obey the space group symmetry of the crystal. The global symmetry must be regained on averaging across the crystal but may not be present on the local scale of a few nanometres, or if one prefers, on the scale of a few unit cells. This renders many of the tools of conventional crystallography of limited use.

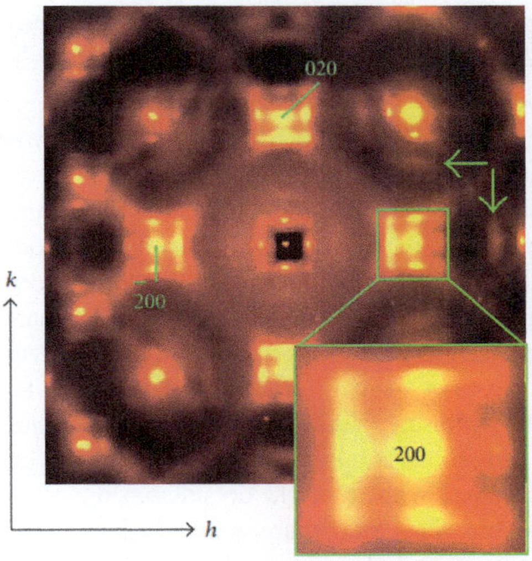

Figure 1: Figure 1: A single exposure from a stationary single crystal of wustite, $Fe_{1-x}O$ (false colour map, white = high intensity). Highenergy X-rays (65.35 keV in this case) give a small wavelength, λ, and an almost flat Ewald sphere. Hence these data closely approximate the $hk0$ section. Every Bragg reflection (and also the origin) is surrounded by a motif of incommensurate scattering. Second-order incommensurate spots have been arrowed and some Miller indices noted for scale [22]..

In particular, the idea of a single, repeating unit cell is not an adequate "solution" of the local structure—indeed, the very idea of what is meant by "structure solution" is thrown into question. Conventional crystallography solves (a better word may be "determines") very important aspects of the structure—those describing its global average. Analysis of SCDS reveals another important aspect—the pair correlations, and thus some

Citation: D. J. Goossens, "Diffuse Scattering from Lead-Containing Ferroelectric Perovskite Oxides" ISRN Materials Science, vol. 2013, Article ID 107178, 17 pages, 2013. doi:10.1155/2013/107178

deeper information on the population of local environments, which will be closely related to the crystal chemistry. But even SCDS is limited to two-body correlations. An experiment that could measure the phase shift of the scattered radiation (X-ray or neutron, in most cases) would be sensitive to higher-order correlations, and so on. At what level of detail one could consider the structure to be "solved" is almost a philosophical question. In practice, the level of detail depends on what is needed to give insight into the properties of the material. The structure is "solved" when the questions being asked can be answered.

Though not the subject of this review, it can be noted that diffuse scattering data collection is difficult.

The weak diffuse scattering must be measured in the presence of strong Bragg reflections, and so the experiments require low noise, high dynamic range, and detectors with good resistance to saturation [32, 33]. Data may require considerable correction and reconstruction [34–37] before they reach a useful format. Dedicated instruments for diffuse scattering are relatively rare, for example DNS at Jülich [20], D7 at the Institut Laue Langevin [19], Corelli at the Spallation Neutron Source [38], a custom-made laboratory X-ray machine [39, 40], the E2 flat cone instrument at the Helmholtz-Zentrum Berlin für Materialien und Energie [41], and so on. However, many other instruments are capable of collecting high quality data, including the powder diffractometer at the Australian Synchrotron [42], SXD at the ISIS neutron source [27, 43–45], Wombat at the OPAL reactor [35, 46], the SNBL beamline at ESRF [47], 11-ID-B at the Advanced Photon Source [48,49], and others [50–52].

The correlation structure is likely to be anisotropic, such that a detailed model of the SRO will often require a three-dimensional (3D) dataset [23, 53–56], in turn implying the need to measure the distribution of weak and delocalised features throughout significant volumes of reciprocal space. This requires long experiments, very carefully characterised instruments, and large crystals—especially for neutron diffuse scattering. Even at a synchrotron a 3D dataset will take hours to collect, inhibiting studies as functions of temperature, pressure, and so on. If the crucial section of reciprocal space can be observed with a single crystal setting, such studies are practicable. This is perhaps made easier by using high X-ray energies that flatten the Ewald sphere such that a single exposure is almost flat in reciprocal space and can be made through sample orientation to coincide with some important symmetry directions. Hence a particular reciprocal space plane can, as long as it contains the origin, be captured in a single exposure if the crystal is appropriately aligned (Figure 1) [57, 58].

The importance of diffuse scattering in examining structures in ferroelectrics has long been recognised [59, 60], and highly structured diffuse scattering has been observed in many ferroelectric materials, including those of commercial importance: KDP (KH_2PO_4) [61], TGS ($(NH_2CH_2COOH)_3H_2SO_4$) [62], benzil ($C_{14}H_{10}O_2$) [2,43, 63, 64], $BaTiO_3$ [59, 65], NBT ($Na_{0.5}Bi_{0.5}TiO_3$) [66], and so on.

Ferroelectric phenomena are inherently related to correlations amongst structural features, resulting as they do from electric dipoles due to ionic displacements [67]. As noted, it is well established that the diffraction images of $BaTiO_3$, a much-studied ferroelectric material, contain diffuse scattering (DS) [59, 65]. The observation of the evolution of the diffuse patterns through the ferroelectric phase transition [68] leads to ongoing debate over the character of the phase transition—for example, order-disorder as against displacive—in this material [69]. DS planes were observed and considered to be the result of displacements correlated in chains, with the correlations being either static or dynamic.

In the modern relaxor ferroelectric materials [70], DS has often been interpreted using the idea of domain-like polar nanoregions (PNRs) embedded in a paraelectric matrix, which are then linked to the outstanding electromechanical properties [71]. While the PNRs offer a convenient and somewhat intuitive model for the behaviour, the actual need for the concept is, according to some recent work, not thoroughly established [72]. It is also subtly different from the idea of ferroelectric nanodomains (FND) essentially "butting up" against each other—the main difference being the latter requires no paraelectric matrix.

Many of the most significant ferroelectric materials rely on lead, Pb, as a key ingredient in their crystal chemistry: PZT ($PbZr_{1-x}Ti_xO_3$) ([73–76] and others), PZN (($PbZn_{1/3}Nb_{2/3}O_3$) and PZN-PT ($PbZn_{1/3}Nb_{2/3}O_3$-$PbTiO_3$) ([76–78] and so on), and PMN ($PbMg_{1/3}Nb_{2/3}O_3$) and PMN-PT (($PbMg_{1/3}Nb_{2/3}O_3$-$PbTiO_3$) ([78–81]) ([78–81] and others). The search for replacement Pb-free materials has often proceeded by analogy, with the obvious replacement for Pb^{2+} being Bi^{3+} ([82, 83] and many many more) and some exploration of Sn^{2+} [84–86], despite its difficult chemistry.

There are commonalities in the SCDS observed in many of the lead-based relaxors, particularly in their cubic phase [30, 87, 88]. A range of microscopic models have been developed to explain the systematics of the distribution of DS. Often these explanations draw on the idea of the PNR, whose geometry relates to the observed diffuse—atomic displacements correlated in two-dimensional sheets when the diffuse scattering is rod-like, one dimensional chains when the DS is sheet-like, and so on [27, 31, 44, 89–92].

A recent report by Bosak et al. [23] indicates that the diffuse scattering from these compounds can be modelled in a thermal-like way; in particular the often-referred-to diffuse "butterfly" (see the motif around the 300 peak in Figure 2) can be modelled using a purely thermal diffuse scattering-like approach that does not imply the existence of PNRs and has more in common with a phonon-based approach. This model uses parameters which are analogous to elastic constants, but are not elastic constants as such. It could be argued that of the two reciprocal space slices modelled in Figure 4 of [23] (reproduced in Figure 2) the second (hkk) is less than convincing; it seems to give streaks that are rather too strong and motifs of diffuse scattering which are too strong in every second horizontal row of features along the vertical axis (an example is arrowed by the current author on Figure 2(b)). Having said that, the model is clearly very powerful with very few parameters and shows that a displacement-wave picture rather than an order/disorder picture has much to recommend it. How the B-site randomness folds in to this is an interesting question, and it would be interesting to apply this approach to, for example, the PZT materials modelled in [25], in which the Pb-Pb ferroelectric domains were found to be of very different shape, being one-dimensional chains, and in which the static strains due to chemical disorder and B-site cation size mismatch are weaker and should therefore make a thermal-like model more applicable.

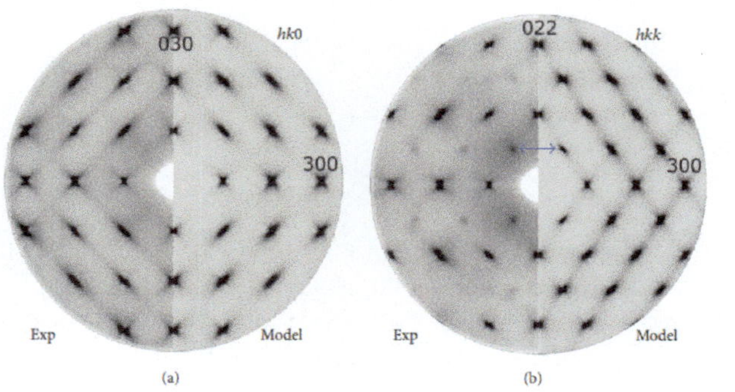

Figure 2: The quality of agreement between experimental data "exp" and calculation "model" is excellent in (a), perhaps less so in (b). (From Figure 4 of [23], used by permission of the IUCr, http://dx.doi.org/10.1107/S0108767311040281.)

SURVEY OF DIFFUSE SCATTERING IN PB-BASED FERROELECTRICS

$PbZr_{1-x}Ti_xO_3$

PZT, $PbZr_{1-x}Ti_xO_3$, ceramics are of great commercial importance and find uses as ultrasonic transducers in medical imaging, in accelerometers, and elsewhere [77, 93–96]. The structural phase diagram is summed up in Figure 3. The physical properties are optimal close to the morphotropic phase boundary, MPB, where the structure is prone to critical fluctuations and therefore responds more strongly to, for example, an external poling field. On the figure are noted coordinates of the measurements discussed below.

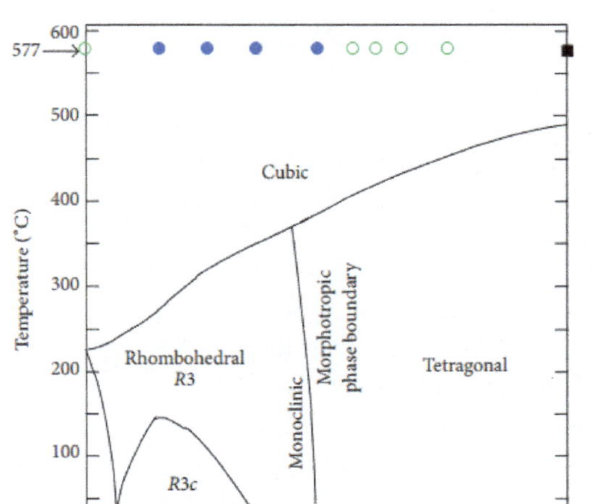

Figure 3: Schematic phase diagram for PZT. The morphotropic phase boundary is noted, along with key structural phases. Temperatures and x values for the various datasets discussed are noted on the figure—results at 577°C (850 K) are taken from [24], while letters (a), (b), and (c) correspond to diffraction patterns shown in Figure 4.

PT, $PbTiO_3$, appears to be a well-ordered perovskite [24, 97] with little diffuse scattering related to the PNR structures discussed here. Antiferroelectric PZ, $PbZrO_3$, on the other hand, does show diffuse scattering. The average structure at high temperatures shows $\langle 110 \rangle_P$ b^{2+} displacements [24]. Antiferroelectrically correlated displacements along one of the $\langle 110 \rangle_P$ directions had been suggested in early diffraction work [98], though ordered—but with unrealistic degrees of octahedral distortion. This was largely remedied in terms of the average structure by introducing two substructures related by a shear of c/2 and allowing some disorder [99].

PDF work has since suggested that the Pb^{2+} are displaced at all temperatures, and the correlations amongst these displacements determine the observed average structure [100], including development of correlated c-axis displacements (compared to the antiferroelectric displacements in the bc plane) as expected to be associated with the weak room temperature piezoelectric effect in PZ [101]. Given the lack of B-site randomness, PNRs are unlikely to form if random strain fields (see below) play a role, and this is observed in PZ.

The average structure of PZT has been the subject of numerous studies (just a few include [24, 102–105]), and the work continues. It has been shown that at room temperature powder diffraction data are best modelled by a multiple-phase approach, in which the sample consists of a rhombohedral R3c component and a monoclinic Cm component, with the proportion of Cm phase increasing with x (Ti content) as the MPB is approached [102]. Whether this is definitive or is an indication of the limits of powder diffraction in studying disordered materials is almost a question of ideology. It does appear that at high temperatures, in the cubic phase, the Pb atoms show disorder [24] that is not present in the Rietveld modelling [102], despite tests that were undertaken. Further, the presence of highly structured diffuse scattering in PZT at room temperature [25, 106–108] unambiguously requires an average crystal structure that permits local ordering to occur. Given that the Rietveld fits in [102] are of extremely high quality and yet do not yield models that permit SRO in Pb displacements to occur, it must be concluded that the powder diffraction is simply not sensitive to crucial aspects of the structure of PZT, despite the excellent fit statistics and the all-but perfect visual quality of the fits. This is a salutary lesson.

In the high-temperature cubic phase, it has been noted that the Pb atoms do show disorder. For $x = 0$ ($PbZrO_3$) they are distributed across 12 $\langle 110 \rangle_c$ -type directions (c-implies indexing on the parent cubic structure); as Ti is doped in this, it switches to the 8 $\langle 111 \rangle_c$-type directions, and until for x>0.5 it switches back to the $\langle 110 \rangle_c$-type directions, until at $x = 1$ ($PbTiO_3$) the Pb shows a single-ordered site [24]. Such a model accords extremely well with the modelling of electron diffuse scattering, which appears to show that the observed diffuse scattering can be explained by one-dimensional chains of Pb displacements that result in planes of diffuse scattering perpendicular to $\langle 111 \rangle_c$ directions [107], and also accords well with neutron diffraction results [109]. Such a model was implemented using Monte Carlo simulation and was indeed found to reproduce many details of the observed electron diffuse scattering [25, 106]. As for other Pb^{2+} -based materials, the chronic underbonding of Pb^{2+} in the environment determined by the B-site cations (whose bonding is in a sense stiffer than the Pb-O bonds [110, 111]) is central to the mechanism of the off-centring.

Figure 5 shows a calculated diffraction pattern corresponding to in Figure 4(a), alongside a representation of the domain structure from the MC model. In this model, the Pb^{2+} are displaced along the eight $\langle 111 \rangle_c$ shifts, but not in equal proportions. For the domain structure shown in Figure 5(a), ~51% of the Pb^{2+} are displaced along $\langle 111 \rangle_c$, and this can be seen in the preponderance of red arrows (pointing up and to the right) in the figure. This imbalance is posited to drive the system away from cubic, and it may be conjectured that the average structure observed at high temperatures [24] is showing a state in which thermal energy is both evening out the distribution of Pb^{2+} displacements across the eight $\langle 111 \rangle_c$ shifts and overwhelming the correlations between displacements, giving a real space picture of the structural phase transition as one in which the types of individual Pb^{2+} displacements do not change—though the proportions along the $\langle 111 \rangle_c$ directions do—but the correlations amongst them break down. Since the crystal chemistry around a given $Pb^{2+}$is not changing with T, this seems plausible.

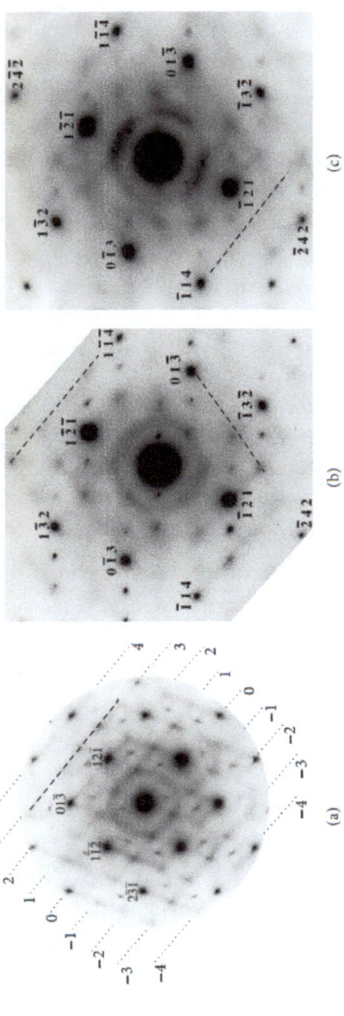

Figure 4: Figure 4: Observed electron diffuse scattering from PZT. (a) $PbZr_{1-x}Ti_xO_3$ with $x = 0.3$ (rhombohedral phase) viewed down the [531] zone axis (c means relative to the cubic parent structure). (b) $x = 0.46$ (monoclinic) viewed down [731]. (c) $x = 0.7$ (tetragonal) viewed down [731]. Some reflections and lines of scattering have been marked.

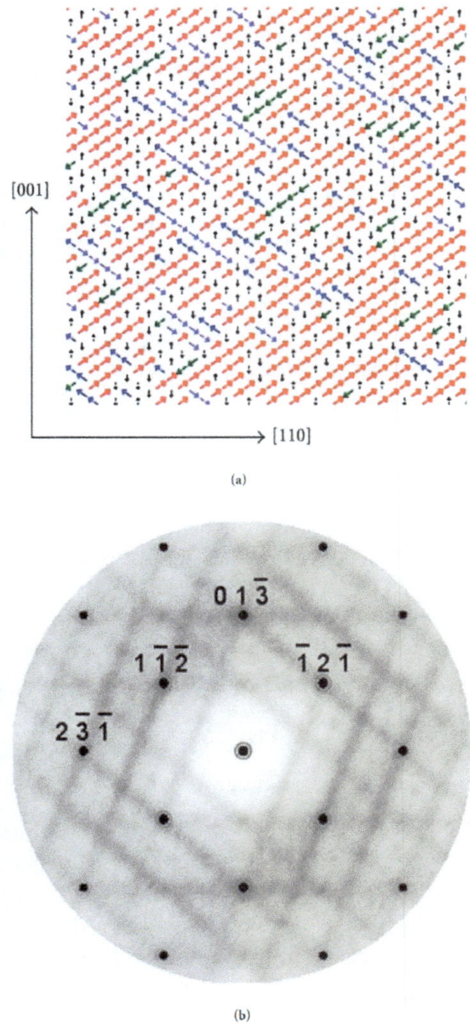

Figure 5: Calculated electron diffuse scattering from PZT. (a) The model domain structure and (b) the resulting diffraction pattern, to be compared to (a) in Figure 4 [25].

Contrary to this atomistic model, a model in which inhomogeneous strains are caused by inclusions of a tetragonal phase within a rhombohedral or monoclinic matrix has been proposed [108]. On the basis of synchrotron diffuse and inelastic scattering the model reproduces some of the structure seen around selected Bragg peaks via these strain-inducing inclusions through a Huang-like formalism. This is qualitatively different from the relaxors like PMN-PT and PZN-PT, in which the Huang formalism cannot reproduce the observed scattering.

It is clear that the diffuse scattering in PZT varies strongly with composition [25, 106], as would be expected if random strain fields induced by B -site doping play a role in inhibiting long-range order in Pb^{2+} displacements and promoting nanodomain growth.

Interestingly, a Huang component has been observed in PMN above the Burns temperature, T_d [112], which is therefore not associated with domains but with "simple, isotropic defects" which are considered to be fluctuations in Nb^{5+}: Mg^{2+} concentrations on the nanoscale. This component of the diffuse scattering in PMN and related materials has been noted elsewhere [28, 113, 114]. It is relatively independent of temperature, being related to the chemical ordering on the B-site.

Earlier work using pair distribution function analysis showed that local structure in PZT and closely related La-doped PLZT ($[(Pb_{1-y}La_y)_{1-\alpha}\square_\alpha][(Zr_{1-x}Ti_x)_{1-\beta}\square_\beta]O_3$ or $(Pb_{1-3x/2}La_x)(Zr_yTi_{1-y})O_3$) was "more complex than that implied by the crystallographic structure" [115], with the PDF showing a wider range of Pb-O distances than the average structure suggests. The existence of short-range-ordered polar clusters has been suggested to explain Raman and other results in PLZT [116]. An electron diffraction study of $(Pb_{1-3x/2}La_x)(Zr_yTi_{1-y})O_3$ [117] draws related conclusions, noting that the "fundamental dipolar units in these materials correspond to highly anisotropic <111> chain dipoles formed from off-centre Pb/La and coupled Ti/Zr displacements."

Hence the overall picture is one in which the Pb^{2+} environments drive atomic Pb^{2+} displacements which interact via other atoms (in fact, via the BO_6 octahedra) and affect those other atoms. This is a common theme across the Pb^{2+} ferroic materials.

$PbZn_{1/3}Nb_{2/3}O_3$

Above T_c $PbZn_{1/3}Nb_{2/3}O_3$ (PZN) is cubic $(Pm\bar{3}m)$ and average structure analysis that shows disordered off-centre displacements of Pb atoms [118]. According to early studies by Kuwata et al. [119, 120], on cooling PZN undergoes a phase transition to the rhombohedral phase [121, 122]. The existence of the R phase in bulk was questioned [91] after measurements of splitting of the (111) reflection, and an unusual bulk structure was posited in which a sample/grain showed a rhombohedral outer shell and a cubic inner core. This has been shown to be unnecessary [121, 122]. That PZN from $PbMg_{1/3}Nb_{2/3}O_3$ (PMN) (PMN) which stays cubic down to 5 K [123]. It has been suggested that the phonon behaviour and diffuse scattering are very similar in the two materials [124].

You [125] recorded two-dimensional X-ray DS for PMN, showing "butterfly" and ellipsoidal-shaped DS, and related it to transverse optic soft modes producing correlations along the cubic <110> directions (later works have shown that a coupling of optic and acoustic-like displacements is needed to explain the intensity variation of DS in PMN [126]).

Some three-dimensional DS studies of PZN [27, 35, 44, 46] have applied $m\bar{3}m$ symmetry to the diffuse features even in the rhombohedral phase, as these appear to obey this symmetry within experimental resolution—the rhombohedral angle ~89.95°.

The diffuse scattering from PZN has been studied extensively. It was suggested that locally polarised "regions (probably of a size of the order of several unit cells)" could explain measurements of the index of refraction of PLZT ($(Pb_{1-3x/2}La_x)(Zr_yTi_{1-y})O_3$) and also the discovery of a second important temperature, the Burns temperature (T_d) well above T_c, above which it was posited that the domains ceased to exist [127]. PZN shows similar behaviour, with diffraction experiments showing that a signature for the posited domains could indeed be observed in the scattering from PZN and closely related PMN [126, 128, 129].

Using single high-energy X-ray exposures and tilting the sample, rather in the manner of a transmission electron microscope diffraction experiment, Xu and coworkers were able to show that the rods of scattering observed in PZN were indeed <110>-type rods and not the intersections of, for example, the sections with out-of-plane rods in directions like <111> [54].

Figure 6 shows X-ray diffuse scattering in the $hk0$, $hk0.5$, and 111 sections of reciprocal space, measured on a single crystal of PZN at 300 K at the advanced photon source. These data clearly show the key diffuse features: the rods of scattering along -type directions (green arrow in Figure 6(a)), and the "butterfly" shapes where the rods intersect (boxed in Figure 6(a)), the <110> -type spots that indicate B-site ordering (box "A" in Figure 6(b)) and that arise from intersection of the streaks with the layer (box "B" in Figure 6(b)). Also apparent is the variation of intensity along the length of the rods.

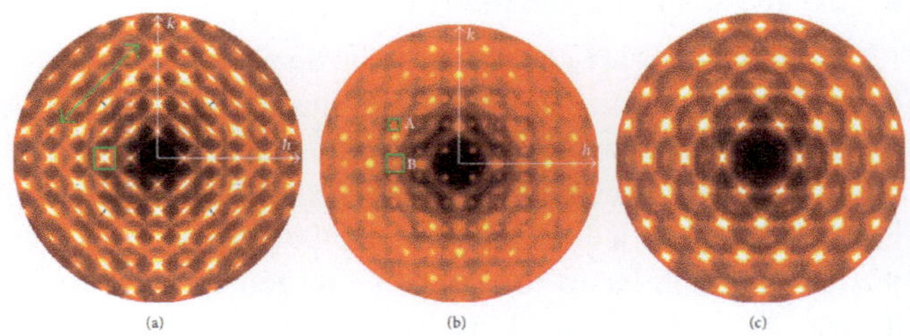

Figure 6: X-ray diffuse scattering from PZN, $PbZn_{1/3}Nb_{2/3}O_3$, showing the (a) $hk0$, (b) $hk0.5$, and (c) 111 sections of reciprocal space. Select features are highlighted. To indicate scale, note that the $\bar{2}00$ is boxed in (a) [26].

Figure 7 shows three slices of diffuse scattering, measured using neutron diffraction at the ISIS-pulsed neutron source. There is no energy analysis undertaken, and so the scattering is treated as wholly elastic. That this induces little or no loss of symmetry—as it does when the scattering is the result of phonons [43]—is indicative that the scattering is not of predominantly thermal origin, or that the thermal fluctuations are slow on the energy scale of thermal neutrons. Again, the key features are visible. More apparent than in Figure 6 is the size-effect scattering around, for example, the 004 Bragg peak and which is also apparent in the intensity of the <110> rods on the inside of a row of Bragg peaks compared to the outside (note green arrows showing direction of intensity transfer).

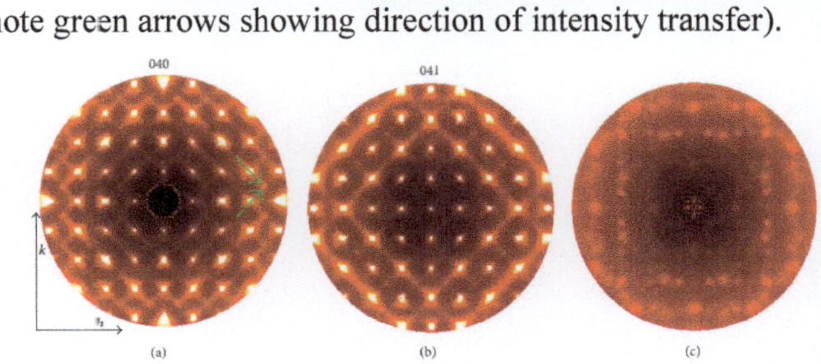

Figure 7: Neutron diffuse scattering from PZN, $PbZn_{1/3}Nb_{2/3}O_3$, showing the (a) $hk0$, (b) $hk1$, and (c) $hk0.5$ sections of reciprocal space. Select features are highlighted [27].

The idea of a relatively static PNR lends itself to a local, atomistic model driven by the crystal chemistry—a real-space chemical picture rather than a reciprocal space phonon-based picture, as it was. Such a model is also reasonable given that local inhomogeneity appears to play a role in relaxor behaviour—relaxor behaviour is often induced by doping when the "pure" compound is a nonrelaxor ferroelectric. This is even the case in multiferroics, where the doping of $BiFeO_3$ with $BaTiO_3$ induces a change to relaxor-like behaviour [130, 131].

Hence, in modelling neutron diffuse scattering, Welberry and coworkers developed models for the diffuse scattering from PZN based on what could be termed chemical principles [27]. For example, the displacing of the Pb^{2+} ions is considered to be driven by the fact that Pb^{2+} is heavily underbonded in the average unit cell of PZN and also possesses an electron lone pair. The suggestion is that an inherently polar atom in a large environment (as parameterised by bond valence sums [132]) will satisfy its bonding requirements by moving off-centre rather than isotropically contracting its environment, and this is especially true of a "soft" atom like Pb^{2+} [133] which can be considered as sitting within a "hard" framework established by the oxygen atoms and the B-site cations.

The Pb atoms are considered to be displaced along <110>-type directions in the unit cell, and these displacements are correlated within two-dimensional domains (see Figure 8), such that the displacement direction is parallel to a <110>-type direction that lies within the plane of the domain. In essence, this is quite similar to the structure posted earlier based on diffuse X-ray scattering [54, 131]—except that there is no paraelectric matrix, but rather the volume is virtually entirely occupied by ferroelectric nanodomains (FND) of a range of sizes and orientations; this has led to the introduction of the term FND in distinction to PNR [72].

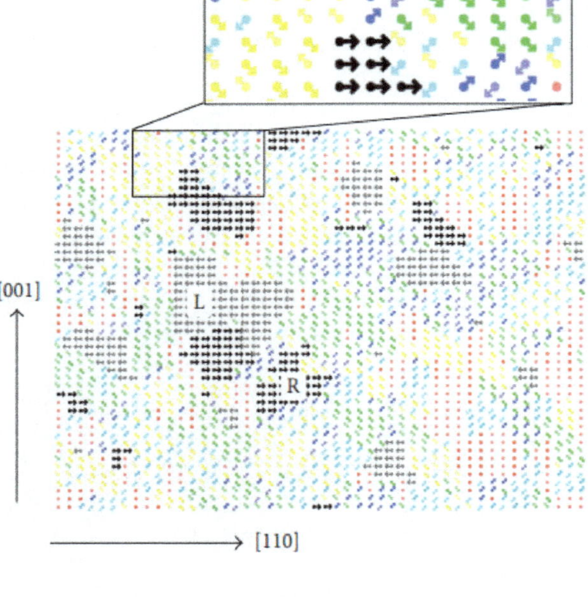

Figure 8: Schematic diagram of posited PNRs (domains) in PZN in the plane of one of the domains, axes as noted (section is normal to $\bar{1}\bar{1}0]$). Grey arrows lie in the plane and point left (L), black point right (R). Red (pink) dots point into (out of) the page. Other directions are intermediate.

This model and its ramifications, including the response of the O ions to the Pb^{2+} framework and the alternating ordering of the Nb/Zn (frustrated by the 2 : 1 ratio of Nb to Zn), give rise to a qualitatively good agreement between observed and calculated data, including subtle details like the fact that every second diffuse rod is weak in the $hk1$ layer (Figure 7(b)). Further, the model gives an empirical chemical basis for the observed behaviour from which one can attempt to generalise in the search for Pb-free compounds. However, no underlying cause for the Pb^{2+} ions to form into the posited {110} domains is apparent.

More significantly, it has been shown that the simple heuristic interpretation of the diffuse scattering that leads to the displacement directions being of $\langle 110 \rangle$ -type directions is misleading [89]. Investigation of these ideas shows that, at least in atomistic models, displacements indeed need not be along $\langle 110 \rangle$ directions, but that the domains themselves do need to be a single atomic layer thick, and with the established $\langle 110 \rangle$-type interfaces [134]. Having said that, high-temperature average structure determinations suggest that when the Pb site disorders, it is split across the 12 $\langle 110 \rangle$ directions [118]. Thus in some sense the initial model remains the best supported.

The question arises as to the power of the diffuse scattering to reveal the underlying structures. It has been shown that the diffuse scattering can be well modeled and quantitatively by a model that dispenses with static planar domains entirely [26]. This is supported by work discussed above [23] that also dispenses with PNRs and does a good job of modelling the scattering with just a handful of free parameters. Having said that, there is evidence that PNRs have been observed in relators using imaging techniques [135, 136] (with appropriate caveats about how representative of the bulk the surface may be), and so perhaps the ability to model the diffuse scattering with and without PNRs says more about the limits of the technique and the need to draw on multiple techniques than it says about PZN.

$PbMg_{1/3}Nb_{2/3}O_3$

Any discussion of $PbMg_{1/3}Nb_{2/3}O_3$-$PbTiO_3$ and $PbMg_{1/3}Nb_{2/3}O_3$ must relate to PZN, as the two materials are highly analogous.

$PbMg_{1/3}Nb_{2/3}O_3$ (PMN) is a heavily studied relaxor. Some work has used a dipole glass model to interpret diffuse scattering from PMN [137]. Earlier again it was suggested based on powder diffraction data that heavily displaced polar clusters of ions, particularly Pb^{2+} and O^{2-}, were required to model the average structure [123,138]. Butterfly-shaped and ellipsoidal diffuse scattering was seen in X-ray diffraction experiments and explained in terms of transverse optic soft modes along the cubic <110> directions [125]. This has been elaborated more recently [126]. As with PZN, the polar nanoregion has been invoked, with the T dependence of the DS being attributed to the growth of PNRs with cooling [139]. The relation of the lattice dynamics in PMN and PNRs (or local polar correlations) has been explored in some detail using a range of experimental techniques [140–144] and also ab initio calculations [145, 146].

As noted in the discussion of PZN, the diffuse scattering from $PbMg_{1/3}Nb_{2/3}O_3$-$PbTiO_3$ appears to be able to be parameterised using a thermal-like model that dispenses with PNRs [23, 147, 148] and also to be calculated from an ab initio approach [149] using recently established potentials [150]. Further molecular dynamics results suggest that Pb atoms show <111>p displacements, as expected [151] and that while the PNR does seem to be a useful model for PMN at low temperatures [152], it is a sense-limiting case resulting from correlations that, while present at higher temperatures, do not necessarily produce domains as such. This is quite plausible; when there is substantial randomness in the system, components of the displacements can be correlated but a random component that increases with temperature may prevent them actually assembling into a physical domain in which all ions share a common displacement direction. Further, diffuse scattering is sensitive to pair correlations—without distinguishing between static and dynamic in the case of X-ray scattering of neutron energies well above the excitation energies—and so correlations amongst components of displacements on pairs of interacting atoms can show structure that is not apparent in the absolute displacements of the atoms. Importantly, these results depend on the local strains due to the B-site ordering of Nb and Mg as a mechanism for inhibiting large domains from forming, providing a strong link between B-site doping [153] and the introduction of relaxor behaviour [152]. An interesting comparison of scattering from PMN and PMN-60PT [28] shows that increasing PT doping greatly reduces the structure in the DS, leaving features that are weakly elongated in the radial direction (Figure 9). These have been interpreted as short-range chemical order scattering due to weak B-site ordering. As the B-site becomes heavily occupied by Ti^{4+}, the strains will grow weaker, removing the barrier to PNR growth and preventing the formation of the characteristic rods of <110>p scattering.

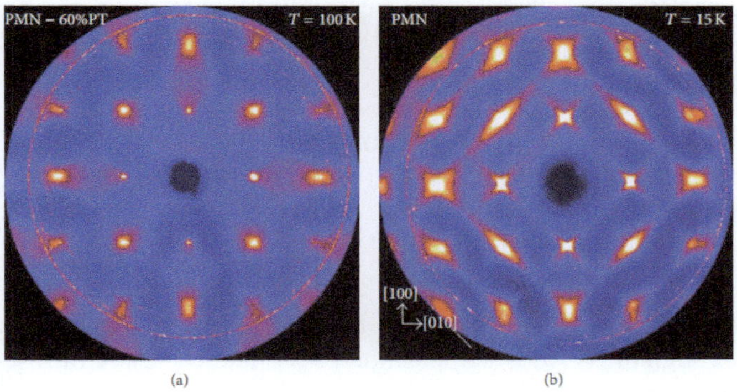

Figure 9: Diffuse scattering in the *hk*0 layer of (a) PMN-60PT at 100 K and (b) PMN at 15 K. The radical change in the form of the scattering with PT fraction (x) and with T is apparent. Reprinted with permission from Figure 4 of [28] (http://link.aps.org/doi/10.1103/ PhysRevB.73.064107). Copyright 2006 by the American Physical Society. (Readers may view, browse, and/or download material for temporary copying purposes only, provided that these uses are for noncommercial personal purposes. Except as provided by law, this material may not be further reproduced, distributed, transmitted, modified, adapted, performed, displayed, published, or sold in whole or part, without prior written permission from the American Physical Society.)

Relatively straight-forward crystal chemical considerations and structural work [154, 155] show that the Pb^{2+}displacements are very large—typically 0.5 Å—which is why the treatment in [23] is "thermal-like" but not truly thermal. One approach to this has been to suggest that atomic shifts possess two components—one from a soft mode and the other a uniform displacement of the PNR [126].

On the assumption that the PNRs are real, diffuse scattering has been used to explore their response to applied field [156], and it is found that for a field applied along [110] the diffuse scattering in the (hk0) plane is affected, indicating a possible redistribution of nanodomains of different polarisations, at least those perpendicular to the field, as some become more energetically favourable than others. Similar effects have been modelled for PZN in the nanodomain picture [44] with good qualitative agreement with observations [92].

Further results, obtained by measuring diffuse scattering with different neutron energies to probe different time scales, suggest that the PNRs in PMN are (relatively) static [143] up to 420 K, and that the Burns temperature, as measured using diffuse scattering, may depend on the energy of the probe used—as the "static" PNRs become more dynamic above 420 K, they fluctuate on the timescale of a slow probe, but not a faster probe, causing different measurements of T_d and implying a switch from a slowly to more quickly fluctuating PNR state. This is similar to but different from molecular dynamics results which, when used to interpret diffuse scattering over the range 10–700 K, suggest that strongly cooperative displacement behaviour sets in at 400 K on cooling [152]. This could also be interpreted as a value for T_d, as the PNRs/FNDs are the result of cooperative displacements.

Direct observation of PNRs through powder diffuse scattering (PDF analysis) has been claimed [151] though what actually manifest in the scattering are displacements and correlations. The work suggests that PNRs are "dispersed in a pseudocubic host lattice" [157] and that the PNRs begin to overlap as the sample cools to close to 200 K and increase in size and volume fraction of the sample but are affected by random fields due to the Mg/Nb disorder and are limited from growing further [158]. Other work also suggests 200 K as an important temperature—where possibly the domains truly manifest [152] and, most crucially, where the Curie temperature, T_c~213 K [123, 138], is found. Hence, the overall picture is one in which on cooling correlations set in at T_d but do not necessarily induce an actual domain structure, at least not one in which the displacements do not possess a significant random component, with both correlated and uncorrelated components fluctuating in time and space; close to 200 K, the domains lock in, possibly as they start interacting with each other more strongly and stabilising their existence, and the Curie temperature is obtained.

$PbSc_{0.5}Nb_{0.5}O_3$ and $PbSc_{0.5}Ta_{0.5}O_3$

There are a range of other Pb-containing ferroic materials. Two significant materials are $PbSc_{0.5}Nb_{0.5}O_3$ and $PbSc_{0.5}Ta_{0.5}O_3$ [159, 160].

Diffuse scattering has been observed in $PbSc_{0.5}Ta_{0.5}O_3$ and derivatives [29, 161, 162] and is of similar form to that seen in PMN and PZN—rods of scattering perpendicular to <110> directions such that "the atomic ferroic shifts correlate within {110} planes of the real space" [29]. Studies as a function of pressure and Ba B -site doping "chemical pressure" suggest that pressure suppresses the diffuse scattering and decouples B-site/Pb displacements and is interpreted in terms of PNRs [161, 162]. The -site stoichiometry of PST allows full anticorrelated ordering, which is seen in a cell doubling along all cubic directions. As a result, at room temperature, the diffuse scattering appears very similar to that from PZN; the diffuse spots which appear at (1/2)(1/2)(1/2)-type positions in PZN are sharp peaks occurring at 111-type positions for the doubled cell (Figure 10). That the increase in pressure suppresses the diffuse scattering accords with the bond valence picture of the driver for Pb^{2+} displacement, in that the Pb^{2+} environment will be reduced in size, lessening the need for displacement off-centre to increase the Pb^{2+}-bonding valence. This would also serve to decrease the -site/Pb displacement correlations, as a Pb^{2+} which is "happier" in its bonding environment will distort the oxygen lattice to a smaller degree, reducing any transmitted effects.

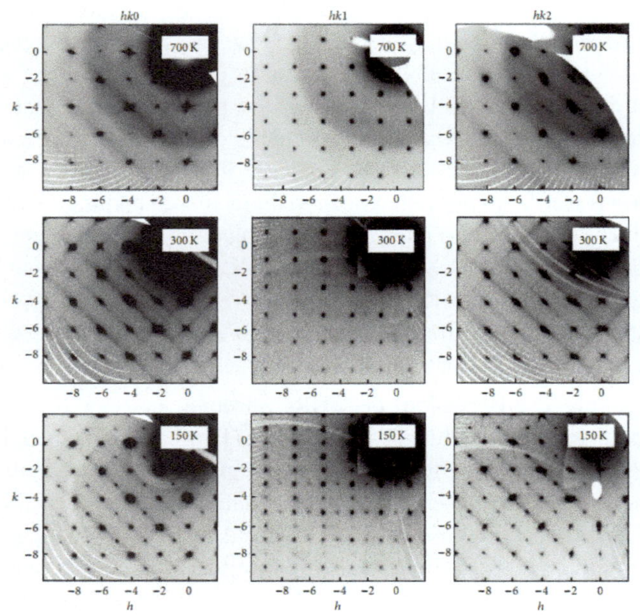

Figure 10: X-ray diffuse scattering from $PbSc_{0.5}Ta_{0.5}O_3$ as a function of temperature. The unit cell here is doubled relative to the parent perovskite cube. Note that the diffuse features at 300 K are very similar to those present in Figures 6 and 7, allowing that the indices here are doubled. Images taken with permission from [29] (http://hasyweb.desy.de/science/annual_reports/2007_report/part1/contrib/42/20848.pdf) and Figure 1 of [30] (http://link.aps.org/doi/10.1103/PhysRevB.79.224108). Copyright 2009 by the American Physical Society (see the following: Readers may view, browse, and/or download material for temporary copying purposes only, provided that these uses are for noncommercial personal purposes. Except as provided by law, this material may not be further reproduced, distributed, transmitted, modified, adapted, performed, displayed, published, or sold in whole or part, without prior written permission from the American Physical Society).

At low T, the diffuse scattering becomes less pronounced because the polar order is quite long ranged. The intersection of the <110> diffuse streaks with the hk1 becomes more apparent though it can be discerned at 300 K.

Diffuse scattering has been observed in $PbSc_{0.5}Nb_{0.5}O_3$ as well [30, 163–165] and again tends to show the characteristic <110>p (p= parent cell) diffuse streaks. These weaken substantially on cooling, rather like PST, whereas in doped PSN (with Ba and Bi in this very thorough study [30]) the diffuse scattering is more strongly maintained on cooling, something also apparent in Ba-doped PST [161]. This is strong evidence that, as discussed in relation to PMN, it is randomness that prevents the PNR from continuing to grow on cooling. In PST and PSN when the B -site cations are in a 1 : 1 ratio and can be strong ordered, such that there is a cell doubling, there is not the randomness that exists in PMN and PZN where the B-site cations are in ratio 2 : 1 and cannot alternate. Hence, on cooling, the PNR growth is less inhibited in PST/PSN, and the diffuse scattering sharpens to the point of almost vanishing—the ferroelectric order becomes virtually long ranged. This can be "remedied" by doping into the A site, introducing greater disorder and restoring the randomness and preventing large domains from forming.

Average structure determination, allowing the Pb to disorder across split sites, has shown that the Pb^{2+} in PSN are indeed displaced (by ~0.40 Å at 2 K), but the displacement directions are not well determined by these neutron powder diffraction data [165]. The observed diffuse scattering in the powder diffraction background suggests that on cooling the "correlation among Pb displacements along <111>pc directions increases to form polar domains" (pc= parent cell). They conclude that in the paraelectric phase the Pb^{2+} is distributed on many split sites around the Wyckoff position and that the minimum energy configuration changes with temperature, appearing to be most likely <111>pc at lower temperatures and <100>pc at the highest temperature considered.

$PbMg_{1/3}Ta_{2/3}O_3$

PMT, $PbMg_{1/3}Ta_{2/3}O_3$, is clearly related to PMN and PZN in terms of its chemistry, and it is perhaps not surprising that the diffuse scattering accords with this. PMT is a relaxor ferroelectric with a mean Curie temperature T_{cm}~170K [166].

Initial neutron diffuse scattering experiments used limited regions of reciprocal space—essentially, linear cuts through the (110) Bragg position [167]. These showed diffuse scattering whose intensity increased and width fell as T was reduced from room temperature to 100 K. These quantities then remained constant down to 20 K. The results suggested that displacements were static at low temperature and gave a low temperature correlation length, ξ , of ξ ~ 39 Å below 100 K, smaller than that has been suggested for PZN [137]. Given that the size difference between Mg and Ta is the largest exhibited in these Pb^{2+} containing compounds (the R_{ij} [132] for relevant species are: Mg^{2+} 1.693 Å, Zn^{2+}: 1.704 Å, Nb^{5+}: 1.911 Å, Ta^{5+}: 1.92 Å, and Sc^{3+}: 1.849 Å). Hence the B -site randomness coupled to the most pronounced size difference gives rise to the strongest random strain field and serves to inhibit large domain growth. In PSN and PST, where the B -site cations are present in a 1 : 1 ratio and a relatively more similar in size, the domains seem able to grow so large, as T is reduced that the diffuse scattering coalesces into sharp features (Figure 10).

Data collected over wider regions of reciprocal space using synchrotron X-rays at 175 K shows similar structure to that observed in PMN and PZN [31] (Figure 11) including evidence for partial -site ordering.

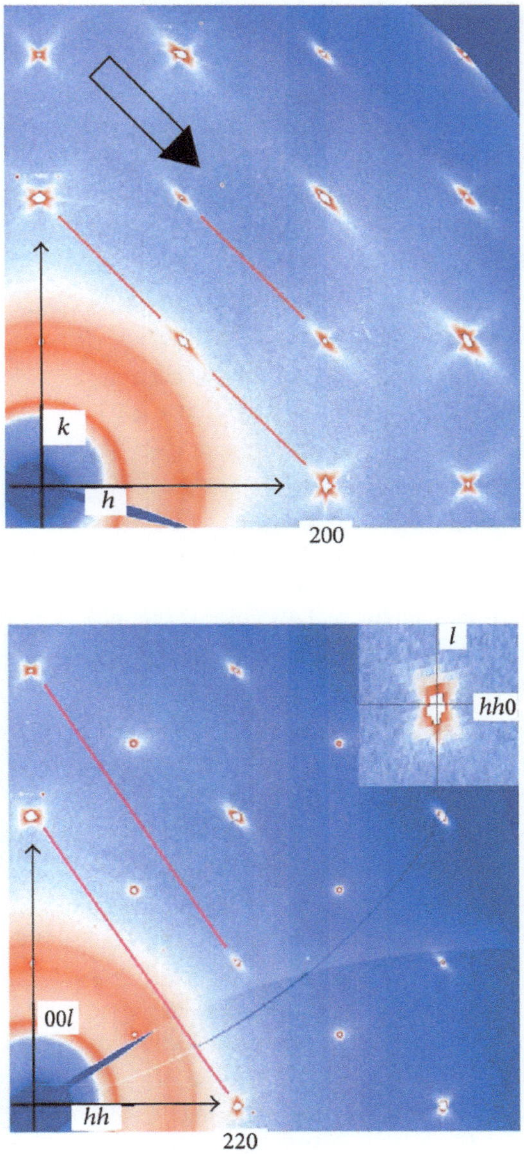

Figure 11: X-ray diffuse scattering from $PbMg_{1/3}Ta_{2/3}O_3$ in the (a) *hk*0 and (b) *hhl* planes, showing similar structure to that observed in PMN and PZN. The inset shows a close view of the 220 Bragg peak and the surrounding diffuse scattering. (Figure from [31], with permission of the IUCr, http://dx.doi.org/10.1107/S0021889811012635.)

These data are modelled with Pb^{2+} displacements along <111>pc. and anisotropic PNRs whose longest dimension is in <110>pc . PNR dimensions were estimated as $\sim 7a_{pc} \times 7a_{pc} \times 20a_{pc}$ (a_{pc}(being parent perovskite cube dimension). This is not greatly larger than the other estimate [167], except for the long axis which is approximately doubled. It should be noted however that the agreement between Figures 2 and 3 in [31] is moderate. The model does not capture the <110>pc rods of diffuse scattering that connect the motifs of scattering around the Bragg peaks. Figure 12 shows a line profile, integrated over a wide band of pixels, taken in the direction of the broad black outline arrow in Figure 11(a). It has been fitted with a Gaussian on a linear background and shows quite clearly that, though weak, the diffuse scattering extends right through the Brillouin zone and is not localised in the vicinity of the Bragg spots. This large extent in reciprocal space strongly suggests that the PNRs do indeed need to be more pancake-like in real space than the model presented in [31] suggests, only enhancing their similarity to those in PMN and PZN.

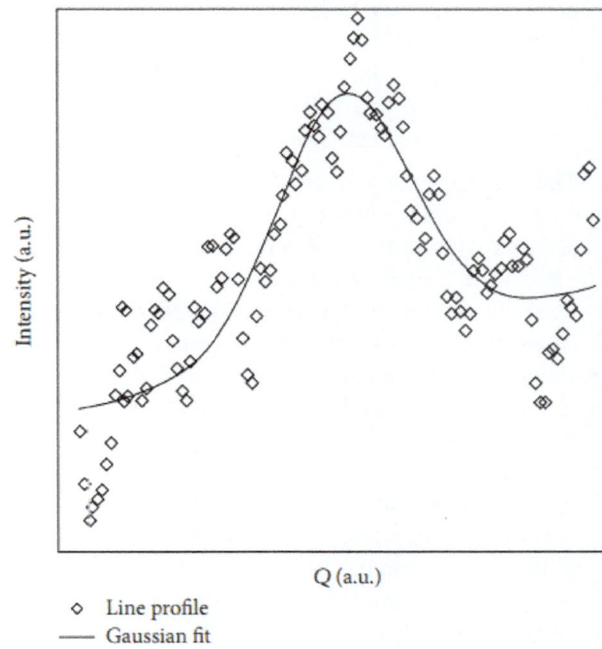

Figure 12: A line profile through the diffuse scattering shown in Figure 11(a), along the direction of the black arrow, demonstrating that the streaks of scattering do not just form "butterflies" and similar shapes around the Bragg spots, but also extend through reciprocal space much like those in PZN (Figure 6).

CONCLUSIONS

It appears that diffuse scattering, both single crystal and powder, does indeed give deep insight into the structure of these materials, but it does not deliver "unique" solutions but constrains the possible solutions. Into the future, energy-analysed diffuse scattering is likely to play an important role, as it provides insights onto the dynamic or static nature of the local correlations, adding important new information to the picture. Further, it is clear that modelling large windows of reciprocal space and preferably three-dimensional volumes are important when the scattering is so anisotropic and extended. Some models work well close to Bragg positions but less well near reciprocal cell boundaries, showing that measuring comprehensive data is necessary if models are to be adequately tested.

The Pb-based relaxors show great commonality of behaviour and similarities in their diffuse scattering. It appears that the idea of the PNR (or FND) may not be mathematically necessary to describe the locus of the diffuse scattering but nevertheless has a strong basis in crystal chemistry, electron imaging, and molecular dynamics modelling and still appears to be a reasonable model for the local order. The questions of whether the domains are static or dynamic and what aspects of their behaviour are crucial in determining the properties are yet to be fully resolved across the family. Is a domain truly an observable polarised volume of the crystal, like a small (often one- or two-dimensional) ferroic crystal embedded in the larger, or is it a relatively evanescent region in which some components of the displacements are correlated while others are random? As yet no definitive picture has emerged although measurements on both PZN and PMN suggest that the PNRs are fluctuating relatively slowly on the time scale of a thermal neutron experiment, which is not to say that random components of the displacements are not fluctuating much faster. It does appear that randomness, whether -site or -site, plays a central role, with the size mismatch and the level of doping being, along with relative to , key in determining the degree of PNR-related scattering.

The overall picture is one in which the Pb^{2+} ion is too small to "sit" in the centre of the O environment determined by the stiffer -site-to-oxygen bonds, and as an inherently polar atom due to its unpaired electrons resolves this by moving off-centre. In systems with chemical disorder, these shifts are prevented from becoming long-range ordered even at low by the random strain fields caused to the size mismatch of the -site cations (or -site when applicable, as in Ba-doped PST). Hence the nature of the local order and that of the relaxor properties appear to be determined by both the average Pb^{2+} environment and by the distribution of local environments—for example, PZ shows Pb^{2+} displacements but not PNRs, while PT, where Ti^{4+} is substantially smaller than Zr^{4+} [168], appears to show an ordered Pb^{2+} site, as might be more likely given that the -site environment is smaller and so the Pb^{2+} is less underbonded; and PZN and its derivatives and close relatives, with significant randomness, do show the PNR scattering.

ACKNOWLEDGMENTS

Thanks are due to the Australian Research Council for support through its Discovery Projects Program. D. J. Goossens thanks Dr. A. P. Heerdegen, Professor T. R. Welberry, Dr. Marek Pas'ciak, Dr. Jessica Hudspeth, and Ross Whitfield of the Research School of Chemistry at the Australian National University for their advice and assistance, but they take no responsibilities for any inaccuracies or opinions expressed herein. The support of the Australian Institute of Nuclear Science and Engineering is appreciated. Thanks are due to Professor Dr. Boriana Mihailova for supplying images used in Figure 10. Use of the Advanced Photon Source, an Office of Science User Facility operated for the U.S. Department of Energy (DOE) Office of Science by Argonne National Laboratory, was supported by the U.S. DOE under Contract no. DE-AC02-06CH11357. The granting of permission to reproduce figures from the noted publications is appreciated. Some aspects of this research were undertaken on the NCI National Facility in Canberra, Australia, which is supported by the Australian Commonwealth Government.

REFERENCES

1. W. Friedrich, "Röntgenstrahlinterferenzen," Physikalische Zeitschrift, vol. 14, pp. 1079–1087, 1913.
2. K. Lonsdale and H. Smith, "An experimental study of diffuse x-ray reflexion by single crystals,"Proceedings of the Royal Society A, vol. 179, pp. 8–50.
3. R. I. Barabash, G. E. Ice, and P. E. A. Turchi, Diffuse Scattering and the Fundamental Properties of Materials, Momentum Press, 1st edition, 2009.
4. M. A. Krivoglaz, Diffuse Scattering of X-Rays and Neutrons by Fluctuations, Springer, Berlin, Germany, 1996.
5. M. A. Krivoglaz, Theory of X-Ray and Thermal-Neutron Scattering by Real Crystals, Plenum Press, New York, NY, USA, 1969.
6. D. W. L. Hukins, X-Ray Diffraction by Ordered and Disorderd Systems, Pergamon Press, New York, NY, USA, 1981.
7. W. A. Wooster, Diffuse X-Ray Reflections from Crystals, Clarendon Press, Oxford, UK, 1962.
8. T. R. Welberry, Diffuse X-Ray Scattering and Models of Disorder, Oxford University Press, 2004.
9. G. Harburn, C. A. Taylor, and T. R. Welberry, Atlas of Optical Transforms, Bell, London, UK, 1975.
10. Th. Proffen, "Analysis of occupational and displacive disorder using the atomic pair distribution function: a systematic investigation," Zeitschrift fur Kristallographie, vol. 215, no. 11, pp. 661–668, 2000. ·
11. W. Schweika, Disordered Alloys: Diffuse Scattering and Monte Carlo Simulations, Springer, 1998.
12. R. B. Neder and Th. Proffen, Diffuse Scattering and Defect Structure Simulations: A Cook Book Using the Program DISCUS, OUP, 2008.
13. V. M. Nield and D. A. Keen, Diffuse Neutron Scattering from Crystalline Materials, OUP, Oxford, UK, 2001.
14. S. J. L. Billinge and M. F. Thorpe, Local Structure from Diffraction, Plenum, New York, NY, USA, 1998.
15. B. D. Butler and T. R. Welberry, "Interpretation of displacement-caused diffuse scattering using the Taylor expansion," Acta Crystallographica A, vol. 49, no. 5, pp. 736–743, 1993.
16. B. E. Warren, B. L. Averbach, and B. W. Roberts, "Atomic size effect in the x-ray scattering by alloys,"Journal of Applied Physics, vol. 22, no. 12, pp. 1493–1496, 1951.
17. T. J. Hicks, "Experiments with neutron polarization analysis," Advances in Physics, vol. 45, no. 4, pp. 243–298, 1996.
18. T. J. Hicks, Magnetism in Disorder, OUP, 1995.
19. J. R. Stewart, P. P. Deen, K. H. Andersen et al., "Disordered materials studied using neutron polarization analysis on the multi-detector spectrometer, D7," Journal of Applied Crystallography, vol. 42, no. 1, pp. 69–84, 2009.
20. W. Schweika and P. Böni, "The instrument DNS: polarization analysis for diffuse neutron scattering,"Physica B, vol. 297, pp. 155–159, 2001.
21. H. Faxén, "Die bei interferenz von röntgenstrahlen infolge der wärmebewegung entstehende streustrahlung," Zeitschrift für Physik, vol. 17, pp. 266–278, 1923.
22. T. R. Welberry and A. G. Christy, "Defect distribution and the diffuse x-ray diffraction pattern of wüstite, $Fe_{1-x}O$," Physics and Chemistry of Minerals, vol. 24, pp. 24–38, 1997.
23. A. Bosak, D. Chernyshov, S. Vakhrushev, and M. Krisch, "Diffuse scattering in relaxor ferroelectrics: true three-dimensional mapping, experimental artefacts and

modelling," Acta Crystallographica A, vol. 68, no. 1, pp. 117–123, 2012.

24. Y. Kuroiwa, Y. Terado, S. J. Kim et al., "High-energy SR powder diffraction evidence of multisite disorder of Pb atom in cubic phase of $PbZr_{1-x}TixO_3$," Japanese Journal of Applied Physics A, vol. 44, no. 9B, pp. 7151–7155, 2005.
25. T. R. Welberry, D. J. Goossens, R. L. Withers, and K. Z. Baba-Kishi, "Monte Carlo simulation study of diffuse scattering in PZT, Pb(Zr,Ti)O 3," Metallurgical and Materials Transactions A, vol. 41, no. 5, pp. 1110–1118, 2010.
26. M. Paściak, A. P. Heerdegen, D. J. Goossens, R. E. Whitfield, A. Pietraszko, and T. R. Welberry, "Assessing local structure in $PbZn_{1/3}Nb_{2/3}O_3$ using diffuse scattering and reverse Monte Carlo refinement,"Metallurgical and Materials Transactions A, vol. 44, pp. 87–93, 2013.
27. T. R. Welberry, M. J. Gutmann, H. Woo et al., "Single-crystal neutron diffuse scattering and Monte Carlo study of the relaxor ferroelectric $PbZn_{1/3}Nb_{2/3}O_3$ (PZN)," Journal of Applied Crystallography, vol. 38, no. 4, pp. 639–647, 2005.
28. C. Stock, D. Ellis, I. P. Swainson et al., "Damped soft phonons and diffuse scattering in 40%Pb $(Mg_{1/3}Nb_{2/3})O_3$-60%$PbTiO_3$," Physical Review B, vol. 73, no. 6, Article ID 064107, 2006.
29. B. Mihailova, B. Maier, C. Paulmann et al., "High-temperature structural transformations in the relaxor ferroelectrics $PbSc_{0.5}Ta_{0.5}O_3$ and $Pb_{0.78}Ba_{0.22}Sc_{0.5}Ta_{0.5}O_3$," Physical Review B, vol. 77, no. 17, Article ID 174106, 2008.
30. B. Maier, B. Mihailova, C. Paulmann et al., "Effect of local elastic strain on the structure of Pb-based relaxors: a comparative study of pure and Ba- and Bi-doped $PbSc_{0.5}Nb_{0.5}O_3$," Physical Review B, vol. 79, Article ID 224108, 2009.
31. A. Cervellino, S. N. Gvasaliya, O. Zaharko et al., "Diffuse scattering from the lead-based relaxor ferroelectric $PbMg_{1/3}Ta_{2/3}O_3$," Journal of Applied Crystallography, vol. 44, no. 3, pp. 603–609, 2011.
32. T. R. Welberry, D. J. Goossens, A. P. Heerdegen, and P. L. Lee, "Problems in measuring diffuse X-ray scattering," Zeitschrift fur Kristallographie, vol. 220, no. 12, pp. 1052–1058, 2005.
33. H. Jagodzinski and F. Frey, International Tables for Crystallography, vol. B, 2006.
34. M. J. Gutmann, SXD2001, ISIS Facility, Rutherford Appleton Laboratory, Oxfordshire, UK, 2005.
35. R. E. Whitfield, D. J. Goossens, A. J. Studer, and J. S. Forrester, "Measuring single-crystal diffuse neutron scattering on the wombat high-intensity powder diffractometer," Metallurgical and Materials Transactions A, vol. 43, pp. 1423–1428, 2012.
36. M. A. Estermann and W. Steurer, "Diffuse scattering data acquisition techniques," Phase Transitions, vol. 67, no. 1, pp. 165–195, 1998.
37. S. Scheidegger, M. A. Estermann, and W. Steurer, "Correction of specimen absorption in X-ray diffuse scattering experiments with area-detector systems," Journal of Applied Crystallography, vol. 33, no. 1, pp. 35–48, 2000.
38. S. Rosenkranz and R. Osborn, "Corelli: efficient single crystal diffraction with elastic discrimination,"Pramana, vol. 71, no. 4, pp. 705–711, 2008.
39. J. C. Osborn and T. R. Welberry, "A position-sensitive detector system for the measurement of diffuse X-ray scattering," Journal of Applied Crystallography, vol. 23, pp. 476–484, 1990.
40. H. J. Lamfers, A Diffuse X-Ray Scattering Diffractometer, Rijksuniversiteit Groningen, 1997.
41. R. Born and D. Hohlwein, "Simultaneous energy analysis in a large angular range: A novel neutron spectrometer, its resolution and applications," Zeitschrift für Physik B Condensed Matter, vol. 74, no. 4, pp. 547–555, 1989.
42. E. J. Chan and D. J. Goossens, "Study of the single-crystal X-ray diffuse scattering in paracetamol polymorphs," Acta Crystallographica B, vol. 68, pp. 80–88, 2012.
43. T. R. Welberry, D. J. Goossens, W. I. F. David, M. J. Gutmann, M. J. Bull, and A. P. Heerdegen, "Diffuse neutron scattering in benzil, C14D10O2, using the time-of-flight Laue technique," Journal of Applied Crystallography, vol. 36, no. 6, pp. 1440–1447, 2003.
44. T. R. Welberry, D. J. Goossens, and M. J. Gutmann, "Chemical origin of nanoscale polar domains in PbZn1 3 Nb2 3 O_3," Physical Review B, vol. 74, no. 22, Article ID 224108, 2006.
45. D. J. Goossens, A. G. Beasley, T. R. Welberry, M. J. Gutmann, and R. O. Piltz, "Neutron diffuse scattering in deuterated para-terphenyl, C 18D14," Journal of Physics Condensed Matter, vol. 21, no. 12, Article ID 124204, 2009.
46. R. E. Whitfield, D. J. Goossens, and A. J. Studer, "Temperature dependence of diffuse scattering in PZN,"Metallurgical and Materials Transactions A, vol. 43, pp. 1429–1433, 2012.
47. A. Cervellino, S. N. Gvasaliya, B. Roessli et al., "Cube-shaped diffuse scattering and the ground state of $BaMg_{1/3}Ta_{2/3}O_3$," Physical Review B, vol. 86, Article ID 104107, 12 pages, 2012.
48. E. J. Chan, T. R. Welberry, D. J. Goossens, and A. P. Heerdegen, "A refinement strategy for Monte Carlo modelling of diffuse scattering from molecular crystal systems," Journal of Applied Crystallography, vol. 43, no. 4, pp. 913–915, 2010.
49. E. J. Chan, T. R. Welberry, D. J. Goossens, A. P. Heerdegen, A. G. Beasley, and P. J. Chupas, "Single-crystal diffuse scattering studies on polymorphs of molecular crystals. I. the room-temperature polymorphs of the drug benzocaine," Acta Crystallographica B, vol. 65, no. 3, pp. 382–392, 2009.
50. N. Ahmed, S. J. Campbell, and T. J. Hicks, "LONGPOL II: analysis of spin-flip and nonspin-flip neutron scattering," Journal of Physics E, vol. 7, no. 3, article no. 318, pp. 199–204, 1974.
51. T. Ersez, S. J. Kennedy, T. J. Hicks, Y. Fei, Th.

Krist, and P. A. Miles, "New features of the long-wavelength polarisation analysis spectrometer—LONGPOL," Physica B, vol. 335, no. 1–4, pp. 183–187, 2003.

52. E. Elkaim, S. Lefebvre, R. Kahn, J. F. Berar, M. Lemonnier, and M. Bessiere, "Diffraction and diffuse scattering measurements in material science: Improvement brought to a six-circle goniometer on a synchrotron beam line," Review of Scientific Instruments, vol. 63, no. 1, pp. 988–991, 1992.

53. M. W. Wall, S. E. Ealick, and S. M. Gruner, "Three-dimensional diffuse x-ray scattering from crystals ofStaphylococcal nuclease," Proceedings of the National Academy of Sciences of the United States of America, vol. 94, pp. 6180–6184, 1997.

54. G. Xu, Z. Zhong, H. Hiraka, and G. Shirane, "Three-dimensional mapping of diffuse scattering in $Pb(Zn_{1/3}Nb_{2/3})O_{3-x}PbTiO_3$," Physical Review B, vol. 70, Article ID 174109, 17 pages, 2004.

55. T. Weber and H.-B. Bürgi, "Determination and refinement of disordered crystal structures using evolutionary algorithms in combination with Monte Carlo methods," Acta Crystallographica A, vol. 58, pp. 526–540, 2002.

56. T. Weber, A. Simon, H. Mattausch, L. Kienle, and O. Oeckler, "Reliability of Monte Carlo simulations of disordered structures optimized with evolutionary algorithms exemplified with diffuse scattering from $La_{0.70(1)}(Al_{0.14(1)}I_{0.86(1)})$," Acta Crystallographica A, vol. 64, no. 6, pp. 641–653, 2008. · ·

57. T. R. Welberry, D. J. Goossens, D. R. Haeffner, P. L. Lee, and J. Almer, "High-energy diffuse scattering on the 1-ID beamline at the advanced photon source," Journal of Synchrotron Radiation, vol. 10, no. 3, pp. 284–286, 2003.

58. A. Gibaud, D. Harlow, J. B. Hastings, J. P. Hill, and D. Chapman, "A high-energy monochromatic laue (monolaue) X-ray diffuse scattering study of KMnF3 using an image plate," Journal of Applied Crystallography, vol. 30, no. 1, pp. 16–20, 1997.

59. G. Honjo, S. Kodera, and N. Kitamura, "Diffuse streak diffraction patterns from single crystals I. General discussion and aspects of electron diffraction diffuse streak patterns," Journal of the Physical Society of Japan, vol. 19, no. 3, pp. 351–367, 1964.

60. C. Randall, D. Barber, R. Whatmore, and P. Groves, "Short-range order phenomena in lead-based perovskites," Ferroelectrics, vol. 76, Article ID 1, pp. 277–282, 1987.

61. S. R. Andrews and R. A. Cowley, "X-ray scattering from critical fluctuations and domain walls in KDP and DKDP," Journal of Physics C, vol. 19, no. 4, 1986.

62. Y. Fujii and Y. Yamada, "X-ray critical scattering in ferroelectric tri-glycine sulphate," Journal of the Physical Society of Japan, vol. 30, no. 6, pp. 1676–1685, 1971.

63. D. J. Goossens, A. P. Heerdegen, T. R. Welberry, and M. J. Gutmann, "Monte Carlo analysis of neutron diffuse scattering data," Physica B, vol. 385-386, pp. 1352–1354, 2006.

64. T. R. Welberry, D. J. Goossens, A. J. Edwards, and W. I. F. David, "Diffuse X-ray scattering from benzil, C14H10O2: Analysis via automatic refinement of a Monte Carlo model," Acta Crystallographica A, vol. 57, no. 1, pp. 101–109, 2001.

65. J. Harada and G. Honjo, "X-ray studies of the lattice vibration in tetragonal barium titanate," Journal of the Physical Society of Japan, vol. 22, no. 1, pp. 45–57, 1967.

66. J. Kreisel, P. Bouvier, B. Dkhil et al., "Effect of high pressure on the relaxor ferroelectrics $Na_{1/2}Bi_{1/2}TiO_3$(NBT) and $PbMg_{1/3}Nb_{2/3}O_3$ (PMN)," Ferroelectrics, vol. 302, pp. 293–298, 2004. ·

67. J. F. Scott, "Prospects for ferroelectrics: 2012–2022," ISRN Materials Science, vol. 2013, Article ID 187313, 24 pages, 2013.

68. R. Comes, M. Lambert, and A. Guinier, "The chain structure of $BaTiO_3$ and $KNbO_3$," Solid State Communications, vol. 6, no. 10, pp. 715–719, 1968.

69. J. Hlinka, T. Ostapchuk, D. Nuzhnyy et al., "Coexistence of the phonon and relaxation soft modes in the terahertz dielectric response of tetragonal $BaTiO_3$," Physical Review Letters, vol. 101, no. 16, Article ID 167402, 2008.

70. L. E. Cross, "Relaxor ferroelectrics," Ferroelectrics, vol. 76, pp. 241–267, 1987.

71. A. A. Bokov and Z.-G. Ye, "Recent progress in relaxor ferroelectrics with perovskite structure," Journal of Materials Science, vol. 41, no. 1, pp. 31–52, 2006.

72. J. Hlinka, "Do we need the ether of polar nanoregions?" Journal of Advanced Dielectrics, vol. 2, no. 2, Article ID 1241006, 2012.

73. C. R. Martin and I. A. Aksay, "Topographical evolution of lead zirconate titanate (PZT) thin films patterned by micromolding in capillaries," Journal of Physical Chemistry B, vol. 107, no. 18, pp. 4261–4268, 2003.

74. R. Bogue, "Energy harvesting and wireless sensors: a review of recent developments," Sensor Review, vol. 29, no. 3, pp. 194–199, 2009.

75. N. Setter, D. Damjanovic, L. Eng et al., "Ferroelectric thin films: rssssssssseview of materials, properties, and applications," Journal of Applied Physics, vol. 100, no. 5, Article ID 051606, 2006. · ·

76. M. Sitti, D. Campolo, J. Yan et al., "Development of PZT and PZN-PT based unimorph actuators for micromechanical flapping mechanisms," in Proceedings of the IEEE International Conference on Robotics and Automation (ICRA ‹01), vol. 4, pp. 3839–3846, May 2001.

77. S. E. Park and W. Hackenberger, "High performance single crystal piezoelectrics: Applications and issues," Current Opinion in Solid State and Materials Science, vol. 6, no. 1, pp. 11–18, 2002.

78. L. A. Ivan, M. Rakotondrabe, J. Agnus et al., "Comparative material study between PZT ceramic and newer crystalline PMN-PT and PZN-PT materials for composite bimorph actuators," Reviews on Advanced Materials Science, vol. 24, no. 1-2, pp. 1–9, 2010.

79. H. Uršič, M. S. Zarnik, and M. Kosec, "Pb($Mg_{1/3}Nb_{2/3}$)O_3-$PbTiO_3$ (PMN-PT) material for actuator applications," Smart Materials Research, vol. 2011, Article ID 452901, 6 pages, 2011.

80. S. H. Baek, J. Park, D. M. Kim et al., "Giant piezoelectricity on Si for hyperactive MEMS," Science, vol. 334, no. 6058, pp. 958–961, 2011.

81. M. Frank, K. S. Moon, and S. Kassegne, "A PMMA coated PMN-PT single crystal resonator for sensing chemical agents," Smart Materials and Structures, vol. 19, no. 3, Article ID 035015, 2010.

82. T. Takenaka, "Piezoelectric properties of some lead-free ferroelectric ceramics," Ferroelectrics, vol. 230, no. 1, pp. 87–98, 1999.

83. E. A. Patterson and D. P. Cann, "Bipolar piezoelectric fatigue of Bi($Zn_{0.5}Ti_{0.5}$)O_3-($Bi_{0.5}K_{0.5}$)TiO_3-($Bi_{0.5}Na_{0.5}$)TiO_3 Pb-free ceramics," Applied Physics Letters, vol. 101, no. 4, Article ID 042905, 2012.

84. J. W. Bennett, I. Grinberg, P. K. Davies, and A. M. Rappe, "Pb-free ferroelectrics investigated with density functional theory: $SnAl_{1/2}Nb_{1/2}O_3$ perovskites," Physical Review B, vol. 83, Article ID 144112, 6 pages, 2011.

85. Y. Uratani, T. Shishidou, and T. Oguchi, "First-principles study of lead-free piezoelectric $SnTiO_3$,"Japanese Journal of Applied Physics, vol. 47, no. 9, pp. 7735–7739, 2008.

86. A. I. Lebedev, "Ab initio calculations of phonon spectra in $ATiO_3$ perovskite crystals (A = Ca, Sr, Ba, Ra, Cd, Zn, Mg, Ge, Sn, Pb)," Physics of the Solid State, vol. 51, no. 2, pp. 362–372, 2009.

87. B. Mihailova, B. Maier, C. Paulmann et al., "High-temperature structural transformations in the relaxor ferroelectrics $PbSc_{0.5}Ta_{0.5}O_3$ and $Pb_{0.78}Ba_{0.22}Sc_{0.5}Ta_{0.5}O_3$," Physical Review B, vol. 77, Article ID 174106, 10 pages, 2008.

88. P. Ganesh, E. Cockayne, M. Ahart et al., "Origin of diffuse scattering in relaxor ferroelectrics," Physical Review B, vol. 81, no. 14, Article ID 144102, 2010.

89. M. Paściak, M. Wołcyrz, and A. Pietraszko, "Interpretation of the diffuse scattering in Pb-based relaxor ferroelectrics in terms of three-dimensional nanodomains of the ⟨110⟩-directed relative interdomain atomic shifts," Physical Review B, vol. 76, no. 1, Article ID 014117, 2007.

90. T. R. Welberry and D. J. Goossens, "The interpretation and analysis of diffuse scattering using Monte Carlo simulation methods," Acta Crystallographica A, vol. 64, no. 1, pp. 23–32, 2007.

91. G. Xu, Z. Zhong, Y. Bing, Z.-G. Ye, C. Stock, and G. Shirane, "Ground state of the relaxor ferroelectric Pb($Zn_{1/3}Nb_{2/3}$)O_3," Physical Review B, vol. 67, Article ID 104102, 5 pages, 2003.

92. G. Xu, Z. Zhong, Y. Bing, Z.-G. Ye, and G. Shirane, "Electric-field-induced redistribution of polar nano-regions in a relaxor ferroelectric," Nature Materials, vol. 5, no. 2, pp. 134–140, 2006.

93. T. R. Gururaja, R. K. Panda, J. Chen, and H. Beck, "Single crystal transducers for medical imaging applications," in Proceedings of the IEEE Ultrasonics Symposium, pp. 969–972, October 1999.

94. Q. F. Zhou, H. L. W. Chan, and C. L. Choy, "PZT ceramic/ceramic 0–3 nanocomposite films for ultrasonic transducer applications," Thin Solid Films, vol. 375, no. 1-2, pp. 95–99, 2000.

95. L. Lebrun, G. Sebald, B. Guiffard, C. Richard, D. Guyomar, and E. Pleska, "Investigations on ferroelectric PMN-PT and PZN-PT single crystals ability for power or resonant actuators," Ultrasonics, vol. 42, no. 1-9, pp. 501–505, 2004.

96. Z. Yang, X. Chao, R. Zhang, Y. Chang, and Y. Chen, "Fabrication and electrical characteristics of piezoelectric PMN-PZN-PZT ceramic transformers," Materials Science and Engineering B, vol. 138, no. 3, pp. 277–283, 2007.

97. B. D. Chapman, E. A. Stern, S.-W. Han et al., "Diffuse x-ray scattering in perovskite ferroelectrics,"Physical Review B, vol. 71, no. 2, Article ID 020102, 2005.

98. E. Sawaguchi, H. Maniwa, and S. Hoshino, "Antiferroelectric structure of lead zirconate," Physical Review, vol. 83, no. 5, p. 1078, 1951.

99. A. M. Glazer, K. Roleder, and J. Dec, "Structure and disorder in single-crystal lead zirconate, $PbZrO_3$,"Acta Crystallographica B, vol. 49, no. 5, pp. 846–852, 1993.

100. S. Teslic and T. Egami, "Atomic structure of $PbZrO_3$ determined by pulsed neutron diffraction," Acta Crystallographica B, vol. 54, no. 6, pp. 750–765, 1998.

101. X. Dai, J. Li, and D. Viehland, "Weak ferroelectricity in antiferroelectric lead zirconate," Physical Review B, vol. 51, no. 5, pp. 2651–2655, 1995.

102. H. Yokota, N. Zhang, A. E. Taylor, P. A. Thomas, and A. M. Glazer, "Crystal structure of the rhombohedral phase of $PbZr_{1-x}TixO_3$ ceramics at room temperature," Physical Review B, vol. 80, Article ID 104109, 12 pages, 2009.

103. J. Frantti, S. Ivanov, S. Eriksson et al., "Neutron diffraction and bond-valence calculation studies of Pb(Zr_xTi_{1-x})O_3 Ceramics," Ferroelectrics, vol. 272, no. 1, pp. 51–56, 2002.

104. J. Ricote, D. L. Corker, R. W. Whatmore et al., "A TEM and neutron diffraction study of the local structure in the rhombohedral phase of lead zirconate titanate," Journal of Physics Condensed Matter, vol. 10, no. 8, pp. 1767–1786, 1998.

105. D. Viehland, Z. Xu, and D. A. Payne, "Origin of F spots and stress sensitivity in lanthanum lead zirconate titanate," Journal of Applied Physics, vol. 74, no. 12, pp. 7454–7460, 1993.

106. K. Z. Baba-Kishi, T. R. Welberry, and R. L. Withers, "An electron diffraction and Monte Carlo simulation study of diffuse scattering in Pb(Zr,Ti)O_3," Journal of Applied Crystallography, vol. 41, no. 5, pp. 930–938, 2008.

107. A. M. Glazer, P. A. Thomas, K. Z. Baba-Kishi, G. K. H. Pang, and C. W. Tai, "Influence of short-range and long-range order on the evolution of the morphotropic phase boundary in Pb($Zr_{1-x}Ti_x$)O_3," Physical Review B, vol. 70, Article ID 184123, 9 pages, 2004.

108. R. G. Burkovsky, A. Yu. Bronwald, A. V. Filimonov et al., "Structural heterogeneity and diffuse scattering

in morphotropic lead zirconate-titanate single crystals," Physical Review Letters, vol. 109, Article ID 097603, 4 pages, 2012.

109. D. L. Corker, A. M. Glazer, R. W. Whatmore, A. Stallard, and F. Fauth, "A neutron diffraction investigation into the rhombohedral phases of the perovskite series PbZr1-xTixO$_3$," Journal of Physics Condensed Matter, vol. 10, no. 28, pp. 6251–6269, 1998. ·

110. S. Adams, "Relationship between bond valence and bond softness of alkali halides and chalcogenides,"Acta Crystallographica B, vol. 57, no. 3, pp. 278–287, 2001. ·

111. I. D. Brown, "Recent developments in the methods and applications of the bond valence model,"Chemical Reviews, vol. 109, no. 12, pp. 6858–6919, 2009. ·

112. R. G. Burkovsky, A. V. Filimonov, A. I. Rudskoy, K. Hirota, M. Matsuura, and S. B. Vakhrushev, "Diffuse scattering anisotropy and inhomogeneous lattice deformations in the lead magnoniobate relaxor PMN above the Burns temperature," Physical Review B, vol. 85, no. 9, Article ID 094108, 2012.

113. S. Vakhrushev, A. Nabereznov, S. K. Sinha, Y. P. Feng, and T. Egami, "Synchrotron X-ray scattering study of lead magnoniobate relaxor ferroelectric crystals," Journal of Physics and Chemistry of Solids, vol. 57, no. 10, pp. 1517–1523, 1996.

114. H. Hiraka, S.-H. Lee, P. M. Gehring, G. Xu, and G. Shirane, "Cold neutron study on the diffuse scattering and phonon excitations in the relaxor $Pb(Mg_{1/3}Nb_{2/3})O_3$," Physical Review B, vol. 70, no. 18, Article ID 184105, pp. 1–7, 2004.

115. S. Teslic, T. Egami, and D. Viehland, "Local atomic structure of PZT and PLZT studied by pulsed neutron scattering," Journal of Physics and Chemistry of Solids, vol. 57, no. 10, pp. 1537–1543, 1996.

116. M. El Marssi, R. Farhi, J.-L. Dellis, M. D. Glinchuk, L. Seguin, and D. Viehland, "Ferroelectric and glassy states in La-modified lead zirconate titanate ceramics: a general picture," Journal of Applied Physics, vol. 83, no. 10, pp. 5371–5380, 1998.

117. R. L. Withers, Y. Liu, and T. R. Welberry, "Structured diffuse scattering and the fundamental 1-d dipolar unit in PLZT (Pb1-yLay)1-α(Zr1-xTix)1-βO$_3$ (7.5/65/35 and 7.0/60/40) transparent ferroelectric ceramics," Journal of Solid State Chemistry, vol. 182, no. 2, pp. 348–355, 2009.

118. Y. Terado, S. J. Kim, C. Moriyoshi, Y. Kuroiwa, M. Iwata, and M. Takata, "Disorder of Pb atom in cubic structure of $Pb(Zn_{1/3}Nb_{2/3})O_3$-$PbTiO_3$ system," Japanese Journal of Applied Physics A, vol. 45, no. 9B, pp. 7552–7555, 2006.

119. J. Kuwata and K. Uchino, "Dielectric and piezoelectric properties of $0.91Pb(Zn_{1/3}Nb_{2/3})O_3$-$0.09PbTiO_3$single crystals," Japanese Journal of Applied Physics, vol. 21, Article ID 1298, 1982. ·

120. J. Kuwata, K. Uchino, and S. Nomura, "Phase transitions in the Pb $(Zn_{1/3}Nb_{2/3})O_3$-$PbTiO_3$ system,"Ferroelectrics, vol. 37, pp. 579–582, 1981.

121. J. S. Forrester, E. H. Kisi, K. S. Knight, and C. J. Howard, "Rhombohedral to cubic phase transition in the relaxor ferroelectric PZN," Journal of Physics Condensed Matter, vol. 18, no. 19, pp. L233–L240, 2006. ·

122. E. H. Kisi and J. S. Forrester, "Crystal structure of the relaxor ferroelectric PZN: demise of the 'X-phase',"Journal of Physics Condensed Matter, vol. 17, no. 36, pp. L381–L384, 2005.

123. P. Bonneau, P. Garnier, E. Husson, and A. Morell, "Structural study of PMN ceramics by X-ray diffraction between 297 and 1023 K," Materials Research Bulletin, vol. 24, no. 2, pp. 201–206, 1989.

124. C. Stock, R. J. Birgeneau, S. Wakimoto et al., "Universal static and dynamic properties of the structural transition in $Pb(Zn_{1/3}Nb_{2/3})O_3$," Physical Review B, vol. 69, no. 9, Article ID 094104, 2004.

125. H. You, "Diffuse X-ray scattering study of lead magnesium niobate single crystals," Physical Review Letters, vol. 79, no. 20, pp. 3950–3953, 1997.

126. K. Hirota, Z.-G. Ye, S. Wakimoto, P. M. Gehring, and G. Shirane, "Neutron diffuse scattering from polar nanoregions in the relaxor $Pb(Mg_{1/3}Nb_{2/3})O_3$," Physical Review B, vol. 65, no. 10, Article ID 104105, 7 pages, 2002.

127. G. Burns and F. H. Dacol, "Crystalline ferroelectrics with glassy polarization behavior," Physical Review B, vol. 28, no. 5, pp. 2527–2530, 1983.

128. D. La-Orauttapong, J. Toulouse, J. L. Robertson, and Z.-G. Ye, "Diffuse neutron scattering study of a disordered complex perovskite $Pb(Zn_{1/3}Nb_{2/3})O_3$ crystal," Physical Review B, vol. 64, Article ID 212101, 4 pages, 2001.

129. D. La-Orauttapong, J. Toulouse, Z.-G. Ye, W. Chen, R. Erwin, and J. L. Robertson, "Neutron scattering study of the relaxor ferroelectric $(1-x)Pb(Zn_{1/3}Nb_{2/3})O_{3-x}PbTiO_3$," Physical Review B, vol. 67, Article ID 134110, 10 pages, 2003.

130. M. Soda, M. Matsuura, Y. Wakabayashi, and K. Hirota, "Superparamagnetism induced by polar nanoregions in relaxor Ferroelectric $(1-x)BiFeO_3$-$xBaTiO_3$," Journal of the Physical Society of Japan, vol. 80, no. 4, Article ID 043705, 2011.

131. Y. Yoneda, K. Yoshii, S. Kohara, S. Kitagawa, and S. Mori, "Local structure of $BiFeO_3$-$BaTiO_3$ mixture,"Japanese Journal of Applied Physics, vol. 47, no. 9, pp. 7590–7594, 2008.

132. N. E. Brese and M. O'Keeffe, "Bond-valence parameters for solids. Bond-valence parameters for solids,"Acta Crystallographica B, vol. 47, pp. 192–197, 1991.

133. S. V. Krivovichev, "Derivation of bond-valence parameters for some cation-oxygen pairs on the basis of empirical relationships between r_o and b," Zeitschrift für Kristallographie, vol. 227, pp. 575–579, 2012. ·

134. T. R. Welberry and D. J. Goossens, "Different models for the polar nanodomain structure of PZN and other relaxor ferroelectrics," Journal of Applied Crystallography, vol. 41, no. 3, pp. 606–614, 2008.

135. L. Xie, Y. L. Li, R. Yu et al., "Static and dynamic polar nanoregions in relaxor ferroelectric $Ba(Ti_{1-x}Sn_x)O_3$system at high temperature," Physical Review B, vol.

85, Article ID 014118, 5 pages, 2012.

136. Z. Guo, R. Tai, H. Xu et al., "X-ray probe of the polar nanoregions in the relaxor ferroelectric $0.72Pb(Mg_{1/3}Nb_{2/3})O_3$-$0.28PbTiO_3$," Applied Physics Letters, vol. 91, no. 8, Article ID 081904, 2007.

137. S. B. Vakchrushev, B. E. Kvyatkovsky, A. A. Nabereznov, N. M. Okuneva, and B. P. Toperverg, "Neutron scattering from disordered perovskite-like crystals and glassy phenomena," Physica B, vol. 156-157, no. C, pp. 90–92, 1989.

138. N. De Mathan, E. Husson, G. Calvarn, J. R. Gavarri, A. W. Hewat, and A. Morell, "A structural model for the relaxor $PbMg_{1/3}Nb_{2/3}O_3$ at 5 K," Journal of Physics, vol. 3, no. 42, article 011, pp. 8159–8171, 1991.

139. G. Xu, G. Shirane, J. R. D. Copley, and P. M. Gehring, "Neutron elastic diffuse scattering study of $Pb(Mg_{1/3}Nb_{2/3})O_3$," Physical Review B, vol. 69, no. 6, Article ID 064112, 2004.

140. J. Hlinka, J. Petzelt, S. Kamba, D. Noujni, and T. Ostapchuk, "Infrared dielectric response of relaxor ferroelectrics," Phase Transitions, vol. 79, no. 1-2, pp. 41–78, 2006.

141. A. Al-Zein, B. Hehlen, J. Rouquette, and J. Hlinka, "Polarized hyper-Raman scattering study of the silentF_{2u} mode in $PbMg_{1/3}Nb_{2/3}O_3$," Physical Review B, vol. 78, no. 13, Article ID 134113, 7 pages, 2008.

142. W. Dmowski, S. B. Vakhrushev, I.-K. Jeong, M. P. Hehlen, F. Trouw, and T. Egami, "Local lattice dynamics and the origin of the relaxor ferroelectric behavior," Physical Review Letters, vol. 100, no. 13, Article ID 137602, 2008.

143. P. M. Gehring, H. Hiraka, C. Stock et al., "Polarized hyper-Raman scattering study of the silent F_{2u} mode in $PbMg_{1/3}Nb_{2/3}O_3$," Physical Review B, vol. 79, no. 22, Article ID 224109, 7 pages, 2009.

144. A. Al-Zein, J. Hlinka, and J. Rouquette, "Soft mode doublet in $PbMg_{1/3}Nb_{2/3}O_3$ relaxor investigated with hyper-raman scattering," Physical Review Letters, vol. 105, no. 1, Article ID 017601, 2010.

145. S. Tinte, B. P. Burton, E. Cockayne, and U. V. Waghmare, "Origin of the relaxor state in Pb(BxB1−x′) O_3perovskites," Physical Review Letters, vol. 97, no. 13, Article ID 137601, 2006.

146. I. Grinberg, P. Juhás, P. K. Davies, and A. M. Rappe, "Relationship between local structure and relaxor behavior in perovskite oxides," Physical Review Letters, vol. 99, no. 26, Article ID 267603, 4 pages, 2007.

147. A. Bosak, D. Chernyshov, S. Vakhrushev, and M. Krisch, "Diffuse scattering in relaxor ferroelectrics: true three-dimensional mapping, experimental artefacts and modelling," Acta Crystallographica A, vol. 68, no. 1, pp. 117–123, 2012.

148. A. Bosak and D. Chernyshov, "On model-free reconstruction of lattice dynamics from thermal diffuse scattering," Acta Crystallographica A, vol. 64, no. 5, pp. 598–600, 2008.

149. M. Paściak and T. R. Welberry, "Diffuse scattering and local structure modeling in ferroelectrics,"Zeitschrift Für Kristallographie, vol. 226, pp. 113–1125, 2011.

150. M. Sepliarsky and R. E. Cohen, "First-principles based atomistic modeling of phase stability in PMN-xPT," Journal of Physics, vol. 23, no. 43, Article ID 435902.

151. I.-K. Jeong, T. W. Darling, J. K. Lee et al., "Direct observation of the formation of polar nanoregions in $Pb(Mg_{1/3}Nb_{2/3})O_3$ using neutron pair distribution function analysis," Physical Review Letters, vol. 94, no. 14, Article ID 147602, 2005.

152. M. Paściak, T. R. Welberry, J. Kulda, M. Kempa, and J. Hlinka, "Polar nanoregions and diffuse scattering in the relaxor ferroelectric $PbMg_{1/3}Nb_{2/3}O_3$," Physical Review B, vol. 85, Article ID 224109, 2012.

153. Q. M. Zhang, H. You, M. L. Mulvihill, and S. J. Jang, "An X-ray diffraction study of superlattice ordering in lead magnesium niobate," Solid State Communications, vol. 97, no. 8, pp. 693–698, 1996.

154. T. Egami, H. D. Rosenfeld, B. H. Toby, and A. Bhalla, "Diffraction studies of local atomic structure in ferroelectric and superconducting oxides," Ferroelectrics, vol. 120, no. 1, pp. 11–21, 1991.

155. H. D. Rosenfeld and T. Egami, "Model of local atomic structure in the relaxor ferroelectric $Pb(Mg_{1/3}Nb_{2/3})O_3$," Ferroelectrics, vol. 150, no. 1-2, pp. 183–197, 1993.

156. J. Wen, G. Xu, C. Stock, and P. M. Gehring, "Response of polar nanoregions in 68%Pb $(Mg_{1/3}Nb_{2/3})O_3$-32%$PbTiO_3$ to a [001] electric field," Applied Physics Letters, vol. 93, no. 8, Article ID 082901, 2008.

157. R. Blinc, V. Laguta, and B. Zalar, "Field cooled and zero field cooled ^{207}Pb NMR and the local structure of relaxor $PbMg_{1/3}Nb_{2/3}O_3$," Physical Review Letters, vol. 91, no. 24, Article ID 247601, 4 pages, 2003.

158. Z.-G. Ye, Y. Bing, J. Gao et al., "Development of ferroelectric order in relaxor (1-x)$Pb(Mg_{1/3}Nb_{2/3})O_3$-x$PbTiO_3$(0LTHEXA~x~0.15)," Physical Review B, vol. 67, Article ID 104104, 8 pages, 2003.

159. B. Mihailova, N. Waeselmann, B. J. Maier et al., "Chemically induced renormalization phenomena in Pb-based relaxor ferroelectrics under high pressure," Journal of Physics, vol. 25, no. 11, Article ID 115403, 2013.

160. B. J. Maier, N. Waeselmann, B. Mihailova et al., "Structural state of relaxor ferroelectrics $PbSc_{0.5}Ta_{0.5}O_3$and $PbSc_{0.5}Nb_{0.5}O_3$ at high pressures up to 30 GPa," Physical Review B, vol. 84, no. 17, Article ID 174104, 2011.

161. A.-M. Welsch, B. J. Maier, J. M. Engel et al., "Effect of Ba incorporation on pressure-induced structural changes in the relaxor ferroelectric PbSc0.5 Ta0.5 O_3," Physical Review B, vol. 80, no. 10, Article ID 104118, 2009.

162. B. Mihailova, R. J. Angel, A.-M. Welsch et al., "Pressure-induced phase transition in $PbSc_{0.5}Ta_{0.5}O_3$ as a model Pb-based perovksite-type relaxor ferroelectric," Physical Review Letters, vol. 101, Article ID 017602, 4 pages, 2008.

163. N. Takesue, Y. Fujii, M. Ichihara, H. Chen, S. Tatemori, and J. Hatano, "Effects of B-site ordering/disordering in lead scandium niobate," Journal of Physics Condensed

Matter, vol. 11, no. 42, pp. 8301–8312, 1999.

164. N. Takesue, Y. Fujii, M. Ichihara, and H. Chen, "X-ray diffuse scattering study of glass-like transition behavior of relaxor lead scandium niobate," Physics Letters A, vol. 257, no. 3-4, pp. 195–200, 1999.

165. C. Perrin, N. Menguy, E. Suard, C. Muller, C. Caranoni, and A. Stepanov, "Neutron diffraction study of the relaxor-ferroelectric phase transition in disordered $Pb(Sc_{1/2}Nb_{1/2})O_3$," Journal of Physics Condensed Matter, vol. 12, no. 33, pp. 7523–7539, 2000.

166. G. A. Smolensky, Ferroelectrics and Related Materials, Academic Press, New York, NY, USA, 1981.

167. S. N. Gvasaliya, B. Roessli, D. Sheptyakov, S. G. Lushnikov, and T. A. Shaplygina, "Neutron scattering study of $PbMg_{1/3}Ta_{2/3}O_3$ and $BaMg_{1/3}Ta_{2/3}O_3$ complex perovskites," European Physical Journal B, vol. 40, no. 3, pp. 235–241, 2004.

168. R. D. Shannon, "Crystal physics, diffraction, theoretical and general crystallography," Acta Crystallographica A, vol. 32, no. 751, 1976.

CHAPTER 3

EXTRINSIC CONTRIBUTION AND INSTABILITY PROPERTIES IN LEAD-BASED AND LEAD-FREE PIEZOCERAMICS

José Eduardo García

Keywords: Piezoelectric materials, nonlinear properties, extrinsic contribution, dielectric properties, piezoelectric properties

INTRODUCTION

Piezoelectric ceramics have been widely used in sensor and actuator applications for more than a half century, ranging from sonar to pressure sensors, ultrasonic transducers, fuel injectors, microphones, resonators, filters, and many others 1 . While piezoceramics are parts of a mature technology 2 , emerging challenges have stepped up the demand for new high performance piezoelectric materials with specific functionalities. For instance, the miniaturization trend in piezoelectric technologies requires ceramics with submicron or even lower grain size. Furthermore, environmental concerns as well as governmental regulations against hazardous substances have led to the development of new environmentally friendly piezoceramic compositions. More recently, further advances in piezoelectric technology for energy harvesting have demonstrated the need for low cost, high sensitive piezoelectric materials in order to achieve large deformations as a response to low applied voltages.

Lead-based microstructured materials with perovskite structure are the most common active elements in piezoelectric devices because of their excellent properties and relatively low manufacturing costs. In fact, polycrystalline lead zirconate titanate (PZT) has for decades been dominant in the market for piezoelectric materials 3 . The PZT system enjoys high flexibility in terms of compositional modifications, thereby providing a wide spectrum of compositions with improved properties for specific applications. PZT-based materials are classified into two groups, known as "hard" and "soft" according to some particular properties 4 . These terms are commonly used for ranking piezoelectric materials in order to specify their ability to be transferred to applications. Hard materials are characterized by low losses and a high quality factor, but also by moderate values of both dielectric constant and piezoelectric coefficients, while soft materials exhibit high values of dielectric constant and piezoelectric coefficient, but also high losses. Hard materials are typically used in resonant devices where a high mechanical quality factor and a high electromechanical coupling factor are needed, while soft materials are usually used in non-resonant devices where their critical figures of merit are determined by both the piezoelectric strain coefficient and the piezoelectric voltage coefficient.

The dielectric and piezoelectric responses of piezoceramics to either an electrical or a mechanical external stimulus are due to two contributions known as intrinsic and extrinsic. The intrinsic contribution concerns the linear lattice distortion and is associated with the change in the polarization of the unit cell. This change in polarization is related to both the polarization extension and the polarization rotation 5 . On the other hand, the extrinsic contribution is easily defined as all responses that are different from the intrinsic response, which is mainly due to domain wall motion in perovskite polycrystals 6 . Both contributions maximize in the region delimiting different polymorphic phases, such as a morphotropic phase boundary (MPB), thereby improving macroscopic properties of materials 7,8 . The MPB is a compositionally-driven structural change region where the coexistence of two ferroelectric phases enhances the macroscopic properties, which is the reason why high performance piezoelectric materials are MPB systems. This type of phase boundary is also stated as structurally bridging low symmetry regions, which have been described

Citation: José Eduardo García, Extrinsic Contribution and Instability Properties in Lead-Based and Lead-Free Piezoceramics; doi: 10.3390/ma8115426.

as monoclinic symmetry regions for PZT and other systems 9,10 .

Functional properties of piezoceramics are undesirable dependents of external parameters. The dielectric and piezoelectric properties generally exhibit aging, frequency dispersion, and nonlinearity 11 . The term nonlinearity commonly refers to the instability of these properties when a sub-coercive alternating electric field or a dynamic mechanical stress is applied to the material 12,13,14,15 . This nonlinear effect postulates that once the applied field is removed, the material recovers its linear, or low-field, properties. Thus, the electric field is applied in the sub-switching regime, whereas the applied mechanical stress is such that it does not produce depoling. Other phenomena occurring in piezoelectric materials also belong to nonlinear behavior, although they are not addressed in this review. For instance, nonlinearity also refers to: (i) the variation of the elastic properties and the electromechanical coupling factor in resonant conditions, which is essential for establishing the working limits of high-power devices 16,17,18,19 ; and, (ii) the decrease in permittivity under a high DC electric field, known as electrical tuning of permittivity, which is the phenomenon on which a large number of microwave devices are based 20,21,22 .

Piezoceramic materials show a noticeable nonlinear behavior at a relatively low electric field or dynamical stress 12,14 . Since the dielectric and piezoelectric properties of piezoceramics are strongly determined by the ferroelectric/ferroelastic domain structure and the dynamics of the domain walls 23 , the nonlinear response will depend on any microstructural aspect that modifies the domain structure and/or the domain wall motion, such as the crystallographic phase, the grain size, and the compositional modifications by doping. The microstructural dependence of nonlinear response has been widely studied in PZT 24,25,26,27 and other ceramic systems such as barium titanate (BT) 28 , lead magnesium niobate—lead titanate (PMN-PT) 29 , and potassium sodium niobate (KNN) 30 . This review focuses on the role of the ferroelectric domain wall motion/dynamics in the nonlinear response in the most workable lead-based and lead-free piezoelectric ceramics with a perovskite structure. Moreover, a description is given of how nonlinear measurements may be used as a tool to study the mechanisms contributing to dielectric and piezoelectric properties in these materials.

EXTRINSIC CONTRIBUTION AND NONLINEAR RESPONSE

The intrinsic and extrinsic contributions to material response can be experimentally separated by taking into account their different phenomenological nature. The intrinsic contribution comes from the polarization rotation and polarization extension phenomena 5 . The first is essentially dependent on the crystal structure, whereas the second is primarily important in regions where phase transitions occur. Hence, intrinsic contribution may be considered temperature-independent in temperature ranges that are sufficiently far from phase transitions. In contrast, the extrinsic contribution related to domain wall motion is markedly temperature-dependent. The motion of the domain walls is thermally assisted, and is generally the main contribution to material response at room temperature 6,29,30,31,32 . Consequently, only intrinsic response is expected at very low temperature as a result of the domain wall freezing phenomenon 33 .

Figure 1 shows the temperature dependence of the dielectric constant for Nb- and Fe-doped $Pb(Zr_{0.60}Ti_{0.40})O_3$(rhombohedral PZT) and $Pb(Zr_{0.53}Ti_{0.47})O_3$ (MPB PZT) from low temperatures (~15 K) to above room temperature (~390 K). In all cases, $\varepsilon'(T)$ increases from an intrinsic low temperature dielectric constant close to 230 for $Pb(Zr_{0.6}Ti_{0.4})O_3$ and 350 for $Pb(Zr_{0.53}Ti_{0.47})O_3$, reaching different values at room temperature depending on the Zr/Ti ratio and on the dopant. As can be observed, the intrinsic contribution depends only on the crystal structure attaining a higher value at MPB. This is because the existence of thermodynamically equivalent phases in the MPB promotes the polarization rotation, improving dielectric response 7 . The extrinsic response also maximizes at MPB 34 , as shown in Figure 1, but is strongly dependent on the dopant. Both the crystal structure and the dopant determine the domain wall contribution. An expected monotonous increasing $\varepsilon'(T)$ dependence is shown for Nb-doped PZT, while a more complex behavior is displayed for Fe-doped PZT. This different extrinsic response is due to the different character of the defects created by donor and acceptor dopants, as is supported later.

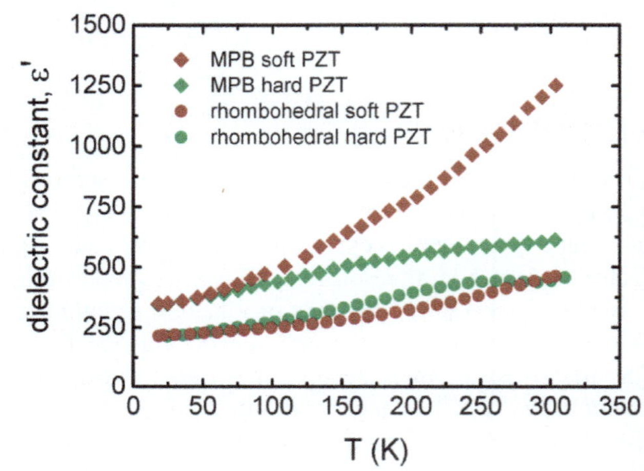

Figure 1: Temperature dependent dielectric constant for $Pb(Zr_{0.60}Ti_{0.40})O_3$(rhombohedral PZT) and $Pb(Zr_{0.53}Ti_{0.47})O_3$ (MPB PZT) doped with Nb^{5+} (soft PZT) or Fe^{3+} (hard PZT) from low temperatures to above room temperature. The data were extracted from the reference.

The nonlinear response concept can be experimentally studied by quantifying the instability of the properties when an alternating external stimulus, either electrical or mechanical, is applied to the material. The increment of a complex property$p = p' - i{\cdot}p''$, which corresponds either to the permittivity or to the piezoelectric coefficient, refers to its rate of change with increasing amplitude of the driving signal and is denoted as $\Delta p = \Delta p' - i{\cdot}\Delta p''$.

Figure 2 shows the increments of the dielectric constant for Nb- and Fe-doped $Pb(Zr_{0.60}Ti_{0.40})O_3$ (rhombohedral PZT), $Pb(Zr_{0.53}Ti_{0.47})O_3$ (MPB PZT) and $Pb(Zr_{0.40}Ti_{0.60})O_3$ (tetragonal PZT). Here, $\Delta\varepsilon$ is calculated from the $D(E)$ response according to the procedure detailed in 36 . As can be observed, the nonlinear response also depends on the crystal structure and the dopant, and is larger in the rhombohedral composition doped with donor ions. These correlations are discussed below.

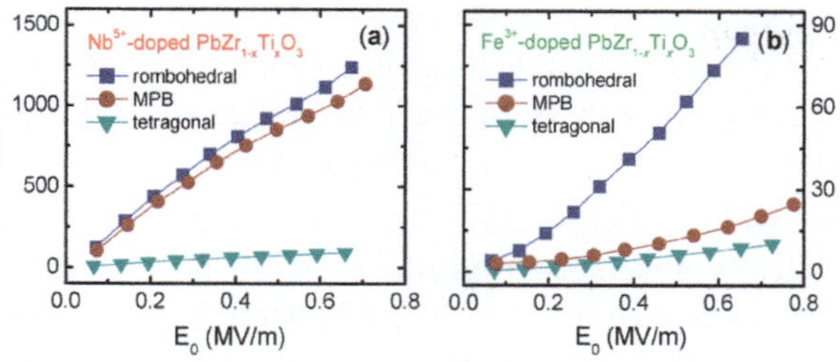

Figure 2: Increments of the dielectric constant as a function of the applied electric field amplitude for $Pb(Zr_{1-x}Ti_x)O_3$ (x = 0.40, 0.47, 0.60) doped with (**a**) softener and (**b**) hardener ions. The data were partially extracted from the reference 26 (Copyright AIP Publishing LCC., 2008).

Extrinsic Origin of the Nonlinear Response

The extrinsic nature of the nonlinear response was first established over twenty years ago 37 . The temperature-dependent hysteretic response related to nonlinear behavior suggests that the origin of nonlinear contribution is a dynamic phenomenon. The increase in the nonlinear behavior supposes an increase in losses, which are generated from the domain wall motion. Since the nonlinear response is extrinsic in nature, it may be reduced by applying a DC electric field or by decreasing the temperature. Here, the extrinsic origin of the nonlinear response is shown by means of the nonlinear dielectric response measurement in temperature.

Figure 3 shows the temperature dependence of the relative dielectric permittivity at 1 kHz as a function of the amplitude of the applied electric field for two well-known piezoceramic compositions. At low temperatures (below 150 K), the nonlinear response almost disappears for both materials, which is as expected as a result of the domain wall freezing phenomenon. This means that when only the intrinsic contribution exits, the nonlinear behavior disappears. The nonlinear dielectric response is significant at temperatures at which the domain wall motion contribution becomes important, usually above 150 K. These observations clearly show that the nonlinear behavior is directly related to the extrinsic effect.

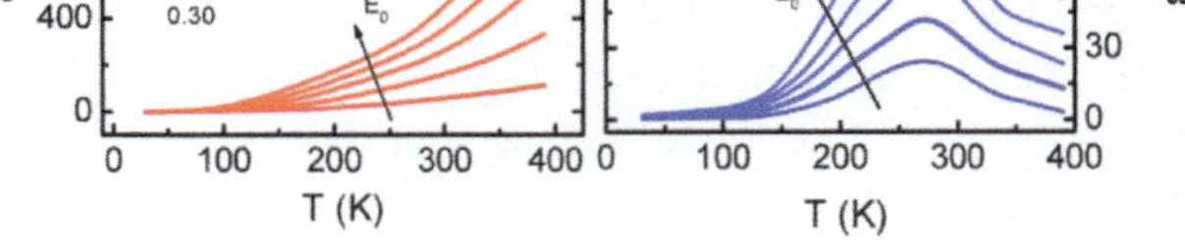

Figure 3: Temperature dependence of real, ε′, and imaginary, ε″, components of the relative dielectric permittivity at 1 kHz as a function of the amplitude of the applied electric field for (**a**) lead-based Nb-doped $Pb(Zr_{0.60}Ti_{0.40})O_3$ (Nb-PZT) and (**b**) lead-free $(K_{0.44}Na_{0.52}Li_{0.04})(Nb_{0.86}Ta_{0.10}Sb_{0.06})O_3$ (KNL-NTS). The data were partially extracted from the reference 38 (Copyright AIP Publishing LCC., 2012).

Domain Wall Dynamics

The dependence of the extrinsic contribution on the material properties has been well studied by using the Rayleigh model. The Rayleigh law, originally stated in order to describe the magnetic response of a ferromagnetic material to the action of an external magnetic field, has been satisfactorily used to describe the dielectric and piezoelectric nonlinear response in a piezoelectric ceramic 12,26,39 . This model can provide valuable information about the dynamics of domain wall motion. One hypothesis sustaining this model is that the nonlinear behavior is exclusively due to the irreversible motion of domain walls 24 . The model assumes that the response of the material is due to the interaction of the domain walls with the defects of the material. These defects act as randomly distributed pinning centres by hindering the movement of the domain walls. Analysis of the dielectric response in terms of the Rayleigh law enables the contribution of the irreversible movement of domain walls to be studied. Therefore, the reversible domain wall motion contribution leads to deviations from the Rayleigh law predictions 40 .

The extrinsic contribution can be evaluated from the increments of the real and imaginary parts of the permittivity as a function of the amplitude of the applied electric field. The Rayleigh model assumes that the increment in the real as well as the imaginary dielectric permittivity linearly depends on the amplitude of the applied electric field, as follows 40 :

$$\Delta\varepsilon' = \varepsilon \times E_0 \tag{1}$$

$$\Delta\varepsilon'' = \frac{4}{3\pi} \times \alpha \times E_0 \tag{2}$$

where α is the Rayleigh coefficient and is directly related to the magnitude of the nonlinear response. Moreover, the ratio between the value of imaginary and real increments of dielectric permittivity is a constant that does not depend on the material:

$$m_\varepsilon = \frac{\Delta\varepsilon''}{\Delta\varepsilon'} = \frac{4}{3\pi} \approx 0.42 \tag{3}$$

Fulfilment of the relations Equations (1), (2), and (3) implies a Rayleigh behavior associated to a preponderant irreversible domain wall motion. Rayleigh behavior involves some hypotheses that are not valid for all materials. Results obtained by different authors show that, for example, in hard PZT the dielectric behavior does not linearly depend on the amplitude of the electric field, and that the relation Δε″/Δε′ does not have the value predicted by the Rayleigh model 40,41,42 . In other cases, even when Δε′ and Δε″ are linearly dependent on E_0, a non-Rayleigh behavior associated with a shift away of m_ε from its theoretical value may occur. Nevertheless, this m_ε shift provides us with information about the preponderant mechanism in the domain wall dynamic, because m_ε relates the ratio between reversible and irreversible domain wall motion processes.

Figure 4a shows the relation between the increments of the imaginary (Δε′) and real (Δε″) parts of the relative permittivity at different temperatures for Fe-doped (hard) rhombohedral $Pb(Zr_{0.60}Ti_{0.40})O_3$. This relation for Nb-doped (soft) $Pb(Zr_{0.60}Ti_{0.40})O_3$ is also shown in the inset. At each temperature, the interval of applied field amplitudes is the same in all cases. As can be observed, there is a linear relation between the real and imaginary parts of the permittivity, which is a common characteristic that has been reported for a wide number of piezoceramics 26,27,29,30,43 , even when $\Delta\varepsilon'$ and $\Delta\varepsilon''$ do not depend linearly on the field amplitude. Two features may be highlighted in the $\Delta\varepsilon''$ *versus* $\Delta\varepsilon'$ graph: the distance between two adjacent points of a plot and the slope of the plot. The distance between the points reveals the significance of the nonlinear effect, while the slope of the plot quantifies the m_ε value. This value indicates that a quantitative relation between the real and imaginary parts of the permittivity exists, *i.e.*, an increment in the dielectric constant implies a given increment of losses.

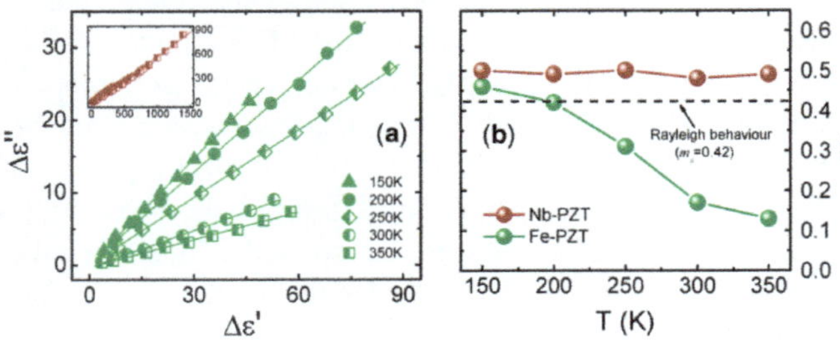

Figure 4: (**a**) Relation between the increments of the imaginary ($\Delta\varepsilon''$) and real ($\Delta\varepsilon'$) parts of the relative permittivity for Fe-doped $Pb(Zr_{0.60}Ti_{0.40})O_3$ (Fe-PZT) at several temperatures. In the inset, this relation is shown for Nb-doped $Pb(Zr_{0.60}Ti_{0.40})O_3$ (Nb-PZT) at the same temperatures. Linear fits are shown as solid lines; (**b**) Ratio between the increments of the imaginary and real parts of the permittivity (m_ε) for both Fe-PZT and Nb-PZT at several temperatures. The data were extracted from the reference 38 (Copyright AIP Publishing LCC., 2012).

The m_ε value as a function of the temperature is shown in Figure 4b for both Fe-doped and Nb-doped $Pb(Zr_{0.60}Ti_{0.40})O_3$. Note that the m_ε value remains almost constant for Nb-PZT, while an evolution towards lower values is observed for Fe-PZT. In terms of the Rayleigh model, the m_ε values for Nb-doped PZT appear to be slightly higher than those expected, thereby revealing the presence of some dissipative processes that have not been taken into account by the model; *i.e.*, processes in which the losses contribute more than that predicted by the Equation (3). Field-induced phenomena, e.g., conductivity effects, may account for the obtained additional dissipation in this material. In any case, and based on the Rayleigh law, the contribution of the irreversible displacement of domain walls appear to be the mechanisms governing the Nb-PZT response for all temperatures. On the other hand, the dielectric response fits the Rayleigh model only for $T = 200$ K in Fe-PZT, thereby showing that the relevant mechanisms governing the dielectric response differ according to the temperature. In this material, there appears to exist a reversible (elastic) domain wall motion mechanism contributing to the dielectric constant, but which has no associated losses 40 .

Analysis of the nonlinear dielectric response undoubtedly indicates a possible change in the dynamics of the domain wall motion in Fe-doped PZT. The different temperature dependence of the extrinsic response due to the different domain wall dynamics determines the room temperature properties in PZT piezoceramics 38 . Changes in the domain wall dynamics may be responsible for other singular properties of emerging materials. For example, a change in domain wall dynamics has been detected in the $(K,Na)NbO_3$ system when crossing the orthorhombic to tetragonal phase transition 44 , which could be decisive for their properties when this phase boundary is located near room temperature.

A wider interpretation of the Rayleigh law was subsequently developed from the Preisach model 45 . The Preisach formalism assumes that a hysteretic system consists of a superposition of independent bi-stable units, each one characterized by two parameters: a bias field, induced in each bi-stable by the action of its environment, and a coercive field, defined as the amplitude of the external field required to switch the bi-stable 46 . From a macroscopic point of view, the system is defined by a distribution function of the parameters allied to each bi-stable.

The Preisach model allows us to describe the Rayleigh empirical relations when the bistable distribution in the two-dimensional space of bias and coercive parameters is taken to be uniform. Then, the Rayleigh law is derived from the Preisach formalism and may be regarded as a particular case of this formalism. The fact that different distribution functions allow us to generate different types of response makes the Preisach model an important tool for describing the nonlinear response of ferroelectric materials 47,48 . However, a restriction exists, and that is that the Preisach formalism is based on a hysteretic system where all contributions to the dielectric or piezoelectric response are directly related to losses. Thus, in this model it is assumed that in any case only irreversible processes contribute to the nonlinear response of the material.

NONLINEARITY IN LEAD-BASED PIEZOCERAMICS

Crystallographic Phase Effect

The formation of the domain configuration in ferroelectrics is mainly determined by the crystal structure. For instance, 90° and 180° domain walls appear in tetragonal materials, while 71°, 109°, and 180° domain walls are allowable in rhombohedral materials. Since the domain wall motion is directly related to the domain configuration, the crystallographic phase has a decisive influence on the extrinsic response and, consequently, on the nonlinear behavior. Figure 5 shows the total value and the increment of the converse piezoelectric coefficient as functions of the amplitude of the applied electric field for donor-doped PZT compositions with different crystallographic phases. As can be observed, the nonlinear response of PZT is strongly dependent on the crystal structure. The increment in dielectric constant and piezoelectric coefficient are greater in rhombohedral materials than in MPB, and in both cases are greater than in tetragonal ones. This is the same observation that can be made from the dielectric response in Figure 2a. Results from the dielectric and piezoelectric response of acceptor-doped PZT with the same nominal compositions show an equivalent tendency 15 . Similar results were also obtained in donor-doped PZT from the direct piezoelectric effect measurements 24 . Analogous observations were reported from the analysis of the dielectric and converse piezoelectric responses of the $(1-x)BiScO_3-xPbTiO_3$ (BS-PT) ceramic system 49 .

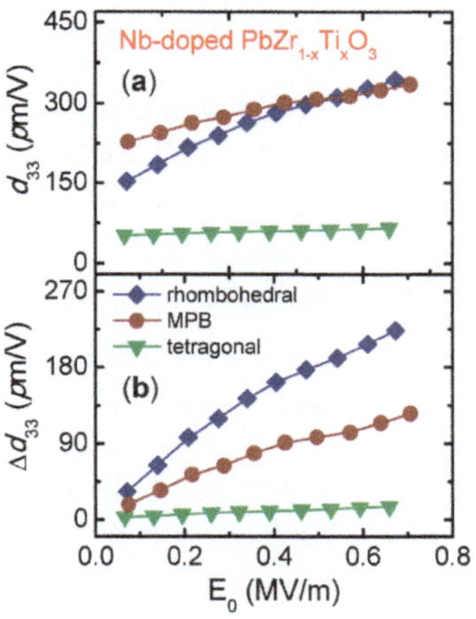

Figure 5: (**a**) Total value and (**b**) increment of the converse piezoelectric coefficient as a function of the applied electric field amplitude for Nb-doped $Pb(Zr_{1-x}Ti_x)O_3$ with $x = 0.40$ (rhombohedral), 0.47 (MPB) and 0.60 (tetragonal). The data were extracted from the reference [26] (Copyright AIP Publishing LCC., 2008).

Two phenomena may contribute to the dependence of the nonlinear response on the crystallographic phase. Firstly, the amount of spontaneous strain of the crystalline structure may govern the domain wall mobility. The spontaneous strain of the rhombohedral elemental cell is less than 0.5%, while the strain is greater than 2% in the tetragonal cell, reaching 6% for Ti-rich PZT compositions [4]. The high distortion of the tetragonal cell may create strong local internal stress, which will make domain wall motion difficult. On the other hand, the rhombohedral domain walls can move with relative freedom due to the low distortion of the lattice, which scarcely causes internal stress. Secondly, the nonlinear response is enhanced at the rhombohedral phase because more directions are available to be oriented, and therefore the material exhibits a higher extrinsic contribution. A further analysis is required to understand why nonlinear behavior is lower in the MPB than in the rhombohedral phase. A complex domain structure is formed at MPB as a result of the coexistence of tetragonal and rhombohedral domains. The domain coexistence may lead to the appearance of an internal stress that reduces domain wall mobility, as was demonstrated in $(K,Na)NbO_3$ based ceramics [50,51]. It is important to point out that the highest value of low-field piezoelectric coefficient is reached at MPB, where the intrinsic contribution plays an important role.

Doping Effects

Compositional engineering by doping is a very active research line for obtaining piezoceramics with enhanced properties. Many papers are published every year on compositional modifications as an approach to improving some specific material property. In particular, the substitution of the A- or B-site of the perovskite structure by lower valence ions generates oxygen vacancies that give rise to the formation of so-called complex defects [52]. These defects operate as pinning centers, thereby making the domain wall movement more difficult [53]. On the other hand, the A- or B-site substitution by higher valence ions generates lead vacancies and reduces oxygen vacancies, thus making domain wall movement easier [52]. So, for instance, in the PZT system the substitution of Zr^{4+} o Ti^{4+} by lower valence ions (e.g., Fe^{3+}) hardens the material properties, while higher valence ions (e.g., Nb^{5+}) soften the properties.

Figure 6 shows the role of the hardener and softener substitutions in nonlinear response. As can be observed, as regards the pure material, a significant increase in the dielectric constant and in the piezoelectric coefficient is produced in donor-doped materials, while a decrease in the same is produced in acceptor-doped ones.

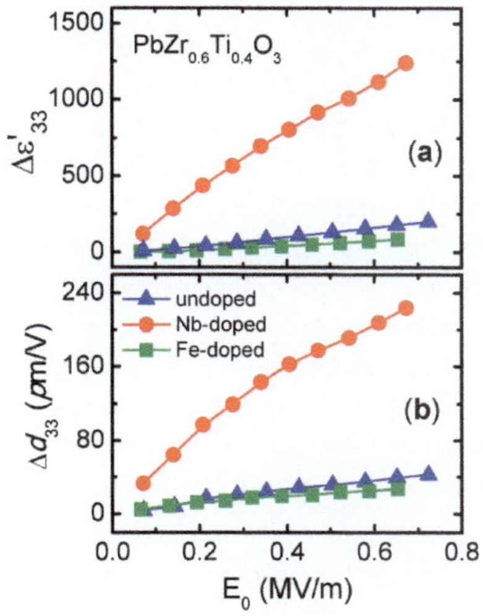

Figure 6: Increments in (**a**) the dielectric constant and (**b**) the converse piezoelectric coefficient as a function of the applied electric field amplitude for undoped, Nb-doped, and Fe-doped $Pb(Zr_{0.60}Ti_{0.40})O_3$. The data were extracted from the reference [26] (Copyright AIP Publishing LCC., 2008).

It should be noted that the effect of the softener ions is notably greater than that of the hardener ones. These results can be explained by taking into account the interaction of the domain walls with the defects formed by oxygen vacancies. It is well known that during the process of synthesis of PZT, oxygen vacancies are formed as a result of the unavoidable evaporation of lead [54]. These vacancies form complex defects resulting in a hard-type behavior for undoped PZT. The addition of Fe^{3+}ions in substitution for Zr^{4+} or Ti^{4+} ions will increase the oxygen vacancies and lead to the creation of new complex defects. These defects can be oriented in the direction of the applied field and thus make the movement of the domain walls difficult, as pointed out above. On the other hand, the addition of Nb^{5+} ions creates lead vacancies by increasing the mobility of the domain walls as a consequence of the low stress of the lattice. Moreover, the insertion of donor impurities in a perovskite structure will reduce the oxygen vacancies by charge compensation [55], so the pinning of the walls is reduced as a result of the decrease in the concentration of complex defects. Definitively, the decrease of oxygen vacancies seems to be the main reason for the increase in domain wall mobility in donor-doped materials. This hypothesis has been successfully employed in order to reduce the nonlinear response of $(K,Na)NbO_3$–based piezoceramics [56]. Furthermore, an adequate softener and hardener co-doping may enable the material properties to be tuned from soft to hard [27].

Recovering Figure 1, different dielectric responses can be detected near room temperature for the same composition, depending on dopant. Anomalous $\varepsilon'(T)$ is observed in acceptor-doped PZT but not in donor-doped PZT for both rhombohedral and MPB compositions; this was also verified for tetragonal composition 35 . This anomaly, which appears to be intimately related with the acceptor dopant, is a direct manifestation of the domain wall pinning effect.

Grain Size Effect

Piezoelectric devices are not oblivious to the current miniaturization trends in ceramic technology for microelectronics. Microelectromechanical systems, for instance, require ceramic layers with a submicron or even a nanometric grain size 57 . A large number of studies exist that address the grain size effect in $BaTiO_3$ (BT), whose properties exhibit a singular grain size dependence that was firstly associated with the evolution of domain structure 58 . A more recent study correlates the grain size with the tetragonal distortion evolution and the ferroelectric properties, revealing an intrinsic size effect 59 . Measurement of the nonlinear piezoelectric response in coarse- and fine-grained BT show that domain wall motion decreases when the grain size is smaller 28 . The published research on grain size effect in BT demonstrates that both intrinsic and extrinsic contributions depend on grain size in perovskite ferroelectrics.

In general, high sensitivity, microstructured piezoceramics undergo a significant functionality reduction when grain size evolves to nanoscale. In soft PZT, for instance, the longitudinal piezoelectric coefficient notably decreases when the grain size moves from micron to submicron size, as a result of the decrease in domain wall mobility 60 . In this context, the scientific challenge is then to obtain high performance piezoceramics with submicron grain sizes close to the nanoscale. Significant advances have been made in this regard in the last decade 61 , and a large number of papers have been published related to piezoceramic properties with submicron grain sizes.

The worsening of the functional properties when grain size is decreased to a submicron scale appears mainly to be due to the reduction of domain wall motion as a consequence of the domain wall clamping effect, which implies a decrease in the nonlinear response. The few reported results about the grain size effect on nonlinear response support this hypothesis 24,43 . Therefore, the reduction of grain size decreases the functionality of piezoceramics, but provides more stability to their properties. In any case, further studies are needed in order to effectively research the grain size effect on the nonlinear behavior of piezoelectric ceramics. Furthermore, the nonlinear response analysis could shed new light on the grain size effect on the dynamics of the domain wall motion.

NONLINEARITY IN LEAD-FREE PIEZOCERAMICS

The discovery of a $(K,Na)NbO_3$–based piezoceramic with a large piezoelectric response 62 triggered a race to replace lead-containing piezoceramics with non-toxic alternatives. This race was also driven by the legislative activity, initially promoted from the European Union, and subsequently followed by other countries, for the purpose of protecting human health as well as the environment by the exclusion or substitution of hazardous substances used in electrical and electronic devices. The intense scientific activity conducted in developing new lead-free piezoceramics has been continuously reviewed in the last decade 63,64,65,66,67,68,69,70,71,72,73,74 . Although the number of the annual refereed publications is extremely high (about 400 refereed papers 74), only few reports can be found related to field-induced instability properties in lead-free compositions. Here, some aspects related to nonlinear behavior of representative lead-free piezoceramics are discussed.

KNN-Based Compositions

$(K,Na)NbO_3$-based systems have attracted an indisputable interest as the most promising candidates for commercially viable lead-free piezoceramics 73,74 . Among these, the $(K,Na)NbO_3$–$LiTaO_3$–$LiSbO_3$, and in particular the $(K_{0.44}Na_{0.52}Li_{0.04})(Nb_{0.86}Ta_{0.10}Sb_{0.04})O_3$ (KNL-NTS) composition 75 is probably the most workable lead-free piezoelectric system known to date. The origin of the high piezoelectric activity in this compound seems to reside in a mixture of a compositional-driven (morphotropic) and temperature-driven (polymorphic) phase transition regions 76 . Unfortunately, its properties are not suitable for all end uses. For instance, KNL-NTS exhibits a relatively high piezoelectric coefficient and coupling factor at room temperature, but also high losses and a low quality factor. Furthermore, the properties are notably temperature- and field-dependent. Figure 7 shows the relative increases of the real ($\Delta\varepsilon'$) and imaginary ($\Delta\varepsilon''$) parts of the relative permittivity as a function of the electric field amplitude for a KNL-NTS piezoceramic obtained by a conventional solid state synthesis process 77 . Both $\Delta\varepsilon'$ and $\Delta\varepsilon''$ show a linear dependence on the electric field amplitude, which corresponds to the Rayleigh-type behavior (Equations (1) and (2)). In addition, the ratio $m_\varepsilon = 0.40$ also verified a Rayleigh-type nonlinear response (Equation (3)) for KNL-NTS. Thus, the irreversible domain wall motion is a mechanism that governs nonlinear response in KNL-NTS ceramics at room temperature, which is similar to that in soft PZT ceramics.

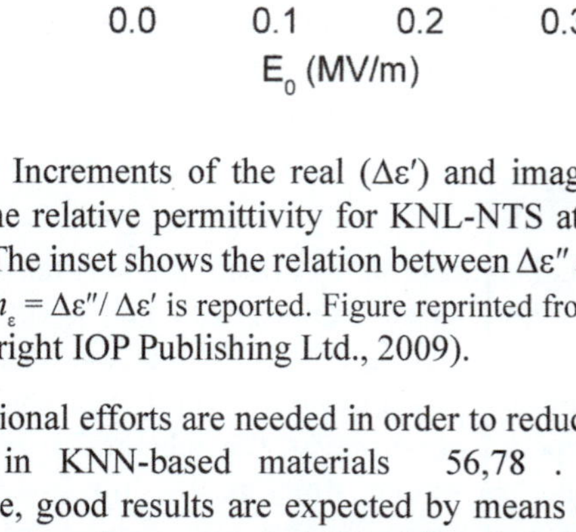

Figure 7: Increments of the real ($\Delta\varepsilon'$) and imaginary ($\Delta\varepsilon''$) parts of the relative permittivity for KNL-NTS at room temperature. The inset shows the relation between $\Delta\varepsilon''$ and $\Delta\varepsilon'$, and the ratio $m_\varepsilon = \Delta\varepsilon''/\Delta\varepsilon'$ is reported. Figure reprinted from reference 30 (Copyright IOP Publishing Ltd., 2009).

Additional efforts are needed in order to reduce nonlinear behavior in KNN-based materials 56,78 . From this perspective, good results are expected by means of hardener substitutions, such as occur in PZT, although some structural and electrical aspects remain controversial as regards the role of dopants in the KNN system. Significant advances have been reported through Cu-doping. It has been shown that Cu-doped KNN-modified compounds may exhibit typical characteristics of hard behaviour 79,80,81,82 . Nevertheless, few reports exist about the stability of the properties of these materials.

Other Lead-Free Piezoceramics

In the race to replace the PZT, $(Bi_{0.5}Na_{0.5})TiO_3$-$BaTiO_3$ (BNT-BT) and $(Ba,Ca)(Zr,Ti)O_3$ (BCZT) based materials have also attracted a great deal of interest. The (1−x)BNT-xBT system sparked interest because a MPB was verified for x = 0.06–0.07 83 . However, the properties of this MPB system are restricted due to their temperature dependence. In particular, a low depolarization temperature is shown as a manifestation of a ferroelectric to antiferroelectric phase transition 83 . Once they have formed ternary systems, BNT-BT-based materials may become candidates for low temperature actuator applications, as in the case of BNT-BT-KNN 74 .

A comparative analysis of the dielectric constant stability of three representative piezoceramics is displayed in Figure 8. The increment of the dielectric constant as a function of the applied electric field for $0.93(Bi_{0.5}Na_{0.5})TiO_3$–$0.07BaTiO_3$ (BNT-7BT) compound is plotted together with the values for KNL-NTS (KNN-modified) and a commercial hard PZT (Pz26, from Ferropern). BNT-7BT is a room temperature MPB BNT-BT obtained by conventional solid state reaction, which exhibits longitudinal piezoelectric coefficient d_{33} around 180 pC/N. A high stability of the dielectric constant is observed in this compound, which is even higher than that found in a commercial hard PZT. The inset data verifies the same insight from the relative increment of the dielectric constant. Results from the piezoelectric response (not shown here and yet to be unpublished) confirm the high stability of this system. Selected compositions, which show moderate but not low piezoelectric coefficient, could be candidates for specific room temperature applications since their properties are stable. In any event, further studies are required in order to elucidate the microscopic origin of the high stability observed in BNT-BT.

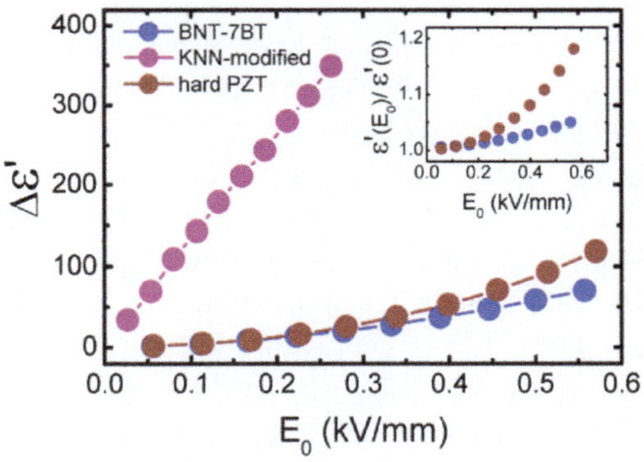

Figure 8. Increments of the dielectric constant as a function of the applied electric field amplitude for $(K_{0.44}Na_{0.52}Li_{0.04})(Nb_{0.86}Ta_{0.10}Sb_{0.04})O_3$ (KNN-modified), $0.93(Bi_{0.5}Na_{0.5})TiO_3$–$0.07BaTiO_3$ (BNT-7BT) and Pz26 (hard commercial PZT, from Ferroperm). The inset shows the relative increment of the dielectric constant for BNT-7BT and hard PZT.

On the other hand, greater research activity is expected to be conducted on the $(Ba,Ca)(Zr,Ti)O_3$ lead-free system in the coming years. Exceptional piezoelectric properties were reported in $(1-x)Ba(Zr_{0.2}Ti_{0.8})O_3$–$x(Ba_{0.7}Ca_{0.3})TiO_3$ 84 that opened a new research era on lead-free piezoelectrics. The BCZT system also shows a low Curie temperature, but deserves a lot of attention for room temperature usage 74 . Nowadays, much research in this system is being carried out. However, the first comprehensive studies of field-induced instability of properties are still to come.

FINAL REMARKS AND OUTLOOK

Nonlinear response analysis is a powerful tool for studying the domain wall dynamics in ferroelectric materials as well as for elucidating the role that extrinsic behavior plays in the material response. This type of characterization may be necessary to evaluate the performance of emerging compositions, in particular, the new lead-free systems. Not only are high values of electromechanical coefficients desirable, but also a high stability of these properties is needed to effectively replace lead-based compositions in high power applications.

The extrinsic nature of the nonlinear response implies that the domain wall motion not only improves piezoelectric properties, but also maximizes the field-induced instability of these properties. The crystal structure determines domains configuration, and consequently the domain wall motion and dynamics. Compositional engineering by doping is an effective method to improve particular properties for specific applications, because this is a real way to modify domain wall dynamics in perovskite ferroelectrics. Further studies related to the grain size effect on the nonlinear response of piezoceramics are required from both the fundamental and applied point of view. From the fundamental point of view, nonlinear response analysis may enable the extrinsic character of the size effect to be assessed, and thus correlate it with micro(nano)structural aspects such as the formation of complex domain structures into submicron grains.

ACKNOWLEDGMENTS

The author's work reviewed in this article has been carried out with the financial support of the projects MAT2004-01341, MAT2007-63445, MAT2010-21088-C03-02, and MAT2013-48009-C4-2-P of the Spanish Government. The author is grateful to many researchers who have closely worked with him, in particular Rafel Pérez and Diego A. Ochoa. The samples used in this review are from the José A. Eiras' group, Universidade Federal de Sao Carlos—Brazil (PZT-based materials), the José F. Fernández' group, Instituto de Cerámica y Vidrio—Spain (KNN-based material) and the Dra. Lourdes Mestres' group, Universitat de Barcelona—Spain (BNT-BT-based material).

REFERENCES

1. Uchino, K. *Ferroelectric Devices*, 2nd ed.; CRC Press: Boca Raton, FL, USA, 2010.
2. *Advanced Piezoelectric Materials: Science and Technology*; Uchino, K., Ed.; Woodhead Publishing Limited: Canbridge, UK, 2010.
3. Jo, W.; Dittmer, R.; Acosta, M.; Zang, J.; Groh, C.; Sapper, E.; Wang, K.; Rödel, J. Giant electric-field-induced strains in lead-free ceramics for actuator applications—Status and perspective. *J. Electroceramics* **2012**, *29*, 71–93.
4. Jaffe, B.; Cook, W., Jr.; Jaffe, H. *Piezoelectric Ceramics*; Academic Press: London, UK, 1971.
5. Damjanovic, D. A morphotropic phase boundary system based on polarization rotation and polarization extension.

Appl. Phys. Lett. **2010**, *97*.

6. Zhang, Q.M.; Wang, H.; Kim, N.; Cross, L.E. Direct evaluation of domain wall and intrinsic contributions to the dielectric and piezoelectric response and their temperature dependence on lead zirconate titanate. *J. Appl. Phys.* **1994**,*75*, 454–459.
7. Bellaiche, L.; Garcia, A.; Vanderbilt, D. Finite-temperature properties of $Pb(Zr_{1-x}Ti_x)O_3$ alloys from first principles.*Phys. Rew. Lett.* **2000**, *84*, 5427–5430.
8. Tutuncu, G.; Li, B.; Bowman, K.; Jones, J.L. Domain wall motion and electromechanical strain in lead-free piezoelectrics: Insight from the model system (1−*x*) $Ba(Zr_{0.2}Ti_{0.8})O_3$−$x(Ba_{0.7}Ca_{0.3})TiO_3$ using in situ high-energy X-ray diffraction during application of electric fields. *J. Appl. Phys.* **2014**, *115*.
9. Noheda, B.; Cox, D.E.; Shirane, G.; Guo, R.; Jones, B.; Cross, L.E. Stability of the monoclinic phase in the ferroelectric perovskite $PbZr_{1-x}Ti_xO_3$. *Phys. Rev. B* **2000**, *63*.
10. Ge, W.W.; Ren, Y.; Zhang, J.L.; Devreugd, C.P.; Li, J.F.; Viehland, D. A monoclinic-tetragonal ferroelectric phase transition in lead-free $(K_{0.5}Na_{0.5})NbO_{3-}x\%LiNbO_3$ solid solution. *J. Appl. Phys.* **2012**, *111*.
11. Damjanovic, D. Ferroelectric, dielectric and piezoelectric properties of ferroelectric thin films and ceramics. *Rep. Prog. Phys.* **1998**, *61*, 1267–1324.
12. Hall, D. Nonlinearity in piezoelectric ceramics. *J. Mater. Sci.* **2001**, *36*, 4575–4601.
13. Zhang, Q.M.; Pan, W.Y.; Jang, S.J.; Cross, L.E. Domain wall excitations and their contributions to the weak-signal response of doped lead zirconate titanate ceramics. *J. Appl. Phys.* **1968**, *64*, 6445–6451.
14. Damjanovic, D. Stress and frequency dependency of the direct piezoelectric effect in ferroelectric ceramics. *J. Appl. Phys.* **1997**, *82*, 1788–1797.
15. Garcia, J.E.; Pérez, R.; Albareda, A.; Eiras, J.A. Non-linear dielectric and piezoelectric response in undoped and Nb^{5+}or Fe^{3+} doped PZT ceramic system. *J. Eur. Ceram. Soc.* **2007**, *27*, 4029–4032.
16. Albareda, A.; Pérez, R.; Casals, J.A.; García, J.E.; Ochoa, D.A. Optimization of elastic nonlinear behavior measurements of ceramic piezoelectric resonators with burst excitation. *IEEE Trans. Ultrason. Ferroelectr. Freq. Control***2007**, *54*, 2175–2188.
17. Jiang, W.; Cao, W. Nonlinear properties of lead zirconate–titanate piezoceramics. *J. Appl. Phys.* **2000**, *88*, 6684–6689.
18. Priya, S.; Viehland, D.; Carazo, A.V.; Ryu, J.; Uchino, K. High-power resonant measurements of piezoelectric materials: Importance of elastic nonlinearities. *J. Appl. Phys.* **2001**, *90*, 1469–1479.
19. Blackburn, J.F.; Cain, M.G. Nonlinear piezoelectric resonance: A theoretically rigorous approach to constant I–V measurements. *J. Appl. Phys.* **2006**, *100*.
20. Ahmed, A.; Goldthorpe, I.; Khandani, A.K. Electrically tunable materials for microwave applications. *Appl. Phys. Rev.***2015**, *2*.
21. Maiti, T.; Guo, R.; Bhalla, A.S. Electric field dependent dielectric properties and high tunability of $BaZr_xTi_{1-x}O_3$relaxor ferroelectrics. *Appl. Phys. Lett.* **2006**, *89*.
22. Curecheriu, L.; Buscaglia, M.T.; Buscaglia, V.; Zhao, Z.; Mitoseriu, L. Grain size effect on the nonlinear dielectric properties of barium titanate ceramics. *Appl. Phys. Lett.* **2010**, *97*.
23. Damjanovic, D. Contributions to the piezoelectric effect in ferroelectric single crystals and ceramics. *J. Am. Ceram. Soc.***2005**, *88*, 2663–2676.
24. Damjanovic, D.; Demartin, M. Contribution of the irreversible displacement of domain walls to the piezoelectric effect in barium titanate and lead zirconate titanate ceramics. *J. Phys. Condens. Matter* **1997**, *9*, 4943–4953.
25. Chaplya, P.M.; Carman, G.P. Dielectric and piezoelectric response of lead zirconate-lead titanate at high electric and mechanical loads in terms of non-180° domain wall motion. *J. Appl. Phys.* **2001**, *90*, 5278–5286.
26. García, J.E.; Pérez, R.; Ochoa, D.A.; Albareda, A.; Lente, M.H.; Eiras, J.A. Evaluation of domain wall motion in lead zirconate titanate ceramics by nonlinear response measurements. *J. Appl. Phys.* **2008**, *103*.
27. Perez-Delfin, E.; García, J.E.; Ochoa, D.A.; Pérez, R.; Guerrero, F.; Eiras, J.A. Effect of Mn-acceptor dopant on dielectric and piezoelectric responses of lead lanthanum zirconate titanate piezoceramics. *J. Appl. Phys.* **2011**, *110*.
28. Demartin, M.; Damjanovic, D. Dependence of the direct piezoelectric effect in coarse and fine grain barium titanate ceramics on dynamic and static pressure. *Appl. Phys. Lett.* **1996**, *68*, 3046–3048.
29. Garcia, J.E.; Guerra, J.D.S.; Araújo, E.B.; Perez, R. Domain wall contribution to dielectric and piezoelectric responses in $0.65Pb(Mg_{1/3}Nb_{2/3})O_3$-$PbTiO_3$ ferroelectric ceramics. *J. Phys. D Appl. Phys.* **2009**, *42*.
30. Ochoa, D.A.; García, J.E.; Pérez, R.; Gomis, V.; Albareda, A.; Rubio-Marcos, F.; Fernández, J.F. Extrinsic contribution and non-linear response in lead-free KNN-modified piezoceramics. *J. Phys. D Appl. Phys.* **2009**, *42*.
31. Gao, J.; Hu, X.; Zhang, L.; Li, F.; Zhang, L.; Wang, Y.; Hao, Y.; Zhong, L.; Ren, X. Major contributor to the large piezoelectric response in (1−*x*)$Ba(Zr_{0.2}Ti_{0.8})O_3$−$x(Ba_{0.7}Ca_{0.3})TiO_3$ ceramics: domain wall motion. *Appl. Phys. Lett.* **2014**,*104*.
32. Ochoa, D.A.; Garcia, J.E.; Pérez, R.; Albareda, A. Influence of extrinsic contribution on the macroscopic properties of hard and soft lead zirconate titanate ceramics. *IEEE Trans. Ultrason. Ferroelectr. Freq. Control* **2008**, *55*, 2732–2736.
33. Zhang, X.L.; Chen, Z.X.; Cross, L.E.; Schulze, W.A. Dielectric and piezoelectric properties of modified lead zirconate titanate ceramics from 4.2 to 300 K. *J. Mater. Sci.* **1983**, *18*, 968–972.
34. Iamsasri, T.; Tutuncu, G.; Uthaisar, C.; Pojprapai, S.; Jones, J.L. Analysis methods for characterizing ferroelectric/ferroelastic domain reorientation in orthorhombic perovskite materials and application to

Li-doped $Na_{0.5}K_{0.5}NbO_3$. *J. Mater. Sci.* **2013**, *48*, 6905–6910.

35. Garcia, J.E.; Gomis, V.; Perez, R.; Albareda, A.; Eiras, J.A. Unexpected dielectric response in lead zirconate titanate ceramics: The role of ferroelectric domain wall pinning effects. *Appl. Phys. Lett.* **2007**, *91*.
36. García, J.E.; Pérez, R.; Albareda, A. High electric field measurement of dielectric constant and losses of ferroelectric ceramics. *J. Phys. D Appl. Phys.* **2001**, *34*, 3279–3284.
37. Li, S.; Cao, W.; Cross, L.E. The extrinsic nature of nonlinear behavior observed in lead zirconate titanate ferroelectric ceramic. *J. Appl. Phys.* **1991**, *69*, 7219–7224.
38. Garcia, J.E.; Ochoa, D.A.; Gomis, V.; Eiras, J.A.; Pérez, R. Evidence of temperature dependent domain wall dynamics in hard lead zirconate titanate piezoceramics. *J. Appl. Phys.* **2012**, *112*.
39. Damjanovic, D.; Demartin, M. The Rayleigh law in piezoelectric ceramics. *J. Phys. D Appl. Phys.* **1996**, *29*, 2057–2060.
40. García, J.E.; Pérez, R.; Albareda, A. Contribution of reversible processes to the non-linear dielectric response in hard lead zirconate titanate ceramics. *J. Phys. Condens. Matter* **2005**, *17*, 7143–7150.
41. Hall, D.A.; Stevenson, P.J. Field-induced desestabilisation of hard PZT ceramics. *Ferroelectrics* **1996**, *187*, 23–37.
42. Mueller, V.; Zhang, Q.M. Shear response of lead zirconate titanate piezoceramics. *J. Appl. Phys.* **1998**, *83*, 3754–3761.
43. Alilat, K.; Thi, M.P.; Dammak, H.; Bogicevic, C.; Albareda, A.; Doisy, M. Grain size effect on electromechanical properties and non-linear response of dense nano and microstructured PIN-PT ceramics. *J. Eur. Ceram. Soc.* **2010**, *30*, 1919–1924.
44. Peng, B.; Yue, Z.; Li, L. Evaluation of domain wall motion during polymorphic phase transition in (K,Na)NbO_3-based piezoelectric ceramics by nonlinear response measurements. *J. Appl. Phys.* **2011**, *109*.
45. Robert, G.; Damjanovic, D.; Setter, N. Preisach distribution function approach to piezoelectric nonlinearity and hysteresis. *J. Appl. Phys.* **2001**, *90*, 2459–2464.
46. Damjanovic, D. Hysteresis in Piezoelectric and Ferroelectric Materials. In *The Science of Hysteresis, Vol III. Hysteresis in Materials*; Bertotti, G., Mayergoyz, I.D., Eds.; Academic Press: Oxford, UK, 2006; pp. 337–465.
47. Robert, G.; Damjanovic, D.; Setter, N. Preisach modeling of ferroelectric pinched loops. *Appl. Phys. Lett.* **2000**, *77*, 4413–4415.
48. Ochoa, D.A.; Pérez, R.; García, J.E. Preisach modelling of nonlinear response in electrically biased lead zirconate titanate-based piezoceramics. *Appl. Phys. A* **2013**, *112*, 1081–1088.
49. Eitel, R.E.; Shrout, T.R.; Randall, C.A. Nonlinear contributions to the dielectric permittivity and converse piezoelectric coefficient in piezoelectric ceramics. *J. Appl. Phys.* **2006**, *99*.
50. Rubio-Marcos, F.; Del Campo, A.; López-Juárez, R.; Romero, J.J.; Fernández, J.F. High spatial resolution structure of (K,Na)NbO_3 lead-free ferroelectric domains. *J. Mater. Chem.* **2012**, *22*, 9714–9720.
51. Rubio-Marcos, F.; Del Campo, A.; Fernández, J.F. Resolution of the ferroelectric domains structure in (K,Na)NbO_3-based lead-free ceramics by confocal Raman microscopy. *J. Appl. Phys.* **2013**, *113*.
52. Eichel, R.-A. Defect structure of oxide ferroelectrics—Valence state, site of incorporation, mechanisms of charge compensation and internal bias fields. *J. Electroceramics* **2007**, *19*, 11–23.
53. Warren, W.L.; Pike, G.E.; Vanheusden, K.; Dimos, D.; Tuttle, B.A.; Robertson, J. Defect-dipole alignment and tetragonal strain in ferroelectrics. *J. Appl. Phys.* **1996**, *79*, 9250–9257.
54. Zhang, Z.; Wu, P.; Lu, L.; Shu, C. Study on vacancy formation in ferroelectric $PbTiO_3$ from ab initio. *Appl. Phys. Lett.***2006**, *88*.
55. Zhang, Z.; Lu, L.; Shu, C.; Wu, P. Computational investigation of B-site donor doping effect on fatigue behavior of lead zirconate titanate. *Appl. Phys. Lett.* **2006**, *89*.
56. Vendrell, X.; García, J.E.; Rubio-Marcos, F.; Ochoa, D.A.; Mestres, L.; Fernández, J.F. Exploring different sintering atmospheres to reduce nonlinear response of modified KNN piezoceramics. *J. Eur. Ceram. Soc.* **2013**, *33*, 825–831.
57. Setter, N.; Waser, R. Electroceramic materials. *Acta Mater.* **2000**, *48*, 151–178.
58. Arlt, G.; Hennings, D.; de With, G. Dielectric properties of fine-grained barium titanate ceramics. *J. Appl. Phys.* **1985**,*58*, 1619–1625.
59. Zhao, Z.; Buscaglia, V.; Viviani, M.; Buscaglia, M.T.; Mitoseriu, L.; Testino, A.; Nygren, M.; Johnsson, M.; Nanni, P. Grain-size effects on the ferroelectric behavior of dense nanocrystalline $BaTiO_3$ ceramics. *Phys. Rev. B* **2004**, *70*.
60. Randall, C.A.; Kim, N.; Kucera, J.-P.; Cao, W.; Shrout, T.R. Intrinsic and extrinsic size effects in fine-grained morphotropic-phase-boundary lead zirconate titanate ceramics. *J. Am. Ceram. Soc.* **1998**, *81*, 677–688.
61. Algueró, M.; Ricote, J.; Jiménez, R.; Ramos, P.; Carreaud, J.; Dkhil, B.; Kiat, J.M.; Holc, J.; Kosec, M. Size effect in morphotropic phase boundary $Pb(Mg_{1/3}Nb_{2/3})O_3$–$PbTiO_3$. *Appl. Phys. Lett.* **2007**, *91*.
62. Saito, Y.; Takao, H.; Tani, T.; Nonoyama, T.; Takatori, K.; Homma, T.; Nagaya, T.; Nakamura, M. Lead-free piezoceramics. *Nature* **2004**, *432*, 84–87.
63. Maeder, M.D.; Damjanovic, D.; Setter, N. Lead free piezoelectric materials. *J. Electroceramics* **2004**, *13*, 385–392.
64. Takenaka, T.; Nagata, H. Current status and prospects of lead-free piezoelectric ceramics. *J. Eur. Ceram. Soc.* **2005**, *25*, 2693–2700.
65. Shrout, T.R.; Zhang, S.J. Lead-free piezoelectric ceramics: alternatives for PZT? *J. Electroceramics* **2007**, *19*, 113–

126.

66. Takenaka, T.; Nagata, H.; Hiruma, Y. Current developments and prospective of lead-free piezoelectric ceramics. *Jpn. J. Appl. Phys.* **2008**, *47*, 3787–3801.

67. Panda, P.K. Review: Environmental friendly lead-free piezoelectric materials. *J. Mater. Sci.* **2009**, *44*, 5049–5062.

68. Rödel, J.; Jo, W.; Seifert, K.T.P.; Anton, E.-M.; Granzow, T.; Damjanovic, D. Perspective on the development of lead-free piezoceramics. *J. Am. Ceram. Soc.* **2009**, *92*, 1153–1177.

69. Aksel, E.; Jones, J.L. Advances in lead-free piezoelectric materials for sensors and actuators. *Sensors* **2010**, *10*, 1935–1954.

70. Xiao, D. Progresses and further considerations on the research of perovskite lead-free piezoelectric ceramics. *J. Adv. Dielectr.* **2011**, *1*, 33–40.

71. Shvartsman, V.V.; Lupascu, D.C. Lead-free relaxor ferroelectrics. *J. Am. Ceram. Soc.* **2012**, *95*, 1–26.

72. Coondoo, I.; Panwar, N.; Kholkin, A. Lead-free piezoelectrics: Current status and perspectives. *J. Adv. Dielectr.* **2013**, *3*.

73. Li, J.-F.; Wang, K.; Zhu, F.-Y.; Cheng, L.-Q.; Yao, F.-Z. (K,Na)NbO_3-based lead-free piezoceramics: Fundamental aspects, processing technologies, and remaining challenges. *J. Am. Ceram. Soc.* **2013**, *96*, 3677–3696.

74. Rödel, J.; Webber, K.G.; Dittmer, R.; Jo, W.; Kimura, M.; Damjanovic, D. Transferring lead-free piezoelectric ceramics into application. *J. Eur. Ceram. Soc.* **2015**, *35*, 1659–1681.

75. Safari, A.; Abazari, M.; Kerman, K.; Marandian-Hagh, N.; Akdo, E.K. $(K_{0.44}Na_{0.52}Li_{0.04})(Nb_{0.86}Ta_{0.10}Sb_{0.04})O_3$ferroelectric ceramics. *IEEE Trans. Ultrason. Ferroelectr. Freq. Control* **2009**, *56*, 1586–1594.

76. Akdoğan, E.K.; Kerman, K.; Abazari, M.; Safari, A. Origin of high piezoelectric activity in ferroelectric $(K_{0.44}Na_{0.52}Li_{0.04})$–$(Nb_{0.86}Ta_{0.10}Sb_{0.04})O_3$ ceramics. *Appl. Phys. Lett.* **2008**, *92*.

77. Rubio-Marcos, F.; Ochoa, P.; Fernandez, J.F. Sintering and properties of lead-free (K,Na,Li)(Nb,Ta,Sb)O_3 ceramics. *J. Eur. Ceram. Soc.* **2007**, *27*, 4125–4129.

78. Vendrell, X.; García, J.E.; Bril, X.; Ochoa, D.A.; Mestres, L.; Dezanneau, G. Improving the functional properties of $(K_{0.5}Na_{0.5})NbO_3$ piezoceramics by acceptor doping. *J. Eur. Ceram. Soc.* **2015**, *35*, 125–130.

79. Matsubara, M.; Yamaguchi, T.; Sakamoto, W.; Kikuta, K.; Yogo, T.; Hirano, S. Processing and piezoelectric properties of lead-free (K,Na)(Nb,Ta)O_3 ceramics. *J. Am. Ceram. Soc.* **2005**, *88*, 1190–1196.

80. Lin, D.; Kwok, K.W.; Chan, H.L.W. Structure, dielectric, and piezoelectric properties of CuO-doped $K_{0.5}Na_{0.5}NbO_3$–$BaTiO_3$ lead-free ceramics. *J. Appl. Phys.* **2007**, *102*.

81. Zhang, S.; Lim, J.B.; Lee, H.J.; Shrout, T.R. Characterization of hard piezoelectric lead-free ceramics. *IEEE Trans. Ultrason. Ferroelectr. Freq. Control* **2009**, *56*, 1523–1527.

82. Lim, J.B.; Zhang, S.J.; Jeon, J.H.; Shrout, T.R. (K,Na)NbO_3-based ceramics for piezoelectric "hard" lead-free materials.*J. Am. Ceram. Soc.* **2010**, *93*, 1218–1220.

83. Takenaka, T.; Maruyama, K.; Sakata, K. $(Bi_{1/2}Na_{1/2})TiO_3$-$BaTiO_3$ system for lead-free piezoelectrics ceramics. *Jpn. J. Appl. Phys.* **1991**, *30*, 2236–2239.

84. Liu, W.; Ren, X. Large piezoelectric effect in Pb-free ceramics. *Phys. Rev. Lett.* **2009**, *103*.

CHAPTER
4

THE GROWTH AND PROPERTIES OF LEAD-FREE FERROELECTRIC SINGLE CRYSTALS

Xiaobing Li, Chao Chen, Hao Deng, Haiwu Zhang, Di Lin, Xiangyong Zhao and Haosu Luo

Keywords: Lead-free ferroelectrics, crystal growth, piezoelectric performance, ferroelectric performance, ion doping

INTRODUCTION

Relaxor-based ferroelectrics are a hot topic at present, in terms of their practical applications and the mechanism that explains their novel properties [1]. Take the lead-based ferroelectrics lead zirconium titanate (PZT) ceramics, for example. They have been used in numerous practical applications such as sensors, actuators, ultrasonic transducers, *etc.* [2,3,4]. Furthermore, the single crystals of these perovskite solid solutions such as $Pb(Mg_{1/3}Nb_{2/3})$-$PbTiO_3$ (PMNT) single crystal exhibit a several times larger piezoelectric effect than conventional ceramics, which make them promising for the next generation of piezoelectric devices [5]. However, due to the toxic nature of PbO, considerable recent studies have been focused on lead-free piezoelectric materials from an environmental protection viewpoint [6].

Among all kinds of lead-free piezoelectric systems, sodium bismuth titanate ($Na_{0.5}Bi_{0.5}TiO_3$, abbreviated as NBT) is a strong ferroelectric and the piezoelectric performance can be enhanced by forming solid solutions with barium titanate ($BaTiO_3$, abbreviated as BT) at the composition of hombohedral-tetragonal morphotropic phase boundary (MPB). Therefore, the $(1-x)(Na_{0.5}Bi_{0.5})TiO_3$-$x$$BaTiO_3$ [abbreviated as NBT-BT $100(1-x)/100x$ or NBBT] solid solutions are considered to be a promising candidate for a lead-free piezoelectric material. Then NBT-BT single crystals will provide an opportunity to conveniently investigate the dielectric and piezoelectric properties [7,8,9,10,11].

However, these properties are still notably inferior to that of lead-based ones. In addition, relatively high leakage currents due to defects are another problem to overcome in NBT-BT, which can interfere with poling and polarization switching. Modest concentrations of dopants have been reported to be an effective method by which to enhance the resistivity of lead-based materials [12,13,14,15,16,17,18]. Then Mn, Eu and Zn doped NBT and NBT-BT crystals were grown and studied, and it have been found that ions substitution can produce local tetragonal distortions and change the domain structures. In addition, the piezoelectric and ferroelectric properties enhanced because the leakage current of NBT-BT crystals decreased notably upon ions substitution [19,20,21,22,23].

Moreover, as another kind of lead free materials, alkaline niobate $(K_xNa_{1-x})NbO_3$ (KNN)-based solid solutions have been attracting much attention as they are believed to provide a relatively high piezoelectric response, a high Curie temperature and a high acoustic velocity. Since the great breakthrough had been made by Saito in textured KNN-based ceramics, preparation of KNN ceramics with better piezoelectric properties has been extensively studied. It provided that texture engineering can lead to an improvement in piezoelectric properties, single crystals of KNN are also expected to have superior properties. In addition, free from the effect of the grain boundary in ceramics, single crystals are more appropriate for investigating the nature of materials.

In the present work, crystal growth, structure and electrical properties of NBT-BT crystal are systemically studied on samples for various contents and crystallographic cuts and it was found that enhanced piezoelectric properties can be obtained in NBT-BT single crystal by using the concept of engineered-domain configuration.

Citation: Xiaobing Li, Chao Chen, Hao Deng, Haiwu Zhang, Di Lin, Xiangyong Zhao and Haosu Luo, The Growth and Properties of Lead-Free Ferroelectric Single Crystals, doi:10.3390/cryst5020172.

GROWTH, PROPERTIES AND PHASE STRUCTURE OF LEAD-FREE FERROELECTRIC SINGLE CRYSTALS

NBT Based Single Crystals

Crystal Growth of NBT-BT

Chemical materials of Na_2CO_3, Bi_2O_3, TiO_2 and $BaCO_3$ with 99.99% were used to synthesize the polycrystalline material of $0.94(Na_{0.5}Bi_{0.5})TiO_3$-$0.06BaTiO_3$ (abbreviated as NBT-BT94/6) by conventional solid-state reaction techniques. Before solid-state reaction, these compounds were weighed according to the following chemical reaction equation and characterized by TG-DTA experiments using a simultaneous thermal analyzer (NETZSCH STA 449C, Sasol, Hamburg, Germany) a heating rate of 10 °C/min in an air atmosphere. The TG-DTA curves of $Na_2CO_3 + Bi_2O_3 + 4TiO_2$ and $BaCO_3 + TiO_2$ are shown in the following Figure 1.

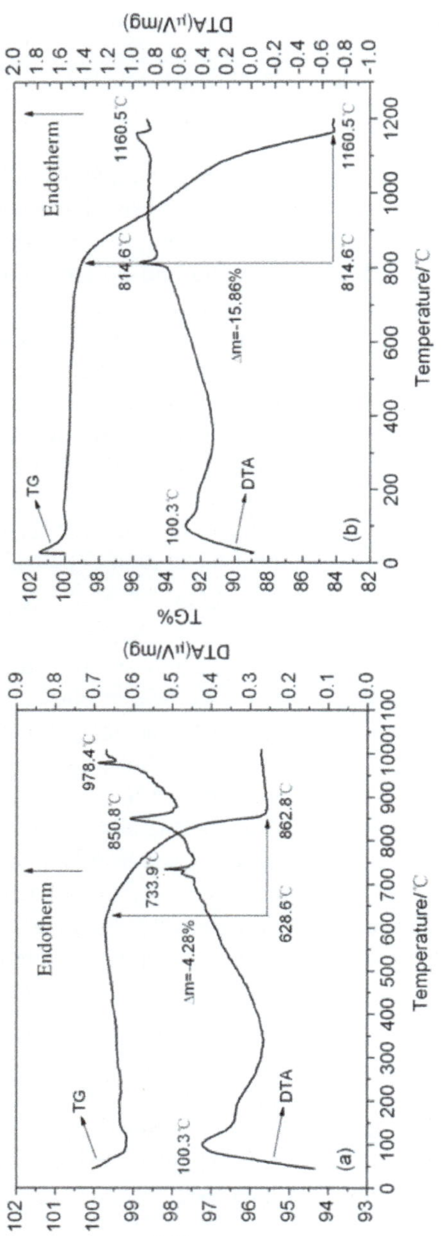

Figure 1. TG-DTA curves of (**a**) $Na_2CO_3 + Bi_2O_3 + 4TiO_2$; and (**b**) $BaCO_3 + TiO_2$.

After these thermal measurements, the proper temperature to synthesize the polycrystalline material of NBT-BT 94/6 crystal was determined:

$$Na_2CO_3 + Bi_2O_3 + 4TiO_2 \xrightarrow{\text{High temperature}} 4(Na_{0.5}Bi_{0.5})TiO_3 + CO_2\uparrow \quad (1)$$

$$BaCO_3 + TiO_2 \xrightarrow{\text{High temperature}} BaTiO_3 + CO_2\uparrow \quad (2)$$

High purity Na_2CO_3, Bi_2O_3, TiO_2 and $BaCO_3$ were weighted according to stoichiometrical ratio of $0.94(Na_{0.5}Bi_{0.5})TiO_3$-$0.06BaTiO_3$. After these compounds were ground and mixed, they were put into platinum (Pt) crucible and heated to 1200 °C for 10 h to decompose the carbonate and form NBT-BT 94/6 polycrystalline material. Then the NBT-BT 94/6 polycrystalline materials were ground and mixed with 20 wt% of excess Na_2CO_3 and Bi_2O_3 as a self-flux for compensating the composition change. These mixtures were heated to 900 °C for an hour [24].

NBT-BT 94/6 single crystal was grown from a Pt crucible, which was heated by using a resistance furnace under air atmosphere. A platinum wire was used as the seed for crystal growth. In order to avoid the formation of poly-crystal in the crystal growth process, a temperature 30–50 °C higher than the melting temperature of polycrystalline material was required initially, to melt the micro-crystal particles in the Pt crucible and keep that temperature for 1 h. The temperature was then lowered to the melting temperature. Initially, a randomly oriented crystal was obtained by spontaneous nucleation on the end of a platinum wire by restricting the diameter of the crystalline material so that only one crystal should be grown. The pulling rate was 2–2.5 mm per day after the crystal diameter reached a certain value; the rotating rate was 10–30 rpm. After growth, the crystal was cooled to room temperature at a rate of 30–50 °C/h. Figure 2 shows the NBT-BT crystal boule grown by the TSSG method. The dimensions of the as-grown crystal are 25 mm in diameter and 10 mm in length [25].

Figure 2. The photograph of as-grown $(Na_{0.5}Bi_{0.5})TiO_3$-$BaTiO_3$ (NBT-BT) crystal. Reprinted with permission from [24]. Copyright 2008 AIP Publishing LLC.

Phases and Domains in NBT

Figure 3a shows the X-ray powder diffraction (XRPD) patterns of the as-grown NBT-BT crystal. In addition, the XRPD patterns of $BaTiO_3$ crystal which was grown by the TSSG method were also given in Figure 3b. It can be seen that these two crystals almost have the same XRPD patterns. Since the barium titanate is a typical perovskite-type structure compounds, it can be deduced that the as grown NBT-BT crystal also possess the perovskite-type structure. The unit-cell parameters were calculated through the TEROR programs according to the XRPD patterns. The unit-cell parameters of the as grown NBT-BT crystal are $a = b = c = 3.8862$ Å and $\alpha = \beta = \gamma = 89.2°$, which are similar to the data of $Na_{0.5}Bi_{0.5}TiO_3$ previously reported in the reference [26]. The diffraction indices corresponding to each diffraction peak are marked on the patterns. There is no splitting of peaks in the XRPD pattern of NBT-BT crystal, while the splitting of (002) and (200) peaks can be observed in the XRPD pattern of $BaTiO_3$ crystal. These results indicate that the as grown NBT-BT crystal belongs to the rhombohedral crystal system and its rhombohedral unit cell are very close to the cubic ones and can be regarded as pseudo-cubic structure [24].

Figure 4 shows domain structures of [001]-oriented NBT and NBT: Mn crystals. The average domain width is estimated to be 50 μm for pure NBT crystal and 25 μm for NBT:Mn crystal, which means that domain size of pure NBT crystal decreases after Mn-doping [27]. It has been reported that the domain configuration of rhombohedral single crystals of PZNT will come into a more stable engineered domain state when it is poled along its nonpolar pseudocubic [001] direction, which can remarkably enhance the piezoelectric properties. The smaller the domain size, the easier the domain reorientation and engineered domain state are obtained. The NBT:Mn crystal also has a rhombohedral structure and possesses small domain size. Thus, we infer that the engineered domain state should be established much easier in NBT:Mn crystal than the case in pure NBT crystal. Consequently, the piezoelectric properties of NBT crystal are obviously improved by Mn doping.

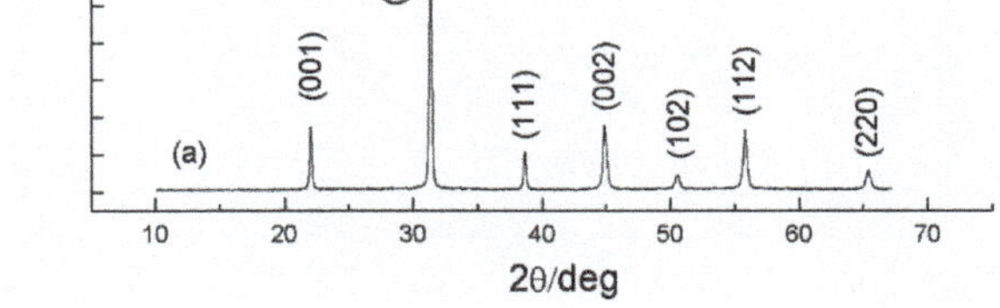

Figure 3. X-ray powder diffraction patterns of (**a**) as grown NBT-BT crystal and (**b**) $BaTiO_3$ crystal. Reprinted with permission from [24]. Copyright 2008 Elsevier.

Figure 4. Room temperature domain structures of [001]-oriented NBT (**a**) and NBT:Mn (**b**) single crystals. Reprinted with permission from [27]. Copyright 2008 Elsevier.

Dielectric, Piezoelectric, Ferroelectric Properties

Figure 5 shows the dielectric constants ε_r as a function of temperature at different frequencies for the poled and unpoled <001> oriented NBT-BT crystal. It can be seen that the dielectric constants show obvious frequency dispersion and exhibit broad dielectric peaks with a maximum at a T_m of 310 °C, which is connected with the transition towards the cubic paraelectric state, as previously reported in NBT-BT system ceramics. These results indicate that the as-grown NBT-BT 92/8 crystal is a relaxor ferroelectric and the T_m is 10 °C lower than that of pure $(Na_{0.5}Bi_{0.5})TiO_3$(NBT). At room temperature, the dielectric constant ε_r of <001> oriented NBT-BT crystal is 770 at 10 kHz and it decreases to 430 after poling. Compared with the unpoled sample, the temperature dependence of dielectric constants for the poled <001> oriented NBT-BT 98/2 crystal shows an obvious hump around 155 °C. The humps in the temperature range of 170–230 °C in the $\varepsilon(T)$ curve were also reported in pure NBT, NBT-BT system ceramics and NBT-BT 94/6 crystal and these researchers all proposed that this hump results from the phase transition from rhombohedral ferroelectric to tetragonal antiferroelectric [24]. However, other experimental results such as X-ray diffraction, Raman and neutron scattering did not indicate the existence of antiferroelectric phase. Since the hump are only observed in the poled sample according to our experimental results, the possible origination of this phenomenon can be attributed to the macro-micro domains switching as it has been discussed in the $Pb(Fe_{1/2}Nb_{1/2})O_3$ and PLZT relaxor ferroelectric ceramics.

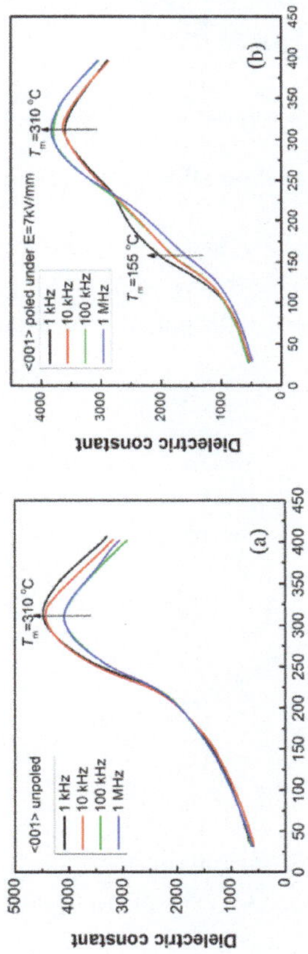

Figure 5. Dielectric constants as a function of temperature and frequency for (**a**) unpoled and (**b**) poled <001>oriented NBT-BT 98/2 crystal. Reprinted with permission from [24]. Copyright 2008 Else-

vier.

Figure 6 shows the piezoelectric constants d_{33} as a function of poling electric field for the <001>, <110> and <111> oriented NBT-BT 98/2 crystal samples. For <001> and <110> oriented NBT-BT 98/2 crystal samples, d_{33} increases firstly as the poling electric field increases and then decrease when the poling electric field exceeds 3 kV/mm. Maximum d_{33} values of 60 and 65 pC/N at the poling electric field of 3 kV/mm were obtained for <001> and <110> oriented NBT-BT 98/2 crystal. For <111> oriented NBT-BT 98/2 crystal samples, the d_{33} values increase moderately as the poling electric field increases. Maximum d_{33}value is 30 pC/N for <111> oriented NBT-BT crystal. Although <111> is the polar direction for rhombohedral NBT-BT crystal, such cuts exhibit low piezoelectric constants d_{33} as a function of poling electric field as shown in Figure 6. Since it has been reported that the piezoelectric in rhombohedral single crystal of $Pb(Zn_{1/3}Nb_{2/3})O_3$-$PbTiO_3$ can be obviously enhanced when it was poled along its nonpolar pseudocubic <001> direction and come into engineered domain state. Perhaps the same mechanism enhances the piezoelectric properties of lead-free rhombohedral NBT-BT crystal as shown in Figure 7, which will be further studied combined with *in situ* domain observation under DC-bias using a polarizing microscope [24].

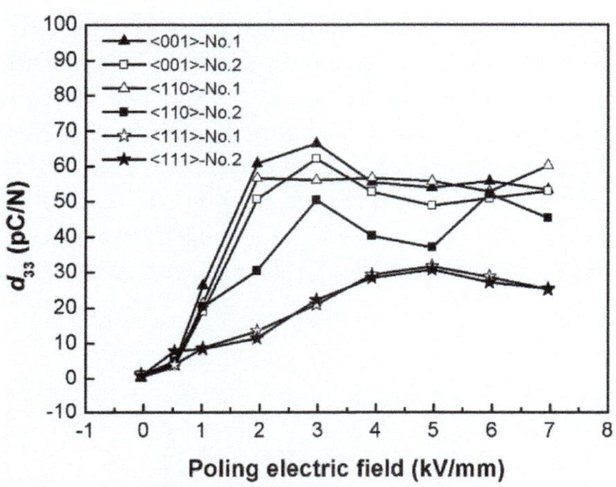

Figure 6. Piezoelectric constants d_{33} as a function of poling electric field for <001>, <110> and <111> oriented NBT-BT 98/2 crystal. Reprinted with permission from [24]. Copyright 2008 Elsevier.

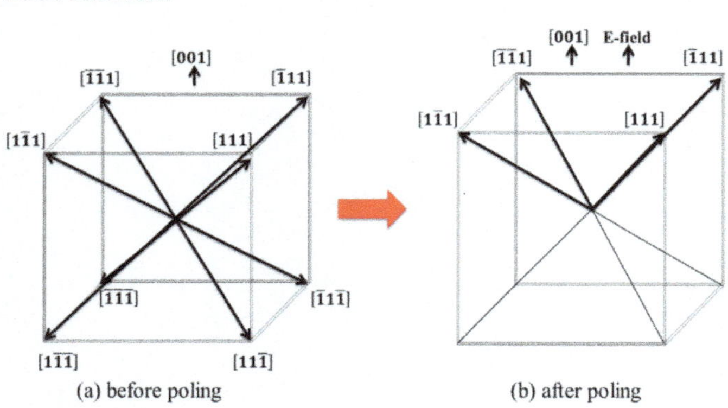

Figure 7. Schematics of domain configuration in the rhombohedral 0.92NBT-0.08KBT crystal: (**a**) before poling, and (**b**) after poling along the <001> direction. Reprinted with permission from [25]. Copyright 2011 AIP Publishing LLC.

In the case of the [001] oriented NBT-BT single crystal, the domain configuration can be graphically represented, as shown in Figure 7. Before poling, the rhombohedral crystal has eight polar vectors with an equivalent magnitude and different [111] directions. After complete poling along one of the pseudocubic [001] directions, an engineered-domain state can be achieved, *i.e.*, each domain has four possible polar directions ($[11\bar{1}]$, $[\bar{1}1\bar{1}]$, $[\bar{1}\bar{1}\bar{1}]$, $[1\bar{1}\bar{1}]$) as shown in Figure 7b. Such an engineered domain state is more stable than the single domain state because the four types of domains are energetically equivalent and are equally populated under the [001] poling electric field. When the [001] oriented rhombohedral 0.92NBT-0.08KBT crystal with the engineered-domain configuration is actuated by an electric field along the [001] direction, the four possible polar directions are expected to incline close to the E field direction in each domain, which results in an increased rhombohedral lattice distortion and a high piezoelectric response. According to Figure 7a, the component of each polar vector along the <001> direction is completely equal to every other one and the remanent polarization of the <001> oriented rhombohedral 0.92NBT-0.08KBT crystal should be 1:3√ that of the <111> oriented crystal. This inference can be confirmed by the polarization hysteresis loop measurements. Figure 8 shows the room-temperature polarization hysteresis loops of <001>, <110> and <111> oriented 0.92NBT-0.08KBT crystals. The ferroelectric parameters taken from Figure 8 are collected inTable 1. It can be seen that <001> oriented crystals have the smallest remnant polarization (P_r = 8.1 μC/cm^2), while the <111> oriented crystals have the largest remnant polarization (P_r = 13.5 μC/cm^2). The remanent polarization ratio ofP_r<001>:P_r<110>:P_r<111> is 1:1.33:1.67; close to 1:2√:3√ [25].

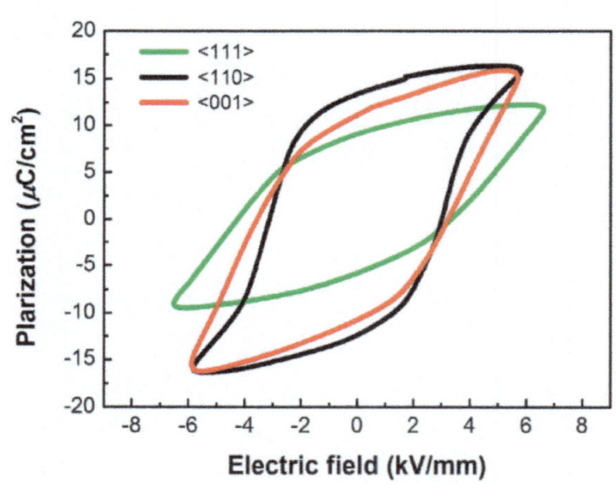

Figure 8. (Color online) Polarization hysteresis loops of the <001>, <110>, and <111> oriented 0.92NBT-0.08KBT crystals at room temperature. Reprinted with permission from [25]. Copyright 2011 AIP Publishing LLC.

Table 1. Dielectric parameters for the unpoled and poled 0.92NBT-0.08KBT single crystal (1 kHz) (poling condition: $E_p = 3$ kV/mm, $T_p = 50$ °C and $t_p = 15$ min). Reprinted with permission from [25]. Copyright 2011 AIP Publishing LLC.

Orientation	Condition	ε_{RM}	tanθ	ε_m	T_m (°C)
<001>	unpoled	805	0.040	4318	316
	poled	683	0.031	4315	316
<110>	unpoled	863	0.047	4160	317
	poled	567	0.030	4154	318
<111>	unpoled	910	0.046	4348	318
	poled	435	0.038	4074	316

KNN Single Crystal

Crystal Growth of KNN

$(K_{0.25}Na_{0.75})NbO_3$ single crystals were grown by a carefully controlled TSSG technique. The raw materials of K_2CO_3(99.99%), Na_2CO_3 (99.99%), and Nb_2O_5 (99.99%) were weighed according to the nominal ratio of $0.70KNbO_3/0.30NaNbO_3$ to obtain KNN25/75 single crystals based on the $KNbO_3$–$NaNbO_3$ phase diagram in ref. [22]. Then the well mixed compounds were put into a platinum (Pt) crucible and calcined at 850 °C for 3 h in air to form the KNN70/30 polycrystalline precursor. Then the polycrystalline materials were ground and mixed with excess ~20 wt% Na_2CO_3 and K_2CO_3 as self-flux. When heated above 850 °C, both Na_2CO_3 and K_2CO_3 decomposed, so the real flux was Na_2O–K_2O. Single crystals were grown using a $[001]_{pc}$ direction seed in a Pt crucible, which was heated to ~1160 °C using a resistance furnace under an air atmosphere. The velocities of rotation and pulling of the rod with the seed crystal were, respectively, ~16 rpm and ~3.0 mm per day. At the end of the crystal growth, the crystals were separated from the flux–melt surface and then slowly cooled down to room temperature [28].

Figure 9a shows the photograph of the as-grown KNN25/75 single crystal pulled along the $[001]_{pc}$ direction, with dimensions of up to Ø30 × 10 mm^2. The as-grown crystal is semi-transparent with no inclusions and exhibits four clear symmetrical growth ridges, indicating the cubic growth habit. A polished $(100)_{pc}$ wafer shown in Figure 9b is transparent with several domain walls visible in it.

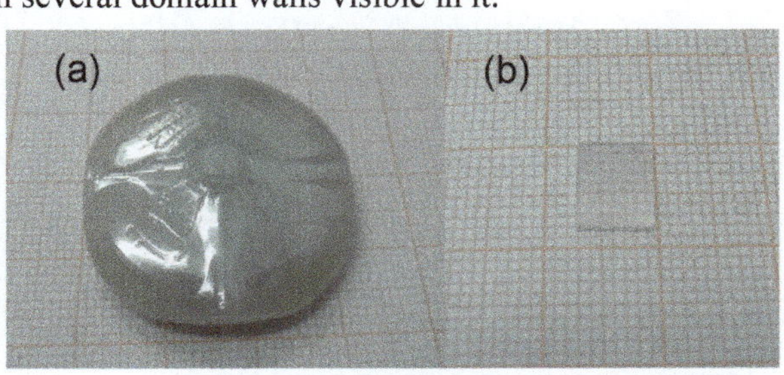

Figure 9. Photographs of (**a**) the as-grown KNN25/75 single crystal and (**b**) the polished KNN25/75 single crystal wafer. Reprinted with permission from [28]. Copyright 2014 Royal Society of Chemistry.

Phases and Domains in KNN

The lattice structure of the as-grown crystal was determined by X-ray diffraction which was carried out using a Huber-G670 X-ray diffractometer (Cu: $K_{\alpha1}$, $\lambda = 1.5406$ Å, Karlsruhe, Germany). The diffractograms of the $(100)_{pc}$ and $(110)_{pc}$ oriented single crystals were obtained using a Bruker D8 Discover HRXRD system. The concentrations of K, Na, and Nb elements in the as-grown crystal were measured by X-ray fluorescence analysis (XRF). Based on the measured results, the effective segregation coefficients of K and Na were calculated.

The X-ray diffraction patterns of the KNN70/30 polycrystalline materials and powders of the as-grown single crystals are presented in Figure 10a. It is clear that both the raw materials and the KNN single crystals are of the perovskite-type structure. The splitting of the (200) cubic indexed peak at ~45°, indicates an orthorhombic symmetry of the KNN crystals. The peaks of the single crystal shifted to a higher angle compared with those of the starting polycrystalline powder, implying the volume decrease of the unit cell. Since the ionic radius of sodium is smaller than that of potassium, it is reasonable to consider that the decrease of the unit cell is induced by the decrease of potassium content in the single crystal caused by segregation. The effective segregation coefficients Keff, (Keff = CS/CL(∞)), of K and Na ions are calculated to be 0.36 and 2.8, respectively. The 2θ–ω scans of the $(100)_{pc}$ and $(110)_{pc}$ oriented single crystals are shown in Figure 10b. The XRD rocking curves of the $(200)_{pc}$ and $(110)_{pc}$ diffraction peaks have full width at half maximum (FWHM) of 2.9 and 1.8 arcmin, respectively, implying relatively high quality of the as-grown crystals [28].

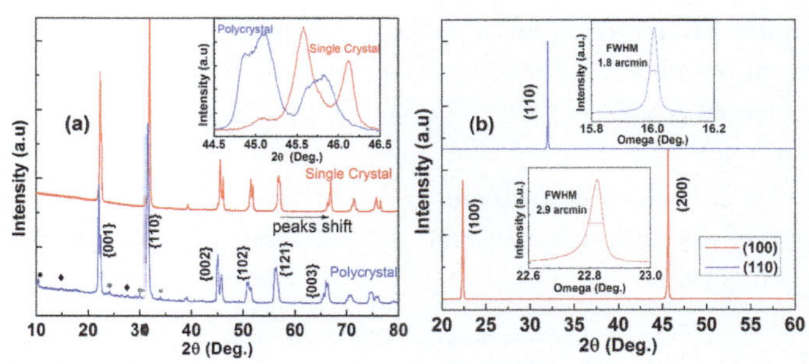

Figure 10. (**a**) XRD patterns of the KNN70/30 polycrystalline material and the powder of as-grown KNN25/75 single crystals. (Minor peaks may be attributed to: "•"—K4Nb6O17 (PDF # 76-0977), "♦"—K3Nb7O19 (PDF # 78-1395), and "*"—$KHCO_3$ (PDF # 12-0292)); (**b**) X-ray diffractograms of the (110)-oriented single crystal (top) and of the (100)-oriented single crystal (bottom). The insets show their respective rocking curves. Reprinted with permission from [28]. Copyright 2014 Royal Society of Chemistry.

Dielectric, Piezoelectric, Ferroelectric Properties

For the electrical property measurements, the as-grown KNN single crystals were sliced into wafers with a thickness of ~0.48 mm, perpendicular to their pseudocubic $[100]_{pc}$ and $[110]_{pc}$ directions. Then, all of the samples were sputtered with gold electrodes. The temperature dependence of the dielectric properties of unpoled samples was analyzed in the temperature range of 30–450 °C using an Agilent 4294 A precision impedance analyzer. For the piezoelectric measurements, the crystal samples were poled in silicon oil at 180 °C for 15 min with a dc electric field of 40 kV/cm and then cooled to room temperature with half of the poling electric field. The d_{33} was tested using a Berlincourt quasistatic meter at 55 Hz. The electric field dependence of the polarization (P–E curves) and strain curves was characterized using a ferroelectric test system (aix ACCT TF analyzer 1000) at room temperature.

Investigation into the dielectric properties of the KNN single crystals along different orientations reveals anisotropic behavior. Figure 11 illustrates the temperature dependence of the dielectric constant (ε) of the $(100)_{pc}$ and $(110)_{pc}$ oriented KNN25/75 crystals at different frequencies. The dielectric constants of both $(100)_{pc}$ and $(110)_{pc}$ plates show a sharp peak at ~396 °C, suggesting the ferro-to-paraelectric phase transition from the tetragonal (P4mm) to the cubic phase (Pm3m). Moreover, another dielectric anomaly peak appears at ~187 °C, which can be attributed to the phase transition from the orthorhombic (Amm2) to the tetragonal (P4mm) phase. The temperatures of every dielectric peak are consistent with each other at different orientations within experimental error. The detailed dielectric and piezoelectric properties of the single crystals are summarized in Table 2 [28].

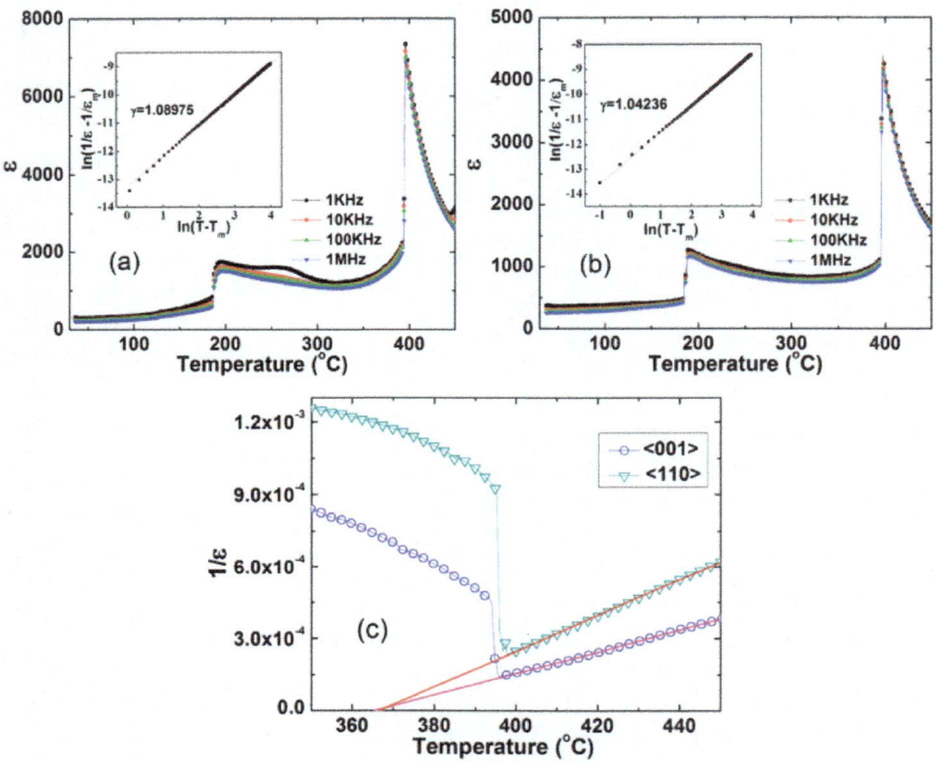

Figure 11. Temperature dependence of the dielectric constants of **(a)** $(100)_{pc}$ and **(b)** $(110)_{pc}$ oriented KNN25/75 crystals at different frequencies. The insets show their respective $\ln(1/\varepsilon - 1/\varepsilon_m)$ *vs.* $\ln(T - T_m)$ curves; **(c)** $1/\varepsilon$ *vs.* T curves of $(100)_{pc}$ and $(110)_{pc}$ oriented KNN25/75 crystals. Reprinted with permission from [28]. Copyright 2014 Royal Society of Chemistry.

Table 2. Dielectric, piezoelectric and ferroelectric properties of $(100)_{pc}$ and $(110)_{pc}$ orientated KNN25/75 single crystals (ε_{rt} represents the dielectric constant at room temperature). Reprinted with permission from [28]. Copyright 2014 Royal Society of Chemistry.

Orientation	ε_{rt}	ε_m	P_r (μC/cm²)	E_c (kV/mm⁻¹)	d_{33} (pC/N)	Strain (%)	k_t
$(100)_{pc}$	375	7407	7.2	0.84	145	0.025	0.69
$(110)_{pc}$	423	4395	10.2	0.61	70	0.05	0.51

It can be seen in Figure 12 that the remanent polarization (P_r) of the $(100)_{pc}$ oriented crystals (7.2 μC/cm^{-2}) is smaller than that of $(110)_{pc}$ (10.2 μC/cm^{-2}). The ratio of remanent polarization along $(100)_{pc}$ to $(110)_{pc}$ is 1:1.42, which is very close to 1:2. On the other hand, the d_{33} value of the $(100)_{pc}$ samples is superior to that of $(110)_{pc}$ ones. Both these phenomena can be explained by the theory of the engineered domain configuration. The schematics of domain configurations for unpoled and poled $(100)_{pc}$ orthorhombic crystals are illustrated in Figure 13. For orthorhombic single crystals, the polar direction is along $<110>_{pc}$. Unpoled orthorhombic single crystals have the configuration wherein the polar vectors are randomly orientated along equivalent $<110>_{pc}$ directions. When poled along $[100]_{pc}$, the crystal reached a domain engineering state in which each domain has one of four possible polar directions ($[110]_{pc}$, $[1\bar{1}0]_{pc}$, $[\bar{1}10]_{pc}$, $[\bar{1}\bar{1}0]_{pc}$). Therefore, the remanent polarization of $[100]_{pc}$ should be the projection of the polarization vector along $[110]_{pc}$. As for the $(110)_{pc}$ orientated crystal, a single domain state can be expected if the complete poling state is achieved. That is why the remnant polarization of $(100)_{pc}$ oriented KNN25/75 crystals should be $1/2\sqrt{}$ of that of the $(110)_{pc}$ oriented crystals. However, the single domain configuration is not stable because of the high elastic and electrostatic energy. The $(110)_{pc}$ oriented crystal would depolarize when the electric field is removed, which results in inferior electromchanical coupling coefficients and piezoelectric properties [28].

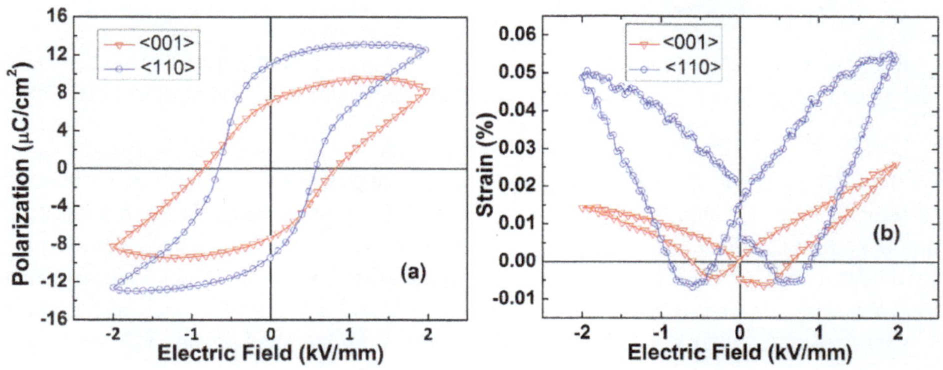

Figure 12. (Color online) (**a**) Polarization and (**b**) strain hysteresis loops of $(100)_{pc}$ and $(110)_{pc}$ oriented KNN25/75 single crystals at room temperature. Reprinted with permission from [28]. Copyright 2014 Royal Society of Chemistry.

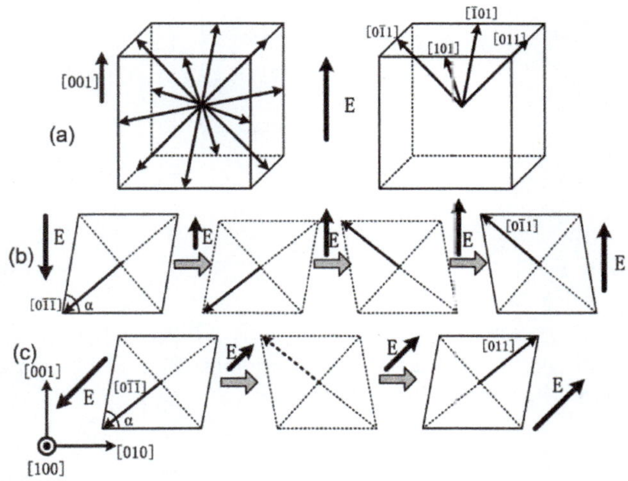

Figure 13. (Color online) (**a**) Schematics of domain configurations for $(100)_{pc}$-oriented orthorhombic KNN25/75 crystals. The left panel: in the unpoled state, the polarization vectors are along twelve equivalent $(110)_{pc}$ directions. The right panel: poled along $(100)_{pc}$, only 4 polarization directions are left; (**b**) The domain switching process of one domain in $(100)_{pc}$ crystals and (**c**) in $(110)_{pc}$ crystals. Reprinted with permission from [28]. Copyright 2014 Royal Society of Chemistry.

THE PROPERTIES MODIFICATIONS BY IONS DOPING AND DEFECTS STRUCTURE IN LEAD-FREE CRYSTALS

Mn-doped NBT-BT Single Crystals and Their Piezoelectric Properties

The electric properties of NBT-BT solid solutions might obviously be enhanced by doping and by optimizing preparation conditions. Values for the piezoelectric constant d_{33} of NBT-BT ceramics and single crystals have been reported to be as high as 205 pC/N and 280 pC/N, respectively [29]. However, these properties are still notably inferior to those of lead-based ones. In addition, relatively high leakage currents due to defects are another problem to overcome in NBT-BT, which can interfere with poling and polarization switching. Modest concentrations of Mn dopants have been reported to be an effective method by which to enhance the resistivity of lead-based materials.

Photos of as-grown Mn-doped crystals are shown in Figure 14. Analysis by ICP-AES of as-grown crystals revealed that the concentrations of Ba and Mn ions were 5.6 and 0.14 at%, respectively. The results indicate that Ba ions experienced notable composition segregation during crystal growth. Pure NBT-BT crystals with the same concentration of Ba ions were grown for comparisons.

Figure 14. (Color online) Photos of Mn-doped NBT-BT single crystals and samples (**a**) grown using <001> oriented seed (Mn:NBT-BT) and (**b**) grown using Pt wire as seed [Mn:NBT-BT]. Reprinted with permission from [29]. Copyright 2009 AIP Publishing LLC.

Figure 15a shows the temperature and frequency dependence of ε_r and tanσ for a poled (001) oriented Mn:NBT-BT crystal. Two abnormal dielectric peaks can be seen, as indicated by T_d and T_m. The depolarization temperature T_d plays an important role with regard to practical applications of NBT-BT single crystals, as it was previously been reported as the temperature of a ferroelectric to antiferroelectric transformation, whereas T_m that of an antiferroelectric → paraelectric transition. In addition, the dielectric peak at T_d is frequency dispersive, implying a relaxorlike character. Recently, other perspectives of the nature of the phase transition at T_d were presented, and clearly further studies concerning the mechanism of depolarization are needed.

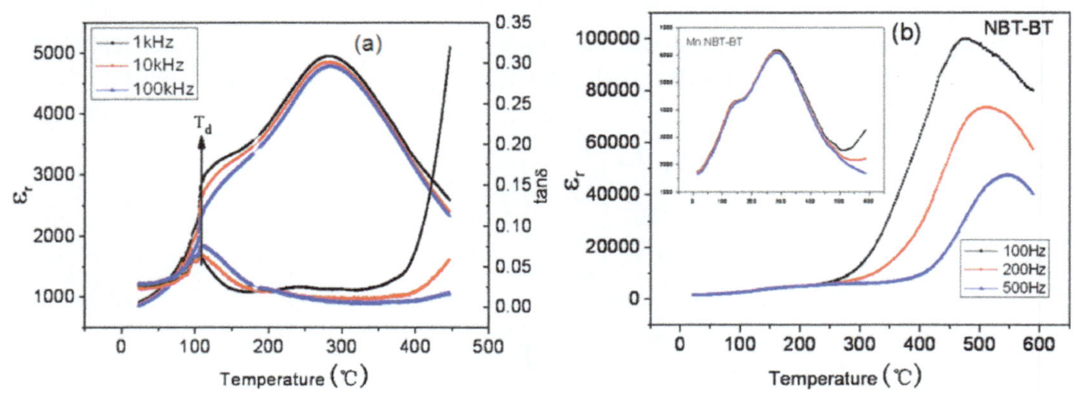

Figure 15. (Color online) Dielectric constant of <001> oriented poled Mn: NBT-BT and NBT-BT crystal samples as a function of temperature. (**a**) Mn:NBT-BT at the frequencies of 1, 10, and 100 kHz; (**b**) Low frequency (100–500 Hz) dielectric constant of Mn:NBT-BT and NBT-BT. Reprinted with permission from [29]. Copyright 2009 AIP Publishing LLC.

Figure 15b shows the low frequency (100–500 Hz) dielectric constant as a function of temperature for NBT-BT, with the inset showing that for Mn:NBT-BT. These data reveal a dramatically enhanced dielectric constant for NBT-BT in the temperature range of 300–600 °C, yielding values in excess of 105, which were extremely frequency dispersive. These data give evidence of the presence of a space charge conduction mechanism at elevated temperatures as previously reported, which results in correspondingly high loss factors. It is important to note in this temperature range of 300–600 °C, that Mn was extremely effective in suppressing such conduction effects, as can be seen in the inset [29].

The *P-E* and *S-E* hysteresis loops at room temperatures under bipolar drive for (001) Mn:NBT-BT crystals are shown inFigure 16a,b, respectively. The remnant polarization was P_r = 45.3 μC/cm^{-2} for Mn:NBT-BT, grown with NBT-BT seeds, which was larger than the 35 μC/cm^{-2} of Mn:NBT-BT, grown with Pt-wire seeds. Because of improvement in crystal quality and a decrease in leakage current density, higher remnant polarizations P_r were found for Mn:NBT-BT relative to Mn:NBT-BT. We also found that the *P-E* hysteresis loops of Mn: NBT-BT had higher saturation polarizations than that previously reported for NBT-BT. Figure 16b shows the *S-E* curve for Mn:NBT-BT, where a maximum strain of 0.15% was found under 25 kV/cm. A comparative summary of the piezoelectric and ferroelectric properties for Mn:NBT-BT crystals with other piezoelectric materials is given in Table 3. This summary clearly shows that Mn:NBT-BT single crystals have notably superior electromechanical properties relative to that previously reported for PZT4 ceramics and for NBT-BT crystals. For our Mn:NBT-BT crystals, values of d_{33}, k_t, and k_{31} of 483 pC/N [29].

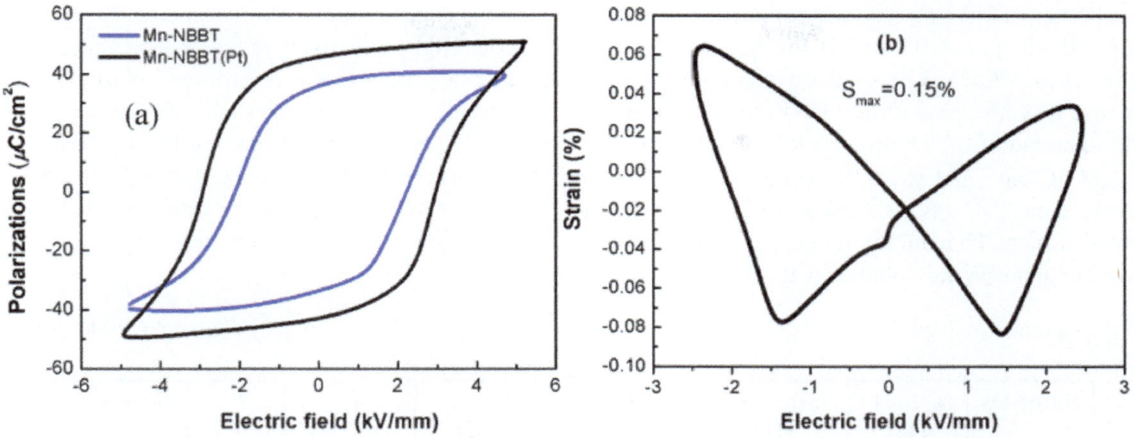

Figure 16. (Color online) (**a**) *P-E* hysteresis loops for Mn:NBT-BT samples compared with Mn:NBT-BT one; (**b**) strain *vs*. bipolar electric field curve for Mn:NBT-BT crystals at room temperatures. Reprinted with permission from [29]. Copyright 2009 AIP Publishing LLC.

Table 3. Piezoelectric and ferroelectric properties of PZT4 ceramics and NBT-BT, Mn:NBT-BT, and Mn:NBT-BT single crystals. Reprinted with permission from [29]. Copyright 2009 AIP Publishing LLC.

Samples	d_{33} (pC/N)	k_t (%)	k_{31} (%)	d_{31} (pC/N)	P_r	E_c (kV/mm)
PZT4	250	48	33	170	–	–
NBT-BT	280	–	–	–	16.44	3.27
Mn:NBT-BT(Pt)	287	56	–	–	35	2.67
Mn:NBT-BT	483	55.6	39.7	155	45.3	2.91

Eu-Doped NBT Single Crystals and Their Electrical Properties

Since the wavelength and positions of photoluminescence excitation and emission bands of Eu^{3+} strongly depend on the local structure, it was considered as a local chemical environment order probe. Therefore, a study on Eu^{3+} doped NBT single crystals will help to better understand the local crystal structure and hence guide the design of high property NBT-based materials. Eu^{3+}-doped NBT single crystals were grown by a top-seeded solution growth (TSSG) method. Photoluminescence excitation and emission were measured. Temperature-dependent $7F_0$-$5D_0$ excitation spectra and decay curves were characterized. The effects of Eu^{3+}-doping on electrical properties were also studied [30].

Figure 17 shows the excitation spectra for the $7F_0$-$5D_0$ transition, which is acquired by monitoring the total luminescence from Eu^{3+} in NBT single crystal at 15, 200 and 300 K. The presence of two peaks around 578 and 581 nm indicates that Eu^{3+}ions are incorporated into two adjacent crystallographic sites *i.e.*, Bi^{3+} [Eu(I)] and Na^{+} [Eu(II)] sites. The strong Eu(I) peak reveals that Eu^{3+} Ions mainly substitute Bi^{3+} ions, while the broad feature of the Eu(II) peak demonstrates that Na^{+} ions displace in a disorder way.

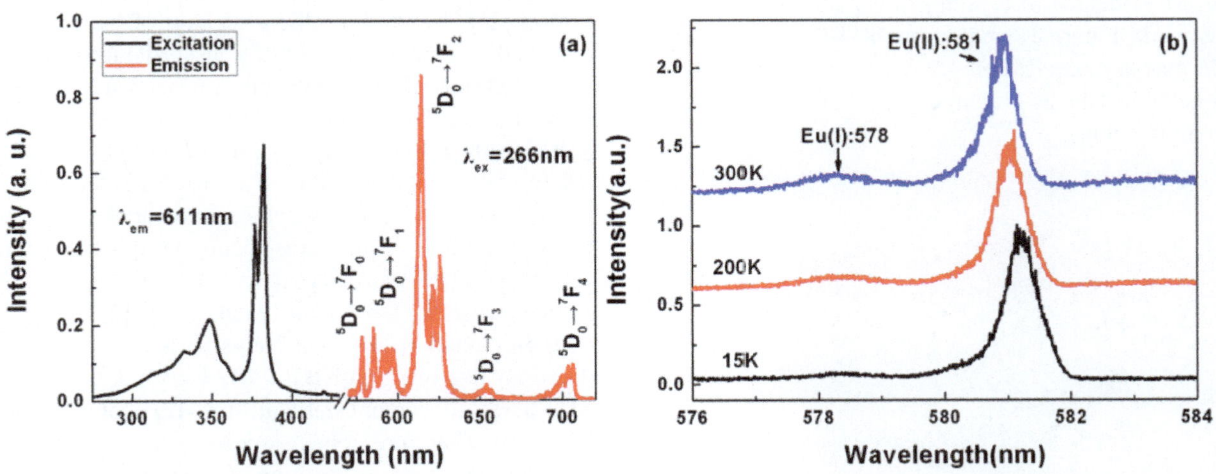

Figure 17. (**a**) Room temperature excitation and emission spectra; (**b**) temperature dependent excitation spectra of 7F0-5D0 transition. Reprinted with permission from [30]. Copyright 2014 Springer.

Electrical properties of NBT crystals are summarized in Table 4 [30]. Compared with pure NBT crystal, the dielectric and ferroelectric properties of Eu: NBT crystal are improved obviously: the dielectric loss decreases from ~0.040 to ~0.030, while the remnant polarization (P_r) and piezoelectric constant (d_{33}) increase to 7.38 μC/cm^{-2} and 86 pC/N, respectively. Based on the photoluminescence results, the effects of Eu^{3+}-doping on electrical properties can be naturally understood. The addition of Eu_2O_3 will compensate the volatilized Bi_2O_3 and hence reduce the concentration of Bi vacancies. Therefore, the leakage current will be effectively reduced, which facilitates the poling process and thus leads to improved piezoelectric properties. The decrease of Bi vacancies will relax the stress caused by lattice distortion. Besides, it also results in the decrease of the extrinsic structure/chemical heterogeneities in nanometer-scale. These mechanisms result in the depressed dielectric dispersion behavior in Eu:NBT single crystals.

Table 4. Electrical properties of NBT and Eu:NBT single crystals. Reprinted with permission from [30]. Copyright 2014 Springer.

Samples	ε_r (@1 Hz)	tanσ (@1 Hz)	E_c (kV/mm)	P_r (C/cm^2)	d_{33} (pC/N)	T_m (°C)	Reference
NBT	700–750	0.037–0.040	1.82	3.0	60–80	358	[21]
Eu:NBT	732	0.027	3.24	7.38	86	333	Present

Zn-doped NBT Single Crystals and Their Electrical Properties

Recent theoretical studies on $Bi(Zn_{1/2}Ti_{1/2})O_3$ (BZT) predicted that it could be a promising lead-free candidate ferroelectric material with a high c/a ratio of 1.21 and large calculated ferroelectric polarization of 150 μC/cm^{-2}, whereas it is not stable in its pure form and can only be stabilized at high pressure. However, it is reported that BZT-containing solid solutions can be stable in atmospheric condition and still have enhanced c/a ratio, ferroelectric polarization, and increasedT_c. Therefore, BZT could be an interesting alternative to some lead-free relaxor ferroelectrics for the purpose of composing binary or ternary solid solutions. Zhang *et al.* [31] reported that a rhombohedral-tetragonal MPB composition exists near $Bi_{0.5}Na_{0.5}TiO_3$-3.75$Bi(Zn_{0.5}Ti_{0.5})O_3$ and the MPB composition shows improved electrical properties.

Raw powders of Na_2CO_3, Bi_2O_3, TiO_2, and ZnO with purity more than 99.99% were used as the starting materials. The well-mixed powders with the composition of NBT-7BZT were calcined to synthesis polycrystalline material at 1000 °C for 10 h in air. Then, the powders mixed with excess 20 wt% Bi_2O_3 and Na_2CO_3 as self-fluxes were put into a platinum crucible. More details of crystal growth can be found in our previous work. Figure 18 shows the NBT-BZT single crystal boule 35 mm in diameter and 12 mm in length grown by the TSSG method. The as-grown crystal looks yellowish and transparent in color.

Figure 18: The as-grown NBT-1.5BZT single crystal (scale in mm). Reprinted with permission from [31]. Copyright 2014 John Wiley and Sons.

Figure 19 shows the dielectric constant ε_r and loss factor tan σ as a function of temperature at 1 kHz for unpoled <001>-oriented NBT and NBT-1.5BZT crystals [31]. It can be seen that the NBT-1.5BZT sample exhibited higher value of ε_r at room temperature. However, the maximum value of ε_r was decreased compared with NBT sample. The phase transition temperature T_m corresponded to the maximum in the dielectric constant. It was found that the addition of BZT into NBT increased T_m slightly from 327 °C to 339 °C. A similar phenomenon was also observed in NBT-BZT ceramics. However, for other BZT-containing system such as BZT–$PbTiO_3$, the large improvement of T_m or Curie temperature (T_c) was obtained when the mole ratio of BZT was up to 40%. This suggested that the T_m could be further improved if the content of BZT was increased in NBT-BZT system.

Figure 20a shows the room-temperature *P-E* hysteresis loops of unpoled <001>-oriented NBT-1.5BZT crystal under different electric fields. It can be seen that the remnant polarization (P_r) and coercive electric field (E_c) under 100 kV/cm. corresponded to 21.5 μC/cm^{-2} and 59.6 kV/cm, respectively. Meanwhile, the *P-E* loop of pure NBT crystal under 60 kV/cm is shown in the inset of Figure 20a. The P_r was determined to be 11.0 μC/cm^{-2}, which was lower than that of NBT-1.5BZT crystal (P_r = 19.1 μC/cm^{-2}) at the same field level. Figure 20b shows the *S-E* curve for NBT-1.5BZT crystal, where a net strain up to 0.23% was observed under 100 kV/cm. In addition, the unipolar strain was measured and is shown in the inset ofFigure 20b. It was found that the maximum strain was 0.12%.The piezoelectric constant d_{33} value measured by a quasistatic Berlincourt meter was 121 pC/N for the <001>-oriented NBT-1.5BZT crystal. It is worth mentioning that the value of d_{33} in NBT-1.5BZT crystal exceeded that of NBT-3.75BZT ceramics around the MPB composition (d_{33} = 92 pC/N). Therefore, a better piezoelectric property could be expected if the real composition of as-grown NBT-BZT crystal is close to the MPB [31].

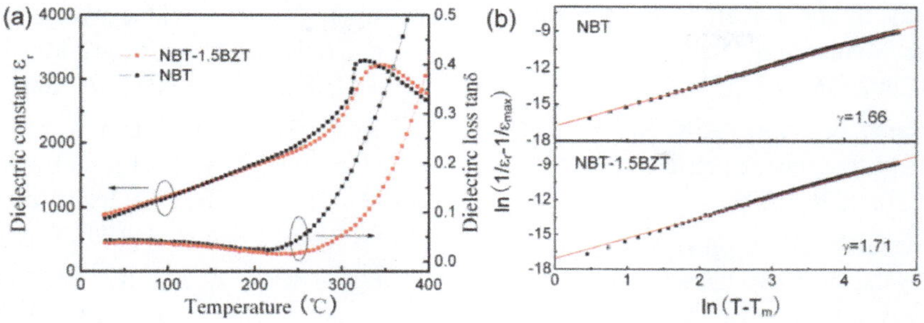

Figure 19. (**a**) Temperature and frequency dependence of dielectric constant and dielectric loss at 1 kHz; and (**b**) a plot of ln($1/\varepsilon_r - 1/\varepsilon_{max}$) *vs.* ln($T - T_m$) for unpoled <001>-oriented NBT and NBT-1.5BZT crystals. Reprinted with permission from [31]. Copyright 2014 John Wiley and Sons.

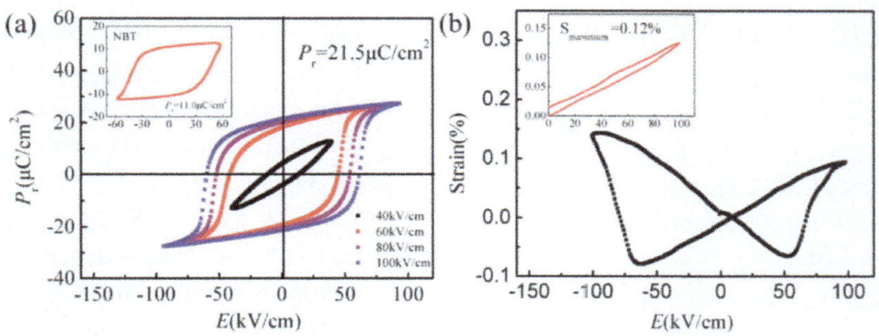

Figure 20. (**a**) *P-E* hysteresis loop and (**b**) *S-E* hysteresis loop for unpoled <001>-oriented NBT-1.5BZT crystal at room temperature. Reprinted with permission from [31]. Copyright 2014 John Wiley and Sons.

CONCLUSIONS

The lead free relaxor-based ferroelectric single crystals NBT-BT with dimensions of Ø25 × 10 mm³ were grown by TSSG technique. The highest piezoelectric constant can be d_{33} ~ 280 pC/N. Mn-doped NBT-BT single crystals had superior piezoelectric and ferroelectric properties. We found values for d_{33} as high as 483 pC/N, which is competitive with conventional PZT ceramics. In addition, the leakage current of NBT-BT crystals decreased notably upon Mn substitution, allowing for a more complete poling. Lead-free NBT-BT crystals are promising as alternatives to conventional PZT piezoelectrics. High quality $(K_{0.25}Na_{0.75})NbO_3$ single crystals with dimensions of up to Ø30 mm × 10 mm were grown by a carefully controlled TSSG method. The piezoelectric constant d_{33} and the thickness electromechanical coupling factor k_t are 145 pC/N and 69% for the $(100)_{pc}$ oriented crystal and 70 pC/N and 51% for the $(110)_{pc}$ oriented crystal, revealing a domain engineering effect.

AUTHOR CONTRIBUTIONS

The paper wrote by Xiaobing Li, the experiments were carried by Chao Chen, Hao Deng and Haiwu Zhang, the analyses are contributed by Di Lin and Xiangyong Zhao, all the paper is directed by Haosu Luo.

REFERENCES

1. Cross, L.E. Relaxor ferroelectrics. *Ferroelectrics* **1987**, *76*, 241–267.
2. Ye, Z.G. Relaxor ferroelectric complex perovskites: Structure, properties and phase transitions. *Key Eng. Mater.* **1998**,*81*, 155–156. [ ]
3. Noheda, B.; Cox, D.E.; Shirane, G.; Gonzalo, J.A.; Cross, L.E.; Park, S.E. A monoclinic ferroelectric phase in the $Pb(Zr_{1-x}Ti_x)O_3$ solid solution. *Appl. Phys. Lett.* **1999**, *74*, 2059–2061.
4. Noheda, B.; Cox, D.E.; Shirane, G.; Guo, R.; Jones, B.; Cross, L.E. Stability of the monoclinic phase in the ferroelectric perovskite $Pb(Zr_{1-x}Ti_x)O_3$. *Phys. Rev. B* **2000**, *63*.
5. Guo, R.; Cross, L.E.; Park, S.E.; Noheda, B.; Cox, D.E.; Shirane, G. Origin of the high piezoelectric response in $Pb(Zr_{1-x}Ti_x)O_3$. *Phys. Rev. Lett.* **2000**, *84*, 5423–5426.
6. Wada, S.; Yako, K.; Kakemoto, H.; Tsurumi, T.; Kiguchi, T. Enhanced piezoelectric properties of barium titanate single crystals with different engineered-domain sizes. *J. Appl. Phys.* **2005**, *98*.
7. Suchanicz, J.; Kwapulinshi, J. X-ray diffraction study of the phase transitions in $Na_{0.5}Bi_{0.5}TiO_3$. *Ferroelectrics* **1995**, *165*, 249–253.
8. Suchanicz, J. Investigation of the phase transitions in $Na_{0.5}Bi_{0.5}TiO_3$. *Ferroelectrics* **1995**, *172*, 455–458.
9. Suchanicz, J. Peculiarities of phase transitions in $Na_{0.5}Bi_{0.5}TiO_3$. *Ferroelectrics* **1997**, *190*, 77–81.
10. Suchanicz, J.; Poparwski, R.; Maryjasik, S. Some properties of $Na_{0.5}Bi_{0.5}TiO_3$. *Ferroelectrics* **1997**, *192*, 329–333.
11. Suchanicz, J. Behavior of $Na_{0.5}Bi_{0.5}TiO_3$ ceramics in the A.C. electric fieled. *Ferroelectrics* **1998**, *209*, 561–568.
12. Isupov, V.; Pronin, I. Temperature dependence of birefringence and opalescence of $Na_{0.5}Bi_{0.5}TiO_3$ crystals. *Ferroelectr. Lett.* **1984**, *2*, 205–208.
13. Jones, G.O.; Thomas, P.A. Investigation of the structure

and phase transitions in the novel A-site substituted distorted perovskite compound $Na_{0.5}Bi_{0.5}TiO_3$. *Acta Crystallogr. Sect. B* **2002**, *58*, 168–178.

14. Takenaka, T.; Maruyama, K.I.; Sakata, K. $Na_{0.5}Bi_{0.5}TiO_3$-$BaTiO_3$ system for lead-free piezoelectric ceramics. *Jpn. J. Appl. Phys.* **1991**, *30*, 2236–2239.

15. Chiang, Y.M.; Farrey, G.W.; Soukhojak, A.N. Lead-free high-strain single-crystal piezoelectrics in the alkaline-bismuth-titanate perovskite family. *Appl. Phys. Lett.* **1998**, *73*, 3683–3685.

16. Zhou, T.S.; Huang, R.X.; Shang, X.Z.; Peng, F.; Gu, H.S. Lead-free In_2O_3-doped $(Bi_{0.5}Na_{0.5})_{0.93}Ba_{0.07}TiO_3$ ceramics synthesized by direct reaction sintering. *Appl. Phys. Lett.* **2007**, *90*.

17. Feng, C.D.; Xiang, P.H.; Li, H.D. Electrical Properties of La-Doped $(Na_{0.5}Bi_{0.5})_{0.94}Ba_{0.06}TiO_3$ Ceramics. *Jpn. J. Appl. Phys.* **2003**, *42*, 7387–7391.

18. Zhu, M.K.; Liu, L.Y.; Hou, Y.D.; Wang, H.; Yan, H. Microstructure and Electrical Properties of MnO-Doped $(Na_{0.5}Bi_{0.5})_{0.92}Ba_{0.08}TiO_3$ Lead-Free Piezoceramics. *J. Am. Ceram. Soc.* **2007**, *90*, 120–124.

19. Wang, X.X.; Chan, H.L.W.; Choy, C.L. Piezoelectric and dielectric properties of CeO_2-added 0.94NBT-0.06BT lead-free ceramics. *Solid State Commun.* **2003**, *125*, 395–399.

20. Yi, J.Y.; Lee, J.K.; Hong, K.S. The role of Cation Vacancies on Microstructure and Piezoelectricity of Lanthanum-Substituted NBT Ceramics. *Jpn. J. Appl. Phys.* **2004**, *43*, 6188–6192.

21. Herabut, A.; Safari, A. Processing and Electromechanical Properties of La-doped NBT Ceramics. *J. Am. Ceram. Soc.* **1997**, *80*, 2954–2958.

22. Nagata, H.; Takenaka, T. Additive effects on electrical properties of $(Bi_{0.5}Na_{0.5})TiO_3$ ferroelectric ceramics. *J. Eur. Ceram. Soc.* **2001**, *21*, 1299–1302.

23. Zhou, X.Y.; Gu, H.S.; Wang, T.Y.; Li, W.Y.; Zhou, T.S. Piezoelectric properties of Mn-doped $(Na_{0.5}Bi_{0.5})_{0.92}Ba_{0.08}TiO_3$ceramics. *Mater. Lett.* **2005**, *59*, 1649–1652.

24. Ge, W.W.; Liu, H.; Zhao, X.Y.; Pan, X.M.; He, T.H.; Lin, D.; Xu, H.Q.; Luo, H.S. Growth and characterization of $Na_{0.5}Bi_{0.5}TiO_3$-$BaTiO_3$ lead-free piezoelectric crystal by the TSSG method. *J. Alloy. Compd.* **2008**, *456*, 503–507.

25. Sun, R.B.; Zhao, X.Y.; Zhang, Q.H.; Fang, B.J.; Zhang, H.W.; Li, X.B.; Luo, H.S. Growth and orientation dependence of electrical properties of $0.92Na_{0.5}Bi_{0.5}TiO_3$–0.08 $K_{0.5}Bi_{0.5}TiO_3$ lead-free piezoelectric single crystal. *J. Appl. Phys.* **2011**,*109*.

26. Smolensky, G.A.; Isupov, V.A.; Agranovskaya, A.I.; Krainik, N.N. New ferroelectrics of complex composition. *Sov. Phys. Solid State* **1961**, *2*, 2651–2654.

27. Liu, H.; Ge, W.W.; Jiang, X.P.; Zhao, X.Y.; Luo, H.S. Growth and characterization of Mn-doped $Na_{1/2}Bi_{1/2}TiO_3$ lead-free ferroelectric single crystal. *Mater. Lett.* **2008**, *62*, 2721–2724.

28. Deng, H; Zhao, X.Y.; Zhang, H.W.; Chen, C.; Li, X.B.; Luo, H.S. Orientation dependence of electrical properties of large-sized sodium potassium niobate lead-free single crystals. *CrystEngComm* **2014**, *16*, 2760–2765.

29. Zhang, Q.H.; Zhang, Y.Y.; Wang, F.F.; Wang, Y.J.; Lin, D.; Zhao, X.Y.; Luo, H.S. Enhanced piezoelectric and ferroelectric properties in Mn-doped $Na_{0.5}Bi_{0.5}TiO_3$-$BaTiO_3$ single crystals. *Appl. Phys. Lett.* **2009**, *95*, 2760–2765.

30. Zhang, H.W.; Zhao, X.Y.; Deng, H.; Chen, C.; Lin, D.; Li, X.B.; Luo, H.S. Photoluminescence and electrical properties of Eu-doped $(Na_{0.5}Bi_{0.5})TiO_3$ ferroelectric single crystals. *Appl. Phys. A* **2014**, *114*, 357–361.

31. Chen, C.; Zhang, H.W.; Zhao, X.Y.; Deng, H.; Li, X.B.; Luo, H.S. Structure, Electrical, and Optical Properties of $(Na_{1/2}Bi_{1/2})TiO_3$–1.5at.%$Bi(Zn_{1/2}Ti_{1/2})O_3$ Lead-free Single Crystal Grown by a TSSG Technique. *J. Am. Ceram. Soc.***2014**, *97*, 1–5.

CHAPTER 5

PREPARATION, STRUCTURAL, ELECTRICAL, AND FERROELECTRIC PROPERTIES OF LEAD NIOBATE–LEAD ZIRCONATE–LEAD TITANATE TERNARY SYSTEM

Rashmi Gupta, Seema Verma, Vishal Singh, and K. K. Bamzai

INTRODUCTION

Materials with high Curie temperature and piezoelectric properties are required for high temperature applications in industries. Barium titanate (BT) or lead zirconate titanate (PZT) is used for many commercial applications like medicine, industry, and research but due to their low Curie temperature, that is, ~−120°C for BT [1] and ~390°C for PZT [2], they are not suitable for high temperature applications. Recently, a great deal of interest has been focused on relaxor ferroelectric materials due to their excellent piezoelectric as well as dielectric properties [3, 4]. Lead metaniobate ($PbNb_2O_6$ or PN) reported by Goodman [5] is a member of tungsten bronze family and is often used in nondestructive testing and medical diagnostic imaging and for deep submergence hydrophones. However, problems such as high level of porosity and relatively low mechanical strength are often encountered in its use and since its discovery no much work has been done on it in spite of the high quality pioneering work [6, 7]. Niobium based perovskite oxides have attracted much attention because they show combined electrical and mechanical properties for use in the electronic industries [8–10]. In spite of the reports of an attractive high Curie temperature (517–570°C) [11, 12], fewer studies [13–15] on PN have been carried out which can be attributed to the preparing of piezoelectric form (the metastable orthorhombic structure) in pure phase. On the other hand, one of the most widely studied ferroelectric materials due to its various potential technological applications is lead zirconate titanate, that is, $PbZr_{1-x}TixO_3$ (PZT) [16]. Aliovalent modifications to PZT which can be either with higher valence substitutions (donors) or with lower valence ions (acceptors) can alter its properties and have been studied by many authors [16, 17]. One of the most used additives is niobium which substitutes either Zr^{4+} or Ti^{4+}, promotes the lead vacancies, and is considered as a donor dopant. Most of the studies have reported on the effect of niobium modification on the dielectric, piezoelectric, pyroelectric, and electrooptic properties of PZT in bulk as well as thin film form [18–25]. However, there are so many reports that due to lead evaporation during processing, both doped and undoped PZT, are seldom observed to be perfect [26–28]. In the present investigation, we have investigated the preparation and properties of lead niobate–lead zirconate–lead titanate (PN–PZ–PT) ternary system. The effects on the electrical properties of the system have been investigated in detail by varying the PZ and PT concentration while keeping the PN concentration steady. To the best of author's knowledge, there is no such report on preparation and structural and electrical properties of such a ternary compound having general formula (x) lead niobate (PN)–(y)lead zirconate (PZ)–(x-y)lead titanate (where $x = 0.5$, $y = 0.15, 0.25, 0.35$).

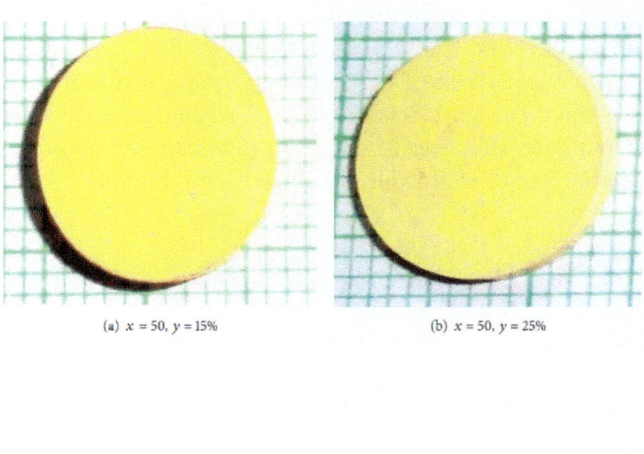

(a) $x = 50$, $y = 15\%$ (b) $x = 50$, $y = 25\%$

Citation: Rashmi Gupta, Seema Verma, Vishal Singh, and K. K. Bamzai, Preparation, Structural, Electrical, and Ferroelectric Properties of Lead Niobate–Lead Zirconate–Lead Titanate Ternary System, doi.org/10.1155/2015/835150.

(c) $x = 50$, $y = 35\%$

MATERIALS AND METHODS

Polycrystalline ceramic of xPN–yPZ–(x-y) PT (where x = 0.5, y = 0.15, 0.25, and 0.35) popularly known as lead niobate zirconate titanate (PNZT) was prepared using the conventional mixed oxide route. Oxide powders of PbO, ZrO_2, TiO_2, and Nb_2O_5 from S D Fine-Chem Limited (99% pure) were used as raw materials. Stoichiometric amount as per the molecular formula was weighed and mixed together in a mortar and pestle first and then in a high energy ball mill, as well as homogenized with triple distilled water for 48 h using zirconia balls as the milling media. The mixture was then dried in an oven at 100°C for 2 h. After drying, the reaction of the uncalcined powders taking place during heat treatment was investigated by thermogravimetric and differential thermal analysis (TGA–DTA, Shimadzu), using a heating rate of 10°C/min in air from room temperature up to 1000°C. Based on the TGA–DTA results, the mixture was calcined at a temperature of 900°C for a dwell time 2 h and heating/cooling rates ranging 4°C/min in closed platinum crucible in order to ensure the proper chemical homogeneity of the constituent elements. The calcined powder was then grinded and mixed with 5 wt% of polyvinyl alcohol as binder and uniaxially pressed into pellets of 13 mm diameter by applying a pressure of 200 MPa in a hydraulic press. These pellets were then sintered at 1100°C using a heating/cooling rate of 4∘ C/min with dwell time of 4 h in a sealed platinum crucible. Figure 1 shows the photographs of the sintered ceramic pellets for all the three compositions considered in the present investigation. X-ray diffraction pattern was collected using a Rigaku X-ray diffractometer with CuKα radiation (λ = 1.5405A) and nickel filter in a wide 2 ˚ θ range of 20–80°C using step scanning with a step size of 0.02∘ and step time of 65.6 s. The surface morphology and microstructures were examined using scanning electron microscope (SEM) of JEOL model number JSM–6390 LV and the presence of all the major elements was confirmed by energy dispersive X-ray analysis (EDAX) of model number JEOL JED-2300. The dielectric measurements have been recorded with the help of automated impedance analyzer (4192A LF model) interfaced with USB–GPIB converter 82357 (Agilent) and further automated by using a computer for data recording, storage, and analysis.The polarization versus electric field (P– E) hysteresis loop at room temperature was traced using an automatic PE loop tracer (Marine India Electrocom Ltd., New Delhi, India).

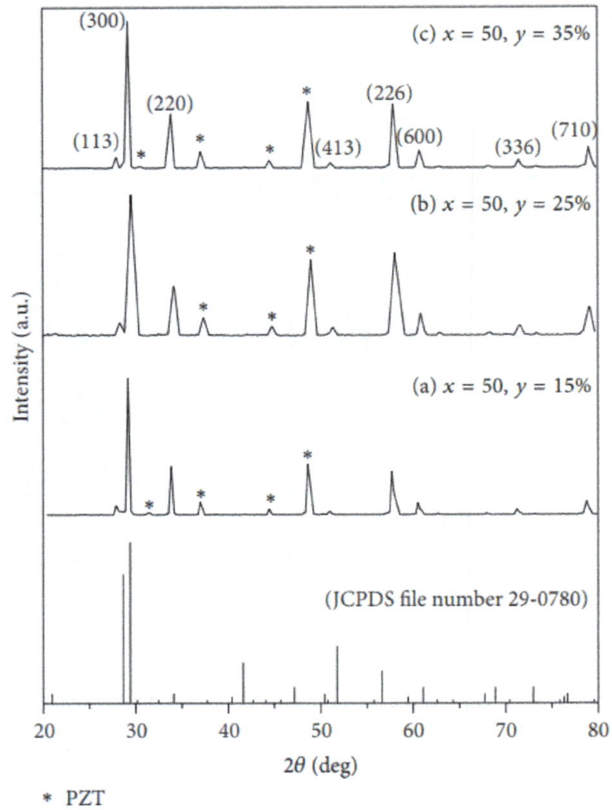

Figure 2: Room temperature X-ray diffraction pattern of xPN–yPZ–(x-y)PT ceramics sintered at 1100∘ C matched with JCPDS file number 29–0780 for the rhombohedral PbNb2O6; (∗) indicates the PZT phase.

RESULTS AND DISCUSSION

Structural Characterization

Figure 2 shows the X-ray diffraction (XRD) patterns of three different compositions of PNZT, namely, xPN–yPZ–(x-y)PT (where x = 0.5, y = 0.15, 0.25, and 0.35) with a well-crystallized structure without the presence of pyrochlore or unwanted phases. The major diffraction peaks observed correspond to rhombohedral phase of $PbNb_2O_6$ (JCPDS 29 - 0780) with lattice parameters a = 10.501, $\alpha = \beta = 90°$, $\gamma = 120°$, and space group R3m (160) [15, 29]. The peaks marked with star (∗) in the figure correspond to PZT, thereby suggesting the presence of mixed phases in PNZT. The peaks were indexed according to the JCPDS card. The peaks with maximum intensity and maximum crystallite size were observed for the composition having 35% zirconium which shows the improved crystallinity of the material. The backscattered scanning electron micrograph (SEM) images are shown in Figures 3(a)–3(c).

The micrographs clearly show the formation of fairly dense grains and the average grain size calculated using linear intercept method was found to be 1.40, 1.46, and 1.18 μm for 15, 25, and 35% of zirconium content, respectively. The elemental composition was analyzed by energy dispersive X-ray analysis (EDAX) (Figures 4(a)–4(c)) which confirms the presence of almost all the constituent elements, that is, Pb, Nb, Ti, Zr, and O in the composition. Table 1 shows the grain size, crystalline size calculated by Debye Scherrer formula [30] as well as Williamson-Hall plot [31], and strain and crystallinity index for all the three compositions. Crystallinity is evaluated from the crystallinity index equation [32]:

$$I_{\text{cry}} = \frac{D_p\,(\text{SEM, TEM})}{D_{\text{cry}}\,(\text{XRD})}, \quad I_{\text{cry}}\,(\geq 1.00), \tag{1}$$

where "I_{cry}" is the crystallinity index; "D_p" is the particle size (obtained from either TEM or SEM morphological analysis); "D_{cry}" is the particle size (calculated from the Scherrer equation). If "I_{cry}" value is close to 1, then it is assumed that the crystallite size represents monocrystalline whereas a polycrystalline has a much larger crystallinity index.

Thermal Analysis

The result of thermogravimetric analysis (TGA) and differential thermal analysis (DTA) of a milled powder having the composition 0.50PN–0.25PZ– 0.25PT is shown in Figure 5. Three endothermic peaks followed by small weight loss were observed in the DTA curve in the temperature range 250–460°C. This weight loss may be attributed due to the presence of adsorbed and trapped solvent. Also, there is a sharp increase of weight loss between 325 and 375°C which is due to the reason that the solvent adsorbed on the outer surface of the particles requires relatively lower temperature to evaporate than those entrapped between the particles during milling. The weight loss at 578° C may be assigned due to loss of lead at high temperature suggesting that the precursor materials should be heated well at 600°C [33]. No significant weight loss was observed for temperature higher than 600°C. The weight loss and the associated endothermic peak in DTA may be due to transformations into metastable phases, which formed prior to the crystallization of the perovskite phase. The last endothermic peak over the temperature range of 800–1000°C is not associated with weight loss in the TGA curve which corresponds to the crystallization of perovskite phase of lead niobate zirconate titanate (PNZT) [34].

Table 1: Variation of grain size, particle size, and strain and crystallinity index for PNZT system

Composition	Grain size (μm)	Particle size (nm) Debye Scherrer method	Particle size (nm) Williamson-hall method	Strain (ε)	Crystallnity index, I_{cry}
0.5PN–0.15PZ–0.35PT	1.40	28.41	34.07	0.00229	49.27
0.5PN–0.25PZ–0.25PT	1.46	19.19	20.85	0.00035	76.08
0.5PN–0.35PZ–0.15PT	1.18	38.47	65.34	0.00108	30.67

(a) $x = 50$, $y = 15\%$

(b) $x = 50$, $y = 25\%$

(c) $x = 50$, $y = 35\%$

Figure 3: Scanning electron micrographs of xPN–yPZ– (x-y)PT (PNZT) composite ($x = 50$, $y = 15$, 25, and 35%).

Electrical Studies

Figures 6 and 7 show the dependence of real part of dielectric permittivity (ε') and dielectric loss (tan δ) on frequency at different temperatures. From the figures, the value of both ε' and tan δ is found to decrease with increasing frequency which is a normal dielectric behaviour followed by all dielectric materials. In the lower frequency region, decrease in the value of ε' and tan δ is observed which is due to the dominance of space charge polarization and interface effects at lower frequencies. However, in the high frequency region, frequency independent behaviour of these parameters is observed. At very low frequencies ($\omega \ll 1/\tau$), dipoles are able to follow the applied field and we have the value of dielectric constant at quasistatic field ($\varepsilon' \sim \varepsilon s$). As the frequency increases, the dipoles began to lag behind the field resulting in a small increase in dielectric constant and as the characteristic frequency is reached, the value of dielectric constant drops which is the relaxation process. At very high frequencies ($\omega \gg 1/\tau$), the dipoles are not able to align with the applied field and hence we have ($\varepsilon' \sim \varepsilon\infty$). The maximum and minimum value of dielectric constant was found to be 111, 145, 128 and 11.7, 0.65, 37.4 for $x = 50$, $y = 15$, 25, and 35% in xPN– yPZ–(x-

) PT, respectively. The temperature dependence of dielectric constant (ε' , the real part of dielectric permittivity) at different frequencies is shown in Figure 8.

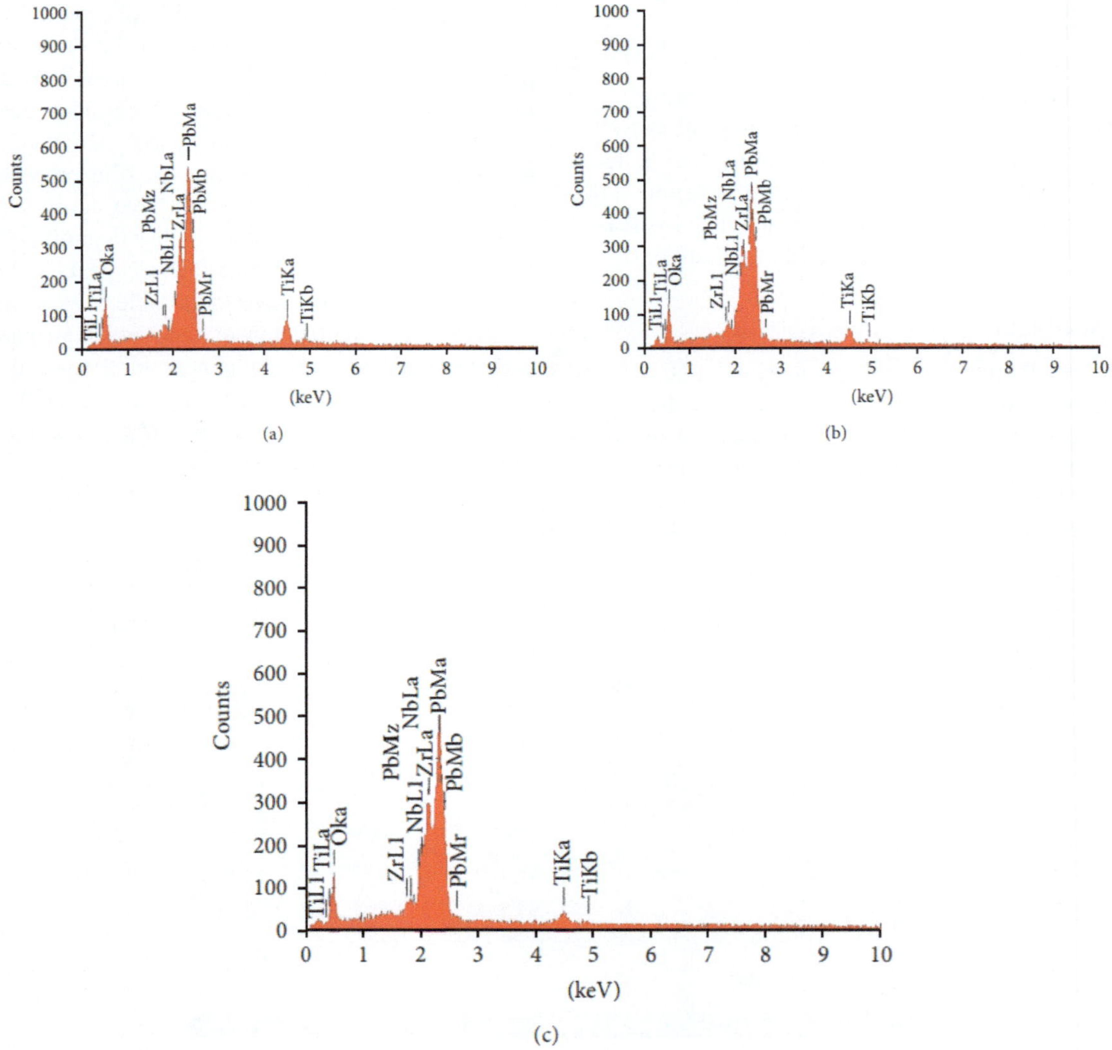

Figure 4: Energy dispersive X-ray analysis results for xPN–yPZ–(x-y)PT (PNZT) composite: (a) $x = 50$, $y = 15\%$; (b) $x = 50$, $y = 25\%$; and (c) $x = 50$, $y = 35\%$.

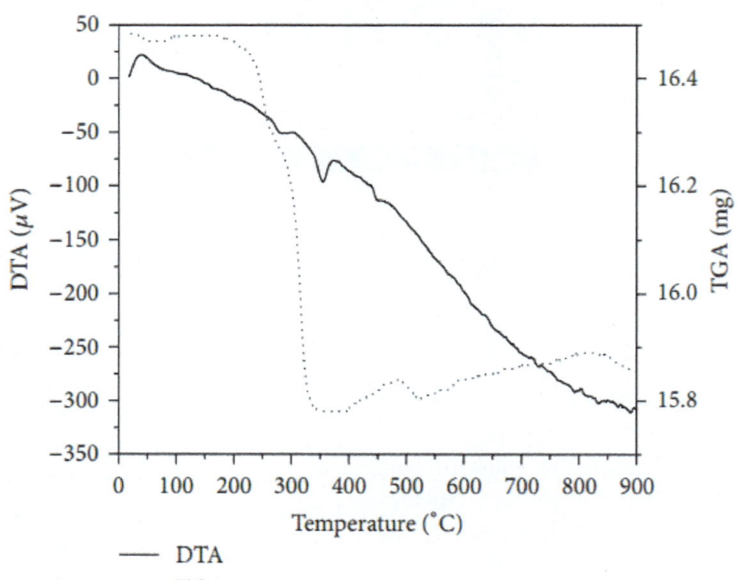

Figure 5: Thermogravimetric analysis along with differential thermal analysis (TGA/DTA) results for milled powder having composition 0.5PN–0.15PZ–0.35PT.

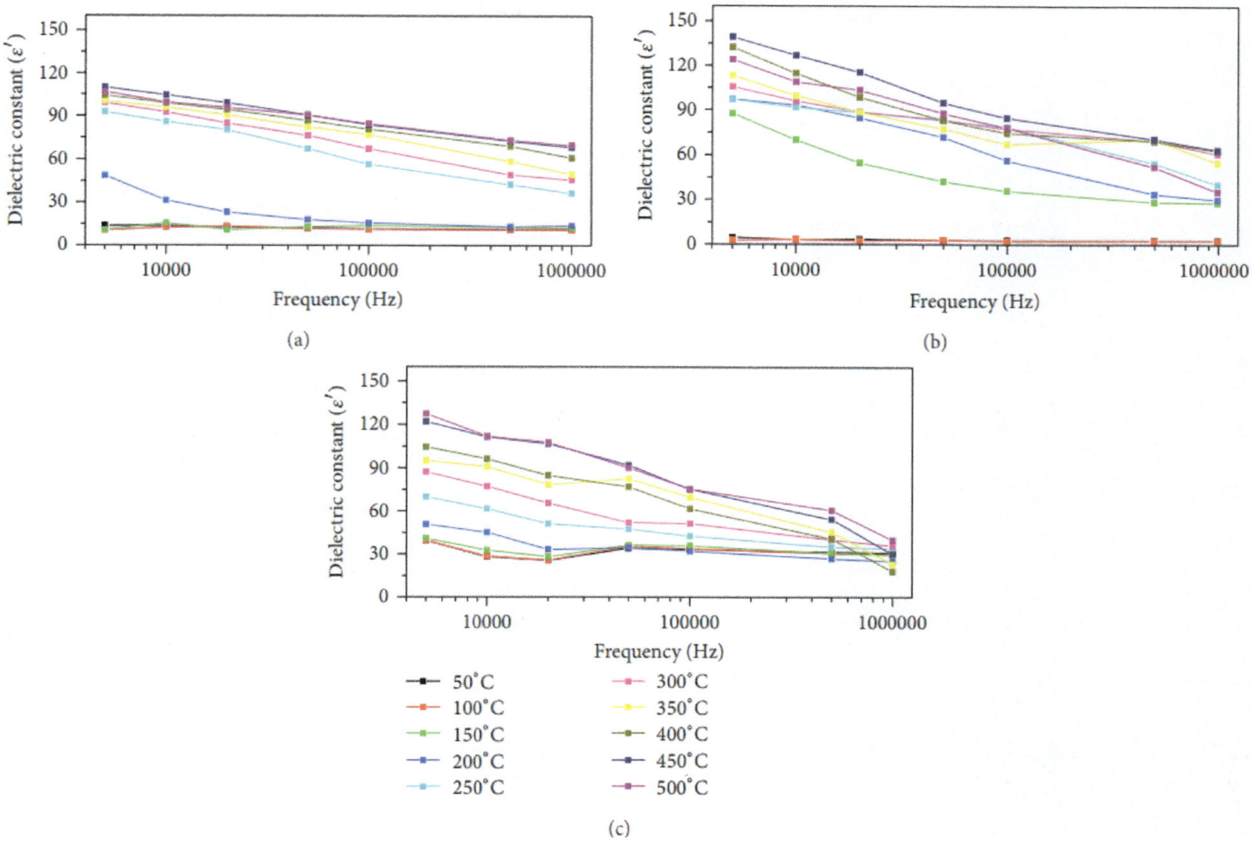

Figure 6: Variation of real part of dielectric constant with frequency for xPN–yPZ–(x-y)PT (a) $x = 50$, $y = 15\%$; (b) $x = 50$, $y = 25\%$; and (c) $x = 50$, $y = 35\%$.

Two types of dielectric anomalies were observed, one below 350°C, which may be due to ferroelectric to antiferroelectric phase transition and another above 350°C, which may be due to antiferroelectric to paraelectric phase transition. Thus, for these three compositions, two dielectric anomalies were seen which is of relaxor type. This relaxor-like anomaly was widely reported in various materials [35] and the anomaly in the temperature range of 400–900°C in oxide materials especially for those containing titanium is related to oxygen vacancies [36]. However, in present case, the formation of two much diffused peak at two different dielectric anomalies may be suggested to be due to the formation of mixed phase. Hence, in these three compositions because of poor peak formation, we are not able to give exact Curie temperature. At a given temperature other than the anomaly regions, the dielectric constant decreases with an increase in frequency and the dependence of dielectric constant with frequency increasingly varies with temperature. With increase in temperature, the peak position of dielectric constant in the curve shifts towards higher frequency side and this indicates frequency dispersion, thereby suggesting the thermally activated nature of dielectric relaxation. Further, all the plots show broad maxima, that is, diffuse phase transitions. In solid solutions, this broadening of peak or diffuse phase transition is a common feature due to the presence of more than one cation in the sublattice which produces some kind of heterogeneity.

The poly-dispersive nature of dielectric relaxation can be analysed quantitatively through complex Argand plane plots of imaginary part of dielectric constant (ε'') versus real part of dielectric constant (ε.) commonly called Cole-Cole plots. For a pure nondispersive Debye process, the plot is a perfect semicircle with its center located on the real axis. However, for poly-dispersive relaxation, the plot is a distorted semicircle with end points on the real axis and whose center lies below the real axis. Mathematically, the Cole-Cole plot obeys the following empirical relation [37]:

$$\varepsilon^* = \varepsilon' - j\varepsilon'' = \varepsilon_\infty + \frac{(\varepsilon_s - \varepsilon_\infty)}{\left[1 + (i\omega\tau)^{1-\alpha}\right]}. \tag{2}$$

Here, "ε_∞" is the high frequency limit of the permittivity, "ε_s" is the static dielectric constant, "$\varepsilon_s - \varepsilon_\infty$" is the dielectric strength, "ω" is the angular frequency, "τ" is the mean relaxation time, and the parameter "α" represents the distribution of relaxation time and can be determined from the angle subtended by the radius of Cole-Cole plot with the real axis passing through the origin of the imaginary axis.

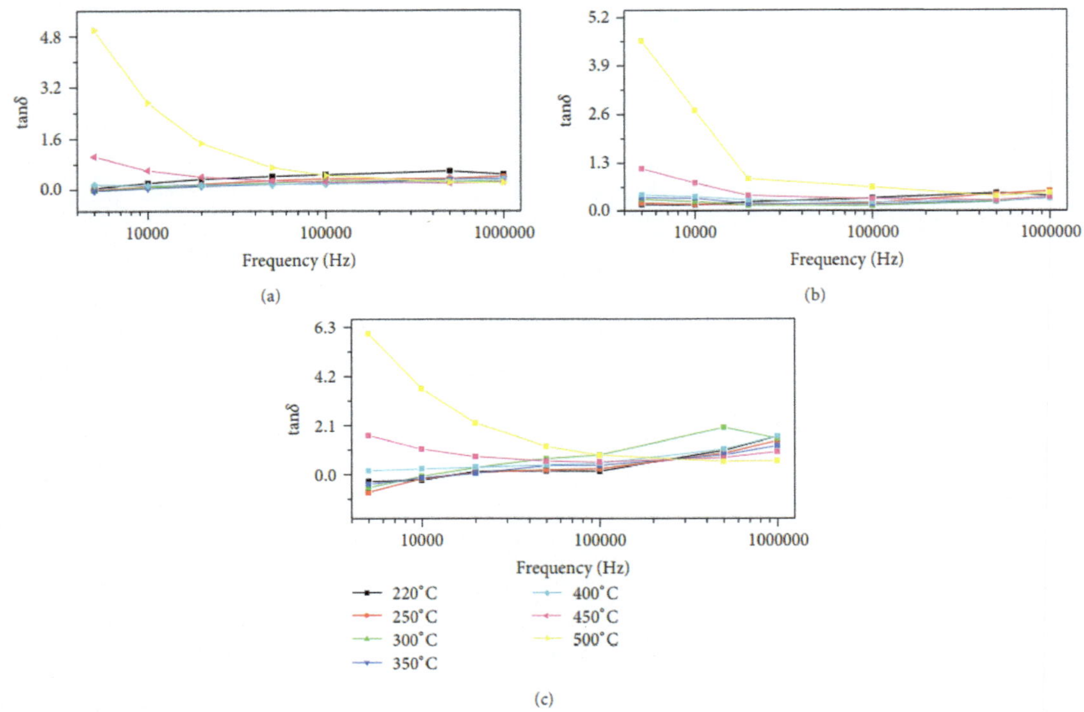

Figure 7: Variation of dielectric loss with frequency for xPN–yPZ–(x-y)PT (a) x = 50, y = 15%; (b) x = 50, y = 25%; and (c) x = 50, y = 35%.

At "ε_∞" and "ε_s", there will be no dielectric loss and maximum loss occurs at the midpoint between the two dielectric values. Figures 9(a)–9(c) depict the Cole-Cole plot representations at T = 200° C for all the three compositions. At low temperatures, that is, less than 200° C, the Cole-Cole plot resembles almost straight lines (not shown here) with large slopes, suggesting the insulating behaviour at low temperatures. However, with increasing temperature, the slopes of these curves decrease and they bend towards the real axis forming semicircles suggesting an increase in conductivity of the material with temperature. The formation of two semicircular arcs in the Cole-Cole plot can be explained on the basis of two parallel RC elements connected in series, one associated with grain (higher frequency range) and the other with grain boundary (lower frequency range) as the relaxation time for the grain boundary is much larger than that of the bulk crystal, thereby indicating negative temperature coefficient of resistance (NCTR) behaviour of PNZT like that of a semiconducting material [38]. As "ε'" is directly related to capacitance, that is, $\varepsilon' = Ct/\varepsilon_0 A$, where "$A$" is area in mm2 and "$t$" is thickness of sample in mm, the capacitances C_g (associated with grain) and C_{gb} (associated with grain boundary) have been calculated from the low frequency intercept of the semicircles on the real ε' -axis. As relaxation time, $\tau_g = R_g C_g$ and $\tau_{gb} = R_{gb} C_{gb}$, the resistances R_g and R_{gb} were obtained from the maximum condition of the semicircles, that is, $\omega_{mac} = 1/\tau = 2\pi f$max, where "$f$max" is the frequency at the maximum of semicircle. The values of R_g, R_{gb}, C_g, and C_{gb} obtained from the Cole-Cole plot are given in Table 2. As seen from table, bulk resistance decreases with increase in grain size. Also, "α" lies below the real axis and its value was found to be 0.42, 0.37, and 0.16 radians for the three compositions, thereby confirming the non-Debye type relaxation in PNZT.

Table 2: Values of grain resistance (R_g), grain capacitance (C_g), grain boundary resistance (R_{gb}), and grain boundary capacitance (C_{gb}) obtained from Cole-Cole plot for PNZT system.

Composition	$R_g \times 10^4$ (Ω)	C_g (pF)	$R_{gb} \times 10^5$ (Ω)	C_{gb} (pF)
0.5PN–0.15PZ–0.35PT	42.3	23.61	24.8	40.26
0.5PN–0.25PZ–0.25PT	30.0	59.54	68.5	29.21
0.5PN–0.35PZ–0.15PT	65.1	5.12	44.5	14.97

ac Conductivity Analysis

ac conductivity (σac) has been calculated from the dielectric loss (tan δ) data using the empirical formula [39]:

$$\sigma_{ac} = \omega\varepsilon_0\varepsilon' \tan\delta = 2\pi f\varepsilon_0\varepsilon' \tan\delta, \tag{3}$$

where $\varepsilon_0 = 8.845 \times 10^{-12}$ Fm^{-1} is the permittivity of free space. Figure 10 shows the variation of ac conductivity as a function of 1000/T.

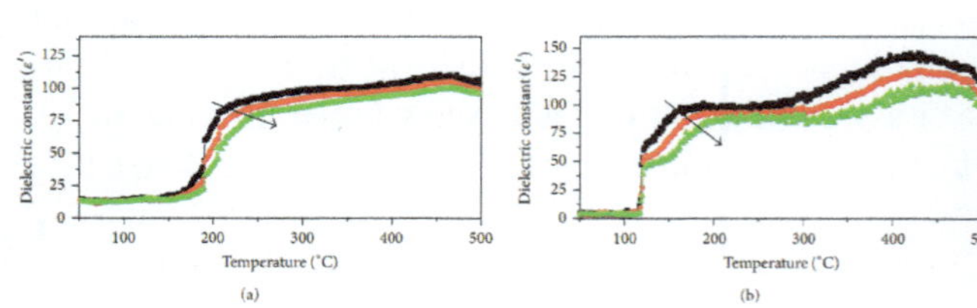

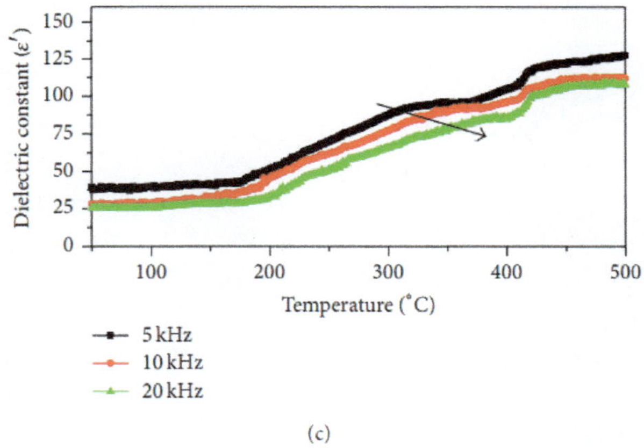

Figure 8: Variation of real part of dielectric constant with temperature at 5, 10, and 20 kHz frequency for xPN–yPZ–(x-y)PT (a) $x = 50$, $y = 15\%$; (b) $x = 50$, $y = 25\%$; and (c) $x = 50$, $y = 35\%$.

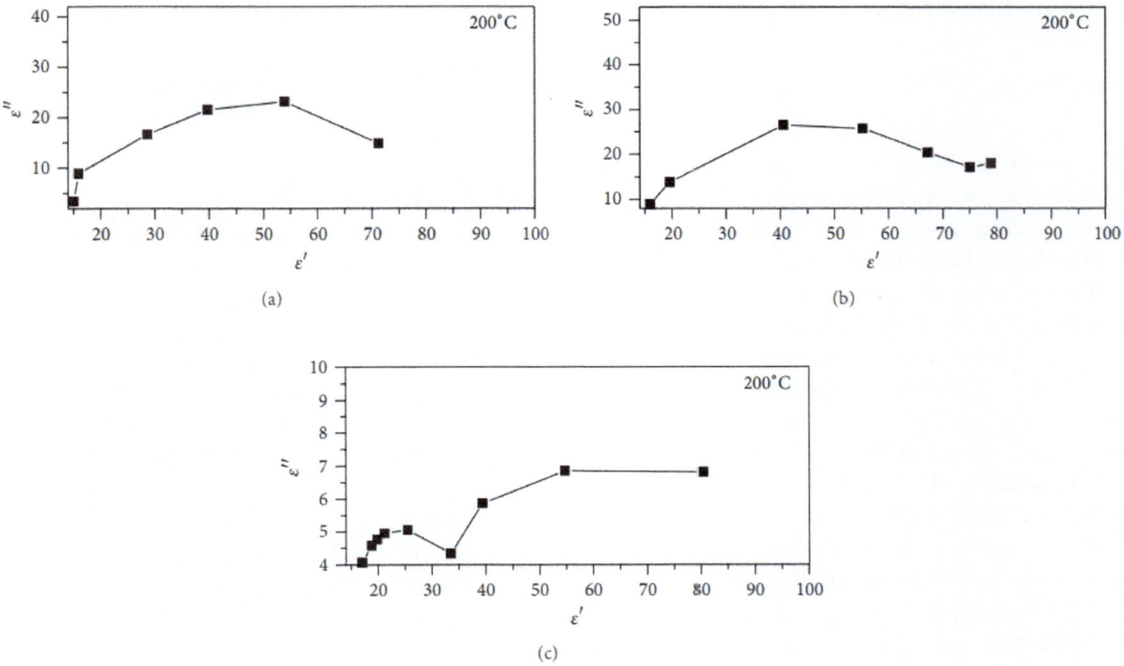

Figure 9: Complex plane Argand plot between ε'' and ε' at 200∘ C for xPN–yPZ–(x-y)PT (a) $x = 50$, $y = 15\%$; (b) $x = 50$, $y = 25\%$; and (c) $x = 50$, $y = 35\%$.

Table 3: Variation of activation energy (Ea) with frequency and exponent (s) with temperature for three different compositions of PNZT system

Composition	Activation energy (E_a) for different frequencies (eV)			Exponent (s) for different temperatures (°C)					
	5 kHz	10 kHz	20 kHz	250	300	350	400	450	500
0.5PN-0.15PZ-0.35PT	0.66	0.46	0.27	1.0	1.1	1.3	1.0	0.7	0.3
0.5PN-0.25PZ-0.25PT	0.24	0.16	0.13	1.0	0.9	0.8	0.8	0.6	0.3
0.5PN-0.35PZ-0.15PT	0.99	0.50	0.23	0.8	1.2	1.0	0.8	0.4	0.1

Figure 10: Temperature dependence of the ac conductivity at 5, 10, and 20 kHz frequency for xPN–yPZ–(x-y)PT (a) $x = 50$, $y = 15\%$; (b) $x = 50$, $y = 25\%$; and (c) $x = 50$, $y = 35\%$.

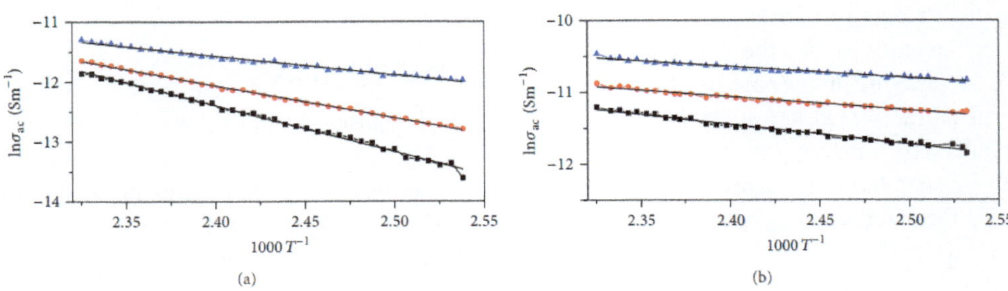

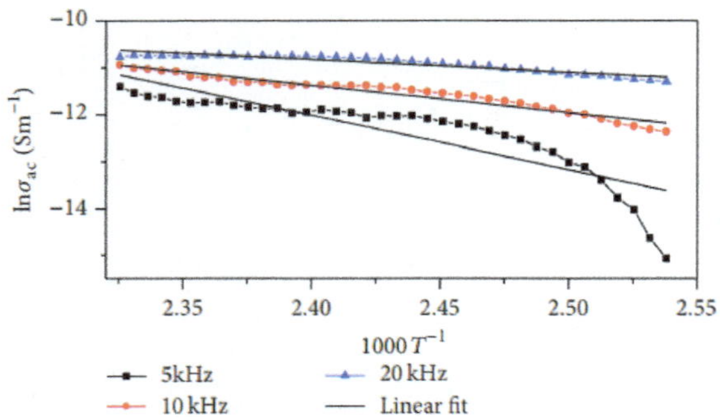

(c)

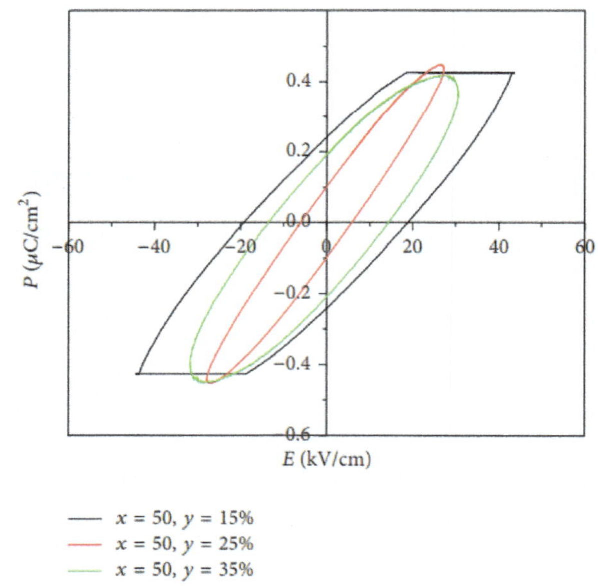

Figure 11: Variation of polarization with electric field for xPN–yPZ–(x-y)PT (x = 50, y = 15, 25, and 35%) at room temperature.

The results indicate an increase in conductivity with rise in temperature, thereby indicating the negative temperature coefficient of resistance (NTCR) character.This type of temperature dependence indicates that the electrical conduction in this material is the thermally activated transport process governed by Arrhenius equation [29]:

$$\sigma_{ac} = \sigma_0 \exp\left(-\frac{E_a}{kT}\right), \tag{4}$$

where "σ_0" is the ac conductivity preexponential factor and "E_a," the ac conductivity activation energy. The activation energy of these three compositions in the temperature range of 394–430°C was calculated from the slope of ln (σ_{ac}) versus 1000/T graph and is given in Table 3. As seen from the table, the activation energy shows a decreasing trend with increase in frequency. This is because of the reason that the increase in the frequency of the applied field results in the enhancement of the charge carriers to jump between localized states which therefore results in the decrease in electrical activation energy with increasing frequency [40, 41].The low value of activation energy may be due to the carrier transport through hopping between localized states in a disordered manner [42, 43]. Also, the activation energy is found to be minimum for the 0.5PN–0.25PZ–0.25PT composition thereby suggesting the decrease in resistivity. The frequency dependence of ac conductivity follows the universal power law [44, 45], that is,

$$\sigma(\omega) = A\omega^s, \tag{5}$$

where "A" is temperature dependent constant, "ω" is the angular frequency of the applied ac field, and the exponent "s" is temperature as well as frequency dependent constant and its value approaches unity at low temperature and decreases with increase in temperature. The values of "s" deduced from the slopes of ln(σ_{ac}) versus ln(ω) curves are given in Table 3, from which one can see that "s" is approaching unity at low temperature which confirms assumption that ac conductivity is predominant due to the relaxation dipole moment. The power law dependence of the ac conductivity on frequency corresponds to the short range hopping of carriers through trap sites separated by an energy barrier of various heights. Since the value of "s" actually scales the extent of localized charge carriers [45], *for example*, in the case of $s = 0$, the above equation shows the usual reciprocal frequency behaviour, and the system is nondispersive transport of free charge carriers process; for $s = 1$, the system has the feature of nearly constant loss relating to strictly localized carriers, while for 0<s

Ferroelectric Studies

The polarizations versus electric field (P-E) hysteresis loops with different Z_r concentrations are shown in Figure 11 whereas the values of remanent polarization and coercivity are given in Table 4. The minimum value of remnant polarization, $P_r \sim 0.10$ μC/cm^2 , and the coercive field, $E_c \sim 6.33$ kV/cm, was observed for the composition 0.5PN–0.25PZ–0.25PT. The saturation polarization was obtained only in the case of 0.5PN–0.15PZ– 0.35PT with a value of $P_s = 0.43$ μC/cm^2 at a maximum applied field of 20 kV/cm. However, in the other two cases, no saturation in polarization in P-E curve was achieved up to the maximum applied field which is due to the low resistivity of the composition [38]. The compositions with 25 and 35% Zr content could not withstand the electric field beyond 14 and 16 kV/cm and further increase in electric field led to the electrical breakdown and thus unsaturated P-E loops were obtained in these two compositions. Low value of coercivity and remanent polarization suggests that these ternary systems are soft ferroelectric.

Table 4: Variation of remnant polarization (Pr) and coercivity (Ec) for PNZT ternary system

Composition	Remnant polarization (P_r) (μC/cm^2)	Coercivity (E_c) (kV/cm)
0.5PN–0.15PZ–0.35PT	0.24	19.15
0.5PN–0.25PZ–0.25PT	0.10	6.33
0.5PN–0.35PZ–0.15PT	0.20	13.61

CONCLUSIONS

Polycrystalline compositions of xPN–yPZ–(x-y)PT (where x = 0.5, y = 0.15, 0.25, and 0.35) known as PNZT ceramic were prepared by conventional mixed oxide route. Room temperature X-ray diffraction analysis of PNZT confirms the well-crystallized rhombohedral phase of the prepared compositions with the presence of both PN and PZT phase.

Scanning electron micrographs depict formation of fairly dense grains with average grain size in the submicrometric range whereas presence of major elements was confirmed by energy dispersive X-ray analysis (EDAX). Thermal analysis of the milled powder suggests the optimum temperature for calcination is 900° C. Electrical properties carried by dielectric spectroscopy on PNZT indicate that the material exhibits two types of dielectric anomalies, one below 350°C and another above 350°C and shifting of maximum value of dielectric constant towards higher temperature side with increase in frequency indicating the relaxor behaviour. The temperature dependence of ac conductivity indicates increase in conductivity with rise in temperature thereby suggesting negative temperature coefficient of resistance (NCTR) character. The poly-dispersive nature of dielectric relaxation has been studied by complex plane Argand plot and its shape indicates the presence of both grain and grain boundary effect in the relaxation process in PNZT. The frequency dependence of ac conductivity obeys the universal power law. The ferroelectric studies suggest these ternary systems to be soft ferroelectric.

ACKNOWLEDGMENTS

The authors express their gratitude to Sophisticated Test and Instrumentation Centre (STIC), Cochin University, for providing XRD and SEM–EDAX facilities. They are also thankful to Dr. Binay Kumar, Department of Physics, University of Delhi, for providing the facility of P–E loop tracer.

REFERENCES

1. M. M. Vijatovic, J. D. Bobi ´ c, and B. D. Stojanovi ´ c, "History and ´ challenges of barium titanate: part II," Science of Sintering, vol. 40, no. 3, pp. 235–244, 2008.
2. N. Vittayakorn, G. Rujijanagul, X. Tan, M. A. Marquardt, and D. P. Cann, "The morphotropic phase boundary and dielectric properties of the $xPb(Zr_{1/2}Ti_{1/2})O_3$-(1-x) $Pb(Ni_{1/3}Nb_{2/3})O_3$ perovskite solid solution," Journal of Applied Physics, vol. 96, no. 9, pp. 5103–5109, 2004.
3. L. E. Cross, "Relaxor ferroelectrics: an overview," Ferroelectrics, vol. 151, no. 1, pp. 305–320, 1994.
4. S.-E. Park, "Relaxor based ferroelectric single crystals for electro-mechanical actuators," Materials Research Innovations, vol. 1, no. 1, pp. 20–25, 1997.
5. G. Goodman, "Ferroelectric properties of lead metaniobate," Journal of the American Ceramic Society, vol. 36, no. 11, pp. 368– 372, 1953.
6. M. H. Francombe and B. Lewis, "Structural, dielectric and optical properties of ferroelectric leadmetaniobate," Acta Crystallographica, vol. 11, no. 10, pp. 696–703, 1958.
7. E. C. Subbarao and G. Shirane, "Nonstoichiometry and ferroelectric properties of PbNb2O6-type compounds," The Journal of Chemical Physics, vol. 32, no. 6, pp. 1846–1851, 1960.
8. L. A. Reznichenko, G. A. Geguzina, and N. V. Dergunova, "Piezoelectric solid solutions based on alkali niobates," Inorganic Materials, vol. 34, no. 2, pp. 167–173, 1998.
9. I. P. Raevski and S. A. Prosandeev, "A new, lead free, family of perovskites with a diffuse phase transition: $NaNbO_3$-based solid solutions," Journal of Physics and Chemistry of Solids, vol. 63, no. 10, pp. 1939–1950, 2002.
10. J. M. Haussonne, G. Desgardin, A. Herve, and B. Boufrou, "Dielectric ceramics with relaxors and a tetragonal tungsten bronze," Journal of the European Ceramic Society, vol. 10, no. 6, pp. 437–452, 1992.
11. M. Venet, A. Vendramini, F. L. Zabotto, F. Guerrero, D. Garcia, and J. A. Eiras, "Piezoelectric properties of undoped and titanium or barium-doped lead metaniobate ceramics," Journal of the European Ceramic Society, vol. 25, no. 12, pp. 2443–2446, 2005.
12. K. R. Sahu and U. De, "Thermal characterization of piezoelectric and non-piezoelectric Lead Meta-Niobate," Thermochimica Acta, vol. 490, no. 1-2, pp. 75–77, 2009.
13. U. De, K. R. Sahu, K. R. Chakraborty, and S. K. Pratihar, "Dielectric and thermal investigations on PbNb2O6 in pure piezoelectric phase and pure non-piezoelectric phase," Integrated Ferroelectrics, vol. 119, no. 1, pp. 96–109, 2010.
14. K. R. Sahu and U. De, "Dielectric properties of PbNb2O6 up to 700° C from impedance spectroscopy," Journal of Materials, vol. 2013, Article ID 702946, 15 pages, 2013.
15. K. R. Sahu and U. De, "Dielectric properties of rhombohedral PbNb2O6," Journal of Solid State Physics, vol. 2013, Article ID 451563, 9 pages, 2013.
16. Y. Xu, Ferroelectric Materials and Their Applications, Elsevier Science, Amsterdam, the Netherlands, 1991.
17. G. H. Haertling, "Ferroelectric ceramics: history and technology," Journal of the American Ceramic Society, vol. 82, no. 4, pp. 797–818, 1999.
18. X. Dai and Y. Wang, "Multistage heat-electric energy conversion working on F(*LT*)-F*R*(*HT*) phase transition in NB doped PZT97/3 ceramics," Ferroelectrics, vol. 109, no. 1, pp. 253–258, 1990.
19. N. Duan, N. Cereceda, B. Noheda, and J. A. Gonzalo, "Dielectric characterization of the phase transitions in $Pb_{1-y/2}(Zr_{1-x}Ti_x)_{1-y}Nb_yO_3(0.03 < x < 0.04, 0.02 < y < 0.05)$," Journal of Applied Physics, vol. 82, article 779, 1997.
20. Z. Ujma, J. Handerek, and G. E. Kugel, "Phase transitions in Nb-doped $Pb(Zr_{0.95}Ti_{0.05})O_3$ ceramics investigated by dielectric, pyroelectric and Raman scattering measurements," Ferroelectrics, vol. 198, no. 1, pp. 77–97, 1997.
21. H. C. Nie, X. L. Dong, N. B. Feng et al., "Quantitative dependence of the properties of $Pb0.99(Zr_{0.95}Ti_{0.05})0.98\ Nb_{0.02}O_3$ ferroelectric ceramics on porosity," Materials Research Bulletin, vol. 45, no. 5, pp. 564–567, 2010.
22. R. D. Klissurska, K. G. Brooks, I. M. Reaney, C. Pawlaczyk, M. Kosec, and N. Setter, "Effect of Nb doping on the microstructure of sol-gel-derived PZT thin films," Journal of the American Ceramic Society, vol. 78, no. 6, pp. 1513–1520, 1995.
23. V. Kayasu and M. Ozenbas, "The effect of Nb doping on dielectric and ferroelectric properties of PZT thin films prepared by solution deposition," Journal of the

European Ceramic Society, vol. 29, no. 6, pp. 1157–1163, 2009.

24. H. Han, S. Kotru, J. Zhong, and R. K. Pandey, "Effect of Nb doping on pyroelectric property of lead zirconate titanate films prepared by chemical solution deposition," Infrared Physics and Technology, vol. 51, no. 3, pp. 216–220, 2008.
25. D. F. Ryder and N. K. Raman, "Sol-Gel processing of Nb-doped Pb(Zr, Ti)O_3 thin films for ferroelectric memory applications," Journal of Electronic Materials, vol. 21, no. 10, pp. 971–975, 1992.
26. R. Gerson, "Variation in ferroelectric characteristics of lead zirconate titanate ceramics due to minor chemical modifications," Journal of Applied Physics, vol. 31, no. 1, pp. 188–194, 1960.
27. R. Gerson and H. Jaffe, "Electrical conductivity in lead titanate zirconate ceramics," Journal of Physics and Chemistry of Solids, vol. 24, no. 8, pp. 979–984, 1963.
28. S. K. Hau, K. H. Wong, P. W. Chan, and C. L. Choy, "Intrinsic resputtering in pulsed-laser deposition of lead-zirconatetitanate thin films," Applied Physics Letters, vol. 66, pp. 245–247, 1995.
29. K. R. Chakraborty, K. R. Sahu, A. De, and U. De, "Structural characterization of orthorhombic and rhombohedral lead meta-niobate samples," Integrated Ferroelectrics, vol. 120, no. 1, pp. 102–113, 2010.
30. B. D. Cullity and S. R. Stock, Elements of X-Ray Diffraction, Prentice Hall, Englewood Cliffs, NJ, USA, 3rd edition, 2001.
31. V. D. Mote, Y. Purushotham, and B. N. Dole, "WilliamsonHall analysis in estimation of lattice strain in nanometer-sized ZnO particles," Journal of Theoretical and Applied Physics, vol. 6, article 6, 2012.
32. M. N. Rifaya, T. Theivasanthi, and M. Alagar, "Chemical cappingsynthesis of nickel oxide nanoparticles and their characterization studies," Nanoscience and Nanotechnology, vol. 2, no. 5, pp. 134–138, 2012.
33. A. Shukla and R. N. P. Choudhary, "Ferroelectric phasetransition and conductivity analysis of La3+/Mn4+ modified $PbTiO_3$ nanoceramics," Physica B: Condensed Matter, vol. 405, no. 11, pp. 2508–2515, 2010.
34. B. Praveenkumar, H. H. Kumar, D. K. Kharat, and B. S. Murty, "Investigation and characterization of La-doped PZT nanocrystalline ceramic prepared by mechanical activation route," Materials Chemistry and Physics, vol. 112, no. 1, pp. 31– 34, 2008.
35. C. C. Wang and S. X. Dou, "Pseudo-relaxor behaviour induced by Maxwell-Wagner relaxation," Solid State Communications, vol. 149, no. 45-46, pp. 2017–2020, 2009.
36. C. Ang, Z. Yu, and L. E. Cross, "Oxygen-vacancy-related lowfrequency dielectric relaxation and electrical conduction in Bi: $SrTiO_3$," Physical Review B—Condensed Matter and Materials Physics, vol. 62, no. 1, pp. 228–236, 2000.
37. K. S. Cole and R. H. Cole, "Dispersion and absorption in dielectrics I. Alternating current characteristics," The Journal of Chemical Physics, vol. 9, no. 4, pp. 341–351, 1941.
38. A. K. Roy, A. Singh, K. Kumari, K. A. Nath, A. Prasad, and K. Prasad, "Electrical properties and AC conductivity of (Bi0.5Na0.5)0.94Ba0.06TiO3 ceramic," ISRN Ceramics, vol. 2012, Article ID 854831, 10 pages, 2012.
39. R. N. P. Choudhary, D. K. Pradhan, G. E. Bonilla, and R. S. Katiyar, "Effect of La-substitution on structural and dielectric properties of $Bi(Sc_{1/2}Fe_{1/2})O_3$ ceramics," Journal of Alloys and Compounds, vol. 437, no. 1-2, pp. 220–224, 2007.
40. A. A. Ebnalwaled, "On the conduction mechanism of p-type GaSb bulk crystal," Materials Science and Engineering B: SolidState Materials for Advanced Technology, vol. 174, no. 1–3, pp. 285–289, 2010.
41. N. A. Hegab, M. A. Afifi, H. E. Atyia, and A. S. Farid, "AC conductivity and dielectric properties of amorphous $Se_{80}Te_{20-x}Ge_x$ chalcogenide glass film compositions," Journal of Alloys and Compounds, vol. 477, no. 1-2, pp. 925–930, 2009.
42. S. Upadhyay, A. K. Sahu, D. Kumar, and O. Parkash, "Probing electrical conduction behavior of BaSnO3," Journal of Applied Physics, vol. 84, no. 2, pp. 828–832, 1998.
43. K. Prasad, C. K. Suman, and R. N. P. Choudhary, "Electrical characterisation of $Pb_2Bi_3SmTi_5O_{18}$ ceramic using impedance spectroscopy," Advances in Applied Ceramics, vol. 105, no. 5, pp. 258–264, 2006.
44. A. G. Hunt, "Ac hopping conduction: perspective from percolation theory," Philosophical Magazine B, vol. 81, no. 9, pp. 875– 913, 2001.
45. A. R. Long, Hopping Transport in Solid, North-Holland, Amsterdam, The Netherlands, 1991.

CHAPTER 6

PREPARATION AND ELECTRICAL PROPERTIES OF $BA_2TIOSI_{2-X}GE_XO_7$ (x = 0.0 AND 0.2) FERROELECTRIC CERAMICS

S. K. Barbar and M. Roy

INTRODUCTION

Fresnoite, a rare barium titanium silicate mineral, $Ba_2TiOSi_2O_7$, was discovered during the geological investigation of the sanbornite deposits at Fresno County, California [1]. The mineral shows a noncentrosymmetric tetragonal crystal structure with space group P4bm [2]. Due to its noncentrosymmetricity, the material shows good ferroelectric, piezoelectric, pyroelectric, optical, and other physical properties useful for different practical devices [3–8]. The single crystal data of fresnoite shows a broad peak in dielectric permittivity at ~433 K but the same peak is absent in dielectric loss curve. It is surprising that the ceramic samples fail to reveal the dielectric anomaly shown in case of single crystal at 433 K [3]. Not only that, but a small reproducible dielectric anomaly has also been reported in the frequency range 0.1– 100 kHz at 805 K. However, the single crystal X-ray data taken at 297 K and 573 K revealed no change in symmetry except small changes in atomic position, the largest of which were less than 0.02A [°8]. Recent studies on oriented crystallization using thermal gradient, crystallization kinetics, and electrical properties through impedance spectroscopy were also carried out on fresnoite and fresnoite like glassy materials [9–11]. The ceramic sample also showed good hysteresis loop at room temperature but the shape of the loop remained unchanged when temperature increased to 875 K. It is surprising that the high temperature X-ray, DTA, and specific heat measurements gave no indication of phase transition. The present paper reports on the synthesis, X-ray diffraction, dc conductivity, and dielectric properties of pure and 10% Ge-doped fresnoite compounds.

Citation: S. K. Barbar and M. Roy, Preparation and Electrical Properties of $Ba_2TiOSi_{2-x}Ge_xO_7$ (x = 0.0 and 0.2) Ferroelectric Ceramics, doi.org/10.1155/2014/585701.

MATERIALS AND METHOD

The Polycrystalline ceramic samples of pure and Gesubstituted fresnoite compounds of general formula $Ba^2TiOSi_{2-x}Ge_xO_7$ (x = 0.0 and 0.2) were prepared by conventional solid state reaction technique using high purity (99.9%) carbonates and oxides ($BaCO_3$, TiO_2, GeO_2 and SiO_2). Stoichiometric mixture of these oxides and carbonates was thoroughly mixed and grinded for several hours and then calcined at 800–900○ C in silica crucible in air atmosphere. The process of mixing and calcination was repeated until a fine homogeneous powder of the material was obtained. The resulting mixture was compressed into pellet form and sintered at desired temperature.

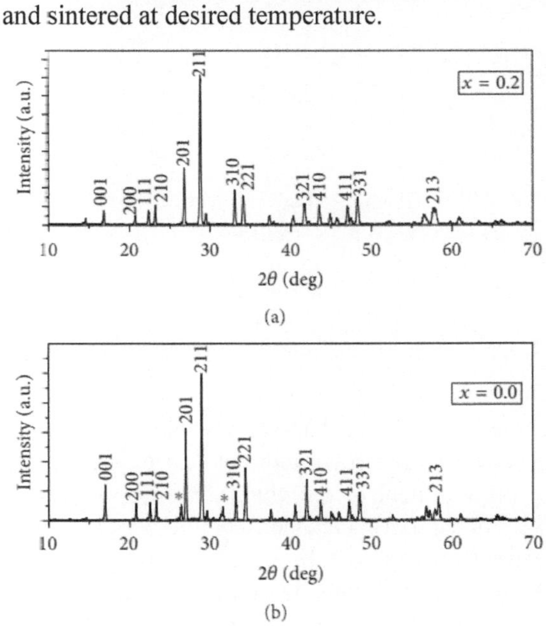

Figure 1: RT X-ray diffraction patterns of $Ba_2TiOSi_{2-x}Ge_xO_7$ (x = 0.0 & 0.2).

The complete synthesis process of both the compounds was reported in our earlier paper [12]. The formation of the single phase compound was checked by X-ray diffraction pattern using the Rigaku X-ray diffractometer with CuKα radiation and nickel filter in a wide range of 2θ from 10 to 70○ with a scanning rate of 2○ per minute. The dc electrical conductivity as a function of temperature was measured by laboratory made setup. The dielectric constant and dissipation factor as a function of frequencies (100 Hz to 2 MHz) and temperature (300 K– 673 K) were measured by Hioki 3532-50 LCR HiTester. The dc conductivity and dielectric measurements were performed on the circular dish-shaped samples with silver paint coated on two sides as the electrodes.

RESULTS AND DISCUSSION

The room temperature (RT) X-ray diffraction patterns of the samples $Ba_2TiOSi_{2-x}Ge_xO_7$ (x = 0.0 and 0.2) are shown in Figure 1. From the X-ray diffraction pattern it is observed that there is a small shift in some of the peaks positions and intensities for the substituted compound with respect to the pure compound. The detailed Rietveld refinement of X-ray diffraction data for both the compounds was reported in our earlier paper [12]. The lattice parameters and some other structural parameters are summarized in Table 1. The temperature dependence of dc electrical conductivity (σ_{dc}) for both the compositions has been evaluated by taking the steady-state values of current and results are plotted for elevated temperature range from 643 to 753 K shown in Figure 2. The resistivities of these samples are so high that they are almost temperature independent from RT to 643 K and hence the results are plotted for an elevated temperature range from 643 to 753 K. The high resistivity of the samples reveals the insulating property of ferroelectrics.

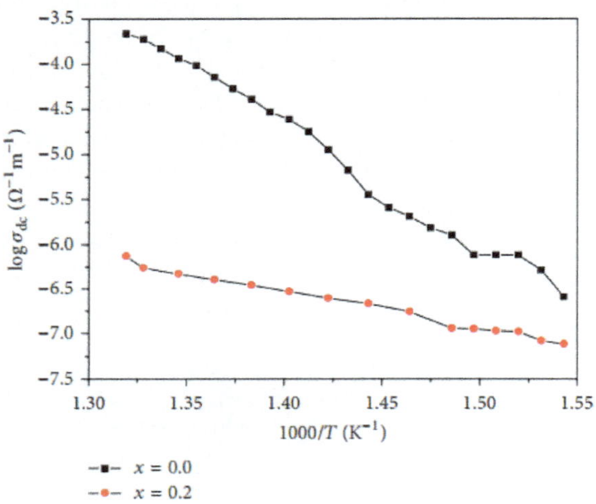

Figure 2: log σ versus 1000/T curves of $Ba_2TiOSi_{2-x}Ge_xO_7$ (x = 0.0 & 0.2).

Not only this, when an electric field is applied to the samples, there is the known reorientation of the dipoles, but also the displacement of the charge carriers. Therefore, the electrical conductivity behavior should be considered for ferroelectricity. After 643 K, the conductivity of both the samples increases up to the measured temperature range. The increase in conductivity may be explained by the fact that it is a consequence of thermally activated processes, which can be described by the Arrhenius relation:

$$\sigma = \sigma_0 \exp\left(\frac{-E_a}{kT}\right), \tag{1}$$

where σ_0 is the preexponential factor, E_a is the activation energy, and k is the Boltzmann constant having the value k = 8.6 × 10–5 eV/K. The Arrhenius equation is an accurate formula for the temperature dependence of reaction rates. In the natural logarithm form, the Arrhenius equation has the same form as an equation for a straight line; hence, we fit the data as a straight line and calculate the activation energy by measuring the slope of this straight line. The activation energies of both the samples (x = 0.0 and 0.2) were calculated from log σ versus 1000/T curve and tabulated in Table 1. It is well known that oxide materials need high temperature for their processing; at temperature above 1000◦ C, the oxygen present in the compound is expected to escape from the sample lattice and create oxygen vacancies in the lattice.

During the cooling process, oxygen again enters in the lattice but these oxygen ions cannot totally compensate the loss of oxygen. This is the case with titanates and other layered structure compounds including present compounds [13, 14]. The high value of activation energy of the pure fresnoite compound and the decrease in E_a of Ge-substituted compound can be explained on the basis of energy band gap (E_g) of silicon (Si) and germanium (Ge). The energy band gap for Ge (0.67 eV) is smaller than energy band gap for Si (1.11 eV); hence, with the substitution of Ge on Si site the activation energy decreases and conductivity increases

Table 1: Structural and electrical parameters of $Ba_2(TiO)_{1-x}Ge_xO_7$ (x = 0.0 and 0.2)

Composition	$a = b$ (Å)	c (Å)	V (Å^3)	Tetragonality (c/a)	E_a (eV)
x = 0.0	8.542	5.220	381.083	0.61	4.74
x = 0.2	8.551	5.222	383.437	0.61	1.47

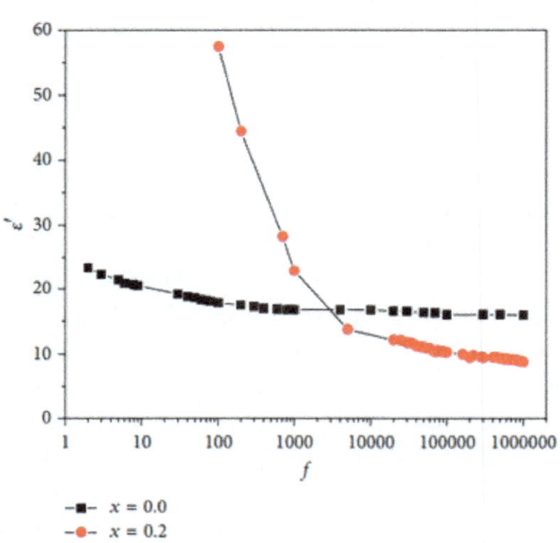

Figure 3: Dielectric constant versus log f curves of $Ba_2TiOSi_{2-x}Ge_xO_7$ (x = 0.0 & 0.2) at room temperature.

This increase in conductivity can also be explained on the basis of dissociation bond energy of Si–O bond and Si–Ge bond. When the Ge-ion is substituted in Si-site, the Si–O bond of high bond energy 452 kJ/mol breaks and makes a new bond Si–Ge of low bond energy 297 kJ/mol hence bond strength reduce and increases the conductivity of substituted compound. The detailed description of the high activation energy value of the pure fresnoite compound has been given on the basis of energy band model in our earlier paper [15]. Although there is a difference in conductivity of both samples, their small but finite conductivities have been taken into account to explain the ferroelectricity in the samples.

Figure 3 shows the variation of dielectric constant (ε') as a function of frequencies (100 Hz to 2 MHz) at RT. From the figure, it is clear that dielectric constant decreases with the increase of frequency. At lower frequencies, the dielectric constant decreases rapidly but it remains almost constant (relaxation behavior) or frequency independent at higher frequencies. This variation of dielectric constant and dissipation factor at lower frequencies is attributed not to the electronic and ionic contribution of polarization, but to the space charge contribution. While with the increase of frequency, the ionic and electronic contribution become dominant and space charge contribution diminishes gradually and hence dielectric constant decreases with the increase of frequency.

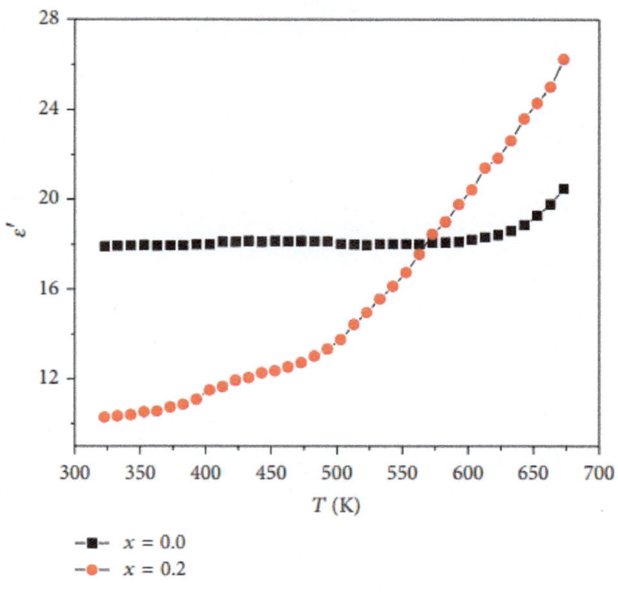

Figure 4: Dielectric constant versus temperature curves of $Ba_2TiOSi_{2-x}Ge_xO_7$ (x = 0.0 & 0.2) at 100 kHz.

The temperature variation of dielectric constant of $Ba_2TiOSi_{2-x}Ge_xO_7$ (x = 0.0 and 0.2) at a frequency of 100 kHz is shown in Figure 4. From the figure it is observed that dielectric constant of pure compound is almost temperature independent up to 600 K and then increases up to the measured temperature range. When the Ge^{4+} ions are substituted on Si4+ ions site, dielectric constant increases rapidly and continuously after a temperature of 400 K. The plot reveals that at lower temperature the magnitude of dielectric constant of the $x = 0.2$ compound is lower than the dielectric constant of pure ($x = 0.0$) compound. But, at higher temperature the dielectric constant increases with respect to the dielectric constant of the $x = 0.0$ compound. This increase in dielectric constant for substituted compound implies that this substitution produces more and more dipole moments in the lattice at higher temperature and hence increases the dielectric constant. But, no dielectric anomaly occurs up to the measured temperature range in both the compounds. Figure 5 shows the variation of dissipation factor (ε'') of both the compounds as a function of temperature at a frequency of 100 kHz. The dissipation factor shows almost similar behavior as the dielectric constant. The decrease in magnitude of dissipation factor for the $x = 0.2$ compound at higher temperature supports the nature of dielectric constant as shown in Figure 4. Not only this, but no dielectric anomalies have also been observed in any case up to the measured temperature range.

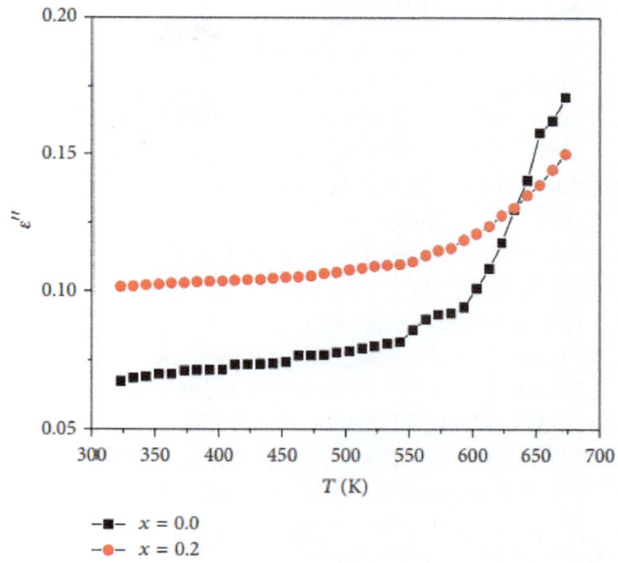

Figure 5: Dissipation factor versus temperature curves of $Ba_2TiOSi_{2-x}Ge_xO_7$ (x = 0.0 & 0.2) at 100 kHz

CONCLUSION

Analyzing the X-ray results, it is concluded that the Rietveld refinement with a structural model of Pseudo-Voigt function using non-centrosymmetric space group P4bm confirmed the tetragonal structure of both the compounds. The dependence of dielectric constant and dielectric loss on temperature could be discussed with the effect of Ge^{4+} concentration on both conductivity and dielectric behaviour. The high resistivity and moderate dielectric constant of both the samples tends to ferroelectricity and responsible for good dielectric and ferroelectric behavior useful for different practical devices. Because of low dielectric constant, both the materials are useful for low frequency resonator.

REFERENCES

1. J. T. Alfors, M. C. Stinson, R. A. Matthews, and A. Pabst, "Seven new barium minerals from eastern Fresno County, California," The American Mineralogist, vol. 50, pp. 314–340, 1965.
2. P. B. Moore and S. J. Louisnathan, "The crystal structure of fresnoite, $Ba_2(TiO)Si_2O_7$," Zeitschrift für Kristallographie, vol. 130, no. 1–6, pp. 438–448, 1969.
3. A. Halliyal, A. S. Bhalla, S. A. Markgraf, L. E. Cross, and R. E. Newnham, "Unusual pyroelectric and piezoelectric properties of fresnoite ($Ba_2TiSi_2O_8$) single crystal and polar glass-ceramics," Ferroelectrics, vol. 62, pp. 27–38, 1985.
4. M. Kimura, Y. Fujino, and T. Kawamura, "New piezoelectric crystal: synthetic fresnoite ($Ba_2Si_2TiO_8$),"Applied Physics Letters, vol. 29, no. 4, p. 227, 1976.
5. M. Kimura, "Elastic and piezoelectric properties of $Ba_2Si_2TiO_8$," Journal of Applied Physics, vol. 48, p. 2850, 1977.
6. S. Haussühl, J. Eckstein, K. Recker, and F. Wallrafen, "Growth and physical properties of fresnoite $Ba_2TiSi_2O_8$," Journal of Crystal Growth, vol. 40, no. 2,

pp. 200–204, 1977.

7. S. A. Markgraf, A. Halliyal, A. S. Bhalla, R. E. Newnham, and C. T. Prewitt, "X-ray structure refinement and pyroelectric investigation of fresnoite, $Ba_2TiSi_2O_8$," Ferroelectrics, vol. 62, no. 1, p. 17, 1985.
8. M. C. Foster, D. J. Arbogast, R. M. Nielson, P. Photinos, and S. C. Abrahams, "Fresnoite: a new ferroelectric mineral," Journal of Applied Physics, vol. 85, no. 4, pp. 2299–2303, 1999.
9. T. Hoche, M. Grodzicki, F. Heyroth, and P. A. van Aken, "Convergence of the Møller-Plesset perturbation series for the fcc lattices of neon and argon," Physical Review B, vol. 82, no. 20, Article ID 205111, 2005.
10. Y. Takahashi, M. Osada, H. Masai, and T. Fujiwara, "Softer region at boundary of supercooled liquid-crystal in glassy fresnoite," Applied Physics Letters, vol. 94, Article ID 241909, 2009.
11. B. Rangarajan, T. Shrout, and M. Lanagan, "Crystallization kinetics and dielectric properties of fresnoite BaO-TiO_2-SiO_2 glass-ceramics," Journal of the American Ceramic Society, vol. 92, no. 11, pp. 2642–2647, 2009.
12. S. K. Barbar and M. Roy, "Synthesis of $Ba_2TiOSi_2O_7$ and $Ba_2TiOSi_{1.8}Ge_{0.2}O_7$ ferroelectric ceramics and the structure determination by Rietveld analysis," Ceramics International, vol. 37, no. 7, pp. 2695–2700, 2011.
13. R. E. Newnham, R. W. Wolfe, and J. F. Dorrian, "Structural basis of ferroelectricity in the bismuth titanate family," Materials Research Bulletin, vol. 6, no. 10, pp. 1029–1039, 1971.
14. P. S. Rama Sastry, G. S. Kumar, T. Bhimashankaram, and G. Prasad, "Pyroelectricity in calcium substituted sodium bismuth titanate layer structured ferroelectric ceramics," Bulletin of Materials Science, vol. 22, pp. 59–64, 1999.
15. M. Roy, S. K. Barbar, P. Dave, S. Jangid, and I. Bala, "X-ray, scanning electron microscopy and electrical properties of synthetic fresnoite ($Ba_2TiSi_2O_8$) ceramics," Applied Physics A, vol. 100, no. 4, pp. 1191–1196, 2010.

CHAPTER 7

OXIDE THIN FILM HETEROSTRUCTURES ON LARGE AREA, WITH FLEXIBLE DOPING, LOW DISLOCATION DENSITY, AND ABRUPT INTERFACES: GROWN BY PULSED LASER DEPOSITION

Michael Lorenz, Holger Hochmuth, Christoph Grüner, Helena Hilmer, Alexander Lajn, Daniel Spemann, Matthias Brandt, Jan Zippel, Rüdiger Schmidt-Grund, Holger von Wenckstern, and Marius Grundmann

INTRODUCTION

Pulsed Laser Deposition (PLD) is a relatively new exploratory growth technique especially suitable for oxide thin films and heterostructures [1,2]. In contrast to other, "older" physical deposition techniques as, for example, thermal evaporation and molecular beam epitaxy (MBE), or cathode sputtering, the development of the "younger" PLD technique was considerably speeded up by the first successful epitaxy of high-*T*csuperconducting thin films by Venkatesan after 1986 [3,4].

The principle of PLD is the use of a small single source target with diameter of about one inch, which may be a single- or a multicomponent compound material [5]. An ns-pulse high-power laser with wavelength in the UV ablates material from the target and excites it into a plasma state. This plasma propagates with particle energies up to about 100 eV perpendicularly to the target surface and condenses as thin film on the substrate. Well-established advantages of PLD are the unique flexibility concerning the deposited film material, the high growth rate, and a high structural quality of the films. Due to the spatial decoupling of the laser as main plasma energy source and the growth chamber, the PLD parameters such as the background gas partial pressure and the growth temperature can be varied in a uniquely wide range. Furthermore, often congruent ablation and deposition of multielement compounds are found, resulting in a stoichiometric transfer of the compound target material into the thin film [5]. However, several drawbacks are generally thought to limit the application of PLD, as for example, the small substrate area of usually not more than one cm^2, sometimes rough surfaces and interfaces, and also particulates and droplets on the films. Sometimes, deviations from the stoichiometric transfer may influence a controlled doping and alloying. Furthermore, because of the highly nonlinear, pulsed nature of the deposition process, very smooth surfaces and interfaces are not always easy to obtain with PLD [6]. In heteroepitaxial growth of films on substrates of different compounds, the misfit and differences in thermal expansion coefficients induce usually high densities of dislocation lines [7].

Therefore, the aim of this paper is to show how we are able to practically overcome some of these above-mentioned limitations in PLD of oxides. First we will demonstrate the possibilities of PLD to grow complex multielement compounds, as for example, the high-*T*csuperconductor $YBa_2Cu_3O_{7-\delta}$, doped ZnO semiconductor layers, or all-oxide optical multilayers, so called Bragg reflectors, with high lateral homogeneity on large area sapphire or silicon substrates with 2-inch, 3-inch, or even 4-inch diameter. A particular challenge concerning the high-*T*c $YBa_2Cu_3O_{7-\delta}$ films was to grow them not only on one substrate side as it is the usual case, but on both front and back sides of a 3-inch or 4-inch diameter sapphire wafer. For the films deposited on the two wafer sides, similar high electrical performance and homogeneity were required.

Next, we will discuss the possibilities to grow doped and alloyed materials. This concern is in particular the II-VI semiconductor ZnO [6,8] which was doped with donors,

Citation: Michael Lorenz, Holger Hochmuth, Christoph Grüner, et al., "Oxide Thin Film Heterostructures on Large Area, with Flexible Doping, Low Dislocation Density, and Abrupt Interfaces: Grown by Pulsed Laser Deposition," Laser Chemistry, vol. 2010, Article ID 140976, 27 pages, 2010. doi:10.1155/2010/140976.

potential acceptor elements, and 3d elements to induce a "diluted" ferromagnetism. The ferroelectric thin film material $BaTiO_3$was doped with several transition metal oxides to reduce dielectric loss and to tune the temperature dependence of the permittivity. Here, also the occurrence of trace impurities in the films will be discussed, which may act as unintentional electrically or magnetically active dopant.

The third main section deals with growth of thin films with reduced structural defects to enhance electrical properties as, for example, the free carrier mobility in the oxide semiconductor ZnO. Here, we used a multistep-PLD process with low-temperature intermediate layers to reduce the strain and by this the dislocation density in heteroepitaxial ZnO thin films. Much lower defect densities are obtained in homoepitaxial ZnO growth on ZnO single crystals, and high Hall mobilities up to 800 cm^2/Vs below 100 K were measured.

Finally, recent results on the growth of very abrupt MgZnO-ZnO quantum well structures with well widths in the range 1–7 nm are shown. These structures grown by a specially optimized PLD process with reduced laser energy density show strong quantum-confined Stark effect which appears as a measure of the abruptness of the embedded quantum well nanolayers. A new PLD setup with an in situ high-pressure RHEED allows the monolayer growth control in $SrTO_3$-$BaTiO_3$multilayers through the observation of RHEED monolayer oscillations.

The presented results give an overview on the state of the art capability of PLD for growth of advanced oxide thin film structures for basic and applied investigations up to the level of fully functional demonstrator devices.

BASICS AND EXPERIMENTAL

In this section selected aspects of modelling of the PLD process (Section2.1) are discussed. Our modular PLD deposition chambers with large-area heaters up to 4-inch diameter, suitable for double-sided deposition, are shown in Section 2.2. Optical emission spectra of the laser plumes of the typically deposited materials are presented in Section2.3.

Modelling of PLD Process

After 1987, the number of publications on basics and applications of PLD increased exponentially, and more and more material systems were deposited successfully with PLD [1,2,5–7]. Beside the bulk of published papers (see [6]), also several PhD theses give more details about basic understanding and application of PLD to selected materials as, for example,(a) ferroelectrics, for example, $BaTiO_3$and $PbZr_xTi_{1-x}O_3$, also $BaTiO_3/SrTiO_3$multilayers, by Curran [12], Klarmann [13], Köbernik [14], Maier [15], Mertin [16], Petraru [17], Siegert [18],(b)dielectric oxides as, for example, Y_2O_3, $SrTiO_3$, MgO, Sc_2O_3, $Y_3Al_5O_{12}$, YSZ, by Bar [19], Goldfuß [20], Hobein [21], Hühne [22], Keiper [23], Kuzminykh [24],(c)high--csuperconducting oxides, by Ferchland [25], Maier [15], Schey [26], Wörz [27],(d)nitrides as BN, AlN, by Hohage [28], Six [29],(e)carbon nanotubes by Gorbunoff [30], nonoxide ternary semiconductors by Roussak [31], Mo/Si multilayers by Braun [32], Nd-Fe-B layers by Hannemann [33], diamond-like carbon by Thärigen [34] and Mercado [35], Ni-C multilayers by Sewing [36],(f)ZnO, by Jin [37], Nobis [38], Rahm [39], Czekalla [40], and Brandt [41]. The fundamental physical processes of PLD and results of plasma diagnostics are discussed in [1,2,5,6] and in addition in [16,19,29,30,42]. Up to now, a comprehensive model of the different processes during PLD with remarkable impact on the experiment does not exist. In the following we mention briefly some results which are important for the description of the innovative PLD processes.

Usually, the PLD process will be divided into the following three spatially separated sections, as shown in Figure1, namely, (1) laser interaction with the source target, (2) plasma expansion, and (3) film growth on the substrate [16,19,43]. The ablation of atoms, ions, and clusters can be considered in many cases as a classical thermal evaporation process [29]. In addition, photoablation, that is, electronic sputtering, may participate as nonthermal ablation process [37]. In case of weakly absorbing target materials as, for example, dielectrics, the ablation will be due to laser-induced decomposition (dielectric breakdown) [19]. The laser absorption of materials with energy band gap above the laser photon energy (as, e.g., MgO, Al_2O_3) can be increased by the use of sintered targets with many grain boundaries and defects [29].

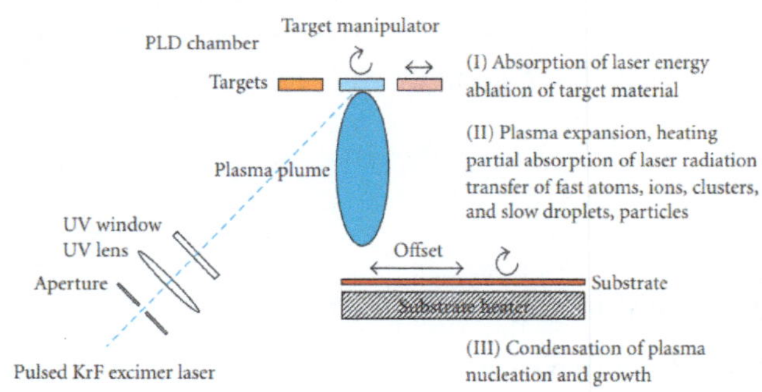

Figure 1: Scheme of a typical PLD setup for large-area "offset" film growth with the main functional components. The three fundamental process steps during (I) target ablation, (II) plasma expansion, and (III) film nucleation and growth are shortly described at the right. The "offset" distance has to be adjusted for laterally homogeneous films growth, as explained in Section3on large-area PLD.

Concerning the energy balance of incoming laser pulse energy (100%), only about 4% will be used to ablate material from the target, as deduced from reflectivity measurements and a heat conduction model for excimer laser ablation of $BaTiO_3$for 2 J/cm^2laser pulse energy [16]. About 64% of incoming laser energy will be absorbed in the plasma above the target surface, another 4% will be reflected from the target surface, and 28% go by heat conduction into deeper regions of the target. As undesired processes, the laser-target interaction includes the emergence of particulates and droplets by exfoliation and hydrodynamic sputtering, respectively. However, several techniques were developed to reduce the droplet density at the deposited films, as for example, perpendicular arrangements of target and substrate surface, velocity filters, or two-laser beam configurations [1,2].

The adiabatic plasma expansion can be described by a hydrodynamic model with the Navier-Stokes equation [16]. The spatial plasma expansion is determined by a certain angular distribution of the form ($a\cos\Theta + b\cos^n\Theta$), with$\Theta$being the angle to the target normal, and the exponent$n$is in between 3 and 12, in maximum also up to 30, in dependence on the laser parameters and focus size, as pointed out by Saenger in [1]. In addition, the background gas pressure influences the angular plasma expansion considerably by scattering processes. Tillack has combined numerical predictions and experimental results to assess the competing effects of ionization and supersaturation [42]. For example, the time-dependencies of laser energy density, of target surface temperature, and of plasma velocity profiles were deduced from a one-dimensional model of the laser plasma. This model includes laser absorption, the thermal surface behavior, mass transfer, the plasma ionization (Saha equation) and recombination, and the hydrodynamics of the gas phase, the heat transport, and the cluster growth [42]. We conclude that all the above-mentioned results on plasma diagnostics in PLD experiments are valid only for very certain experimental conditions [29] and generalizations are difficult.

The growth mode and the film properties at PLD are mainly determined by the substrate temperature and the supersaturation $\Delta m = kT\ln(/R_e)$[37], R being the real and R_e the equilibrium deposition rate at temperature T. Due to the extremely high growth rate, the supersaturation of plasma flow may be up to 10^5 J/mol, which often stimulates a 2-dimensional growth with monolayer steps [37]. In heteroepitaxy, the growth will be lattice matched (pseudomorphic), domain-matched, or relaxed in dependence on the lattice mismatch, the film thickness, and the materials combination. In domain-matched epitaxy with in-plane lattice mismatch above 7%, for the ZnO(0001) growth on c-plane sapphire, there is a fit of 5 to 6 ZnO-(2110) lattice planes to 6 or 7 α-Al_2O_3(3030) planes, respectively [37,45].

Up to now, the experimental variation of the PLD parameters is the only feasible method for the optimisation of the experimental PLD growth because, as explained above, a comprehensive modelling of the strongly nonlinear processes in PLD including ablation of target material, plasma emergence, plasma expansion, condensation, and film growth is not possible so far. However, specific parts of the PLD process can be treated by present models in good correlation to the experimental behavior.

Modular PLD Chambers for Large-Area Deposition

From the year 1990, the Leipzig PLD group has developed, designed, built, and optimized their PLD chambers in-house according to the requirements of the running projects [43]. Up to 2003, 3-inch and 4-inch diameter, double-sided high--csuperconducting $YBa_2Cu_3O_{7-\delta}$ thin films for microwave applications were developed together with Robert BOSCH GmbH Stuttgart and other partners. From 2000, other oxide materials as the ferroelectric $BaTiO_3$ and the semiconducting ZnO [6] replaced the $YBa_2Cu_3O_{7-\delta}$ as main material under investigation. Within the last 20 years, 8 complete PLD systems including modular large-area substrate heaters with maximum substrate diameter of 3-inch or 4-inch (see Figures 2 and 3) and target manipulators were developed and built.

In particular, the large-area substrate heater as shown in Figure 2 is the result of a series of prototype heaters with different designs of the heater element and the substrate rotation. The current design allows a homogeneous and reproducible radiative substrate heating with low trace element contamination and therefore determines the state of the art of heater development [46]; see also [6] which shows further details of our PLD development. For higher growth temperatures above 800°C, much more expensive laser or SiC heaters are necessary [47]. In addition to the hardware part, a PLD process control software was written in Pascal (Delphi 2007 for Win32) which runs on normal personal computers. This PLD software controls all external devices of the PLD system including the Coherent (Lambda Physik) LPX 305 KrF excimer lasers and was adapted recently to Windows XP.

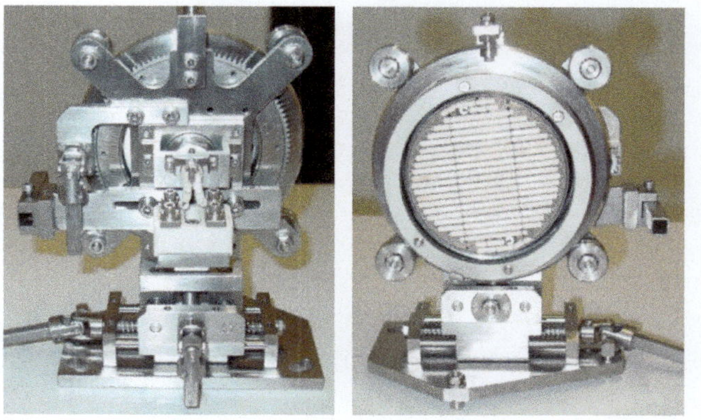

(a) (b)

Figure 2: Back side (a) and front side (b) of the modular PLD substrate heater (design and construction D. Natusch, Leipzig) for 3-inch diameter substrates, suitable for homogeneous double-sided deposition of oxide thin films. The active heater element is KANTHAL wire, embedded in alumina ceramic tubes. To reduce the temperature gradient to the substrate edge, an additional front side heater element is applied (not included in the photos).

(a) (b)

Figure 3: Inner (a) and outer (b) view of our modular medium-size PLD chamber with in-situ-spectroscopic ellipsometer Woollam M 2000 and modular 3-inch diameter heater (see Figure 2) and target manipulator. The laser entrance window is top left under 45° incidence on the target. Additional shields (not included here) may reduce the unavoidable coating of the SUPRASIL plan parallel plate used as laser entrance. The position of the whole substrate heater (left photo) can be adjusted parallel to the substrate surface in order to change the offset distance for the large-area "offset" PLD (see Figure 1).

Table 1 summarizes typical conditions of our PLD processes for thin films and nanostructures. For more details of the the high-pressure "nanowire" PLD process, see [6,48]. Typical for our low-pressure film growth processes is the target to substrate distance around 100 mm. In combination with the offset distance of projected center of target (that is also the plasma plume axis in propagation direction) to the center of the substrate, a laterally homogeneous deposition on large areas of 2-inch, 3-inch, or even 4-inch diameter is possible, as shown in paragraph 3.

Table 1: Typical parameters for PLD growth of thin film-like structures (at low partial gas pressure) and of nanowires (at high pressure). The offset is the distance from projected center of the target to center of the substrate, to ensure the homogenous large-area deposition with continuous substrate rotation.

Parameter	Low-pressure "film" PLD	High-pressure "nanowire" PLD
Grown materials	Highly flexible for oxides, for example, YBCO, $BaTiO_3$, ZnO, $ZnFe_2O_4$,. . .	ZnO, ZnMgO
Used substrates	Single-crystalline oxides, for example, Al_2O_3 ZnO, $SrTiO_3$ . . ., or SiO_2 glass	Sapphire Al_2O_3 with gold colloides for nucleation.
Dopants	Highly flexible to the host oxide	P, Li, Al, Mn, Co
Background gas	O_2, N_2, N_2O	Ar
Background pressure	10^{-5} to 3 mbar	50 to 200 mbar
Gas flow rate	N/A	50 sccm @ 100 mbar
Growth temperature	20 to 770 °C	420–940 °C
Laser fluence	1 to 5 J/cm^2	2 J/cm^2
Target-substr.distance	100 mm	5 to 40 mm
Offset distance	0 to 35 mm	N/A

Plasma Diagnostics with Optical Emission Spectroscopy

To get information on the optically excited species in the PLD plasma plume, we acquired optical emission spectra of the plasma plumes of several oxide materials [49] using a compact miniaturized optical spectrometer "Maya 2000 Pro" from Ocean Optics with a Hamamatsu back-thinned area CCD detector and USB interface. The spectrometer is equipped with a fixed HR composite grating with 300 lines per mm with variable-order filter to cover a wide spectral range from 200 to 1100 nm. The fixed spectrometer slit width is 10 µm, and for light collection we use a 400 µm UV-VIS optical SMA fiber and a Dynasil collimator lense. This collimator lense was combined with suitable viewports (7056 glass) of the PLD chambers (Figure3) with direct view to the plasma plume (see Figure4) for light collection. The viewport glass has usable transmission from about 250 nm to 2.5 µm. The spectra were taken always at a fixed position in the plasma plume of about 20 mm from the target surface. The spectra were averaged over 200 scans each with 100 ms integration time with a laser pulse repetition frequency of 10 Hz. With that, the resulting spectra are averaged over about 200 laser pulses. The software "Spectra Suite" was used to export the acquired spectra.

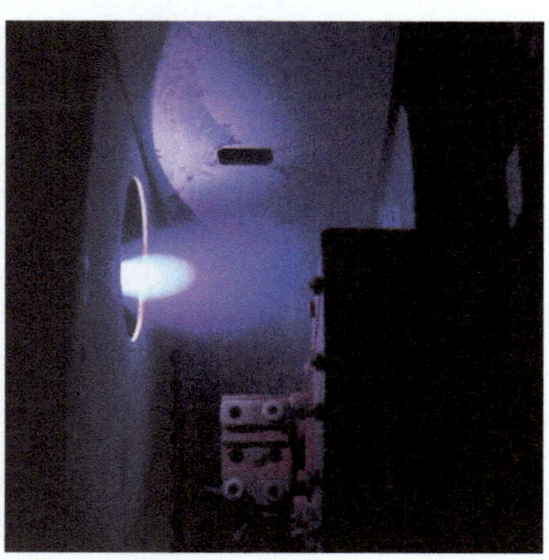

Figure 4: Photograph of the PLD plasma plume of ZnO in 0.16 mbar O_2background gas in the PLD chamber shown in Figure3. The light for the OES spectra was collected from the bright emission plume at 20 mm from the target surface at the left side. Photo was taken by C. Hanisch, University of Leipzig.

The aim of these investigations [49] was also to check if the optical emission spectra are suitable as input for a further sophisticated PLD process control to further enhance the stability and reproducibility of the PLD process. The dominating optical excitation mechanisms in the laser plasma are fluorescence (resonance) excitation due to absorption of the laser light, thermal excitation due to the high plasma temperature of several 1000 K, and shock excitation due to collisions with electrons and other atoms.

In the following we show typical optical emission spectra of our used oxide target materials. The peaks were identified by comparison with published data bases [51–53]. We identified emission lines of neutral atoms (called for Zn-atoms: Zn I), of single-ionized atoms (Zn II), and double-ionized atoms (called for oxygen atoms: O III). However, in our PLD plasma, similar to arc plasma with plasma temperature below 10,000 K, the emission from neutral atoms seems to dominate, as shown in Figures5,6,7, and8.

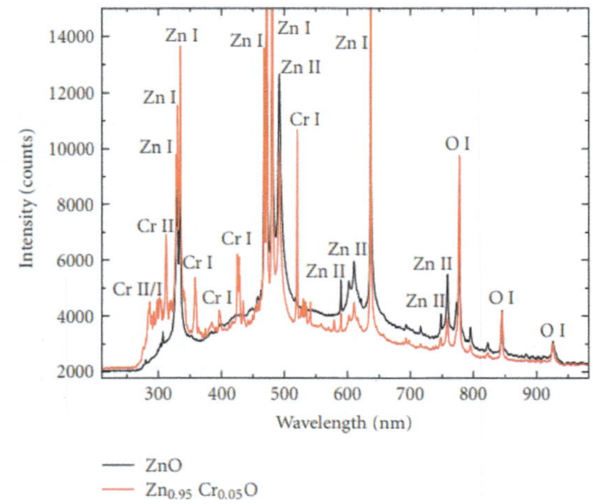

Figure 5:OES spectra taken from PLD plumes of pure ZnO (black curve) and of ZnCrO with 5% Cr (red curve), both in 0.1 mbar O_2. The emission intensities of ZnO and the fluorescence background in the Cr-doped ZnO are remarkably reduced in relation to the pure ZnO emission.

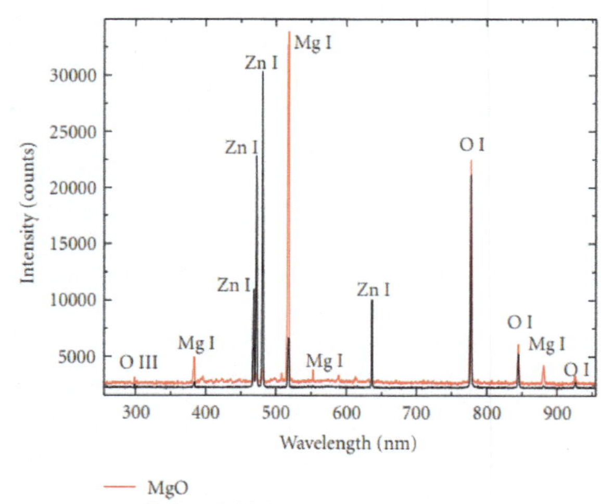

Figure 6:OES spectra taken from PLD plumes of pure MgO (red

spectrum) and of ZnO with 4% MgO (black spectrum) in 0.1 mbar O_2.

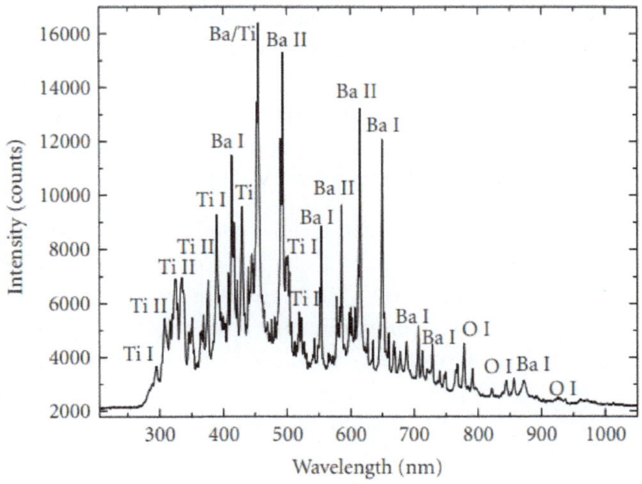

Figure 7: OES spectrum taken from PLD plume of $BaTiO_3$ in 0.1 mbar O_2. We have identified about 21 Ti and 19 Ba lines.

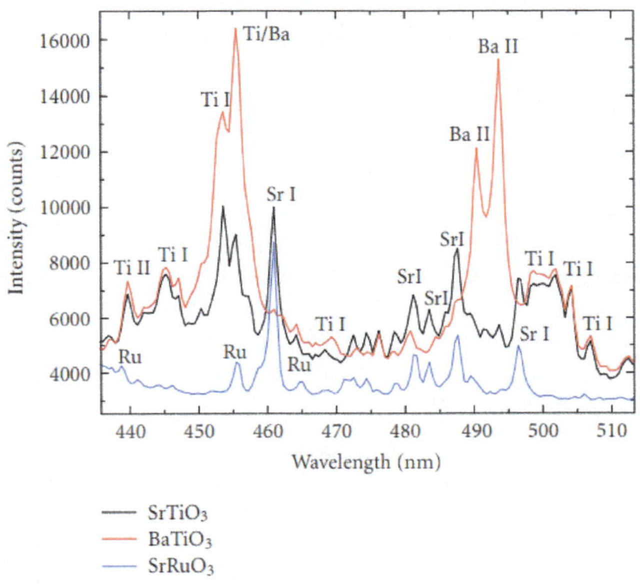

Figure 8: OES spectra taken from PLD plumes of $SrTiO_3$, $BaTiO_3$, and $SrRuO_3$ in 0.1 mbar O_2, in the wavelength range from 435 to 515 nm. For ruthenium only 2 lines could be clearly identified, while for Sr, Ti, and Ba about 20 lines were found from 250 to 1,000 nm wavelength.

The process gas was found to have minor influence on the measured spectra, as we have shown for ZnO ablated in 0.1 mbar O_2 and N_2 (see [49], not shown here). Obviously, the optical plasma emission is mainly determined by the target material itself. The interaction with the background gas may become important at later stages of plasma expansion and with the film nucleation and growth. We checked also the dependence of plasma emission from the background gas pressure. However, at lower pressures of 0.01 to 0.0005 mbar the spatial extension of plasma changes considerably, and the emitted light intensities decrease. Therefore, all spectra shown here are taken at 0.1 mbar oxygen partial pressure, and the laser energy density at the target was 2 J/cm^2 using our KrF laser with 248 nm wavelength. Spectral lines from trace impurities in the targets were not found, as we used mostly 5N powders for target preparation. Figure 9 shows the OES spectra of undoped and Mn-doped zirconia which are surprisingly almost identical and which show a very high spectral density of lines.

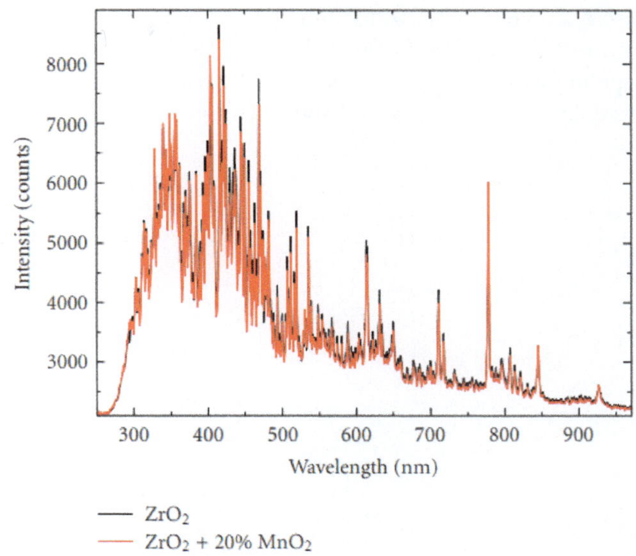

Figure 9: OES spectra taken from PLD plumes of pure ZrO_2 and of ZrO_2 with 20% MnO_2 in 0.1 mbar O_2. Only one additional peak from Mn I was found at 476.61 nm. In contradiction, Mn-doped ZnO shows 10 peaks from Mn. Zirconium and hafnium show together 1497 lines in the OES data base. A strong fluorescence background is visible here.

PLD ON LARGE-AREA SUBSTRATES

Large-area PLD is demonstrated here for double-sided $CeO_2/YBa_2Cu_3O_{7-x}$ films on both sides of r-plane sapphire wafers up to 71×75 mm^2 size for microwave filter applications (Section 3.1.), for thin ZnMgO films on 50×50 mm^2 a-plane sapphire substrates for field effect transistor demonstrators (3.2.), and for large-area Bragg reflectors built from 10 double layers Al_2O_3/yttria stabilized zirconia (YSZ) with total thickness around 1 µm (3.3.). In addition, large-area ZnO- and $BaTiO_3$-based films are discussed in Sections 4.1. and 4.2., respectively.

3.1. Double-Sided High-T_c Superconducting $YBa_2Cu_3O_{7-x}$ Thin Films on 71×75 mm^2 Sapphire Substrates

The Leipzig development of a PLD process for double-sided deposition of $CeO_2/YBa_2Cu_3O_{7-x}$ (YBCO) on both sides of r-plane sapphire wafers began in 1992 and is described in [9–11,43,55–62]. The possible approaches for large-area PLD are demonstrated by Greer in [1, page 293 ff], and we adopted the so-called "off-axis" PLD, where the center of the large-area substrate holder and heater is laterally moved by the "off-axis" distance from the center of the PLD target and the axis of the plasma plume, as shown schematically in Figure 1 and in [61]. The "off-axis" distance has to be optimized together with the oxygen partial pressure to achieve the desired lateral homogeneity of film thickness and other physical properties. We have chosen the critical current density j_c at 77 K [56,58] as measure for optimization of the large-area YBCO process. A high j_c is a necessary precondition for a low microwave surface resistance and a high microwave power handling capability [63]; see also [10]. Figure 10 shows the critical current density for two 71×75 mm^2 YBCO films on both sides of*r*-plane sapphire wafers, demonstrating a very good lateral homogeneity of j_c as the most important quality criterion of the high-T_c superconducting thin films (Table2).

Table 2:Typical microstructural and superconducting properties of double-sided 3-inch diameter PLD-YBCO films on sapphire with CeO_2 buffer layer [10]. Microwave measurements were done in Wuppertal [63] and Karlsruhe [70,71].

Property	Typical value
YBCO-thickness d_{YBCO}.	(230 ± 30) nm
C-axis lattice constant c.	(11.659 ± 0.006) Å
A-axis lattice constant a.	(3.829 ± 0.003) Å
Orthorhombic splitting (b-a)/a.	(1.52 ± 0.05)%
Percentage of a-axis oriented grains.	0 to 0.5%
In-plane epitaxially oriented film area to total illuminated area in Raman spectroscopy.	75 to 94 %
Critical temperature T_c.	87.5 to 90 K
Critical current density j_c at 77 K and $B = 0$.	3.5 to 5.5 MA/cm²
Surface resistance R_s at 10 GHz and 77 K.	500 to 600 μΩ
Microwave surface magnetic field B_s at 8.5 GHz and 77 K for R_s + 20% [63].	8 to 10 mT
Surface resistance R_s at 145 GHz and 77 K [70].	40 to 60 mΩ

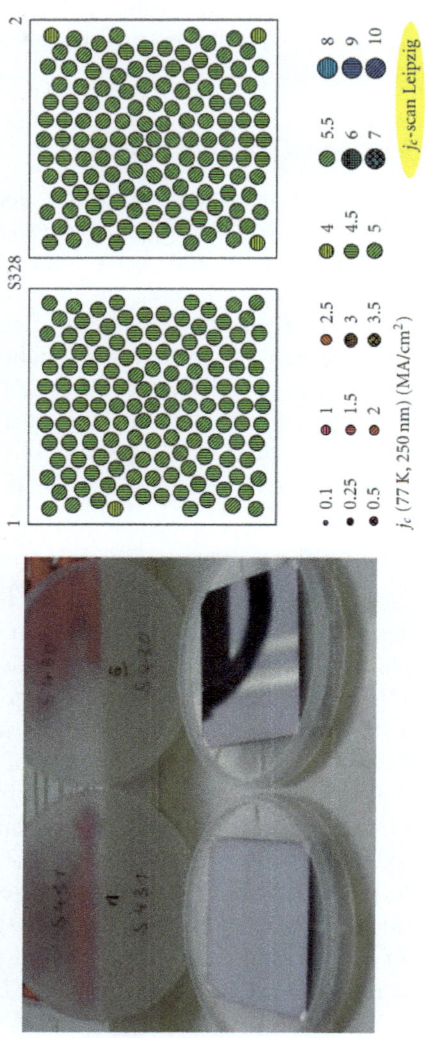

Figure 10:(a) Photograph of two double-sided YBCO wafers on71×75 mm² *r*-plane sapphire substrates with CeO_2 buffer layers. Obvious is the mirror-like, homogeneous appearance of the YBCO films. (b) Mapping of the critical current density (77 K) of such a71×75 mm² YBCO wafer with two YBCO films at both wafer sides, demonstrating the very good homogeneity of j_c using the "j_c-scan Leipzig." As much as 15 consecutively deposited71×75 mm² YBCO wafers showed such excellent j_c-homogeneity, indicating a high reproducibility of the "space qualified" large-area PLD process [9–11].

Using these YBCO films (for sputtered and evaporated films see [64–67]), applications as low-loss microwave band-pass filters for multiplexers in future satellite communication systems or mobile communication base stations could be successfully demonstrated by the industrial project partners Robert BOSCH GmbH Stuttgart and Cryoelectra Wuppertal, respectively [68,69]. Beside the development of the PLD growth itself, simple and effective characterization techniques are important. The inductive scan of the critical current density of high-T_c superconducting thin films called "j_c-scan Leipzig" [56,58] is internationally successful with this respect.

ZnMgO Thin Films on50×50 mm²Sapphire for Reproducible Field Effect Transistors

Large-area ZnMgO thin films are deposited on double-side polished50×50 mm² a-plane sapphire substrates. The layers are intended for the channel in transparent metal-semiconductor field-effect transistors (T-MESFETs) [72,73]. The PLD target was pressed from ZnO (5N) with 0.25 wt.% MgO (99,998%) and was sintered at 1150°C for 12 hours. The deposition was carried out using a 248 nm KrF excimer laser with pulse energy of 600 mJ. The substrate was heated to 670°C, and the oxygen pressure was kept constant at 0.02 mbar. The axis of the plasma plume touches the substrate at an "offset" distance of about 25 mm from the center.

In the following, investigations regarding the electrical homogeneity of films grown by PLD on such50×50 mm² substrates are presented. The substrate was cleaved after the deposition into 25 pieces of10×10 mm² size supported by a predefined line array at the wafer back side, as shown in Figure11. The relative positions of investigated samples are given from the center of the small sample to the center of the uncleaved wafer.

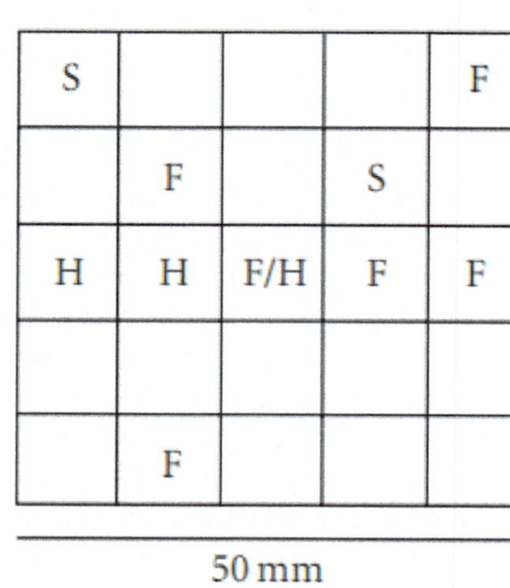

Figure 11:Schematic overview of the detailed sample positions for the presented measurements (S: SEM thickness, H: Hall effect, F: field effect mobility and transconductance) on the50×50 mm² a-plane

sapphire substrate. The film thickness was determined ex situ by field-emission microscope cross sections prepared by focused ion beam milling using Ga-ions (see Figure12). There is a slight reduction of the MgZnO layer thickness from the center to the edge of the wafer.

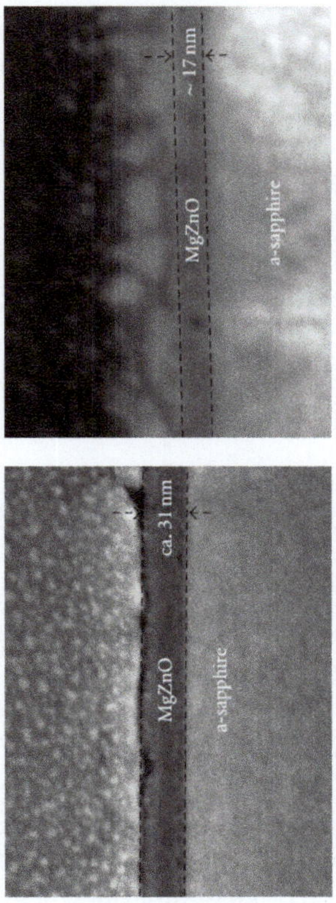

Figure 12:Cross-sectional microscopic image of MgZnO thin film in vicinity of the substrate center (a) and at the substrate edge (b). For the detailed positions of these samples within the50×50mm^2 wafer, see Figure11.

Hall-effect measurements reveal a homogeneous free carrier concentration of about3×1018cm^{-3}in the MgZnO thin film (Figure13). Please note that such a high value is due to the diffusion of aluminium from the substrate, which acts as shallow donor [74,75]. The Hall mobility does not vary significantly over the substrate radius and exhibits a value of about 20 cm^2/Vs (Figure13). From this we conclude that the microstructural properties of the sample should be also laterally homogeneous.

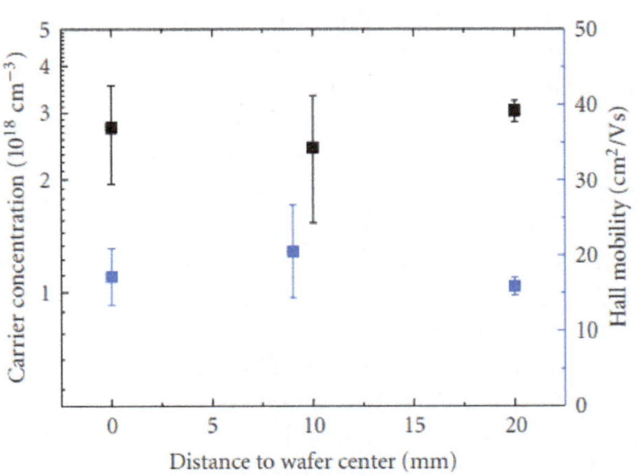

Figure 13:Variation of the free charge carrier density and the mobility determined by Hall-effect measurements over the substrate, as shown in Figure11. The data point of the carrier concentration at 10 mm distance is slightly shifted horizontally to avoid overlap.

The homogeneity of the MgZnO film is further demonstrated by the electrical properties of MESFETs fabricated on samples cleaved from different positions of the substrates as shown in Figure11. For all MESFETs, the transfer characteristic (drain current versus gate voltage) was recorded for a drain voltageVD=2V, and the maximum transconductance (max(dID/dVG)) was determined [72,73]. For the purpose of comparison, the transconductance was normalized with the width to length ratio of the different MESFET channels (Figure14). The best value of the maximum transconductance for each sample is depicted in Figure14, revealing an almost constant value of about$1\times10-5$S.

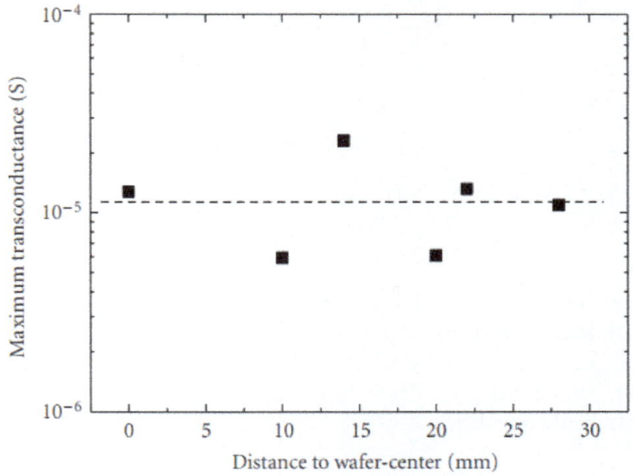

Figure 14:Variation of the best value of the maximum transconductance normalized to a width to length ratio of the different MESFET channels of 10.75 distributed over the50×50substrate of Figure11.

Bragg Reflectors on 3-Inch Diameter Si Substrates

Bragg reflectors (BRs) are wavelength selective highly reflective mirrors, made of periodic stacks of two transparent materials with different refractive indices but optical thickness of a quarter wavelength and are commonly used as one of the building blocks of microresonator structures [78,79]. For various applications, it is desirable to vary the energetic distance between the optical and electronic mode in such resonators on one sample, but to maintain the properties of the Bragg reflector, especially its central wavelength.

For this purpose, BRs consisting of 10 layers pairs of alumina and yttrium stabilized zirconia have been grown by PLD on 3″-Si wafers, applying an oxygen background pressure of 0.002 mbar and heating the substrate to 650°C. For more details of the growth and optical properties of such Bragg reflectors, see [43,78,80]. To achieve homogeneous large-area BR, the substrate holder was placed far off the center of the plasma plume, near the wafer edge, and was rotated with 9 rpm. The average thickness of one-layer pair and of the whole BR structure within 2 cm distance from the wafer center is 93 nm and 930 nm, respectively.

Figure15shows reflectivity spectra, calculated for normal incidence from a layer stack model obtained by transfer-matrix analysis of the ellipsometry data. The maximal reflectivity of the BR stop-bands is 99.7%. The central wavelength respective photon energy shifts from 3.32 eV in the wafer center via 3.31 eV to 3.36 eV at the radial distance of 2 cm (Figure16). In an area of diameter of 3 cm, the central wavelength varies less than 0.4%. The energy shift corresponds to the changing layer thickness at the wafer edge, as visualized by the color gradient in the photograph in Figure16.

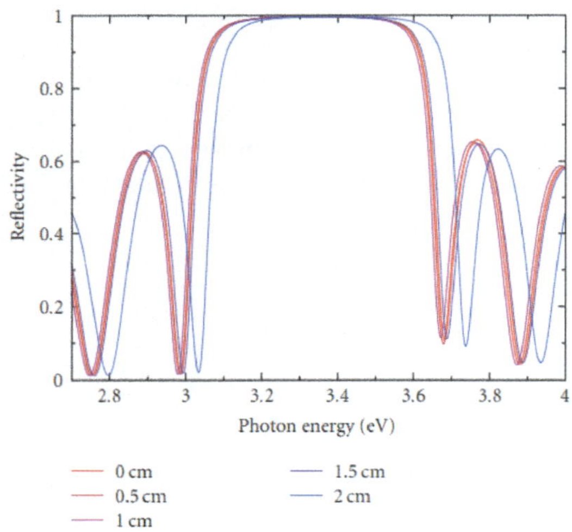

Figure 15:Five reflectivity spectra calculated using a layer stack model from the experimental ellipsometry data taken at the indicated radial positions on the 3-inch diameter Bragg structure on Si(100) wafer. The maximum reflectivity at the BR stop band energy with perpendicular incidence is 99.7%.

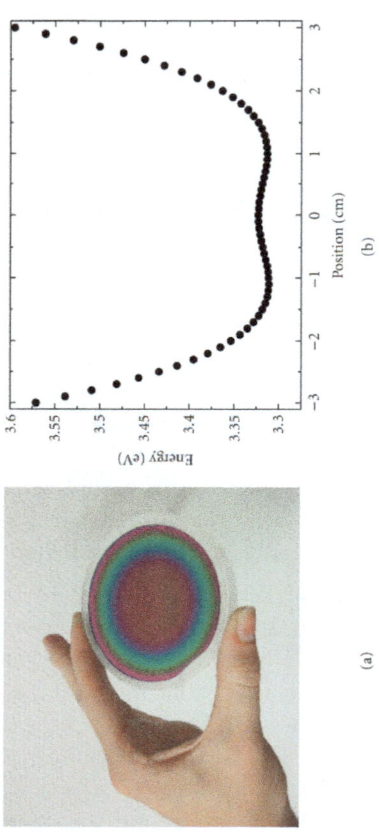

Figure 16: Photograph of the 3-inch diameter Al_2O_2/YSZ Bragg mirror on Si (100) (a) and radial scan of the center energy of the Bragg reflector stop bands (b). The position 0 cm is the wafer center. Within a diameter of 3 cm, the center energy is shifted by only 0.4% and within 4 cm diameter by 1.3%. The thickness of one layer pair is at the wafer center (0 cm) 93.3 nm, at 1 cm distance from center 93.6 nm, and at 2 cm distance 91.8 nm.

CONTROLLED DOPING OF OXIDE THIN FILMS

As mentioned above, the unique flexibility of PLD concerning the deposited material is one of its outstanding advantages. Here we show examples for doping and alloying of ZnO thin films for future optoelectronic applications (Section4.1), and for doped $BaTiO_3$thin films with optimized dielectric properties within a certain temperature range (Section4.2). Because trace impurities may influence considerably the desired effect of doping as, for example, the demanding p-type conductivity of the ZnO semiconductor thin films, we include trace element analyses of PLD targets and films in the discussion in Section4.1.

Doping of ZnO Thin Films to Control Their Electrical Properties

A variety of dopants have been investigated in PLD ZnO films, as shown in Table5and in [44,81–92]. The addition of the dopants is done simply by mixing the dopant element in form of the appropriate oxide into the ZnO powder used for PLD target preparation [6]. By ball milling the powder mixture, a homogenous distribution of the dopant will be achieved. Because the mill beaker and balls are one of the possible sources of unintentional trace impurities, we investigated the effect of three different beaker and ball materials, namely, silica (agate), zirconia and tungsten carbide (Table6). After pressing the powder in a hardened stainless steel compression mould, the ZnO-based targets will be sintered at 900 to 1,150°C in air for 12 h in a furnace. In the PLD process, deviations of the film composition from the initial target composition may occur because of the particular chemical properties of ZnO and the dopants, as for example, evaporation temperature and dopant solubility in the ZnO matrix. Table5summarizes the average composition transfer factors from the PLD target into the grown film. The film compositions were analyzed by ion beam analysis using combined Rutherford backscattering spectrometry (RBS) and particle-induced X-ray emission (PIXE) with 2 MeV $He^+$and 1.2 MeV H^+ions, respectively [93]. For most dopant elements, a transfer factor above 1 was determined (Table5). That means that the chemical composition of the dopant element in relation to Zn is increased by the PLD process. The element with the lower evaporation temperature (mostly Zn) may evaporate preferably from the heated substrate due to the lower sticking coefficient, thus yielding a lowered composition of this particular element in the film. However, some dopant elements, as P [86], Ni, and Cu, and Cd [44,92], show transfer factors considerably below 1. Here, P and Cd show lower evaporation temperatures as compared to Zn, thus explaining the low transfer factors of these elements into the ZnO-based films. In addition, a low solubility of CdO in ZnO of a few at. % is reported [44,92], thus explaining the extremely low transfer factor of Cd. However, the correlation of low transfer factor and low evaporation temperature does not apply for Ni and Cu as neighbor elements to Zn in the periodic system.

In contrast, Mg has the highest transfer factor of all dopant elements [44,87–89]. Probably, the preferential ablation of ZnO out of the MgZnO target is an additional reason for the Mg-enriched films because of the low optical absorption coefficient of MgO with$E_g$around 8 eV for the excimer laser light at 248 nm wavelength. The observed continuous shift of the MgZnO film composition with the ablation state of the MgZnO target [41] supports this assumption of the preferential target ablation.

Figure17shows the resistivity and carrier concentration range of nearly 7 orders of magnitude that can be reached by doping of ZnO films with Mg, Ga, Cd, and Al, in addition to undoped ZnO thin films. The optimized undoped ZnO films show lower intrinsic carrier concentration. The carrier concentration and resistivities were determined by Hall measurements in Van der Pauw geometry with sputtered Au-contacts at the corners of the10×10 mm^{2}films.

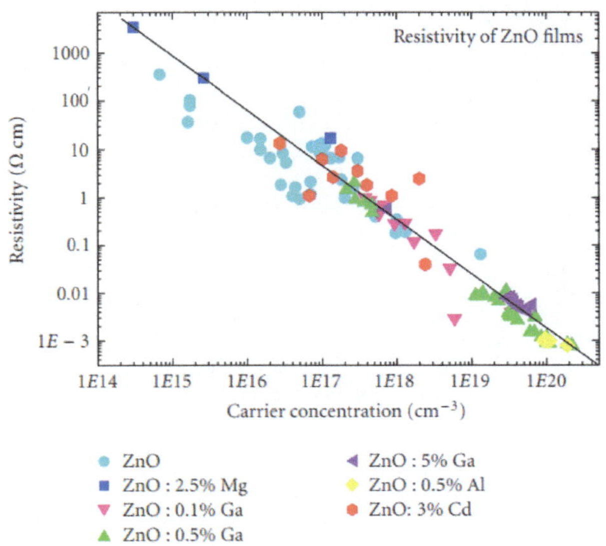

Figure 17: Typical resistivity and free carrier concentration of n-type conducting undoped ZnO, and Al-, Ga-, Mg-, and Cd-doped ZnO thin films on10×10 mm^2 sapphire substrates taken from Hall measurements at 300 K and 0.4 Tesla. The given dopant concentrations are the initial concentrations in the PLD source targets. The line is drawn to guide the eye only. See [44] for additional information.

To get a first estimate on the trace element contamination of the PLD ZnO films, we analyzed three MgZnO targets prepared in different ball mills (Table6) and two undoped ZnO thin films (Table7) for their trace element concentration by PIXE/RBS [93]. The results of Table6show the minor effect of the material of the grinding beaker of the used ball mill for homogenising the initial powders in the PLD target preparation. Within the detection limits of the PBS-PIXE analysis, no remarkable differences for the three beaker materials were found. The MgZnO films grown from these three targets are investigated by capacitance-voltage spectroscopy and thermal admittance spectroscopy, but no remarkable difference of free carrier concentrations and defect levels is obtained up to now.

For most elements listed in Table7, the detected trace element concentrations in the large-area ZnO films are below the given detection limits of the ion beam analysis. At the edges of the 32.8 mm diameter wafers, remarkable concentrations of the elements K, Ca, Ti, Cr, and especially Fe were found. Most of these elements are constituents of stainless steel, and the stainless steel substrate holder and the hot parts of the substrate heater in the PLD chamber are most probably the source of the Fe and Cr contamination at the film edges. In the center of both investigated films, the impurity concentrations of nearly all elements are near or below the detection limit of the ion beam analysis, thus demonstrating the high potential of PLD to deposit films with low concentration of unintentional trace elements. The concentration of Fe, Cr, and Ni can be further reduced by the use of a substrate holder made of molybdenum instead of stainless steel, as we have shown in recent investigations of defect-induced ferromagnetism in ZnO [94].

More detailed information about the very broad ZnO research activities of the Semiconductor Physics Group in Leipzig which is based on PLD ZnO thin films and nanostructures can be found in Sections3.2,5.1, and5.2, and6.1. of this paper, and at the homepagehttp://www.uni-leipzig.de/~hlp/, and in [6,43].

Doped $BaTiO_3$Thin Films with Optimized Temperature Dependence of Dielectric Permittivity

Doped and undoped ferroelectric $BaTiO_3$(BTO) and $Ba_xSr_{1-x}TiO_3$(BSTO) thin films were investigated in cooperation with the industrial partners Robert Bosch GmbH (Stuttgart) and Marconi Communications (Backnang) to demonstrate electrically tuneable ferroelectric microwave devices for automotive distance radars and for directional radio applications, respectively. Voltage-controlled varactors may tune oscillators, filters, phase shifters, delay lines, and antennas [95,96]. As substrate material for the film growth we had to use polycrystalline, Al_2O_3-based microwave ceramic plates with50×50 mm^2 (CoorsTek, see Figure18). The working temperature range of the device structure was set to −35°C up to +85°C. The homogeneity of the dielectric properties within this temperature range was controlled on one hand by the Ba- to Sr-composition ratio in the mixed compound $Ba_xSr_{1-x}TiO_3$(e.g., BSTO-0.8 stands for$x$=0.8) [97], and on the other hand by doping of the BTO films by metal oxides. There are some indications in the literature that doping of BTO with Mg, Al, La, and Ta reduces the losses (Q-factor) but also the relative permittivityεr[98].

Figure 18:(a) PLD $Ba_xSr_{1-x}TiO_3$capacitor structures on microwave ceramic substrates (CoorsTek SUP 996 Hirel,$50\times50\times0.635$ mm^3, A-face:Ra~21 nm) with DC sputtered 3 mm diameter gold (G1072) or platinum (G1017) electrodes~150 nm thick. The nominal thickness of the dielectric film is 1 to 2 µm. (b) Self-made temperature controller platform with adjustable top contacts for the large-area50×50mm^2 capacitor structures. The water cooled Peltier element enables a temperature sweep from −35°C up to +85°C.

In our work, we have partially substituted the Ba and Ti within the $BaTiO_3$by Sr and Zr, respectively, to obtain BSTO-0.6, BSTO-0.8, and $BaTi_{0,8}Zr_{0,2}O_3$. Furthermore, we have doped $BaTiO_3$with 1% and 2% MgO, 1% and 2% Y_2O_3, 3% TiO_2, and 0,5%, 1%, 2%, and 5% Fe_2O_3, employing the unique materials flexibility of PLD. All substitutes and dopants were mixed as oxides directly into the PLD targets. The dielectric permittivity$\varepsilon r$and the loss tangent$\tan\delta$were measured at low frequency of 1 kHz using thin film capacitors in the temperature range −35°C up to +85°C. The capacitors consist of a DC-sputtered Au or Pt ground film (200 nm) at the smoother side of the above-mentioned microwave ceramic substrates. The doped BTO and BSTO films were PLD grown [97], and on top was sputtered the upper Au or Pt capacitor electrode using shadow masks, as shown in Figure18(a). An RCL meter Fluke 6306 and a temperature platform with a Peltier element (see Figure18(b)) were used for the capacitor measurements at 1 kHz with 20 mV or 50 mV AC-amplitude and DC bias 0–10 V. This measurement gives also the$Q$-factor as the ratio of capacitive to Ohmic resistance, which is roughly equal to the inverse loss tangent$\tan\delta$.

The results of the low frequency investigations are shown in Figures19,20,21,22,23, and in Table8. Figure19shows the effect of Ba substitution by Sr on the temperature dependence of the capacity and permittivity at 1 kHz. The permittivity maximum shifts to higher temperature with decreasing substitution. For typical capacity of$C$=100 nF, typical thickness of the dielectric of$d$=1 μm, and 3 mm circular top electrode the relative permittivity$\varepsilon r$of the dielectric films is above 1,600; see Figures19,20,21,22, and23, and for corresponding microwave data, see Table3. Because of the rough substrate surface, the effective thickness of the dielectric is hardly to determine for each particular capacitor. Figure20shows the capacity and$Q$value of a $BaTiO_3$thin film capacitor with Pt electrodes in dependence on temperature and DC bias voltage (1 kHz). The tunability of capacity by the DC voltage of the same structure is shown in Figure21and is nearly independent of temperature from −35°C up to +85°C.

Table 3:Relative permittivity and loss tangent of typical large-area BaxSr1−xTiO3thin films grown by PLD or sputtering on the indicated substrates (size50×50 mm2or 2-inch diameter), measured at 35 GHz and 300 K with an open Fabry-Perot resonator by R. Heidinger at Karlsruhe Research Center. The BSTO films grow with polycrystalline structure on the ceramic substrates and with oriented epitaxial structure on the single-crystalline substrates.

Growth method	Substrate	$Ba_xSr_{1-x}TiO_3$-film	Thickn.	ε_r	$\tan\delta$
PLD (U Leipzig)	CoorsTec *	$Ba_{0.45}Sr_{0.55}TiO_3$ (poly)	500 nm	1140–1640	(0.3 bis 0.8) $\times 10^{-2}$
PLD (U Leipzig)	CoorsTec *	$Ba_{0.6}Sr_{0.4}TiO_3$ (poly)	500 nm	720	5×10^{-2}
PLD (U Leipzig)	LTCC *	$Ba_{0.7}Sr_{0.3}TiO_3$ (poly)	390 nm	500	30×10^{-2}
PLD (U Leipzig)	r-sapphire	$Ba_{0.7}Sr_{0.3}TiO_3$ (110)	130 nm	900–1150	35×10^{-2}
PLD (U Leipzig)	$LaAlO_3$	$Ba_{0.7}Sr_{0.3}TiO_3$ (100)	180 nm	1300–1600	50×10^{-2}
sputtering (FZK)	$LaAlO_3$	$Ba_{0.7}Sr_{0.3}TiO_3$ (100)	200 nm	500–550	(7 to 9) $\times 10^{-2}$

* Al_2O_3-based polycrystalline microwave ceramics. The surface roughness R_a of the LTCC-substrates was 95 to 145 nm, and R_a of the CoorsTec substrates 21–27 nm (A-face HIREL quality).

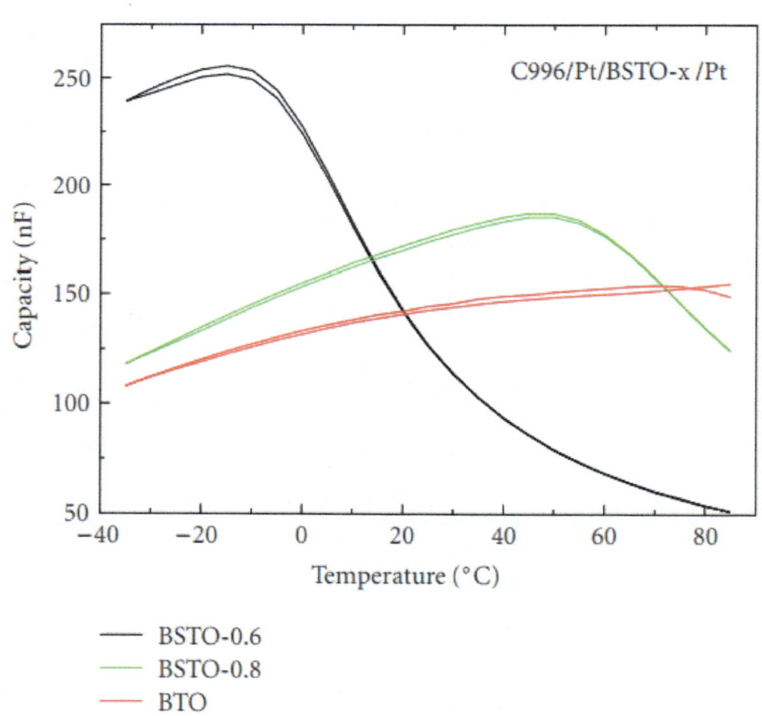

Figure 19:Temperature dependence of capacityC=□0□rA/d (A is electrode area and d thickness of dielectric) of thin film capacitors, measured with 50 mV AC amplitude at 1 kHz and zero DC-bias. The dielectric was Ba0.6Sr0.4TiO3(BSTO-0.6, black line), BSTO-0.8 (green), and BTO (red). With decreasing Ba substitution, the ferroelectric Curie temperature (maximum of□r) increases. This structural phase transition is remarkably broadened due to the thin film nature of the dielectric. ForC=100 nF,d=1 μm, and 3 mm circular electrodes, the relative permittivity□rof the dielectric is above 1,600.

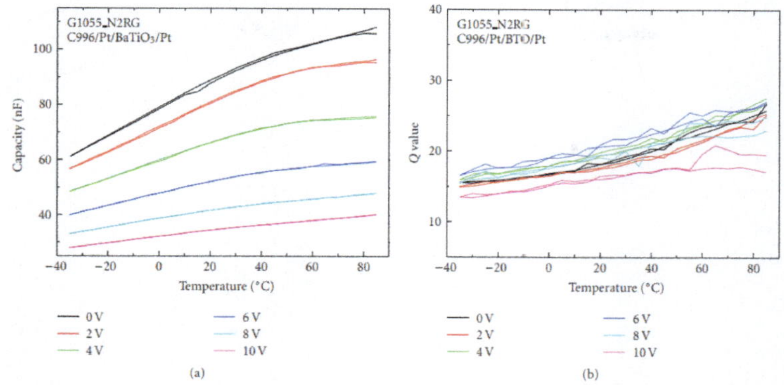

Figure 20:Capacity and□value of an optimized BaTiO3thin film capacitor with Pt electrodes on CoorsTec microwave ceramic substrate in dependence on temperature and DC bias voltage (1 kHz). The relative tunability of capacity is shown in Figure21. The□value is maximum for 6 V DC bias.

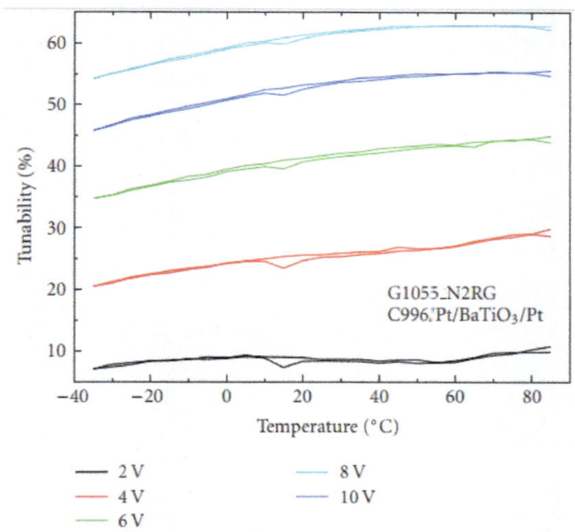

Figure 21:Relative tunability of capacity$(C_0-C)/C_0$in dependence on temperature and DC bias voltage for the BTO sample of Figure20. The temperature dependence is weak as desired for applications. The tunability is 60% for electrical field strength in the dielectric of~5 V/μm with an assumed thickness of 2 μm.

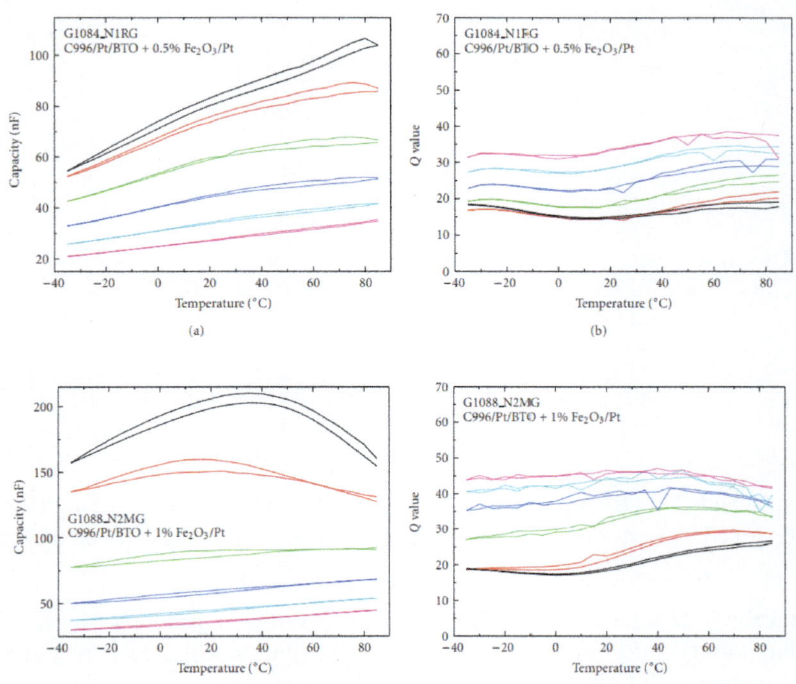

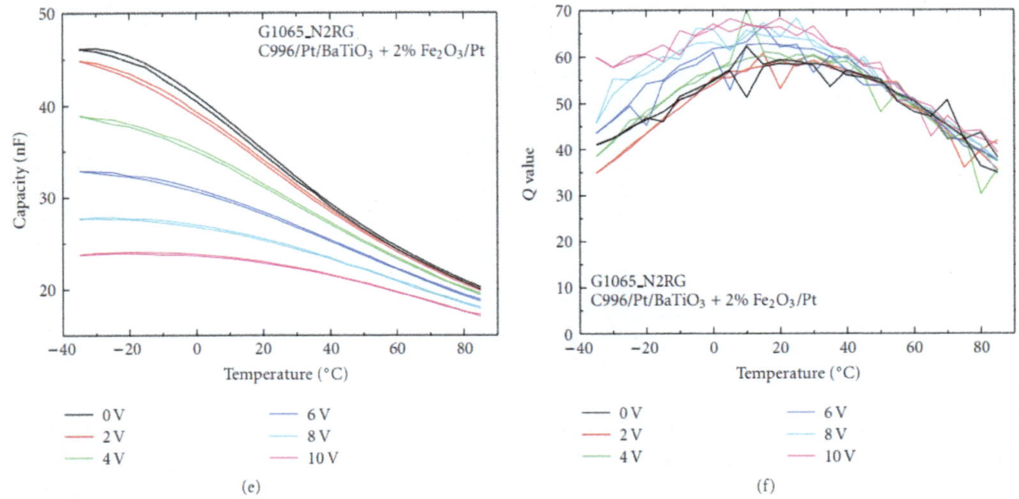

Figure 22: Effect of Fe doping of the BTO dielectric on capacity (a, c, e) and Q value (b, d, f) of Pt/BTO: Fe/Pt thin film capacitors for 0.5% (a, b), 1% (c, d), and 2% (e, f) Fe_2O_3 in the BTO PLD target. The dopant concentration influences the temperature dependence of permittivity considerably.

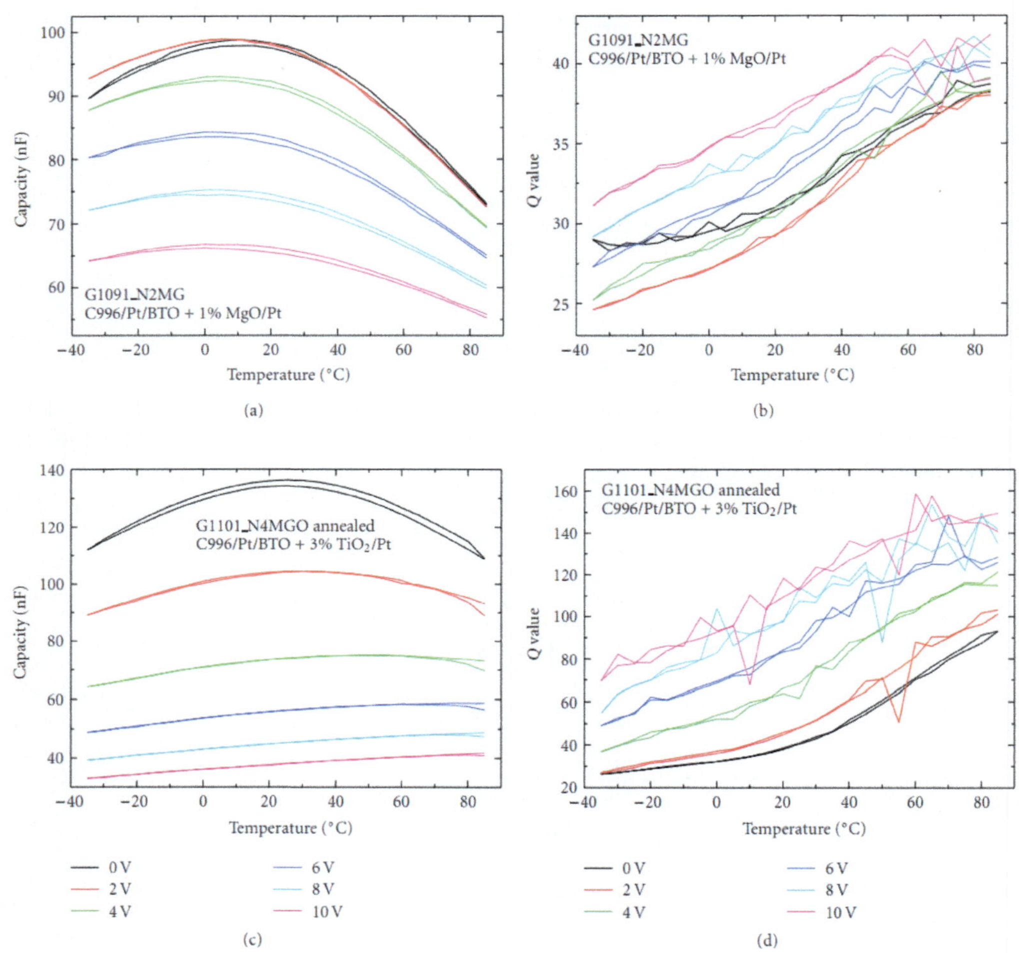

Figure 23: Effect of Mg (a, b) and Ti (c, d) doping of the BTO dielectric on capacity (a, c) and Q value (b, d) of Pt/BTO: Mg,Ti/Pt thin film capacitors. The Ti-doped capacitor (c, d) is postannealed in oxygen which improved the Q value by at least a factor of four.

Doping and postannealing effects on the dielectric properties of BTO thin films are demonstrated in Figures 22 and 23. Increased doping with Fe reduces the losses, but influences considerably the temperature dependence of permittivity; see Figure 22 and [97]. Other metal dopants change the temperature dependence of capacity (εr) and of the Q-value in different ways, as Figure 23 and Table 8 show. Annealing steps reduce remarkably the loss (Q-value) but the permittivity remains nearly unchanged; see Figure 23 and Table 8. A summary on all investigated substitutes and dopants is given in Table 8, together with the low frequency dielectric properties of the thin film capacitors. The highest values of permittivity (capacity), Q value, and tunability are marked in red in Table 8. Obviously, high Sr-substitution and Fe-doping [97] promote a high permittivity.

In addition, the dielectric properties were measured at microwave frequency of 35 GHz for selected large-area BSTO films without any metal electrode by Heidinger using an open Fabry-Perot-resonator at Karlsruhe Research Center similar to that in [70,71]. Table 3 shows the results for $Ba_xSr_{1-x}TiO_3$ films with x=0,45, 0,6, and 0,7. For these measurements, the BSTO films were grown without any metal electrode on different substrates as mentioned in Table 3. With decreasing Ba-content (increasing substitution by Sr), the relative permittivity increases, and the losses are reduced simultaneously. In addition, the growth parameters influence the dielectric microwave properties. Large-area PLD films show high microwave permittivities clearly above 1,000. Based on these investigations, the PLD BSTO thin films were used for planar microwave components and for phase shifter demonstrators (see Figure 24) in cooperation with Marconi Communications Backnang.

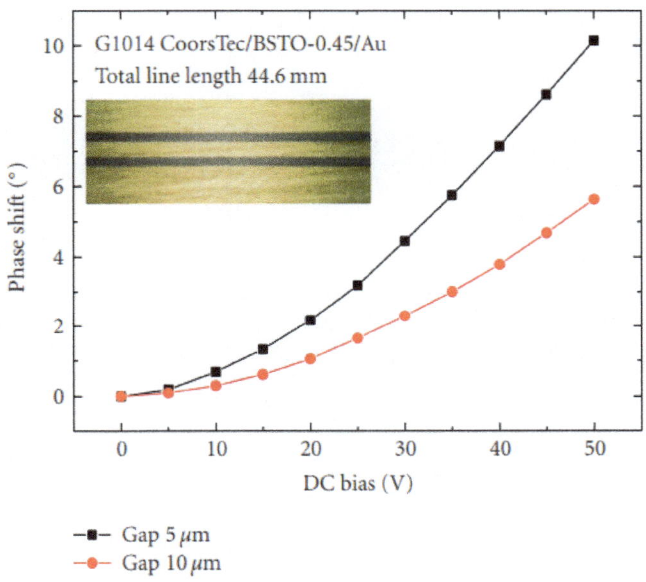

Figure 24: Phase shift at 10 GHz by transmission through a coplanar microwave line (made of gold) on BSTO-0.45 thin film at −25°C. The line width is 200 µm, and the gap to the grounded areas is 10 µm (red curve) and 5 µm (black curve); see inset. The Curie temperature of BSTO-0.45 is below −50°C and therefore shows a weak tunability of phase shift at +25°C.

Microwave investigations of ferroelectric thin films are of current international interest; see for example the overview on tunability and dielectric losses of STO, BSTO, and BTO thin film heterostructures and bulk materials [99]. A "616 element scanning phased array antenna" based on 48 two-inch diameter BSTO-0.5 films on LAO was presented by the NASA Glenn Research Center [100]. This antenna system includes 616 identical phase shifters for 19 GHz. The average relative permittivity of the BSTO-0.5 films was 2,129 with a standard deviation of 149 [100]. A current research subject is the strain engineering of the dielectric properties of ferroelectric BTO films [101].

ZnO FILMS WITH LOW DISLOCATION DENSITY

By the introduction of low-temperature interlayers, we were able to reduce the dislocation density in heteroepitaxial ZnO thin films on c-sapphire to achieve high Hall mobilities above 150 cm²/Vs at 300 K (Section 5.1.). ZnO homoepitaxy is a further possibility to grow films with very low dislocation density and corresponding high Hall mobilities up to 800 cm²/Vs below 100 K (Section 5.2.).

Multistep PLD of ZnO on C-Sapphire with High Hall Mobility

A multistep PLD process was developed to grow nominally undoped ZnO thin films on c-plane sapphire with thickness of 1 µm to 2 µm using low-temperature nucleation and intermediate layers between the high-temperature grown ZnO main layers [102]. The multi-step grown ZnO thin films on c-plane sapphire exhibit high Hall mobilities of more than 150 cm²/Vs at 300 K. The key issue of this multistep PLD process is the insertion of 30 nm thin ZnO relaxation layers grown at reduced substrate temperature as shown in Table 4; see also [102] for further details.

Table 4: XRD ω-scan width of ZnO (0002) peak, PL-D^0X_A FWHM, Hall mobility μ_H, and carrier concentration n of a series of ZnO thin films grown by the indicated multi-step PLD process on c-plane sapphire substrate. The nucleation and intermediate layers are always 30 nm, and the main layers are 1,200 nm thick. The thin nucleation and interlayers are grown at 600°C and the thick ZnO layers at $1.5\times10-2$ mbar oxygen partial pressure and 750°C growth temperature [102].

PLD steps	XRD ω-scan ZnO(002) (arsec)	FWHM PL-D^0X_A(2 K) (meV)	μ_H (cm²/Vs) (300 K)	n (cm⁻³) (300 K)
One-step	446	3.2	30	5×10^{17}
Two-step high temp. *	320	2.4	60	1.4×10^{16}
Two-step	338	2.2	141	5.4×10^{16}
Three-step	241	2.0	119	2.8×10^{16}
Four-step	256	1.9	101	7.5×10^{15}

* 750°C growth temperature for the thin nucleation layer, instead of 600°C at the other processes.

Table 5: Selection of dopants for ZnO and the aim of doping, together with transfer factors of dopant composition from the PLD target into the thin film, as determined by RBS and PIXE analysis of at least five different films for each dopant [6,93] DMS stands for diluted magnetic semiconductor which is still challenging, as well as stable and reproducible p-type conductivity in ZnO.

Dopant in ZnO	Aim of doping	Transfer factor from target into thin film
Li	p-type conductivity	1.37 ± 0.72
N	p-type conductivity	N/A
Mg	higher band gap	1.86 ± 0.49
Al	n-type conductivity	1.56 ± 0.44
P	p-type conductivity	0.50 ± 0.01
Ti	DMS	1.15 ± 0.43
Mn	DMS	1.03 ± 0.28
Fe	DMS	1.47 ± 0.16
Co	DMS	1.23 ± 0.09
Ni	DMS	0.15 ± 0.02
Cu	DMS	0.74 ± 0.25
Ga	n-type conductivity	1.54 ± 1.04
Cd	lower band gap	0.09 ± 0.12
Sb	p-type conductivity	1.71 ± 0.32

Table 6: Typical RBS and PIXE trace element concentrations in ppm (2,0 MeV H^+; Ø 0,4 mm; 50 μC) of three ZnO+2%MgO PLD targets prepared from 99.995% ZnO and 99.998% MgO powders. The initial powders were homogenized in ball mills with three different grinding beakers, as indicated in the first column. Obviously, there is nearly no influence of the beaker material on the trace element impurities within the detection limits. Concentrations with "<" are below the detection limits of the analysis; for Si the detection limit is ~300 ppm, similar to tungsten.

ZnO+2% MgO targets milled in:	Ti	Cr	Fe	Co	Ni	Zr	W
WC beaker	17 ± 5	15 ± 4	36 ± 6	<14	<24	<49	<317
ZrO_2 beaker	15 ± 5	<5	42 ± 6	<14	<23	<46	<309
SiO_2 beaker front side*	12 ± 4	<5	47 ± 7	<14	<24	<49	<321
SiO_2 beaker back side*	16 ± 5	<5	52 ± 7	<14	<24	<47	<319

*Front and back side refers to the side of the PLD target. Front side was ablated already for film growth.

Table 7: Typical atomic concentrations of trace elements from combined PIXE/RBS analysis with 1.2 MeV protons (all values given in ppm) of two large-area 32.8 mm diameter ZnO thin films on sapphire. The PLD target was sintered from 99.9995 at.% ZnO powder. Films were grown with stainless steel substrate holder at growth temperature of about 650°C. Therefore, the Fe concentration is much higher at the wafer edges. Most elements are below the detection limits.

Sample*	Si	K	Ca	Ti	V	Cr	Mn	Fe	Ni	Cu
E391 c	1,342	<16	24	<11	<14	<15	20	35	<67	<153
E391 e	<901	30	28	<13	<16	28	<19	1,069	<69	<166
E392 c	<971	<30	<24	<21	<24	50	<26	<33	<104	<215
E392 e	<1,272	54	51	27	<29	85	41	206	<103	<243

*Sample no. E391 and E392, c and e stand for center and edge analysis position on the wafer, respectively.

Table 8: Overview of doping dependence of dielectric properties of planar $Ba_xSr_{1-x}TiO_3$:M (M = Fe, Mg, Y, Ti, Zr) thin film capacitors on microwave ceramic substrates. Maximum values of capacity at zero DC bias, Q value, and tunability are highlighted by bold type. See Figures 19, 20, 21, 22, and 23 for the sample design and selected, more detailed graphs of C and Q as functions of T for DC bias from 0 to 10 V.

PLD thin film dielectric in Pt-BSTO:X-Pt capacitor	Capacity [nF] @ 0 V and 20°C	Q value @ 0 V and 20°C	Tunability ΔC/ΔU [nF/V] for ΔU = 10 V @ 20°C	T-coefficient ΔC/ΔU [nF/K] @ 0 V, ΔT = 25°–15°C
Undoped BSTO:				
BTO	89	18	5.5	0.44
BSTO-0.8	**172** (5 V)*	17 (5 V)*	**20.7**	0.77
BSTO-0.8 annealed	**116**	**42**	7.8	0.46
BSTO-0.6	**195** (5 V)*	29 (5 V)*	**24.7**	−4.6
BSTO-0.5 to 0.85 Gradient H. Christen, ORNL, USA	**112** (5 V)*	22 (5 V)*	**10.5**	−1.14
Doped BTO:				
BTO:Fe_2O_3(0.5%)	83	15	5.6	0.41
BTO:Fe_2O_3(1%)	**206**	18	**17.0**	0.49
BTO:Fe_2O_3(2%)	35	**58**	1.2	−0.3
BTO:Fe_2O_3(5%)	24	**49**	0.4	−0.11
BTO:MgO(1%)	98 (5 V)*	31 (5 V)*	6.6	−0.08
BTO:MgO(2%)	40 (5 V)*	14 (5 V)*	0.5	−0.08
BTO:Y_2O_3(1%)	49	17	2.9	0.19
BTO:Y_2O_3(2%)	**149**	18	**12.2**	0.39
BTO:Y_2O_3(2%) annealed	**118**	**26**	**9.4**	0.43
BTO:TiO_2(3%)	90 (5 V)*	23 (5 V)*	9.0	−0.003
BTO:TiO_2(3%) annealed	**136**	**38**	**9.8**	0.09
$BaTi_{0.8}Zr_{0.2}O_3$	78	24	5.5	−0.6

*For these samples the maximum DC bias was 5 V, because of the leackage currents.

The high-mobility samples show atomically flat surface structure with grain size of about 0.5–1 μm, whereas the surfaces of low-mobility films consist of hexagonally faceted columnar grains of only 200 nm size in the atomic force microscopy images [102]. Structurally optimized PLD ZnO thin films show narrow photoluminescence line widths of donor bound exciton of 1.7 meV at 2 K. The concept of nucleation and intermediate layers grown at lower temperature as compared to the main functional layer was first proposed by Amano et al. for GaN films [50] and was transferred successfully to the ZnO system [102]. Obviously, the low temperature layers compensate lateral strain and reduce the number of dislocation lines in growth direction, as shown in the TEM cross section in Figure25.

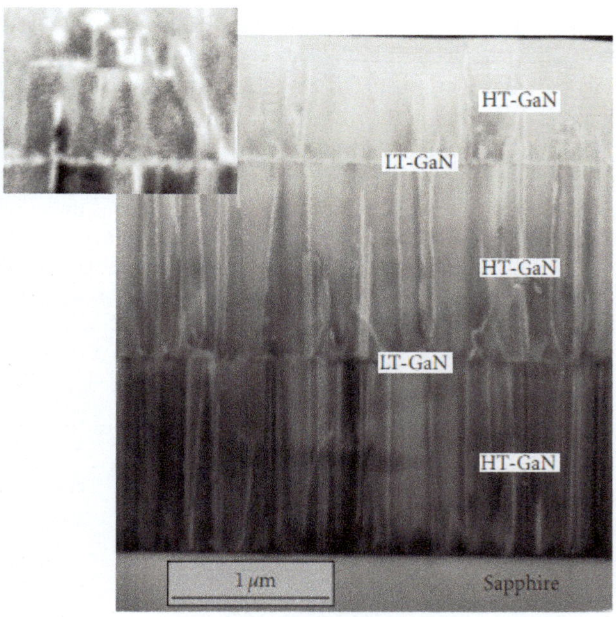

Figure 25: TEM dark field cross-section of an MOCVD GaN structure on c-sapphire with low temperature LT-GaN buffer layer at the sapphire/GaN interface and two low temperature GaN interlayers in between the high-temperature HT-GaN layers. The number of dislocations clearly decreases in growth direction from bottom to top. The inset at the top left shows a dislocation half loop formed in the HT-GaN layer in the vicinity of an LT-interlayer. Obviously, the interlayers act as barriers for the propagation of dislocations. Adapted from Benamara et al. [50].

ZnO and ZnMgO Thin Films on Lattice Matched Substrates

High-quality light emitters and other electronic and optoelectronic devices require ZnO thin films with very high structural quality. One way to reduce the dislocation density in thin films is to reduce the lattice mismatch of film and underlying substrate. Sapphire and ZnO show in-plane lattice mismatch above 10%, and these ZnO films show a high mosaicity with small-angle tilted grains and a high density of dislocations which decrease the electrical performance of the films. A better lattice-matched substrate to ZnO is $ScAlMgO_4$(SCAM) with nearly identical in-plane a-lattice constant of 3.236 Å (aZnO=3.2501 Å). Figure26shows an XRD 2Θ−ωscan of a PLD ZnO thin film on SCAM. Due to the much better lattice match, ZnO on SCAM (006) shows considerably reduced ZnO(002) rocking curve width as compared to ZnO on sapphire (Table9). However, SCAM is an expensive material, and the SCAM substrates show a strongly anisotropic layered morphology with laterally not homogeneous structural properties, as we have found for the tilt mosaicity.

Table 9: Full width at half maximum (FWHM) of HR-XRD ZnO(002) rocking curves of PLD ZnO and ZnMgO thin films on the indicated substrates (best values). ω-scans were taken with Cu K_{c1} radiation and with open detector at the 2Θ position of ZnO(002). For details see [54,104].

Film on substrate (sample no.)	FWHM of HR-XRD rocking curve of ZnO(002) film peak
ZnO on a-sapphire (E1105)	369″
ZnO on n-GaN/c-sapphire (E1111)	305″
ZnO on c-sapphire (multistep PLD)	151″
ZnO on $ScAlMgO_4$(006) (E1108)	71″
ZnMgO on ZnO(0001) (E2064)	27″
ZnO on ZnO(0001) (E1340)	23″ (no peak separation)

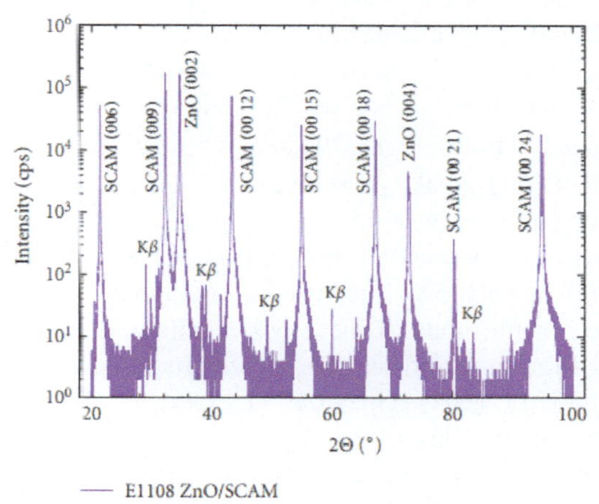

Figure 26: XRD wide angle 2Θ−ωscan (Cu $K_α$) of a PLD ZnO thin film on a $ScAlMgO_4$(006) substrate.

Only recently, ZnO single crystals became commercially available for a reasonable price, and ZnO homoepitaxy could be performed. We have demonstrated the PLD of undoped ZnO thin films with high structural quality [103] on thermally pretreated hydrothermal O-face ZnO single crystals. Details of the ZnO substrate preparation, the homoepitaxial PLD growth, and the preparation of the TEM cross sections can be found in [103–105]. The electrical and optical properties of homoepitaxial ZnO films doped with about 0.01% phosphorus are reported in [105,106]. Interestingly, such as-grown ZnO:P and ZnMgO:P thin films are highly n-type conductive [54,105–107], which is in partial agreement with [108–110]. More structural properties, in particular detailed HR-XRD and TEM investigations of undoped and phosphorus doped homoepitaxial ZnO and ZnMgO thin films, were reported in [54,104].

Selected MgZnO:P films showed an annealing-induced increase of the resistivity of nearly 10^6; however, the films were still n-type. ZnO:P films (without MgO) showed a much smaller annealing effect [54,107]. This is in contradiction to heteroepitaxial ZnO:P films with more structural defects, which could occasionally be switched from n-type into p-type [108–110]. Preliminary evidence for a formation of a two-dimensional electron gas in homoepitaxial MgZnO-ZnO-MgZnO quantum well structures is shown in [111].

In [54], we extended the detailed structural investigations in [104] to homoepitaxial ZnO films alloyed with Mg (0 to 4%) and doped with P (0 to 1%). In particular, a complete survey on Hall data of grown MgZnO:P films is given. The ZnMgO films show atomically flat surfaces with monolayer steps of c/2 or c in the atomic force microscopy (AFM) images (Figure27). Two-dimensional growth with terrace-like surface structure is most prominent for the Mg-alloyed films without P. Transmission electron microscopy (TEM) and high-resolution X-ray diffraction (HR-XRD) confirm a very low dislocation density (Figure28) and narrow rocking curves of undoped homoepitaxial ZnO films grown by PLD (Table9). ZnO(002) rocking curves of $Zn_{95.99}Mg_4P_{0.01}O$ films on ZnO(001) were as narrow as 27 arcsec with an FWHM of the substrate peak of 23 arcsec. Extensive HR-XRD triple-axis scans are reported with special attention to samples which show a clear splitting of film and substrate peaks in XRD [54]. For these samples, c-axis and a-axis lattice constants were determined independently for homoepitaxial film and underlying ZnO substrate. In-plane lattice-matched, pseudomorphic growth with compressive or tensile strain was confirmed for all investigated samples. Figure29shows triple axis $2\Theta-\omega$scans of the ZnO(002) peaks of two selected samples which show tensile or compressive film strain. The results show the balance between tensile strain induced by Mg and compressive strain by P in ZnO [54]. The in-plane lattice-matched, pseudomorphic growth mode of these two tensile or compressively strained films is demonstrated by reciprocal space maps of the asymmetric (104) reflections in Figure30. High electron mobilities up to 190 cm²/Vs at 300 K and up to 800 cm²/Vs at 70 K were found in the homoepitaxial MgZnO:P thin films [54,105–107]. The effect of a further increase of the Mg concentration in homoepitaxial ZnMgO films up to 22 at% on the structural and optical properties is demonstrated in [112]. The corresponding HR-XRD RSMs show an increasing tilt mosaicity and the loss of lattice matched growth with increasing Mg content [112]. However, the homoepitaxial film with highest Mg content shows the highest Hall mobility of~930 cm²/Vs at 65 K [112].

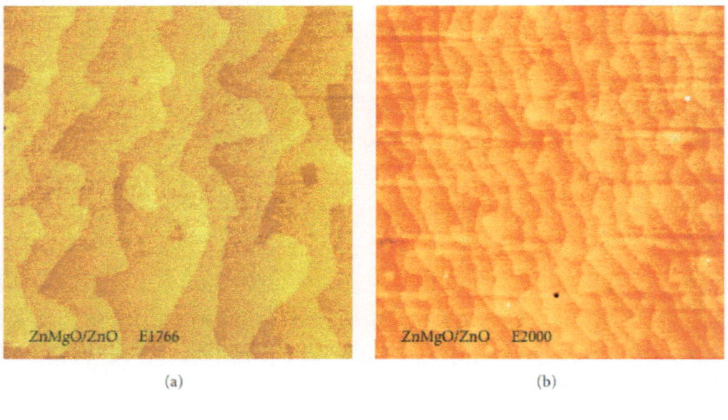

Figure 27:AFM surface images of $Zn_{0.96}Mg_{0.04}O$/ZnO quantum well/$Zn_{0.96}Mg_{0.04}O$ films on ZnO (0001). The scan size is$2\mu m \times 2\mu m$, and the$z$-scales are 2.16 nm (E 1766) and 2.5 nm (E 2000). ZnMgO monolayer terraces with step height c or c/2 corresponding to two-dimensional layer by layer growth are visible. The total film thickness and the out-of-plane strain are 380 nm and 330 nm and −0.18% and −0.22% for E1766 and E2000, respectively [54].

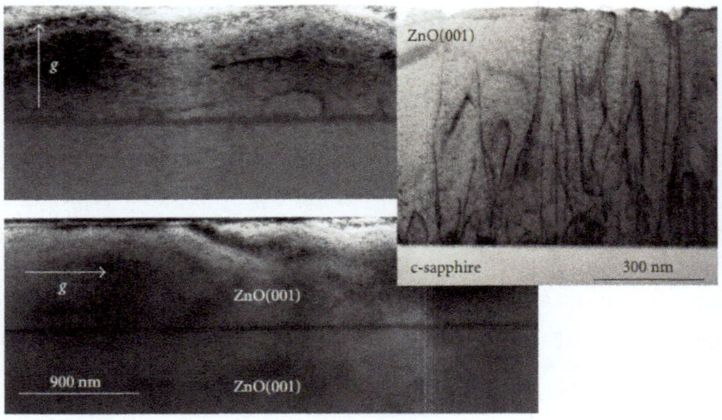

Figure 28:TEM bright field (11.0) cross sections of homoepitaxial undoped ZnO film E1289 taken with two different g-vectors (left), taken by Wagner (see [54]). The few dislocations for$g$-vector in c-direction are out of contrast for$g$=[−11.0], therefore their Burgers vectors are of c-type. The film thickness is893±5 nm, as determined from the intensity oscillations in the HR-XRD $2\Theta-\omega$scan (not shown here). The right inset (provided by W. Mader, Bonn) shows for comparison a heteroepitaxial PLD ZnO film on c-plane sapphire E435 with much higher dislocation density, in particular near the interface.

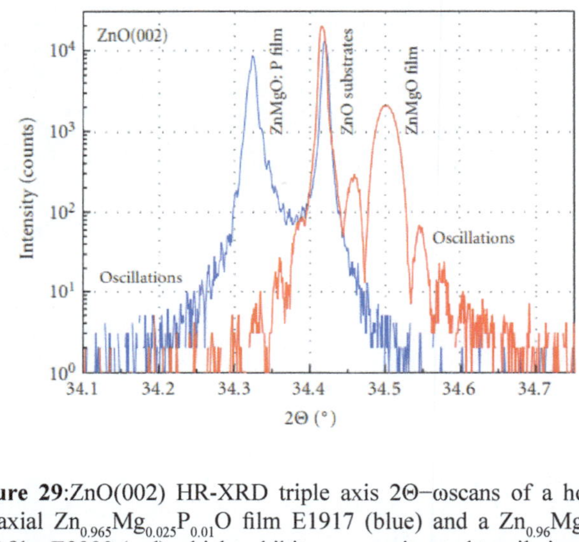

Figure 29:ZnO(002) HR-XRD triple axis $2\Theta-\omega$scans of a homoepitaxial $Zn_{0.965}Mg_{0.025}P_{0.01}O$ film E1917 (blue) and a $Zn_{0.96}Mg_{0.04}O$ QW film E2000 (red) which exhibit compressive and tensile in-plane strain in the pseudomorphic growth mode, respectively. The total film thickness (calculated from the intensity oscillations) and the out-of-plane strain were 1,030 nm and 330 nm and 0.27% and −0.22% for E1917 (blue) and E2000 (red), respectively.

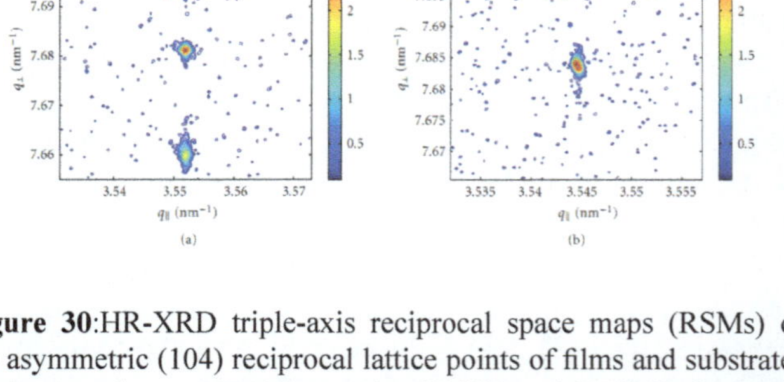

Figure 30:HR-XRD triple-axis reciprocal space maps (RSMs) of the asymmetric (104) reciprocal lattice points of films and substrates of the two homoepitaxial samples E1917 and E 2000 (compare Figure29). In the center of the maps is the ZnO (104) substrate lat-

tice point with higher intensity, and below or above is the weaker ZnMgO:P and ZnMgO (104) film lattice point. The perfect vertical alignment of (104) film and substrate peaks demonstrates the in-plane lattice match and the pseudomorphic growth mode.

OXIDE MULTILAYERS WITH SMOOTH INTERFACES

The interface smoothness of ZnO quantum wells (QWs) with MgZnO barriers could be considerably improved so that the QW photoluminescence gives evidence of the quantum-confined Stark effect, which was up to now observed only in MBE and MOCVD grown ZnO QWs (Section6.1). By in-situ RHEED control, we are able to grow monolayer controlled $SrTiO_3/BaTiO_3$single and multilayers with atomically flat interfaces and surfaces (Section6.2.).

Improved Interface Abruptness of MgZnO/ZnO Quantum Wells Grown by PLD

Due to the polar nature of ZnO and MgZnO, both materials possess a spontaneous polarization parallel to their crystallographicc-axis. The magnitude depends on the Mg content, and a sheet charge forms in MgZnO/ZnO heterostructures if the heterointerface intersects thec-axis. An electric field arises from the interface charge in MgZnO quantum wells. Holes and electrons will be separated leading to the quantum-confined Stark effect (QCSE). An abrupt interface is required for the occurrence of the QCSE.

Quantum confinement effects in $Mg_xZn_{1-x}O$/ZnO quantum wells (QWs) grown by PLD were previously published [77]. We were able to control the thickness of the QWs in the range from 4.8 nm down to 1.2 nm, and the energetic position of the QW emission shifted accordingly. However, even for high QW widths no significant influence of the QCSE on the 2 K photoluminescence was detected [77].

In recent experiments, the laser fluence in the PLD process was controlled in order to modify the energy of the particles in the laser plasma. This had a tremendous influence on the sharpness of the heterointerfaces. For a reduced laser fluence of 1.8 J/cm^2, a systematic redshift of the QW luminescence with increasing well width of up to 230 meV (Figure31) below the emission of the free exciton in bulk ZnO was observed [76]. For samples grown at a high fluence (2.4 J/cm^2), the luminescence peak barely shifts below the emission of bulk ZnO, indicating the absence of a strong electric field in the quantum well, as mentioned above. Additionally, an increase in the exciton lifetime by two orders of magnitude with increasing thickness is observed for the samples grown with the low fluence, while no change is observed for the samples grown with the high fluence, as shown in Figure32.

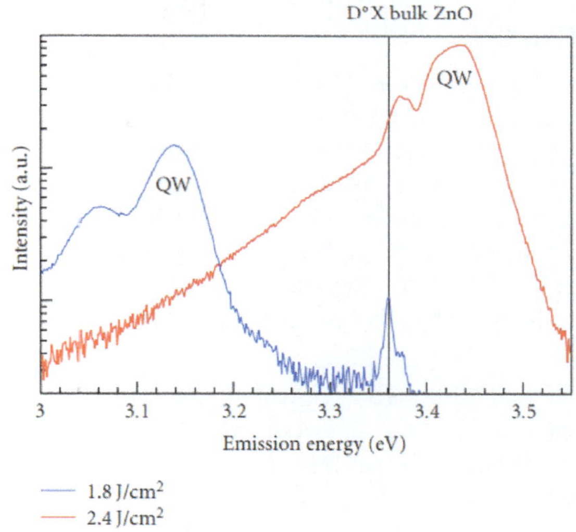

Figure 31:PL-spectra of MgZnO/ZnO QW grown by PLD at high fluence (2.4 J/cm^2, red line) and low fluence (1.8 J/cm^2, blue line). The donor bound excitonic emission (D^0X) of bulk ZnO is indicated by a black line, and the QW emission is marked in the spectra. The quantum well grown at low laser fluence shows the quantum confined Stark effect due to improved interface sharpness, while the other does not.

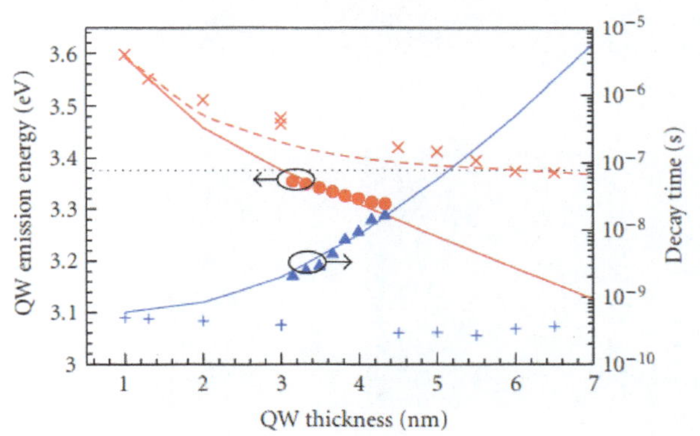

Figure 32:Dependence of the photoluminescence energy (left scale) and decay time (right scale) taken at the maximum of the QW luminescence of PLD grown QW samples (Figure31), on the thickness of the ZnO quantum wells. Closed symbols give data obtained in [76] with remarkable QCSE. Crosses symbolize data from PLD grown samples without remarkable QCSE [77]. The black dotted line shows the free exciton transition energy in bulk ZnO. Solid lines give the expected dependence from the variational calculations with the presence of an electrical polarization mismatch. The dashed red line gives the transition energy as from variational calculations without a mismatch in the electrical polarization.

These results were explained by a difference in the kinetic energies of the constituents of the laser plasma. For high laser fluence, the ions in the plasma have sufficient energy to penetrate the topmost layers of the growing film, resulting in an intermixing between the ZnO QW and the MgZnO barrier, which leads to a reduction of the electric field in the QW [76]. A similar diminishing of the QCSE was seen after ion bombardement of MBE grown MgZnO/ZnO QWs that previously showed the QCSE [113].

PLD of $SrTiO_3$ and $BaTiO_3$ Multilayers with In Situ RHEED

We report the growth of $SrTiO_3$(STO) and $BaTiO_3$(BTO) thin films using PLD combined with high-pressure Reflection High Energy Electron Diffraction (RHEED) system (STAIB "Torr RHEED" with double differentially pumped 35 keV gun) to control the growth rate as well as the surface quality of the thin films with submonolayer resolution. References [114–119] describe the state of the art of the so-called laser MBE technique which is a common name for PLD with in-situ RHEED.

The thin films were grown on STO (100) substrates supplied from Crystal GmbH Berlin. A thermal annealing step has been applied in order to obtain a TiO_2-terminated, monolayer surface quality of the substrates, as shown in Figure33. The substrates were pretreated for 30 seconds with NH_4F-HF solution and subsequently annealed for 2 hours at 870°C in 700 mbar oxygen ambient.

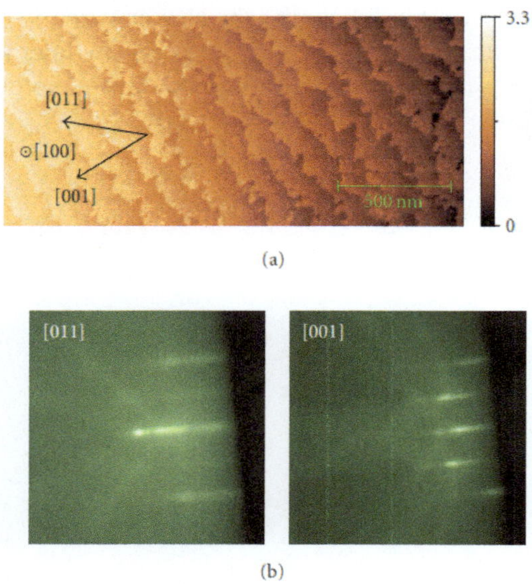

Figure 33:Atomic Force Microscopy (AFM) image of a typical STO(100) substrate surface after etching and heat treatment, with unit cell steps of 3.9 Å height. Below are the corresponding RHEED surface images taken in[011]and[001] directions.

For the growth of the thin films, the substrate temperature T_{sub}, the oxygen partial pressure$p(O_2)$ and the laser fluence at the target$F$are varied between 580°C and 720°C, 0.01 mbar and 0.0003 mbar, and between 1.6 J/cm^{2}and 2.5 J/cm^2, respectively. During the PLD process, RHEED monitors the film growth on line by high speed recording a continuous movie file with up to 50 RHEED patterns per second which reflect the changes of surface morphology. This technique is very surface sensitive and enables to study, for example, whether a two-dimensional layer by layer (or step-flow) growth or a three-dimensional island growth dominates under the chosen PLD parameters; see for comparison [117,118]. By growing STO homoepitaxially on pretreated STO substrates, we are able to observe intensity oscillations of the specular RHEED (0,0) spot indicating a layer by layer growth. The characteristic of the RHEED spot intensity as a function of time for the STO film growth is shown in Figure34(a). The oscillations are visible over the whole recording time interval of 400 sec. and even remain intense and clearly resolved when the repetition rate of the laser is changed from 1 Hz to 3 Hz. The RHEED images recorded at the marked time positions show streaky lines supporting the conclusion of a layer by layer growth. As to the pulsed nature of PLD, the RHEED oscillations show a substructure related to the time regime of the single laser pulses, as shown in Figure34(b).

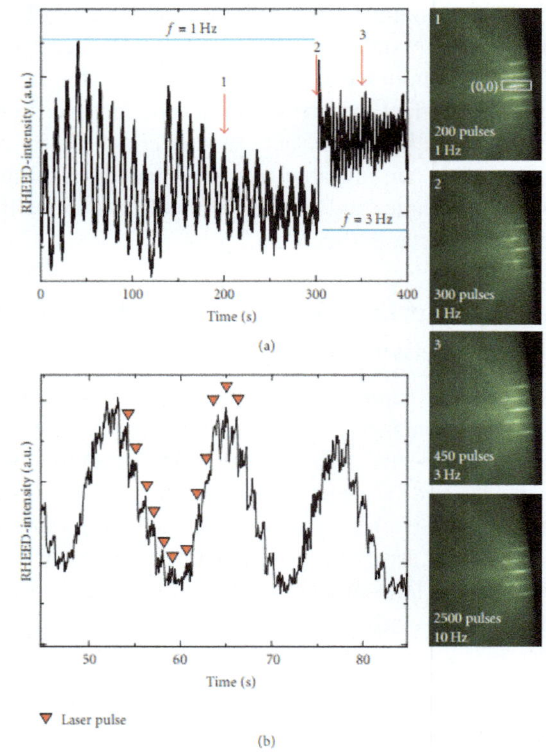

Figure 34:(a) RHEED oscillations during the growth of an STO thin film at$T$sub≈700°C,p(O_2) = 0.002 mbar, and$F$≈1.9 J/cm^2. The oscillations are visible over the whole measurement time even for a laser repetition frequency of 3 Hz. The time positions 1, 2, 3 correspond to the recording of the RHEED images marked 1, 2, 3 (right). In RHEED image no. 1, the specular (0,0) spot is framed which is used for recording the oscillations throughout this section. (b) A detailed look on three RHEED oscillations during growth with 1 Hz. A substructure related to the single laser pulses (marked by red triangles) is resolved. One STO monolayer is grown with 12 pulses.

By looking on the AFM image of the STO film surface after 3000 laser pulses, obviously monoatomic steps as predetermined by the substrate are still visible; see Figure35.

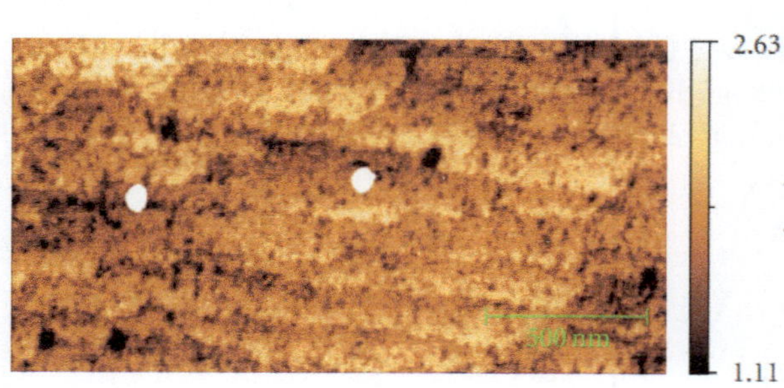

Figure 35:AFM image of an STO thin film at$T$sub≈700°C,p(O_2) = 0.002 mbar, and$F$≈1.9 J/cm^{2}on a pre- treated STO (100) substrate. The thickness of the STO thin film is about 95 nm.

By growing BTO on STO(100), RHEED oscillations are still visible. We observe that in the case of a too small laser flux the BTO growth mode changes from layer by layer growth to a three dimensional island growth as shown in Figure36. The spotty RHEED images clearly indicate the formation of 3-dimensional BTO islands. After finding optimal BTO growth parameters, we were able to preserve the layer by layer growth regime also for BTO, which enables us to grow heterostructures of alternating STO and BTO thin film multilayers. As shown in Figure37, the RHEED-oscillations remain visible even after growing two double layers of STO and BTO, thus enabling the growth of high quality STO/BTO superlattices.

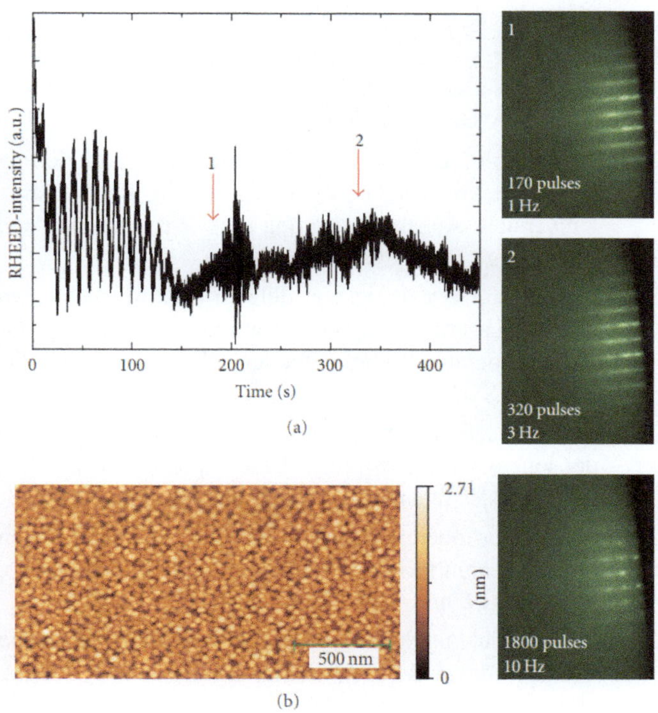

Figure 36:(a) RHEED intensity as function of time for a BTO thin film grown at$T$sub≈700°C,p(O_2) = 0.002 mbar, and$F$≈1.9 J/cm^2. Still in the 1 Hz regime, the growth mode changes from layer by layer growth with clear oscillations up to 150 sec., to an island growth from 150 to 450 sec. as indicated by the more and more spotty RHEED images (right). (b) AFM image of the sample after 3,000 pulses. Clearly the granular surface with BTO islands is visible as implied by the RHEED images (right).

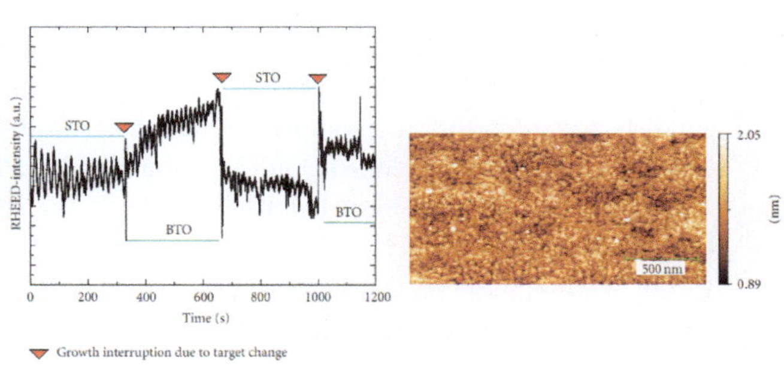

Figure 37:(a) RHEED (0,0) spot intensity as function of time for a BTO/STO heterostructure grown at$T$sub≈700°C,p(O_2) = 0.002 mbar and$F$≈2,5 J/cm^{2}on STO(100). The oscillations are visible over the whole deposition time. The first STO layer is 16, and the following BTO layer is 21 monolayers thick. (b) AFM image after growth of two double layers of STO and BTO.

SUMMARY

Recent experimental progress of several advanced PLD techniques for growth of oxide films and multilayers is reported. Large-area PLD is demonstrated for double-sided CeO_2/$Yba_2Cu_3O_{7-\delta}$films on both sides of$r$-plane sapphire wafers of71×75 mm^{2}size, corresponding to a diagonal of 4-inch, for microwave filter applications, and for thin ZnMgO films on50×50 mm^{2}a-plane sapphire substrates for field effect transistor demonstrators. Large-area Bragg reflectors built from 10 pairs Al_2O_2/YSZ with total thickness around 1 µm show high optical homogeneity within 40 mm diameter. The unique flexibility of PLD concerning the deposited material was utilized for extended doping and alloying of ZnO thin films for future optoelectronic applications, and for doped $BaTiO_3$thin films with optimized dielectric properties within a certain temperature range. The amount of trace impurities in the PLD grown films is discussed, and process-related sources of impurities as the ball mills for PLD target preparation were considered in detail. By the introduction of low-temperature interlayers, the dislocation density in heteroepitaxial ZnO thin films on c-sapphire was reduced to achieve high Hall mobilities above 150 cm^2/Vs at 300 K. ZnO homoepitaxy on single crystalline ZnO substrates is a further possibility to grow films with very low dislocation density and corresponding high Hall mobilities of more than 900 cm^2/Vs at 65 K. The interface abruptness of PLD grown ZnO quantum wells with MgZnO barriers could be considerably improved so that the optical QW emission gives evidence of the quantum-confined Stark effect. By in-situ RHEED control, we are able to grow monolayer controlled $SrTiO_3$/$BaTiO_3$single- and multilayers with atomically flat interfaces and surfaces.

ACKNOWLEDGMENTS

M. Lorenz and all coauthors thank D. Natusch and G. Ramm for their outstanding long-term engagement in development of the modular PLD growth systems and for SNMS depth profiling and PLD target preparation, respectively. They thank R. Heidinger, Karlsruhe Research Center for the microwave characterization of the BTO films within the industrial integrated BMBF Project FKZ 13N8158. G. Wagner (Leipzig) and W. Mader (Bonn) prepared TEM cross sections of ZnO films on ZnO and sapphire, respectively. Florian Schmidt was involved in the CV and TAS investigations of the ZnMgO films from targets prepared with different ball mills. J. Sellmann and S. Schöche optimized the growth of the Bragg reflectors and the STO substrate pretreatment, respectively. M. Thorwart worked at the PLD with in-situ RHEED monitoring. Current financial support of the Deutsche Forschungsgemeinschaft within the SFB 762 “Functionality of oxide interfaces,” of the European Social Fund (ESF), and the Leipzig School of Natural Sciences BuildMoNa (GS 185) is gratefully acknowledged.

REFERENCES

1. D. B. Chrisey and G. H. Hubler, Eds.,Pulsed Laser Deposition of Thin Films, Wiley, New York, NY, USA,

1994.

2. R. Eason, Ed.,Pulsed Laser Deposition of Thin Films—Applications-Led Growth of Functional Materials, Wiley-Interscience, Hoboken, NJ, USA, 2007.

3. D. Dijkkamp, T. Venkatesan, and T. Venkatesan, "Preparation of Y-Ba-Cu oxide superconductor thin films using pulsed laser evaporation from high Tc bulk material,"Applied Physics Letters, vol. 51, no. 8, pp. 619–621, 1987.·

4. B. Roas, L. Schultz, and G. Endres, "Epitaxial growth ofYBa2Cu3O7−xthin films by a laser evaporation process,"Applied Physics Letters, vol. 53, no. 16, pp. 1557–1559, 1988.·

5. D. Bäuerle, "Laser chemical processing," inLandolt-Börnstein New Series, Group VIII Advanced Materials and Technologies, Vol. 1 Laser Physics and Applications, Subvolume C Laser Applications, R. Poprawe, H. Weber, and G. Herziger, Eds., pp. 311–354, Springer, Berlin, Germany, 2004.

6. M. Lorenz, "Pulsed laser deposition of ZnO-based thin films," inTransparent Conductive Zinc Oxide. Basics and Applications in Thin Film Solar Cells, K. Ellmer, A. Klein, and B. Rech, Eds., vol. 104, chapter 7, pp. 303–358, Springer Series in Materials Science, Berlin, Germany, 2008.

7. S. B. Ogale, Ed.,Thin Films and Heterostructures for Oxide Electronics, Multifunctional Thin Films Series, edited by O. Auciello and R. Ramesh, Springer Science + Business Media, New York, NY, USA, 2005.

8. C. Jagadish and S. Pearton, Eds.,Zinc Oxide Bulk, Thin Films and Nanostructures: Processing, Properties and Applications, Elsevier, Amsterdam, The Netherlands, 2006.

9. M. Lorenz, H. Hochmuth, D. Natusch, et al., "Ag-doped double-sided PLD-YBCO thin films for passive microwave devices in future communication systems,"IEEE Transactions on Applied Superconductivity, vol. 9, no. 2, pp. 1936–1939, 1999.·

10. M. Lorenz, H. Hochmuth, and H. Hochmuth, "High-quality Y-Ba-Cu-O thin films by PLD—ready for market applications,"IEEE Transactions on Applied Superconductivity, vol. 11, no. 1, pp. 3209–3212, 2001.·

11. M. Lorenz, H. Hochmuth, D. Natusch, and M. Grundmann, "High-quality reproducible PLD Y-Ba-Cu-O:Ag thin films up to 4 inch diameter for microwave applications,"Physica C, vol. 372–376, no. 2, pp. 587–589, 2002.·

12. C. Curran,Gepulste Laserabscheidung von Aurivillius-Phasen für ferroelektrische dunne Schichten, Ph.D. thesis, Martin-Luther-Universität Halle-Wittenberg, 1997.

13. R. Klarmann,Laserdeposition ferroelektrischer $(Pb_{0.92}La_{0.08})(Zr_{0.65}Ti_{0.35})O_3$-Schichten auf Edelstahlsubstraten, Ph.D. thesis, Universität Augsburg (Shaker, Aachen), 2000.

14. G. Köbernik,Präparation und Charakterisierung ferroelektrischer perowskitischer Multilagen, Ph.D. thesis, Technische Universität Dresden, 2004.

15. M. Maier,In-situ Präparation von komplexen supraleitenden und ferroelektrischen Heterostrukturen mittels gepulster Laser Deposition, Ph.D. thesis, Universität Mainz, 2001.

16. M. Mertin,Modellierung des PLD-Prozesses zur Abscheidung elektrokeramischer Dünnschichten, Ph.D. thesis, D82 RWTH Aachen (Shaker Aachen), 1996.

17. A. I. Petraru,Optical and electro-optical properties of $BaTiO_3$thin films and Mach-Zehnder waveguide modulators, Ph.D. thesis, Universität Köln, 2003.

18. M. Siegert,Wachstumsuntersuchungen an Bariumtitanat-Dünnschichten, hergestellt mit gepulster Laserdeposition, Ph.D. thesis, Universität Köln, 2001.

19. S. Bar,Crystalline, rare-earth-doped sesquioxide PLD-films onα-alumina, Ph.D. thesis, Universität Hamburg, Göttingen, Germany, 2004.

20. J. Goldfuß,Herstellung und Modifizierung heteroepitaktischer Oxidschichten auf Si, Ph.D. thesis, Universität Augsburg (Verlag für Wissenschaft und Forschung), Berlin, Geramny, 2005.

21. B. Hobein,Herstellung von dünnen Elektrolytschichten mittels Laserablation und Kathodenzerstaubüng für Hochtemperatur-Brennstoffzellen, Ph.D. thesis, RWTH Aachen, 2003.

22. R. Hühne,Textur- und Mikrostrukturentwicklung bei der ionenstrahlunterstutzten Laserdeposition von MgO, Ph.D. thesis, Technische Universität Dresden, 2001.

23. B. Keiper,Ionengestützte Laserpulsabscheidung von Oxidschichten für optische Anwendungen, Ph.D. thesis, Technische Universität Chemnitz, 1997.

24. Y. Kuzminykh,Crystalline, Rare-Earth-doped Sesquioxide and YAG PLD -Films, Ph.D. thesis, Universität Hamburg, 2006.

25. L. Ferchland,Untersuchungen zum Wachstum und zur Passivierung von dünnen Schichten aus Hochtemperatur-Supraleitern für die Anwendung bei Sensorbauelementen, Ph.D. thesis, Fortschritt- Berichte VDI-Reihe 9 Elektronik/Mikro- und Nanotechnik, Nr. 320 (VDI-Verlag Düsseldorf), 2000.

26. B. Schey,Laserdeposition großflachiger $YBa_2Cu_3O_{7-x}$-Schichten auf einkristallinen und technischen Substraten mit Textur-Untersuchungen zur ionenunterstutzten Laserablation von Ceroxid, Ph.D. thesis, Universität Augsburg (Shaker, Aachen), 1998.

27. B. Worz,Wachstum biaxial texturierter NiO-Schichten auf chemisch behandelten Ni-5%W-RABiTS und gepulste Laserdeposition epitaktischer Schichten für die Herstellung supraleitender $YBa_2Cu_3O_{7-\delta}$—Bänder, Ph.D. thesis, Universität Augsburg (Mensch & Buch Verlag Berlin 2004), 2003.

28. J. Hohage,Synthese von Bornitridschichten mittels PLD-Verfahren, Ph.D. thesis, Technische Universität Dresden, 2004.

29. S. Six,Synthese epitaktischer Aluminiumoxidschichten mittels ionenstrahlgestützter Laserablation, Ph.D. thesis, Universität Augsburg (Mensch & Buch Verlag Berlin), 2001.

30. A. Gorbunoff,Laser-Assisted Synthesis of

Nanostructured Materials, Habilitation thesis, TU Dresden, Fortschritt-Berichte VDI-Reihe 9 Elektronik/ Mikro- und Nanotechnik, Nr. 357 (VDI-Verlag Düsseldorf), 2002.

31. L. Roussak,Herstellung und Charakterisierung von epitaktischen $2(ZnD_x(CuInD_2)_{1-x}$—Hetero- und Doppelheteroschichtstrukturen für D = S, Se und Te mittels Pulsed Laser Deposition (PLD), Ph.D. thesis, Universität Leipzig, 2005.

32. S. Braun,Gefüge- und Grenzflächenbeschaffenheit von Mo/Si-Multischichten, synthetisiert mittels Puls-Laser- und Magnetron-Sputter-Deposition, Ph.D. thesis, Universität Bielefeld, 2004.

33. U. Hannemann,Wachstum, Mikrostruktur und hartmagnetische Eigenschaften von Nd-Fe-B-Schichten, Dissertation, Technische Universität Dresden, 2004.

34. T. Thärigen,Makroskopische Eigenschaften und Mikrostruktur von Kohlenstoff-Dünnfilmen aus Laser-Plasma- Reaktionen von Graphit mit Stickstoff-Radikalen, Siliciumnitrid oder Bornitrid, Ph.D. thesis, Universität Leipzig, 2000.

35. Y. A. P. Mercado,Diamond-like carbon and ceramic materials as protective coatings grown by pulsed laser deposition, Ph.D. thesis, RWTH Aachen, 2003.

36. A. Sewing,Präparation von Ni-C-Multischichten und Mischsystemen mit dem PLD-Zweistrahlverfahren und Untersuchung der thermischen Stabilität der Schichtsysteme, Ph.D. thesis, Technische Universität Dresden, 2002.

37. C. Jin,Growth and Characterization of ZnO- and ZnO-based alloys $Mg_xZn_{1-x}O$ and $Mn_xZn_{1-x}O$, Ph.D. thesis, North Carolina State University, Raleigh, 2003.

38. T. Nobis,Beobachtung und Modellierung des optischen Flustergallerie-Effekts in hexagonalen Zinkoxid-Nanoresonatoren, Ph.D. thesis, Universität Leipzig, 2006.

39. A. Rahm,Pulsed laser deposition growth and characterization of ZnO-based nanostructures, Ph.D. thesis, Universität Leipzig, 2007.

40. C. Czekalla,Ortsaufgelöste Lumineszenz von Zinkoxid-Mikronadeln, Ph.D. thesis, Universität Leipzig, 2009.

41. M. Brandt,Influence of the electric polarization on carrier transport and recombination dynamics in ZnO-based heterostructures,, Ph.D. thesis, Universität Leipzig, 2010.

42. M. S. Tillack, D. Blair, and S. S. Harilal, "The effect of ionization on cluster formation in laser ablation plumes," Report UCSD-ENG-103, University of California, San Diego, Calif, USA, August 2003.

43. M. Lorenz,Gepulste Laser-Plasmaabscheidung (PLD) von oxidischen Dünnfilm- und Heterostrukturen, Habilitation thesis, Universität Leipzig, 2008.

44. M. Lorenz, E. M. Kaidashev, and E. M. Kaidashev, "Optical and electrical properties of epitaxial(Mg,Cd) xZn1-xO, ZnO, and ZnO:(Ga,Al) thin films on c-plane sapphire grown by pulsed laser deposition,"Solid-State Electronics, vol. 47, no. 12, pp. 2205–2209, 2003.·

45. J. Narayan and B. C. Larson, "Domain epitaxy: a unified paradigm for thin film growth,"Journal of Applied Physics, vol. 93, no. 1, pp. 278–285, 2003.·

46. M. Lorenz, H. Hochmuth, and D. Natusch, "Waferheizer für Saphirsubstrate, Deutsche Patentanmeldung," Az. DE 102 55 453.7 vom 28.11.2002 (über Robert BOSCH GmbH Stuttgart).

47. S. Ohashi, M. Lippmaa, N. Nakagawa, H. Nagasawa, H. Koinuma, and M. Kawasaki, "Compact laser molecular beam epitaxy system using laser heating of substrate for oxide film growth,"Review of Scientific Instruments, vol. 70, p. 178, 1999.

48. M. Lorenz, A. Rahm, and A. Rahm, "Self-organized growth of ZnO-based nano-and microstructures,"Physica Status Solidi B, vol. 247, no. 6, pp. 1265–1281, 2010.·

49. C. Grüner,Untersuchung des PLD-Laserplasmas mittels Optischer Emissionsspektroskopie, M.S. thesis, University Leipzig, Institute for Experimental Physics II, 2009.

50. M. Benamara, Z. Liliental-Weber, and Z. Liliental-Weber, "The role of the multi buffer layer technique on the structural quality of GaN,"MRS Internet Journal of Nitride Semiconductor Research, vol. 5S1, W5.8, 2000.

51. Y. Ralchenko, A. E. Kramida, J. Render, and the NIST ASD Team, "NIST Atomic Spectra Database version 3.1.5.," National Institute of Standards and Technology, Gaithersburg, Md, USA, September 2009,http://physics.nist.gov/asd3.

52. J. E. Sansonetti and W. C. Martin,Handbook of Basic Atomic Spectroscopic Data, 2000.

53. D. R. Lide,Handbook of Chemistry and Physics, CRC Press, Boca Raton, Fla, USA, 73rd edition, 1993.

54. M. Lorenz, M. Brandt, G. Wagner, H. Hochmuth, G. Zimmermann, H. von Wenckstern, and M. Grundmann, "MgZnO:P homoepitaxy by pulsed laser deposition: pseudomorphic layer-by-layer growth and high electron mobility," inZinc Oxide Materials and Devices IV, vol. 7217 ofProceedings of SPIE, San Jose, Calif, USA, January 2009.·

55. M. Lorenz, H. Hochmuth, H. Börner, D. Natusch, and K. Kreher, "Large area pulsed laser deposition of YBCO thin films and buffer layers on 3-inch wafers," inMaterials Research Society Symposium Proceedings, vol. 341 ofEpitaxial Oxide Thin Films and Heterostructures, pp. 189–194, San Francisco, Calif, USA, April 1994.

56. H. Hochmuth and M. Lorenz, "Inductive determination of the critical current density of superconducting thin films without lateral structuring,"Physica C, vol. 220, no. 1-2, pp. 209–214, 1994.

57. G. Kästner, D. Hesse, M. Lorenz, R. Scholz, N. D. Zakharov, and P. Kopperschmidt, "Microcracks observed in epitaxial thin films ofYBa2Cu3O7−δandGdBa2Cu3O7−δ,"Physica Status Solidi A, vol. 150, no. 1, pp. 381–394, 1995.·

58. H. Hochmuth and M. Lorenz, "Side-selective and non-destructive determination of the critical current density of double-sided superconducting thin films,"Physica C, vol. 265, no. 3-4, pp. 335–340, 1996.

59. M. Lorenz, H. Hochmuth, D. Natusch, H. Börner, G.

Lippold, K. Kreher, and W. Schmitz, "Large-area double-side pulsed laser deposition ofYBa2Cu3O7−xthin films on 3-in. sapphire wafers,"Applied Physics Letters, vol. 68, no. 23, pp. 3332–3334, 1996.·

60. M. Lorenz, H. Hochmuth, and H. Hochmuth, "Large-area and double-sided pulsed laser deposition of Y-Ba-Cu-O thin films applied to HTSC microwave devices,"IEEE Transactions on Applied Superconductivity, vol. 7, no. 2, pp. 1240–1243, 1997.

61. M. Lorenz, H. Hochmuth, D. Natusch, and K. Kreher, "Highly reproducible large-area and double-sided pulsed laser deposition of HTSC YBCO:Ag thin films for microwave applications,"Applied Physics A, vol. 69, no. 7, pp. S905–S911, 1999.·

62. M. Lorenz, H. Hochmuth, M. Grundmann, E. Gaganidze, and J. Halbritter, "Microwave properties of epitaxial large-area Ca-dopedYBa2Cu3O7−δthin films onr-plane sapphire,"Solid-State Electronics, vol. 47, no. 12, pp. 2183–2186, 2003.·

63. T. Kaiser,Nichtlineare Hochfrequenzeigenschaften von Hochtemperatursupraleiter-Filmen, Ph.D. thesis, Bergische Universität Gesamthochschule Wuppertal, 1998.

64. R. Semerad, B. Utz, P. Berberich, W. Prusseit, and H. Kinder, "Coevaporation of YBaCuO films up to 9-inches diameter," inApplied Superconductivity, Inst. Phys. Conf. Ser. No. 148, p. 847, IOP, Bristol, UK, 1995.

65. J. Geerk, A. Zaitsev, and A. Zaitsev, "A 3-chamber deposition system for the simultaneous double-sided coating of 5-inch wafers,"IEEE Transactions on Applied Superconductivity, vol. 11, no. 1, pp. 3856–3858, 2001.·

66. Y. Lemaître, D. Mansart, B. Marcilhac, J. C. Mage, and J. Siejka, "Double-sided sputtering deposition ofYBa2Cu3O7−δthin films on2″LaAlO3wafers for microwave applications,"IEEE Transactions on Applied Superconductivity, vol. 9, no. 2, pp. 2363–2366, 1999.·

67. D. W. Face, F. M. Pellicone, R. J. Small, L. Bao, M. S. Warrington, and C. Wilker, "Tl2Ba2CaCu2O8andYBa2Cu3O7films on large area MgO and sapphire substrates for high power microwave and rf applications,"IEEE Transactions on Applied Superconductivity, vol. 9, no. 2, pp. 2492–2495, 1999.

68. M. Klauda, T. Kässer, and B. Mayer, "Superconductors and cryogenics for future communication systems,"IEEE Transactions on Microwave Theory and Techniques, vol. 48, no. 7, pp. 1227–1239, 2000.·

69. T. Kässer, M. Viertel, R. Bölter, C. Neumann, and F. Schnell, "Superconductors and cryotechnology for space communications,"Physica C, vol. 372–376, no. 1, pp. 489–492, 2002.·

70. R. Schwab, R. Heidinger, J. Geerk, F. Ratzel, M. Lorenz, and H. Hochmuth, "Surface impedance mapping at cryogenic temperatures of HTSC wafers at 145 GHz," inProceedings of the 23th International Conference on Infrared and Millimeter Waves, Essex, UK, September 1998.

71. R. Schwab,Ortsaufgelöste Untersuchung des Oberflachenwiderstandes von epitaktischen $YBa_2Cu_3O_{7-x}$- Dünnschichten mit Millimeterwellen, Dissertation, FZ Karlsruhe/Universität Karlsruhe, 1999.

72. H. Frenzel, A. Lajn, H. von Wenckstern, et al., "Recent progress on ZnO-based metal-semiconductor field-effect transistors and their application in transparent integrated circuits,"Advanced Materials.··

73. H. Frenzel, A. Lajn, H. von Wenckstern, and M. Grundmann, "Ultrathin gate-contacts for metal-semiconductor field-effect transistor devices: an alternative approach in transparent electronics,"Journal of Applied Physics, vol. 107, no. 11, Article ID 114515, 2010.·

74. B. K. Meyer, H. Alves, and H. Alves, "Bound exciton and donor-acceptor pair recombinations in ZnO,"Physica Status Solidi B, vol. 241, no. 2, pp. 231–260, 2004.·

75. H. von Wenckstern, S. Weinhold, G. Biehne, R. Pickenhain, H. Schmidt, H. Hochmuth, and M. Grundmann, "Donor levels in ZnO,"Advances in Solid State Physics, vol. 45, pp. 263–274, 2005.

76. M. Brandt, M. Lange, and M. Lange, "Control of interface abruptness of polar MgZnO/ZnO quantum wells grown by pulsed laser deposition,"Applied Physics Letters, vol. 97, no. 5, Article ID 052101, 2010.·

77. S. Heitsch, G. Zimmermann, and G. Zimmermann, "Interface and luminescence properties of pulsed laser depositedMgxZn1−xO/ZnO quantum wells with strong confinement," inMaterials Research Society Symposium Proceedings, vol. 957, pp. 229–234, Boston, Mass, USA, November 2007.

78. C. Sturm, H. Hilmer, R. Schmidt-Grund, and M. Grundmann, "Observation of strong exciton-photon coupling at temperatures up to 410 K,"New Journal of Physics, vol. 11, Article ID 073044, 2009.·

79. H. Hilmer, C. Sturm, R. Schmidt-Grund, B. Rheinländer, and M. Grundmann, "Observation of strong light-matter coupling by spectroscopic ellipsometry,"Superlattices & Microstructures, vol. 47, no. 1, pp. 19–23, 2010.·

80. R. Schmidt-Grund, H. Hilmer, and H. Hilmer, "Two-dimensional confined photonic wire resonators—strong light-matter coupling,"Physica Status Solidi B, vol. 247, no. 6, pp. 1351–1364, 2010.·

81. C. Bundesmann, N. Ashkenov, and N. Ashkenov, "Raman scattering in ZnO thin films doped with Fe, Sb, Al, Ga, and Li,"Applied Physics Letters, vol. 83, no. 10, pp. 1974–1976, 2003.·

82. Q. Xu, L. Hartmann, and L. Hartmann, "Metal-insulator transition in Co-doped ZnO: magnetotransport properties,"Physical Review B, vol. 73, no. 20, Article ID 205342, 2006.·

83. Q.Xu,L.Hartmann,andL.Hartmann,"Magnetoresistance in pulsed laser deposited 3d transition metal doped ZnO films,"Thin Solid Films, vol. 515, no. 4, pp. 2549–2554, 2006.·

84. H. Schmidt, M. Diaconu, and M. Diaconu, "Electrical and optical spectroscopy on ZnO:Co thin films,"Applied Physics A, vol. 88, no. 1, pp. 157–160, 2007.·

85. M. Ungureanu, H. Schmidt, and H. Schmidt, "A comparison between ZnO films doped with 3d and 4f magnetic ions,"Thin Solid Films, vol. 515, no. 24, pp.

8761–8763, 2007.·

86. H. von Wenckstern, G. Benndorf, and G. Benndorf, “Properties of phosphorus doped ZnO,”Applied Physics A, vol. 88, no. 1, pp. 125–128, 2007.·

87. C. Bundesmann, A. Rahm, M. Lorenz, M. Grundmann, and M. Schübert, “Infrared optical properties ofMgxZn1−xOthin films (0≤x≤1): long-wavelength optical phonons and dielectric constants,”Journal of Applied Physics, vol. 99, no. 11, Article ID 113504, 2006.·

88. R. Schmidt-Grund, A. Carstens, and A. Carstens, “Refractive indices and band-gap properties of rocksaltMgxZn1−xO(0.68≤x≤1),”Journal of Applied Physics, vol. 99, no. 12, Article ID 123701, 2006.·

89. S. Heitsch, G. Zimmermann, and G. Zimmermann, “Luminescence and surface properties ofMgxZn1−xOthin films grown by pulsed laser deposition,”Journal of Applied Physics, vol. 101, no. 8, Article ID 083521, 2007.·

90. Q. Xu, H. Schmidt, and H. Schmidt, “Room temperature ferromagnetism in Nd- and Mn-codoped ZnO films,”Journal of Physics D, vol. 41, no. 10, Article ID 105012, 2008.·

91. H. von Wenckstern, H. Schmidt, and H. Schmidt, “Anionic and cationic substitution in ZnO,”Progress in Solid State Chemistry, vol. 37, no. 2-3, pp. 153–172, 2009.·

92. M. Lange, C. P. Dietrich, C. Czekalla, et al., “Luminescence properties ofZnO/Zn1−xCdxO/ZnOdouble heterostructures,”Journal of Applied Physics, vol. 107, no. 9, Article ID 093530, 8 pages, 2010.

93. D. Spemann, E. M. Kaidashev, M. Lorenz, J. Vogt, and T. Butz, “Ion beam analysis of epitaxial(Mg,Cd)xZn1−xOand ZnO:(Li, Al, Ga, Sb) thin films grown onc-plane sapphire,”Nuclear Instruments and Methods in Physics Research, Section B, vol. 219-220, no. 1-4, pp. 891–896, 2004.·

94. M. Khalid, M. Ziese, and M. Ziese, “Defect-induced magnetic order in pure ZnO films,”Physical Review B, vol. 80, no. 3, Article ID 035331, 2009.·

95. R. Waser, Ed.,Nanoelectronics and Information Technology—Advanced Electronic Materials and Novel Devices, Wiley-VCH, Weinheim, Germany, 2003.

96. J. F. Scott, “Applications of modern ferroelectrics,”Science, vol. 315, no. 5814, pp. 954–959, 2007.··

97. M. Lorenz, H. Hochmuth, M. Schallner, R. Heidinger, D. Spemann, and M. Grundmann, “Dielectric properties of Fe-dopedBaxSr1−xTiO3thin films on polycrystalline substrates at temperatures between -35 and +85∘C,”Solid-State Electronics, vol. 47, no. 12, pp. 2199–2203, 2003.·

98. L. C. Sengupta and S. Sengupta, “Electrical properties of doped BSTO-0.6 thin films,”Materials Research Innovations, vol. 2, p. 278, 1999.

99. S. S. Gevorgian and E. L. Kollberg, “Do we really need ferroelectrics in paraelectric phase only in electrically controlled microwave devices?”IEEE Transactions on Microwave Theory and Techniques, vol. 49, no. 11, pp. 2117–2124, 2001.·

100. R. R. Romanofsky, et al., “616 element phased array antenna,” inMaterials Research Society Symposium Proceedings, vol. 720, H 4.1.1, 2002.

101. K. J. Choi, M. Biegalski, and M. Biegalski, “Enhancement of ferroelectricity in strainedBaTiO3thin films,”Science, vol. 306, no. 5698, pp. 1005–1009, 2004.··

102. E. M. Kaidashev, M. Lorenz, and M. Lorenz, “High electron mobility of epitaxial ZnO thin films on c-plane sapphire grown by multistep pulsed-laser deposition,”Applied Physics Letters, vol. 82, no. 22, pp. 3901–3903, 2003.·

103. H. von Wenckstern, H. Schmidt, and H. Schmidt, “Homoepitaxy of ZnO by pulsed-laser deposition,”Physica Status Solidi, vol. 1, no. 4, pp. 129–131, 2007.·

104. M. Lorenz, G. Wagner, and G. Wagner, “Homoepitaxial ZnO thin films by PLD: structural properties,”Physica Status Solidi C, vol. 5, no. 10, pp. 3280–3287, 2008.·

105. M. Brandt, H. von Wenckstern, and H. von Wenckstern, “High electron mobility of phosphorous-doped homoepitaxial ZnO thin films grown by pulsed-laser deposition,”Journal of Applied Physics, vol. 104, no. 1, Article ID 013708, 2008.·

106. H. von Wenckstern, M. Brandt, and M. Brandt, “Properties of homoepitaxial ZnO and ZnO:P thin films grown by pulsed-laser deposition,” inZinc Oxide Materials and Devices III, vol. 6895 ofProceedings of SPIE, San Jose, Calif, USA, January 2008.·

107. M. Brandt, H. von Wenckstern, C. Dietrich, et al., “Dopant activation in homoepitaxial MgZnO:P thin films,”Journal of Vacuum Science & Technology B, vol. 27, p. 1604, 2009.

108. Y. W. Heo, S. J. Park, K. Ip, S. J. Pearton, and D. P. Norton, “Transport properties of phosphorus-doped ZnO thin films,”Applied Physics Letters, vol. 83, no. 6, pp. 1128–1130, 2003.·

109. S. J. Pearton, D. P. Norton, K. Ip, Y. W. Heo, and T. Steiner, “Recent progress in processing and properties of ZnO,”Progress in Materials Science, vol. 50, no. 3, pp. 293–340, 2005.·

110. H. von Wenckstern, G. Benndorf, and G. Benndorf, “Properties of phosphorus doped ZnO,”Applied Physics A, vol. 88, no. 1, pp. 125–128, 2007.·

111. M. Brandt, H. von Wenckstern, G. Benndorf, H. Hochmuth, M. Lorenz, and M. Grundmann, “Formation of a two-dimensional electron gas in ZnO/MgZnO single heterostructures and quantum wells,”Thin Solid Films, vol. 518, no. 4, pp. 1048–1052, 2009.·

112. M. Lorenz, M. Brandt, M. Lange, G. Benndorf, H. von Wenckstern, D. Klimm, and M. Grundmann, “HomoepitaxialMgxZn1−xO(0≤x≤0.22) thin films grown by pulsed laser deposition,”Thin Solid Films, vol. 518, no. 16, pp. 4623–4629, 2010.·

113. J. A. Davis and C. Jagadish, “Ultrafast spectroscopy of ZnO/ZnMgO quantum wells,”Laser and Photonics Reviews, vol. 3, no. 1-2, pp. 85–96, 2009.·

114. T. M. Shaw, A. Gupta, M. Y. Chern, P. E. Batson, R. B. Laibowitz, and B. A. Scott, “Atomic scale oxide

superlattices grown by RHEED controlled pulsed laser deposition,"Journal of Materials Research, vol. 9, no. 10, pp. 2566–2573, 1994.

115. T. Matsumoto, H. Tanaka, K. Kouguchi, T. Kawai, and S. Kawai, "A scanning tunneling microscopy study of laser molecular beam epitaxy onSrTiO3(100) surface,"Surface Science, vol. 312, no. 1-2, pp. 21–30, 1994.

116. D. H. A. Blank, G. J. H. M. Rijnders, G. Koster, and H. Rogalla, "In-situ monitoring during pulsed laser deposition using RHEED at high pressure,"Applied Surface Science, vol. 127–129, pp. 633–638, 1998.

117. G. Koster,Artificially layered oxides by Pulsed Laser Deposition, Ph.D. thesis, Universiteit Twente, Enschede, The Netherlands, 1999.

118. J. Klein,Epitaktische Heterostrukturen aus dotierten Manganaten, Ph.D. thesis, Universität Köln, 2001.

119. H. Y. Hwang, "Atomic control of the electronic structure at complex oxide heterointerfaces,"MRS Bulletin, vol. 31, no. 1, pp. 28–35, 2006.

CHAPTER 8 BROADBAND ELECTROMAGNETIC EMISSION FROM PZT FERROELECTRIC CERAMICS AFTER SHOCK LOADING

Fiodoras Anisimovas, Steponas Ašmontas, Oleg Kiprijanovič, Andrius Maneikis, Bonifacas Vengalis

INTRODUCTION

Nowadays attention is paid to investigations of materials with complex internal structure. The investigations include analysis of their improved mechanical properties considering scale and nonlinear effects [1 – 3]. Sometimes these materials are expected to possess necessary properties in relation to certain electromagnetic (EM) and chemical phenomena. In ferroelectric ceramics and composite dielectrics mechanical-electrical and mechanical-EM transformations after shock loading are also investigated for these purposes [4, 5].

Pulsed mechanical loading of piezoelectric crystals [6], layered and composite dielectric materials [7, 8], induce the pulsed EM radiation. It was found that duration of the EM pulses varies up from microseconds down to nanoseconds, which hints at differences in origins of such generations. Generation of broad band signals and their application, including medical, is of great interest. Investigation of super wideband EM radiation using antennas is difficult, especially of single pulses. Wideband antennas and wideband oscilloscopes are needed for the investigation. Now creation of super wideband antennas and wideband digitized oscilloscopes makes the investigations easier [9]. Both theoretical studies [10] and experimental research [11] keep uncovering new types of super wideband signals. Previously our experimental investigation of EM radiation after pulsed compressive stress of lead zirconate titanate (PZT) ferroelectric ceramic cylinders was carried out using wideband television antennas [12]. It was found series of short EM pulses (30 ns – 50 ns) spanning ~1 μs. The pulses in this series have no indication of period or constant value of amplitude. It was proposed that the stress leads to appearing of high voltage electric pulse series. The electrical pulses have ns duration and during their fast rising fronts the EM pulses are generated. The processes that linked the stress and the generation of electrical pulse series were not sufficiently explained.

Piezoelectric ignition mechanisms (PIM) with PZT ceramics are used for high voltage generation, which works on shock loading for example in piezoelectric cigarette lighters. The aim of this work is to register pulsed EM radiation during action of PIM in the range from 1 GHz up to 80 GHz using different horn antennas with coaxial or waveguide outputs. We also analyze cleaved PZT samples under SEM and try to clarify what causes the series of electrical pulses to appear.

EXPERIMENTAL

Conventional PIM with PZT cylindrical components (cylinders) of 4 mm high and 2 mm in diameter is used in experiments. Such cylinders are fabricated as follows. Firstly PZT powder is calcined at temperatures (1000 – 1300)°C for (1 – 2) h until the required PZT phase is acquired. Then the samples are shaped using pressing or hot-pressing methods. Different piezoelectric properties are imparted depending on how long the pressure is applied at maximal temperature [13]. It is worth noting that during growth of PZT crystallites impurities concentrate in crystallite boundaries whose thickness increases causing changes of their electrical properties. Key problem is in controlling the properties of the boundaries during the fabrication.

Crystalline structure of the PZT cylinders was examined by X-ray diffractometer Bruker AXS D8 using CuKα radiation. We should investigate the PZT ceramics internal structure to establish if it is linked with EM pulse series generation along with its intenced function. In this work there were investigated surfaces formed after cleaving the cylinders along its axes.

Citation: Fiodoras anisimovas, Steponas ašmontas, Oleg kiprijanovič, Andrius maneikis, Bonifacas vengalis, Broadband Electromagnetic Emission from PZT Ferroelectric Ceramics after Shock Loading, doi.org/10.5755/j01.ms.19.4.3137.

To investigate the surfaces we employed scanning electron microscope (SEM) Hitachi 3000. The backscattered electrons are detected to form an image of the surfaces. Such procedure also allows determining composition of surface material.

Schematic diagram of high voltage block of the PIM is presented in Fig. 1. There are two PZT cylinders (1) of 130 mg each with Ag metalized bases (2) and high voltage electrode (3) placed between them. They are rested against a metallic foundation of 800 mg (6). On another cylinder base there is a metallic disk with a bulge of 20 mg (5) which is being hit by metallic hammer of 300 mg (4). The hammer is accelerated by spring which has stiffness 2.75·103 N/m. Energy stored in the spring before the shock is equal to 34·10–3 J. The purpose of the disk is to align shock force evenly on the cylinder base. Both cylinders are in two-layer insulated cases: inner layer is plastic (7) and outer layer is epoxy (8). Under the action of a spring (not shown) the hammer (4) hits the disk (5) and high voltage discharge through the air between (3) and (6) is initiated. The length of the discharge spark channel reaches 6 mm.

Voltage dynamics induced by the hammer are investigated inserting high voltage resistive divider 1 : 470 between the electrode (3) and foundation (6). Schematic diagram of this investigation including oscilloscope having 1 MΩ input and (0 – 100) MHz bandwidth corresponds to the one presented in [12]. Appearing voltage is divided and reduced signal is registered by the oscilloscope.

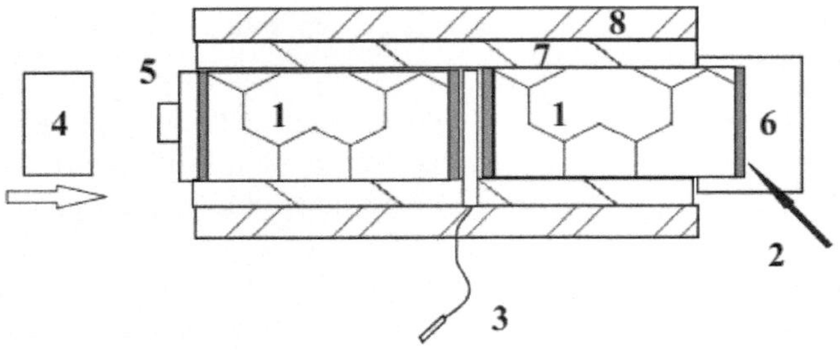

Figure 1: Schematic diagram of high voltage block of the PIM. 1 – PZT cylinder, 2 – cilinder base metallization, 3 – high voltage electrode, 4 – hammer, 5 – disk, 6 – foundation, 7 – plastic insulator, 8 – epoxy insulator

The EM emission of PIM after the shock loading was investigated using wideband antennas and digital and real time oscilloscopes. Schematic diagram of the experimental arrangement for EM pulses registration is presented in Fig. 2. The EM emission registration occurs in a following way. The PIM (1) is placed at small distance from the horn antenna aperture (2) and is excited by pressure. We have used the pyramidal horn with coaxial output. The horn has aperture (34 × 26 cm2) and (1 – 18) GHz operational frequency range. The EM emission is collected by the horn and electrical signal is then transmitted directly to wideband digital oscilloscope (4) by 50 Ω impedance cable. Since the signal was powerful enough, to protect sensitive, 1 V max, oscilloscope input we had to include the 10 dB attenuation (3).

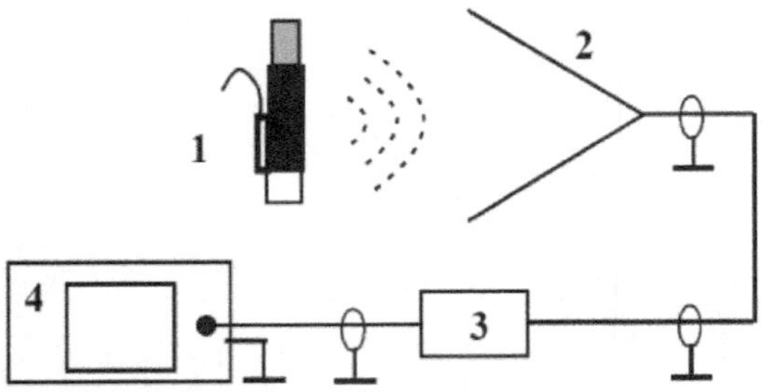

Figure 2: Schematic diagram of experimental set-up for EM radiation registration using digital oscilloscope. 1 – PIM, 2 – horn antenna operating in (1 – 18) GHz frequency range, 3 – 10 dB attenuator, 4 – digital oscilloscope of (0 – 6) GHz bandwidth

Figure 3: Surface images of PZT cylinder cleaved along its axis obtained by SEM. Full vertical scales are 200 μm (above) and 30 μm (below)

Detection of the higher than 18 GHz frequency harmonics was realized using a number of different horn antennas with waveguide outputs. In these measurements it was used (0 – 7) GHz bandwidth real time oscilloscope. Schematic diagram of experimental set-up is structurally similar to that as in Fig. 2. Difference between the diagrams is that now synchronization channel is added. The signal is branched by a tee and a delay line is placed before the oscilloscope input (see in Ref. [12]). The horn waveguides are loaded by detector heads. Signals from the heads are amplified in (0.8 – 2.4) GHz range and transmitted to the oscilloscope input using short cables. The frequency range can be broadened employing waveguide-to-waveguide transition to one with smaller cross section

There were employed following horns in the investigations. Horn with aperture (10 × 8 cm^2) with waveguide output of (11 × 5.5) mm^2 cross section corresponding to (17.44 – 25.95) GHz range; horn with aperture (5 × 4.5 cm^2) with output waveguide 7.2 × 3.4 mm^2 cross section corresponding to (25.95 – 37.5) GHz range; horn with aperture (3.5 × 3 cm^2) with waveguide output of 3.6 × 1.8 mm^2 cross section corresponding to (53.6 – 78.3) GHz range. We have used transition from 7.2 × 3.4 mm^2 to 5.2 × 2.6 mm^2 waveguide which corresponds to (37.5 – 53.6) GHz range.

EXPERIMENTAL RESULTS

Information obtained from X-ray diffraction patterns is essentially the same as presented in Ref. [12]. The cylinders are $PbZr_{0.52}Ti_{0.48}O_3$ compound with tetragonal symmetry and some of the samples had more developed (111) preferred orientation. The Zr: Ti proportion of this composition, near morfotropic transition Zr: Ti = 55: 45, indicates on the PZT ceramics having high piezoelectric effect. Composition analysis using SEM shows somewhat distorted results because of the presence of carbon and additional oxygen at the surface of the cleaved samples

The experiments on high voltage dynamics when discharge is initiated through high ohmic resistive divider show results, which mainly repeat those obtained in [12]. The experiments reveal that the shock induces voltage growth, which reaches its maximum after (1 – 3) μs. Discharge dynamics has zero sign inversion point, which appears in time range from 4 μs to 5 μs. Afterwards in span of (10 – 15) μs voltage decreases, and reaches negative minimum comparable by absolute value with positive one. However important part of the response at the first μs could not be observed mainly because of divider bandwidth. After reaching negative minimum the voltage has lengthy relaxation part of ms order.

Such behaviour of the voltage-time dependence curve with a sharp drop from positive to negative values is observed when investigating surface charge density changes in ferroelectrics. Except in our case at the beginning of the relaxation part there is series of relatively high voltage spikes of microsecond duration with decreasing amplitude (see Fig. 2 in Ref. 12).

The PZT cylinder was split along its axis and the split surface images made by SEM can be seen in Fig. 3. There is chaotic distribution of crystallites having different sizes and forms. This is clearly seen on both pictures despite them having different magnification. The distances between the crystallites boundaries differ from tightly merged to containing nanometer size air filled openings. Crystallites of large size, higher than 2μm indicate on calcining at 1300 °C for 1 h following by non-hot pressing.

Using broadband digital oscilloscope reveals additional results during registration of EM pulse series.

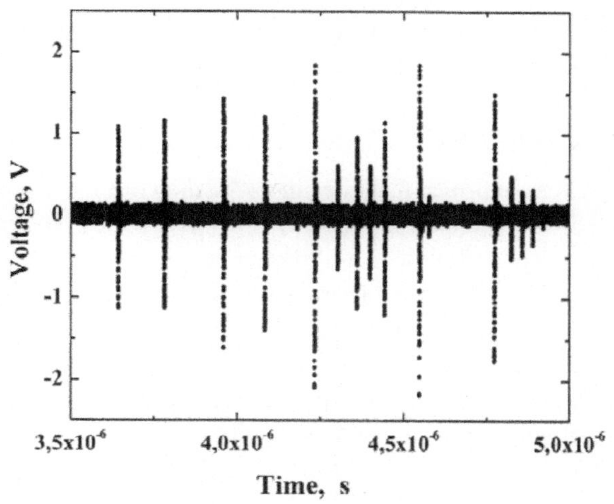

Figure 4: Oscillogram of the half of EM pulse series obtained employing the (1 – 18) GHz horn and digital oscilloscope with (0 – 6) GHz bandwidth. Zero time corresponds to the shock of the hammer

Half of one of the registered series is presented in Fig. 4 where every pulse appears as dashed vertical line. There is a delay between the moment of shock, which is also accompanied by pulse or several pulses, and the first series pulse. The pulse induced by the shock is zero moment of the oscillograms and is not displayed in Fig. 4. The delay ranges from 1.2μs to 4μs depending on the sample. Series duration is longer than 1μs as it was proposed in [12] and can be up to 4μs. The delay is comparable with the time needed for voltage to reach the maximum.

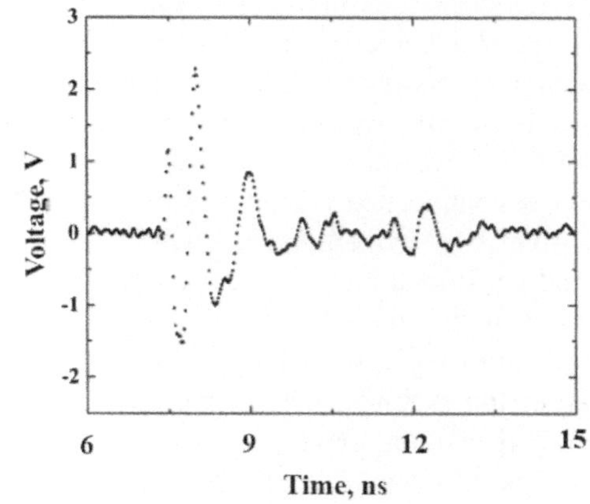

Figure 5: Typical stretched EM pulse from the presented in Fig. 4 series. Higher harmonics are cut by (0 – 6) GHz bandwidth oscilloscope

The oscilloscope allows us to stretch every pulse of the series. They are short wideband pulses, which differ mainly in amplitude. The typical stretched pulse is presented in Fig. 5 and they have durations of (3 – 5) ns. However the form of this pulse is slightly distorted because higher than 6 GHz harmonics are cut off by both the (0 – 6) GHz bandwidth oscilloscope and the cable. Such pulse received on wideband TV antenna with upper frequency about 1 GHz has duration of (30 – 50) ns as shown in [12].

This is an example of oscillogram change depending on the receiving antenna operational bandwidth. Oscillogram of the several pulses from the series detected by detector head of (37.5 – 53.6) GHz range is presented in Fig. 6. The forms of spikes on oscillogram differ depending on semiconductor type in the heads. They can be negative and stretched in time because of their sluggishness.

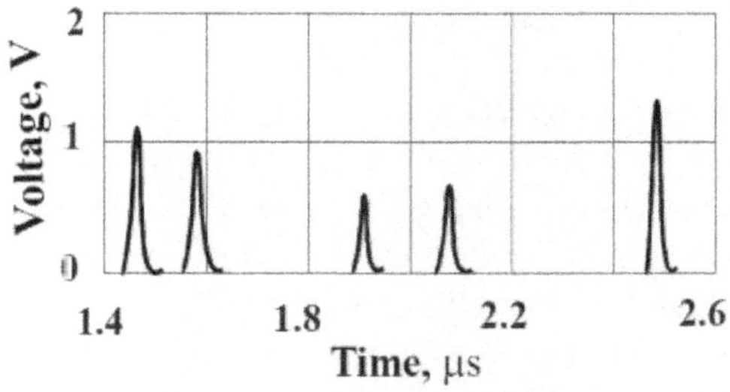

Figure 6: Typical oscillogram of the part of EM pulse series obtained employing the waveguide horn antennas loaded by detector heads and amplification in (0.8 – 2.4) GHz range

DISCUSSION

So, usage of the broadband digital oscilloscope allows for more extensive measurements of the EM series. It is followed from Fig. 4 that there is 3.6μs delay, which can be (1 – 4)μs in other cases, between the hammer shock and the first pulse of the EM series. The correct evaluation of the EM series duration was not possible in [12] because the fact of this delay was not known. The experiments reveal that series duration is longer and is at least (3 – 4) μs instead of ~1 μs.

The EM pulse series with its complex structure has certain features indicating on re-reflections of pulses. Sound speed in the crystallite structure with boundaries has relatively small value. This means that applied hammer shock with a pulse of 4.5 kg·m/s can induce deformation wave and even shock wave if the sound speed is low enough. Such pulses can be soliton-like deformation waves appearing after the hammer shock [14], and electrical pulse, which causes EM pulse, is formed during reflection of this wave from the cylinder base. Clarifying of the situation is complicated because overlapping EM pulses generated from two cylinders are undistinguishable. The waves should be formed after the shock from the two bases, one of them near the disk and another near the foundation. However the hammer shock is relatively long, comparing with ps laser pulse, for example, and interaction of the cylinder with the PIM foundation allows us to propose that cylinders were compressed as a whole rather than in a way which allows forming the soliton-like deformation waves.

It is worth to mention that the high voltage spikes can be understood as a result of elastic oscillations of the cylindrical sample during relaxation. It would mean that mechanical behaviour of the cylinders during the oscillations is as if they were solid.

As one can see from Fig. 3 the cylinder consists of varied form and size crystallites each having a surface and invisible internal domains with their own walls. Electrical properties of the crystallite content and its surface are different. There are Schottky barrier-like surface regions and surface levels, which play important role in polarization screening [15]. When we also take into account switching of the domains and movement of their walls it becomes clear how complex the cylinder structure is from electrical point of view. Processes and their response times and relaxation time constants in such media at applied electric field vary in duration from 10–12 s to ~10–2 s. But high dielectric constant of PZT, which is changeable during repolarization, and entwined ferroelectric and piezoelectric properties could lead to fast, ps or ns, unforeseen processes.

Series of EM pulses can be generated due to generation of electrical pulses of ns durations with fast rising fronts during surface charge changes at the cylinder bases. Therefore it is possible sudden significant weakening of polarization and then returning to its initial state. Polarization switching and accompanying processes including electron emission and shielding are investigated in variety of works [16 – 19]. Investigators in the mentioned works used configuration with application of external electrical pulses to induce the repolarization, following by the polarization switching. In this work we met the problem of fast polarization switching while clarifying the EM radiation origin.

In our case self-repolarization process begins at the maximal voltage value during dynamic changes of cylinder polarity. The velocity of voltage inversion is of the order of 20 kV in 15μs, which means that self-repolarization is quite fast indeed. The EM series beginning coincides with the start of self-repolarization process and a spark discharge in external circuit occurs after the start. We do not, however, exclude possibility that the series begins before the voltage maximum is reached, which indicates that fast reversible polarization switching occurs during the rapid voltage changes. Most probably we have experimental confirmation of the reversible metastable process including fast polarization switching and electron emission [20]. The polarization recovery to the initial state became possible due to differences in fast ferroelectric response times and slower adjusting times of surface state electrons. This process is also important in organizing an efficient emission process. We suppose that in our case fast reversible polarization switching occurs in regime of selfexcitation and is accompanied by electrical pulse series. Our observed EM radiation occurred during these fast polarization changes.

Harmonics above 20 GHz point on the presence of electron emission during recovering and their deceleration, and therefore on the EM radiation. However we have no experimental confirmation whether the spikes in Fig. 6 correspond to the every pulse whose lower frequency harmonics registered and shown in Fig. 4 or to the every second.

Most of the classic models of polarization switching are based on the so called Kolmogorov-Avrami-Ishibashi approach, which considers the switching process to occur in two basic stages: nucleation and growth of reversed domains. This approach cannot accurately describe the fast switching in our experiment because of nonhomogeneous PZT ceramics with complex structure and entwined ferroelectric-piezoelectric properties. In modern theories such media is considered an ensemble of many regions with independent switching dynamics, governed exclusively by a local electric field [20, 21]. If it is true, the self-generation of fast polarization switching mostly appear randomly and, possibly, could have different amplitude and durations of up to hundreds of nanoseconds. This is partially confirmed in our experiments as seen from Fig. 4 where the EM pulses of different amplitudes are spread randomly.

The mean value of amplitude of EM pulse series should depend on density of surface states in crystallites and on domain structure both of which can be regulated during manufacturing. Therefore using different PZT ceramics in devices with shock loading allows generating complex broadband signals which can be amplified and exploited in various fields including medical applications [22].

CONCLUSIONS

- It is confirmed that after shock loading of PIM cylinders it is generated broadband EM pulse series. Usage of broadband digital oscilloscope uncovered that the first pulse of the series appears after a delay and the duration of the series reaches 4 μs.
- The series pulse is extremely broadband and its form on the oscillogram depends on antenna operational frequency band and on bandwidth of oscilloscope.
- Ferroelectric PZT cylinder has complex structure with interacting subsystems and therefore its reaction on external disturbances is characterized by different response times and by relaxation times constants of huge time range. This facilitates processes with

screening effects which damping fast electric field changes. Applying temporary mechanical structure displacements in such media induces not only piezoeffect but also can cause unforeseen fast, ps or ns, processes.

- It is explained that unforeseen presence of such broadband EM emission suggests presence of self-exited generation of fast reversible polarization switching which is accompanied by generation of electrical pulse series of ns duration which begins at voltage maximum and lasts about 4 μs. 5) Presence of EM harmonics in the pulse up to 80 GHz indicates on deceleration processes of emitted electrons during fast polarization switching

REFERENCES

1. Mehta, A., Barker, G., Luck, J.-M. Heterogeneities in Granular Materials Physics Today May 2009: pp. 40 – 44.
2. Goriacheva, I. P., Morozov, N. F. Mechanics – Science of XXI century Vestnik of Russian Academy of Science 82(7) 2012: pp. 638 – 647 (in Russian).
3. Dimitrienko, Ju., Sokolov, A. Numerical Modeling of Composites with Multiscale Microstructure Bulletin of the Russian Academy of Sciences. Physics 75 (11) 2011: pp. 1457 – 1461.
4. Rouquette, J., Hines, J., Bornard, V., Pintard, M., Papet, Ph., Bousquet, C., Konczewicz, L., Gorelli, F., Hull, S.
5. Pressure Tuning of the Morphotropic Phase Boundary in Piezoelectric Lead Zirconate Titanate Physical Review B 70 2004: 014108 p.
6. Millett, J., Bourne, N., Deas, D. The Response of Two Ferroectric Ceramics to One-Dimensional Shock Loading Journal of Physics D: Applied Physics 45 2012: 165203 p.
7. Sud'enkov, Yu. V. Electromagnetic Radiation Induced by the Failure of Piezoelectrics under the Action of Submicrosecond Stress Pulses Technical Physics 46 (12) 2001: pp. 1588 – 1590.
8. Fursa, T., Surzhikov, A., Horsov, N., Osipov, K. Interrelation between the Stressed State in Composites and the Parameters of the Electromagnetic Response to a Mechanical Impulse Technical Physics 51 (4) 2006: pp. 519 – 521.
9. Surzhikov, A., Fursa. T. Mechanoelectrical Transformations upon the Elastic Impact Excitation of Composite Dielectric Materials Technical Physics 53 (4) 2008: pp. 462 – 465.
10. Reid, A., Judd, M. Ultra-wide Bandwidth Measurements of Partial Discharge Current Pulses in SF6 Journal of Physics D: Applied Physics 40 2007: pp. 2948 – 2953.
11. Kravchenko, V., Lazorenko, O., Pustovoit, V., Chernogor, L. F. A New Class of Fractal Ultra-Wideband
12. Signals Doklady Physics 52 (3) 2007: pp. 129 – 133. http://dx.doi.org/10.1134/S1028335807030019
13. Anisimovas, F., Ašmontas, S., Kiprijanovič, O., Matuzas, J. Radiation of Large Bandwidth and Long Duration Signals by Pulsed Excitation of Monopole Antennas Proc. Intern. Conf. "Radiation Interaction with Material and its Use in Technologies" May 14 – 17, Kaunas, Lithuania Kaunas, Technologija, 2012: pp. 574 – 577.
14. Ašmontas, S., Anisimovas, F., Dapkus, L., Gradauskas, J., Kiprijanovič, O., Senulis, M. Electromagnetic Radiation of Ferroelectric PZT Ceramics under Pulsed Compressive Stress Materials Science (Medžiagotyra) 16 (3) 2010: pp. 264 – 267.
15. Okazaki, K. Ceramics Engineering for Dielectrics. Energija, 1976 (in Russian). Kudriavtsev, E., Lotkova, E., Zotov, S., Autric, M., Emel'yanov, V., Rogacheva, A. Temperature Dependence of the Velocity of a Slow Soliton Wave of Changes in Optical Reflection and Conductivity Excited by a Laser Pulse in Solids Laser Physics 10 (3) 2000: pp. 733 – 740.
16. Fridkin, V. Ferroelectrics-Semiconductors. Nauka, Moscow, 1976 (in Russian). Zhukov, S., Genenko, Yu., Hirsch, O., Glaum, J., Granzow, T., Seggern, H. Dynamics of Polarization Reversal in Virgin and Fatigued Ferroelectric Ceramics by Inhomogeneous Field Mechanism Physical Review B 82 2010: 014109 p.
17. Rabkin, L. M., Ivanov, V. N. Electron Energy in Exoelectronic Emission from a Ferroelectric Technical Physics Letters 24 (7) 1998: pp. 565 – 566.
18. Sadykov, S., Borodin, V., Agalarov, A. Reversible Polarization of Ferroelectric Ceramics in a Rapid-Growing Electric Field Technical Physics 45 (6) 2000: pp. 776 – 779. http://dx.doi.org/10.1134/1.1259719
19. Rezinkin, O., Axelsson, A., Vytrishko, V. Test Bench for Investigating the Pulsed Polarization Dynamics of Nonlinear Dielectrics Instruments and Experimental Techniques 53 (5) 2010: pp.755 – 760.
20. Pavlov, A. N., Raevski, I. P., Sakhnenko, V. P. Electron Emission during Pulsed Polarization Switching of Ferroelectric Ceramics Technical Physics 44 (7) 1999: pp. 782 – 785.
21. Tagantsev, A., Stolichnov, I., Setter, N. NonKolmogorov-Avrami Switching Kinetics in Ferroelectric Thin Films Physical Review B 66 2002: 214109 p.
22. Isogawa, T., Kumashiro, T., Song, Ho-Jin, Ajito, K., Kukutsu, N., Iwatsuki, K., Nagatsuma, T. Tomographic Imaging Using Photonically Generated Low-Coherence Terahertz Noise Sources IEEE Transactions on Terahertz Science and Technology 2 (5) 2012: pp. 485 – 492.

CHAPTER 9

POLYMER-STABILIZED FERROELECTRIC LIQUID CRYSTALS

Ingo Dierking

Keywords: *Liquid crystal, polymer network, ferroelectric, antiferroelectric, photopolymerization, material properties*

INTRODUCTION

Liquid Crystals [1–3]

Liquid crystals are anisotropic fluids, which are thermodynamically located between the isotropic liquid and the three-dimensional crystal. Two broad classes of liquid crystals are distinguished, lyotropic phases [4,5], which are formed by variation of the concentration of amphiphilic molecules in a suitable solvent, and thermotropic phases, which are observed by temperature variation. The former class of liquid crystals will be disregarded in this review, while the latter is further distinguished by the molecular shape of the constituent molecules. Calamitic phases are formed by cylinder-like mesogens, discotic phases by disk-shaped molecules, and bent-core phases [6,7] by so called banana liquid crystals.

Numerous liquid crystal phases can be observed, depending on the order of the constituent mesogens [8]. The simplest of the phases is the nematic phase, which solely exhibits an average orientational order of the long axes of the molecules (called the director **n**), while their centres of mass are isotropically distributed (Figure 1a). The nematic phase is also the one observed at the highest temperature. On cooling, fluid smectic phases are formed, which in addition to orientational order also exhibit a one-dimensional positional ordering, *i.e.*, a layered structure. Within a smectic layer the centres of mass of the mesogens is again isotropically distributed. For the smectic A (SmA) phase the layer normal **k** is parallel to the director **n**, while in the smectic C (SmC) phase it is inclined by a temperature dependent angle called the tilt angle, θ (Figure 1b,c, respectively). Further liquid crystal phases are the hexatic phases, which exhibit short and long range two-dimensional positional order, *i.e.*, an additional feature of order within individual smectic layers.

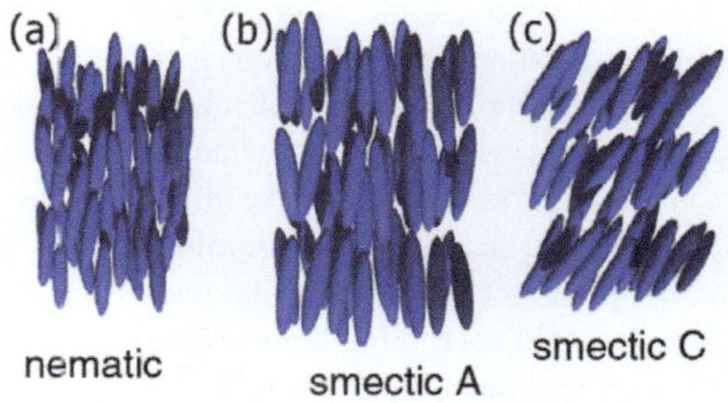

Figure 1. Schematic representation of the (**a**) nematic; (**b**) smectic A (SmA) and (**c**) smectic C (SmC) phase of calamitic (rod-like) mesogens. Molecular order increases on cooling from the left to the right side of the diagram.

In order to identify different liquid crystalline phases, a variety of experimental techniques is generally employed, among them differential scanning calorimetry (DSC), small angle and wide angle X-ray diffraction (SAXS, WAXS), and texture observation by polarizing optical microscopy (POM) [8]. An example of the latter is shown in Figure 2, where the characteristic defects allow a verification of the phase under investigation, in this case point singularities of a nematic phase (Figure 2a, Schlieren texture), Dupin cyclides of a fluid smectic A phase (Figure 2b, focal conic texture), and the broken fan-shaped texture of SmC (Figure 2c).

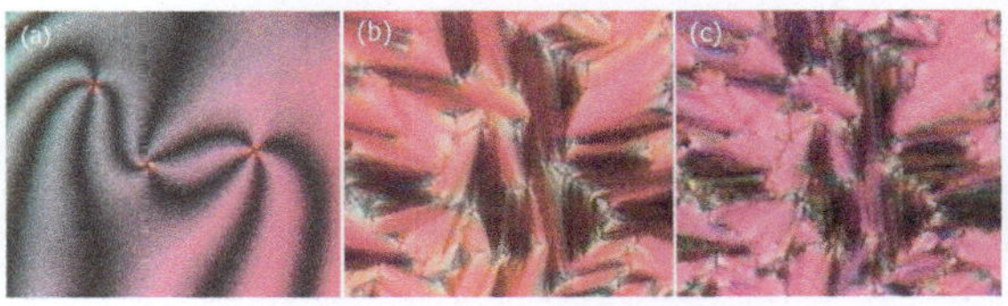

Figure 2. Characteristic polarizing microscopic defect textures of

Citation: Ingo Dierking; Polymer-Stabilized Ferroelectric Liquid Crystals, doi:10.3390/ma7053568.

(**a**) the nematic Schlieren texture; (**b**) SmA focal conic and (**c**) broken focal conic texture of the SmC phase. (Reproduced with permission from [8]. Copyright 2003 Wiley-VCH Verlag GmbH&Co. KGaA).

Ferroelectric Liquid Crystals [9,10]

Introduction of chirality or "handedness" to liquid crystalline materials leads to exciting changes of the phases. Novel structures are observed, like helical superstructures, novel phases, the so called frustrated phases (Blue Phases [11], Twist Grain Boundary phases [12]), which only occur in chiral systems, are formed, and novel effects can be seen. The latter are, for example, selective reflection leading to photonic bandgaps [13], or extremely fast electro-optic modulation via the electroclinic effect [14,15]. However, the most pronounced observation is that of ferroelectricity in fluid systems [16,17]. Indeed, ferroelectric liquid crystals (FLC) are the only fluid ferroelectrics known to man.

The occurrence of a spontaneous polarization in the SmC* phase can easily be understood by following the symmetry arguments of Meyer [16]. He deduced that all chiral titled smectic phases exhibit a local spontaneous polarization and are thus pyroelectric. In the SmC* phase, as in some others, this polarization is switchable between two stable states, thus ferroelectric. As depicted inFigure 3a, the structure of the achiral SmC phase contains three symmetry elements, namely a two-fold rotation axis, a mirror plane and therefore a resultant inversion centre. Application of those local symmetry elements to a molecular dipole moment $\mu = (\mu_x, \mu_y, \mu_z)$ gives $\mu_i = (-\mu_x, -\mu_y, -\mu_z)$, giving a net dipole moment of the achiral SmC phase of $\mu_{SmC} = (0, 0, 0)$, a vanishing spontaneous polarization. For chiral molecules the mirror plane and the inversion centre are lacking and a dipole moment $\mu = (\mu_x, \mu_y, \mu_z)$, which is subjected to the only remaining symmetry element, the two-fold rotation axis, becomes $\mu_{rot} = (-\mu_x, -\mu_y, -\mu_z)$ and the resultant dipole moment of the chiral SmC* phase is $\mu_{SmC}{}^* = (0, \mu_y, 0)$. The spontaneous polarization is the sum of the non-vanishing lateral dipole moment components per unit volume. Reversal of an applied electric field switches the spontaneous polarization between two stable states, and the smectic C* structure is ferroelectric.

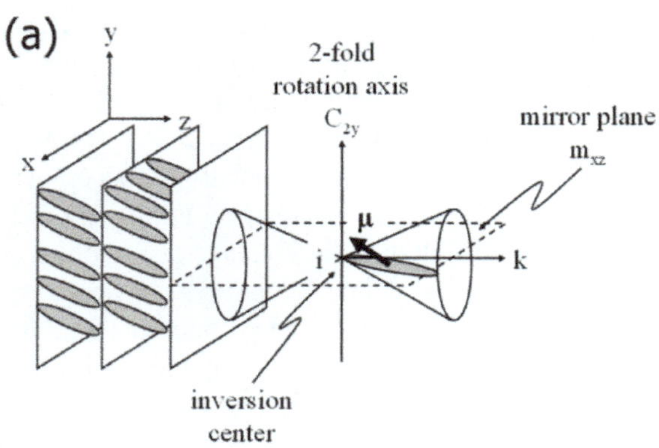

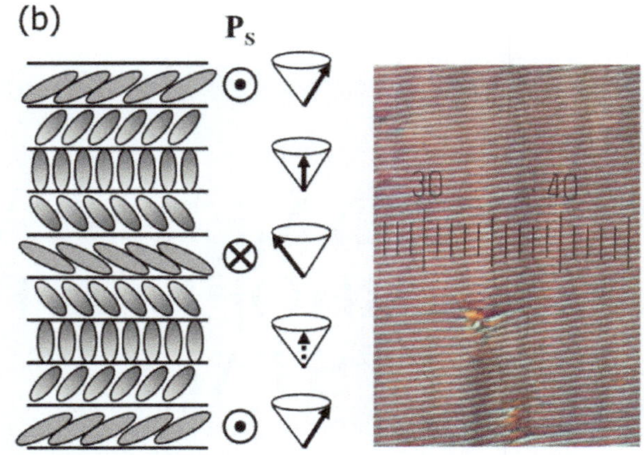

Figure 3. (**a**) Symmetry elements of the achiral SmC phase, explaining the absence of a spontaneous polarization; (**b**) schematic helical superstructure of the chiral SmC* phase (**left**) and associated polarized microscopic image of the equidistant line pattern (**right**). (Reproduced with permission from [8]. Copyright 2003, Wiley-VCH Verlag GmbH&Co. KGaA).

This is not the whole truth though. For a bulk sample the spontaneous polarization compensates by forming a helical superstructure, and the SmC* phase is more correctly called helielectric. The helical structure of SmC* manifests itself in polarizing microscopy through the appearance of an equidistant line pattern, as shown in the inset of Figure 3b Subjecting the SmC* phase to thin cells with strong surface interactions and suppressing the helical superstructure leads to the so called surface stabilized ferroelectric liquid crystal (SSFLC) [17]. This state is indeed ferroelectric and from a virgin sample a domain structure is observed, which represents the two ferroelectric domains of polarization up and polarization down. The latter is shown in Figure 3b.

Polymer Stabilization [18,19]

Polymer stabilisation of liquid crystals refers to a method first used on nematic and cholesteric phases [20–24] in the development of reflective displays, or electronic paper [25–31]. It is now widely used also for other liquid crystal phases and purposes. A small amount, generally less than 10% by weight, of a photoreactive monomer is mixed into the liquid crystalline phase, while care has to be taken to stay below the solubility limit of the monomer. These photoreactive monomers can be liquid crystalline by themselves, but do not need to be, as long as they exhibit a similar elongated shape and form anisotropy as the liquid crystal host molecules. A minute amount of a photoinitiator, often benzoin methyl ether (BME) is also added to the mixture. The latter will not be incorporated into the actual structure, but is simply present as a catalyst to facilitate the chemical reaction, which will take place when the mixture is subjected to UV illumination.

The mixture is filled into suitable cells, which promote a certain desired zero electric field alignment. Since the monomers exhibit the same shape anisotropy as the liquid crystal molecules, they will align along the local director of the liquid crystal. UV illumination then causes an open polymer network to be formed by the photoreactive monomers. This polymer network will follow the local director field in which it was formed, thus acting as a template of the liquid crystal structure [32,33] (Figure 4). The network is phase separated from the liquid crystal, and the interaction between both is thus of elastic nature. The liquid crystal may then be forced out of its equilibrium orientation for example by electric or magnetic fields, and on turning the outside stimuli off, the polymer network drives the liquid crystal back to its original director configuration. Other scenarios may be envisioned. The liquid crystal can be washed out by a suitable solvent and be replaced by a different material. A left handed helical polymer network structure may be filled with a right handed cholesteric material of the same or a different pitch [34]. Polymerisation can be carried out in the orthogonal SmA phase, and the transition into the tilted SmC phase can be investigated, a scenario which we will come back to below. In all cases the fundamental idea is the same: a polymer network is formed which stabilises the liquid crystal director configuration in which it was formed. Elastic interactions between the large surface of the polymer network and the liquid crystal will aim to drive the system back into its equilibrium orientation.

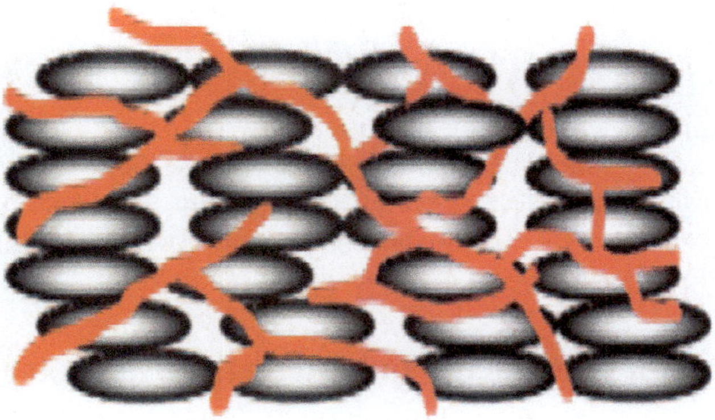

Figure 4. Schematic of a polymer stabilized liquid crystal. The polymer network acts as a template of the self organised liquid crystalline phase that it was formed in.

A few examples of polymer networks formed in different liquid crystals are presented in Figure 5 [33,35–37]. The images are taken by scanning electron microscopy, after coating the polymer network with a thin layer of gold. In part (a) a s = +1 point defect is shown (left), which clearly represents the expected director field from a defect of a Schlieren texture [35] (right) as introduced in Figure 2a. Figure 5b demonstrates the helical superstructure of the cholesteric or chiral nematic phase by an oblique cut through the helix from the top to the bottom substrate plates. The pitch was adjusted to 10 μm so that three half turns of the helix are visible in the 15 μm thick cell [33] (left). The imaged structure is equivalent to a so called Bouligand cut (right), which is used to demonstrate helical superstructures in biological systems. Similarly, the individual smectic blocks of a twist grain boundary (TGB) phase can be imaged [36] (Figure 5c left), including the discontinuous twist which is mediated via boundaries of screw dislocations (right). Also such structures can be found in biological systems. In the SmA phase the polymer network is well oriented along the director of a uniform sample (Figure 5d left) [36]. It can be seen that the individual features of the polymer strands are on average about 0.1 μm in size (Figure 5d right). The size of the voids depends on various preparation conditions, such as monomer concentration, polymerisation temperature, and UV dose [29]. In general, the polymer network formation process is completed within about 10–15 min.

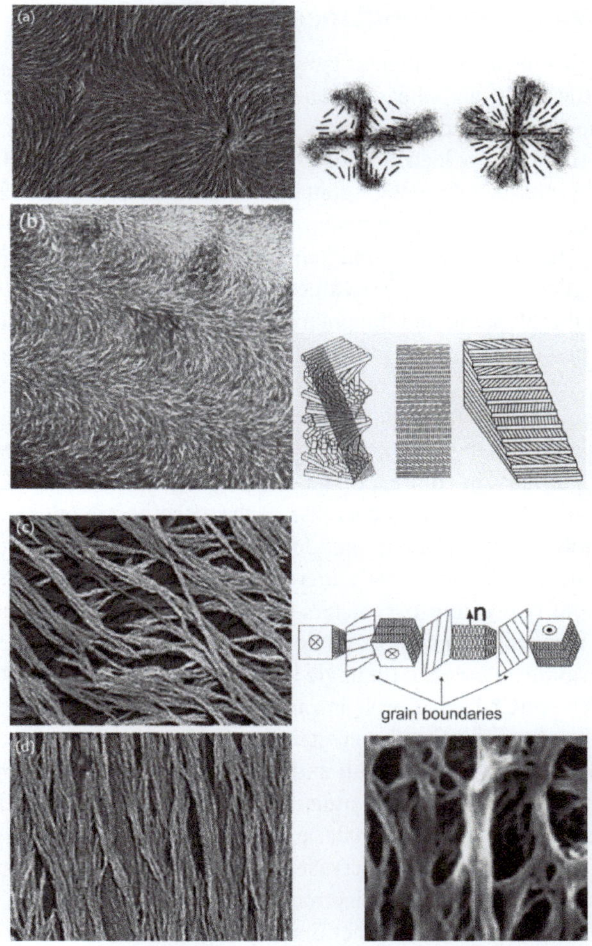

Figure 5. Exemplary scanning electron microscopy (SEM) images of polymer networks formed in different liquid crystalline phases: (**a**) nematic s = ±1 defect pair (**top**) and schematic director field (**bottom**); image side length ~50 μm (Reproduced with permission from [35]. Copyright 2013 Royal Society of Chemistry); (**b**) helical superstructure of the chiral nematic, N* phase (**top**), which can be illustrated by an oblique cut called a Bouligand cut (**bottom**); image side length ~50 μm (Reproduced with permission from [33]. Copyright 1997 American Physical Society); (**c**) different director orientation in SmA* twist grain boundary (TGB) blocks (**top**) and schematic representation (**bottom**); image side length ~5 μm. (Reproduced with permission from [36]. Copyright 2009 Royal Society of Chemistry) and (**d**) well oriented SmA phase (**left**) and close-up (**right**); image side length ~5 μm. The material used is the well known RM257 from Merck, and the liquid crystal has been removed by a suitable solvent after polymerization. For the SEM investigations the polymer network was then coated with a very thin layer of gold.

EFFECTS ON SMC MATERIAL PARAMETERS

The first polymer stabilized ferroelectric liquid crystal (PSFLC) was made by Hikmet and Lub [38] from Philips in 1995, in an effort to overcome the mechanical problems of surface stabilized ferroelectric liquid crystals. Samples were polymerized in the SmC* phase and an oriented polymer network along the smectic layer normal was observed. It was found [39] that polymer network stabilization has a considerable influence on the electro-optic performance of FLCs and the physical parameters as compared to the non-stabilized sample.

Tilt Angle and Spontaneous Polarization

The two most fundamental parameters of the SmC* phase are the tilt angle and the spontaneous polarization. The former is the primary order parameter for the transition between the non-tilted, paraelectric SmA* and the tilted, ferroelectric SmC* phase, while the spontaneous polarization represents the secondary order parameter of the transition, as it is coupled to the tilt angle. While maintaining their general temperature dependence of increasing values for decreasing temperatures, both the tilt angle and the polarization decrease for increasing polymer concentration [40–42]. This is demonstrated in Figure 6a,b, respectively. The reason for this behaviour can be found in regions in the vicinity of the network that are strongly dominated by the polymer, thus not switching at all, or at least only partially, in contrast to the saturated switching of the bulk regions. There is thus a decrease of the effective tilt angle and the polarization. The polymer dominated regions become more and more pronounced with increasing polymer concentration, and the two prime parameters of the SmC* phase decrease for increasing polymer content. The electroclinic effect, describing an induced tilt angle in the very close vicinity of the SmA*–SmC* transition is in our case practically not affected by the process of polymer stabilization, most likely due to the fact that it is very small and only observed over a narrow temperature regime. The situation appears to be different for large electroclinic effect materials, as shown by Petit *et al.* [43] in Figure 7. Here the increase in polymer network density causes a decrease of the electroclinically induced tilt angle, as would be expected. The fact that the respective investigations were carried out on short-pitch FLCs in contrast to the normally investigated surface stabilized FLCs may also play a role for the interpretation of these results.

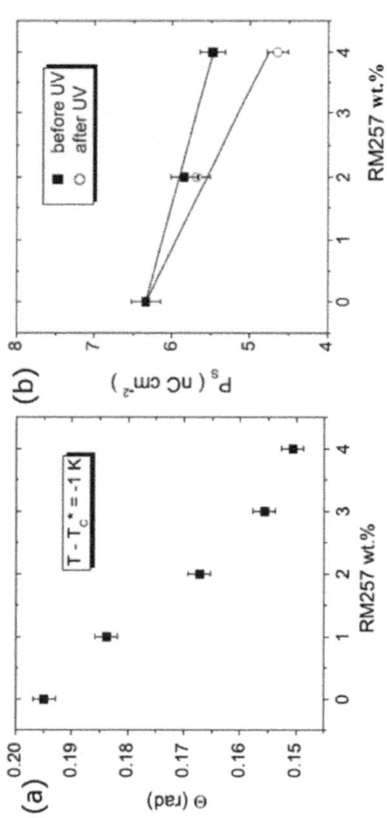

Figure 6. Effect of the polymer network on the principle order parameters of the SmA*–SmC* transition, (**a**) tilt angle (Reproduced with permission from [42]. Copyright 2008, Institute of Physics); and (**b**) spontaneous polarization (Reproduced with permission from [44]. Copyright 2009 Institute of Physics). Both quantities decrease for increasing polymer network concentration.

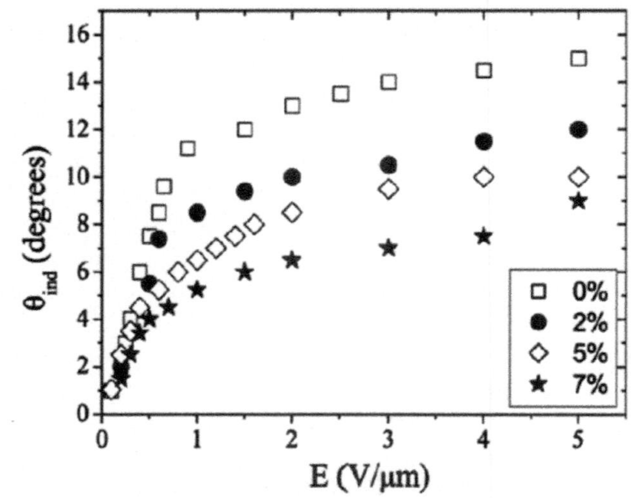

Figure 7. Effect of the polymer network on the electroclinically induced tilt angle in the vicinity of the SmA*–SmC* transition. The electroclinic effect is reduced for increasing polymer content. (Reproduced with permission from [43]. Copyright 2008 Taylor & Francis).

Landau Coefficients and Interaction Parameter

In physical terms, this behaviour can be described by a generalized Landau model of the SmA* to SmC* transition [45,46], which is extended by an additional factor, describing the interaction between the liquid crystal and the polymer network. In a non-helical FLC device subjected to an electric field, E, the difference in the free energy density between the SmA* and SmC* phase, g–g_0, is given by:

$$g-g_0=\frac{1}{2}\alpha(T-T_C)\Theta^2+\frac{1}{4}b\Theta^4+\frac{1}{6}c\Theta^6+\frac{P^2}{2\varepsilon_0\chi_0}-C\Theta P-\frac{\Omega P^2\Theta^2}{2}-PE \tag{1}$$

where Θ is the tilt angle, *P* is the total polarisation, $a = \alpha(T-T_C)$, *b* and *c* are the first three Landau expansion coefficients, C is the bilinear coupling coefficient, χ_0 is the high frequency dielectric susceptibility in the direction of the electric field, Ω is the biquadratic coupling coefficient and T_C is the transition temperature of the related achiral SmA to SmC transition. It should be noted that the Landau description is only valid in the vicinity of the phase transition. Nevertheless, this is the region where the significant changes are observed. For PSFLC systems an additional term is required in the generalised Landau model to take into account the interaction between the polymer network and the liquid crystal. Employing the simple model of Li *et al.* [47] the interaction between the polymer network and the liquid crystal acts to restore the local liquid crystal director to the orientation of the polymer network. They introduced an elastic coupling interaction term, 12WPsin2Θ≈ 12WPΘ212WPsin2Θ≈12WPΘ2, where W_P is the interaction coefficient between the polymer network and the liquid crystal. The free energy density then reads as:

$$g-g_0=\frac{1}{2}\alpha(T-T_C)\Theta^2+\frac{1}{2}W_P\Theta^2+\frac{1}{4}b\Theta^4+\frac{1}{6}c\Theta^6+\frac{P^2}{2\varepsilon_0\chi_0}-C\Theta P-\frac{\Omega P^2\Theta^2}{2}-PE \tag{2}$$

Minimisation of the free energy density (Equation (2)) with respect to the total polarisation P leads to the relation [48,49]:

$$P = \frac{C\Theta + E}{\frac{1}{\varepsilon_0 \chi_0} - \Omega\Theta^2} \tag{3}$$

Simultaneous fitting of the total polarisation as a function of tilt angle with respect to varying applied electric field amplitude allows the determination of C, χ_0 and Ω. Minimisation of Equation (2) with respect to Θ and resubstitution of Equation (3) leads to a temperature-tilt relationship of:

$$T(\Theta, E) = T_C - \frac{1}{\alpha}\left[W_P + b\Theta^2 + c\Theta^4 - \frac{(C\Theta + E)\left(\frac{C}{\varepsilon_0\chi_0} + \Omega\Theta E\right)}{\Theta\left(\frac{1}{\varepsilon_0\chi_0} - \Omega\Theta^2\right)^2}\right] \tag{4}$$

For $T_C \approx T_C^*$ Equation (4) can be directly employed to determine α, b, c and W_P via a second set of simultaneously fitted curves $T(\Theta,E)$, using the previously determined parameters C, χ_0 and Ω. One finds that the Landau coefficients are largely unaffected by the introduction of the polymer network, while the elastic coupling coefficient linearly increases for increasing polymer concentration [42,50] (Figure 8). A more thorough analysis, taking into account the physical dimensions of the network shows, that the elastic coupling coefficient depends on polymer concentration and the penetration depth of the elastic forces into the bulk of the liquid crystal. A linear increase of the elastic coupling coefficient with values of the same order of magnitude was also reported by Petit *et al.* [51].

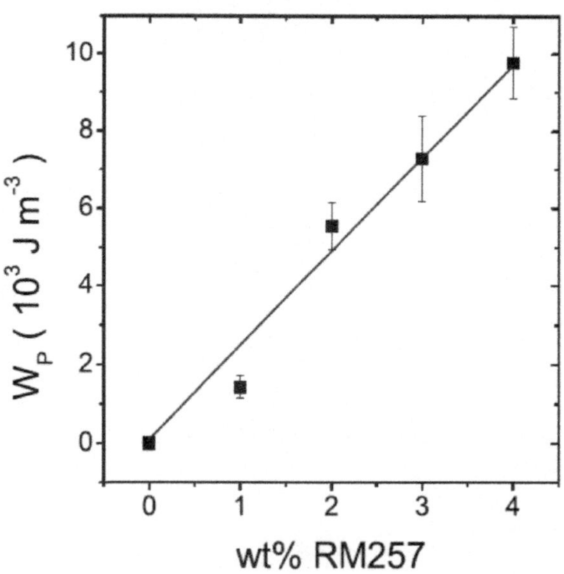

Figure 8. For small concentrations the elastic coupling coefficient between polymer network and liquid crystal increases linearly with increasing network density. (Reproduced with permission from [50]. Copyright 2008 American Physical Society).

Response Times and Effective Viscosity

In the ferroelectric SmC* phase the response times exponentially increase for decreasing temperature, which is a plain viscosity effect when observed at equal applied voltages. This behaviour is not changed for PSFLCs. It is generally found that mixing the monomer into the FLC material increases the response times the more monomer is added. This effect is most likely due to the relatively large size of the monomers, despite themselves being liquid crystalline over a certain range of temperature, with the monomer increasing the viscosity of the FLC-monomer mixtures. Subsequent UV polymerization of the monomers and formation of the polymer networks then reduces the effective viscosity and the response times of the PSFLC become shorter, even shorter than those observed for the neat FLC material [44]. It is further found that after polymerization the response times decrease for increasing polymer network concentration, as depicted in Figure 9a. Since the response time τ is directly proportional to the effective viscosity η via $\tau = \eta/P_S E$, the equivalent trend is also observed for the viscosity as a function of polymer network content (Figure 9b). Such behaviour has been reported by several authors [52–55], but it should be pointed out, that results of the opposite behaviour have also been published, *i.e.*, an increase in viscosity with increasing polymer concentration [56].

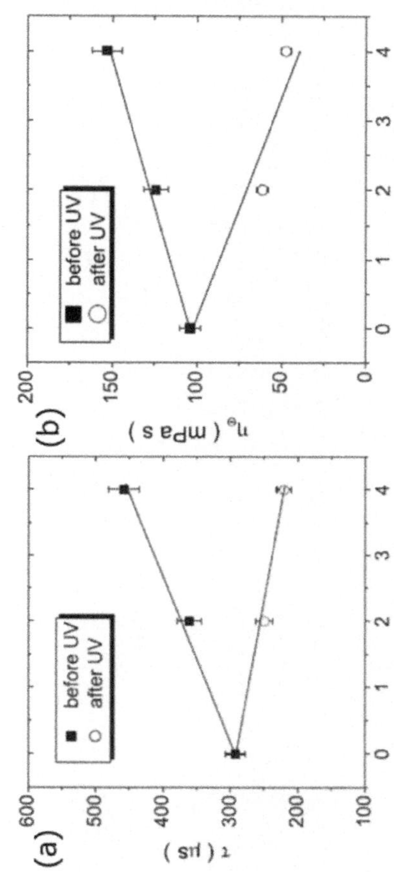

Figure 9. (**a**) The response time after formation of the polymer network decreases for increasing network density and (**b**) for the effective viscosity a comparable trend is observed. (Reproduced with permission from [44]. Copyright 2009 Institute of Physics).

Dielectric Spectroscopy

The ferroelectric SmC* phase exhibits two major collective dielectric modes, the Goldstone mode and the softmode. The former is related to tilt fluctuations on the cone, while the latter is due to tilt fluctuations changing the value of the tilt angle. It should be noted that the effect of a polymer network on the softmode is in general relatively small [57], while pronounced effects are observed on the Goldstone mode. We will thus focus our discussion on the Goldstone mode, and the collective fluctuations of the director on the tilt cone [58]. Dielectric spectroscopic data is often illustrated in a frequency independent plot of the dielectric absorption ε″ as a function of the dielectric permittivity ε′, the so called Cole-Cole plot, as it is shown in Figure 10 for different polymer network concentrations. Already here it can be seen that an increase of polymer network density suppresses the Goldstone mode, or more precisely, it decreases the dielectric strength ($\Delta\varepsilon = \varepsilon_0 - \varepsilon_\infty$) for all temperatures in the SmC* phase (Figure 11a), while the relaxation frequency ν_R is increased (Figure 11b). This general behaviour has been reported by numerous authors [53,57–60] and seems to be noncontroversial. It can be explained by the reduction of collective fluctuations due to an increasing density of polymer network, and thus an increase of elastic interactions between liquid crystal and network. The effect of the polymer network is equivalent to that of an application of an electric bias field [58]. The dielectric strength decreases for increasing polymer concentration, because more and more liquid crystal molecules are elastically coupled to the polymer, reducing their collective fluctuations. The relaxation frequency is proportional to the effective elastic constant K_{eff} and inversely proportional to the effective viscosity, $\nu_R \sim K_{eff}/\eta_{eff}$, and thus displays a behaviour as expected with increasing network content.

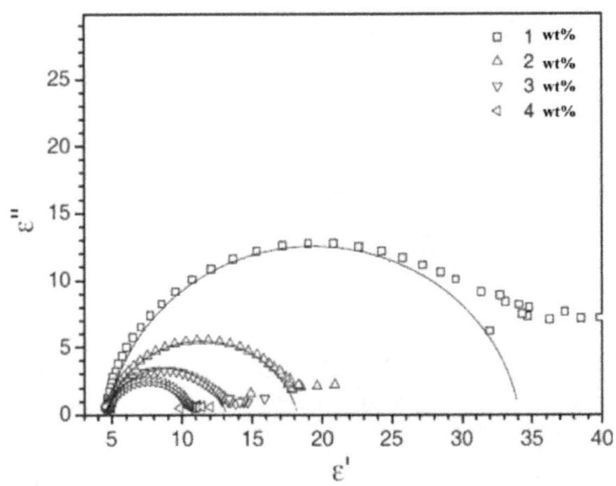

Figure 10. Cole-Cole plot, ε″ *versus* ε′, of the dielectric properties for varying polymer network concentration. (Reproduced with permission from [58]. Copyright 2009 EDP Sciences).

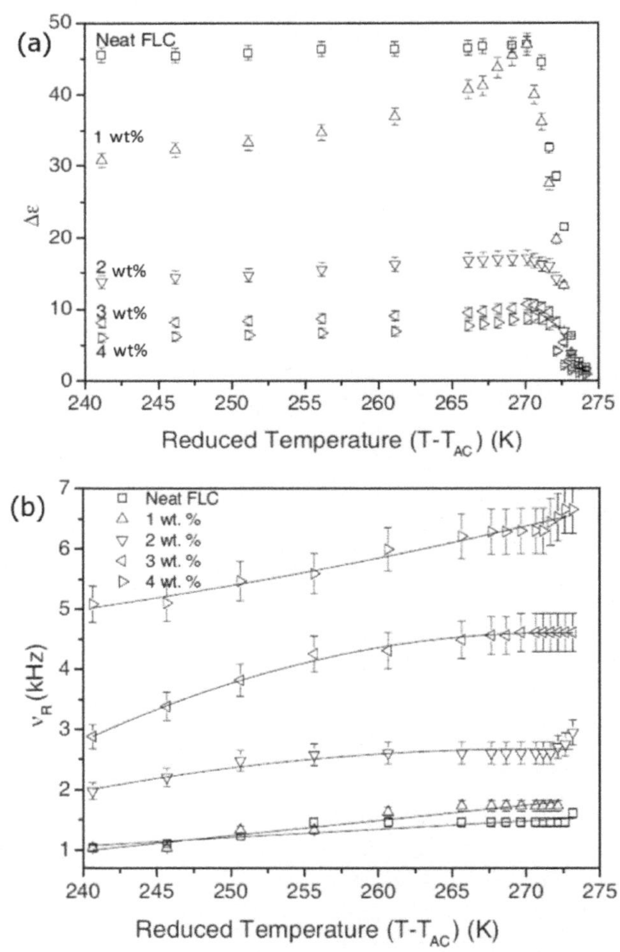

Figure 11. (**a**) The dielectric strength of the Goldstone mode is suppressed for increasing polymer concentration; while (**b**) the

relaxation frequency increases with increasing network density. (Reproduced with permission from [58]. Copyright 2009 EDP Sciences).

As mentioned above, the effect of the polymer network on the softmode of a ferroelectric liquid crystal appears to be much less pronounced [57]. The softmode appears at frequencies above the Goldstone mode, and also here a slight decrease of the dielectric strength is observed for increasing network density at equal reduced temperatures, as depicted in Figure 12a, while the relaxation frequency slightly increases (Figure 12b).

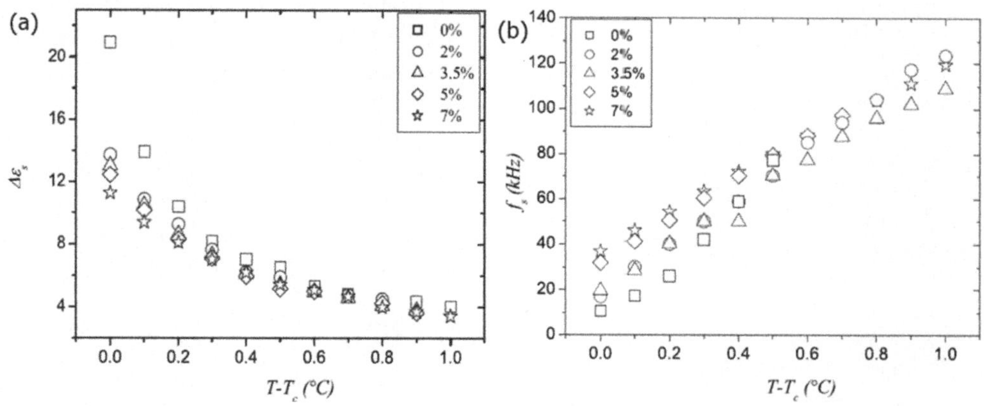

Figure 12. (**a**) Also the softmode dielectric strength is slightly suppressed for increasing polymer concentration; while (**b**) the softmode relaxation frequency slightly increases. (Reproduced with permission from [57]. Copyright 2009 American Physical Society).

In the context of display applications of FLCs, also the stabilization of the so called V-shaped switching mode, which basically exhibits a thresholdless, linear, hysteresis-free electro-optic response to an applied electric field, is of much interest [61,62]. Stabilizing the texture and electro-optic response of such V-shaped switching has been the focus of numerous papers [47,63–66] on polymer stabilized antiferroelectric liquid crystals. Hysteresis may be observed in certain PSFLC materials, especially for low monomer concentrations, which limits application aspects of these materials. Nevertheless, quite often the hysteresis is small and not of significant relevance.

ANTIFERROELECTRIC MATERIALS

SmC* liquid crystals can not only exhibit a ferroelectric structure with all molecules tilted in the same direction (synclinic), but also an antiferroelectric modification, where molecules of adjacent smectic layers are tilted in the opposite direction (anticlinic, disregarding any helical superstructure in a simplified view). The electro-optic switching of these systems is monostable, as compared to the bistable switching of the ferroelectric phase. The antiferroelectric phase switches from the stable E = 0 V state to either one of the ferroelectric states, depending on the polarity of the applied voltage.

Calamitic Antiferroelectrics

Antiferroelectricity in liquid crystals is attributed to have been discovered first for rod-like, elongated molecules [67–70]. The director configuration in its simple form is depicted in Figure 13a, together with the standard electro-optic response (Figure 13b). Introduction of a polymer network makes the switching transition between antiferroelectric to ferroelectric states more continuous, which allows for easy gray scale generation [71,72] (see Figure 14), an advantage of the antiferroelectric SmC^*_A phase over the ferroelectric SmC* phase, when display applications are considered. Of special interest for display applications are the orthoconic antiferroelectric liquid crystal materials, systems with a tilt angle of 45°, or very close to that value [73,74]. These systems exhibit a very good dark state due to the zero birefringence at these tilt angles. Also in this case, polymer stabilization was used to improve the mechanical stability [75].

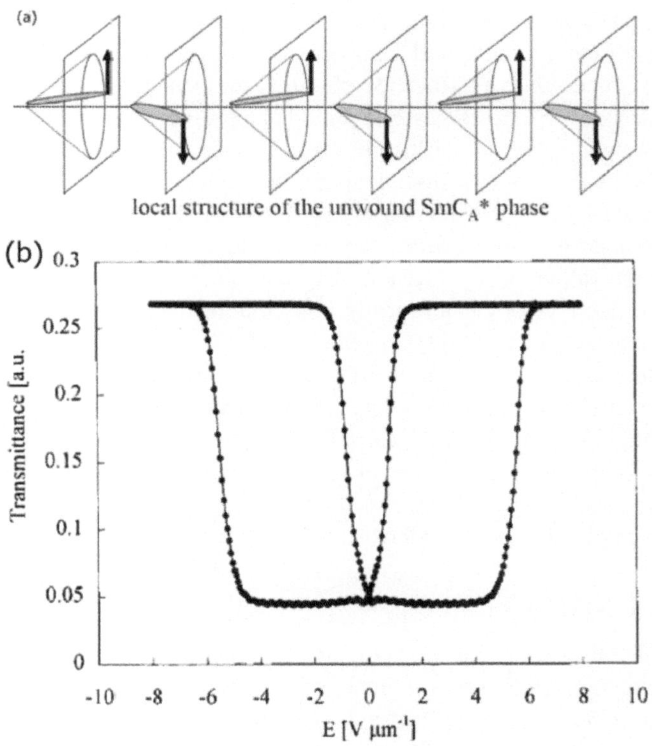

Figure 13. (**a**) Schematic structural drawing of an antiferroelectric liquid crystal, without the helical superstructure (Reproduced with permission from [8]. Copyright 2003 Wiley-VCH) and (**b**) standard electro-optic response of the SmC_A^* phase ((Reproduced with permission from [71]. Copyright 2001 American Chemical Society).

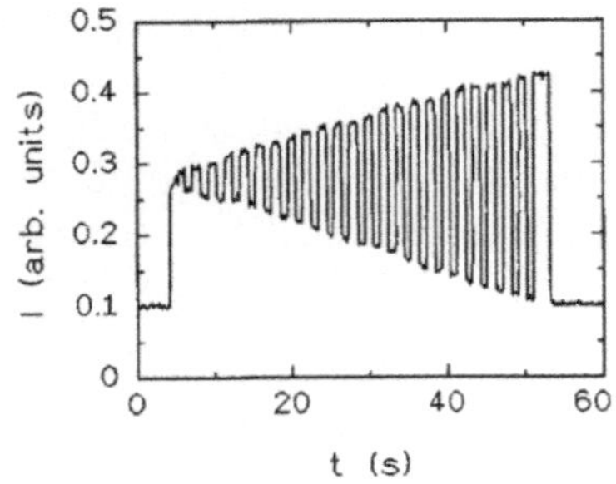

Figure 14: Electro-optic response of a polymer stabilized antiferroelectric liquid crystal, which allows for easier gray scale generation. (Reproduced with permission from [70]. Copyright 1996 American Institute of Physics).

In general, like for ferroelectric liquid crystals, also in antiferroelectric phases the polymer network has an influence on the fundamental material parameters. The contrast of the switching process reduces for increasing polymer concentration, as does the spontaneous polarization. This is due to an effective tilt angle, which is smaller than that observed for the neat liquid crystal, as an increasing amount of molecules contributes less and less to the switching process for increasing network content, due to elastic interactions between liquid crystal and polymer network.

Bent-Core Antiferroelectrics

Bent-core mesogens form quite remarkable liquid crystal phases [6,7], in so far as structural chirality and chiral effects can be observed from achiral molecules [76]. Alternatively, addition of an achiral bent-core material to a chiral matrix has been shown to increase chirality, for example through the decrease of the cholesteric pitch [77], the occurrence [78] or widening [79] of a Blue phase, and the increase of the chiral bilinear tilt-polarization coupling coefficient [80,81]. As neat materials, bent-core phases mostly exhibit antiferroelectric properties. One of the largest obstacles in their use for applications is the fact that bent-core phases are notoriously difficult to orient uniformly, thus diminishing their electro-optic response quality. A recent study [82] of the switching performance of polymer stabilized antiferroelectric bent-core phases has revealed a large temperature region, improved alignment ability, and enhanced contrast for such systems. In addition, it was found that the response times decrease for increasing polymer network content, in agreement with results on polymer stabilized ferroelectric liquid crystals.

SUMMARY

The basic principles of polymer stabilization of ferroelectric liquid crystals were outlined, and their influence on the most important physical parameters characterized. It was shown that both the tilt angle and the spontaneous polarization effectively decrease with increasing polymer concentration. This was explained by the liquid crystal being partially dominated by the network, not contributing to the switching process, due to elastic binding to the polymer. A Landau analysis provides evidence for this interpretation, because an introduced interaction parameter between liquid crystal molecules and stabilizing polymer network increases linearly with increasing network density at small polymer concentrations. The response times of the polymer stabilized systems are found to be shorter than those of the neat liquid crystal, and decrease with increasing network concentration. This also implies that the effective viscosity of the stabilized system decreases. While the process of polymer stabilization does not exhibit a pronounced effect on the soft mode dielectric relaxation, it is clearly observable in the lower frequency Goldstone mode relaxation. The Goldstone mode in general gets suppressed and the dielectric strength reduces with increasing polymer concentration, while the relaxation frequency increases. This behaviour conforms to the interpretation above: elastic binding of the molecules to the network reduces director fluctuations, while it enhances their relaxation frequency. A similar effect of the polymer network as on the parameters of the ferroelectric SmC* phase, can be observed for the antiferroelectric state, independent of whether this state is formed by calamitic or by bent-core mesogens.

REFERENCES

1. Collings, P.J.; Hird, M. *Introduction to Liquid Crystals: Chemistry and Physics*; Taylor & Francis: London, UK, 1997.
2. Chandrasekhar, S. *Liquid Crystals*, 2nd ed.; Cambridge University Press: Cambridge, UK, 1992.
3. De Gennes, P.G.; Prost, J. *The Physics of Liquid Crystals*, 2nd ed.; Clarendon Press: Oxford, UK, 1993.
4. Figueiredo Neto, A.M.; Salinas, S.R.A. *The Physics of Lyotropic Liquid Crystals*; Oxford University Press: Oxford, UK, 2005.
5. Petrov, A.G. *The Lyotropic State of Matter*; Gordon and Breach: New York, NY, USA, 1999.
6. Pelzl, G.; Diele, S.; Weissflog, W. Banana-shaped compounds—A new field of liquid crystals.*Adv. Mater* 1999, *11*, 707–724.
7. Takezoe, H.; Takanishi, Y. Bent-core liquid crystals: Their mysterious and attractive world.*Jpn. J. Appl. Phys* 2006, *45*, 597–625.
8. Dierking, I. *Textures of Liquid Crystals*; Wiley-VCH: Weinheim, Germany, 2003.
9. Lagerwall, S.T. *Ferroelectric and Antiferroelectric Liquid Crystals*; Wiley-VCH: Weinheim, Germany, 1999.
10. Musevic, I.; Blinc, R.; Zeks, B. *The Physics of Ferroelectric and Antiferroelectric Liquid Crystals*; World Scientific: Singapore, 2000.
11. Crooker, P.P. Chapter 7. Blue Phases. In *Chirality in Liquid Crystals*; Kitzerow, H.-S., Bahr, C., Eds.; Springer Verlag: Berlin, Germany, 2001.
12. Kitzerow, H.-S. Chapter 10. Twist Grain Boundary Phases. In *Chirality in Liquid Crystals*; Kitzerow, H.-S., Bahr, C., Eds.; Springer Verlag: Berlin, Germany, 2001.
13. De Vries, H. Rotatory power and other optical properties of certain liquid crystals. *Acta Crystallogr* 1951, *4*, 219–226.
14. Garoff, S.; Meyer, R.B. Electroclinic effect at the ac phase change in a chiral smectic liquid crystal. *Phys. Rev. Lett* 1977, *38*.
15. Garoff, S.; Meyer, R.B. Electroclinic effect at the ac phase change in a chiral smectic liquid crystal. *Phys. Rev. A* 1979, *19*.

16. Meyer, R.B.; Liebert, L.; Strzelecki, L.; Keller, P. Ferroelectric liquid crystals. *J. Phys. Lett* 1975,*36*, 69–71.
17. Clark, N.A.; Lagerwall, S.T. Submicrosecond bistable electro-optic switching in liquid crystals. *Appl. Phys. Lett* 1980, *36*, 899–901.
18. Crawford, G.P.; Zumer, S. *Liquid Crystals in Complex Geometries: Formed by Polymer and Porous Networks*; Taylor & Francis: London, UK, 1996.
19. Broer, D.J.; Crawford, G.P.; Zumer, S. *Cross-Linked Liquid Crystalline Systems*; CRC Press: Boca Raton, FL, USA, 2011.
20. Hikmet, R.A.M. Electrically induced light scattering from anisotropic gels. *J. Appl. Phys* 1990,*68*, 4406–4414.
21. Fung, Y.K.; Yang, D.-K.; Ying, S.; Chien, L.-C.; Zumer, S.; Doane, J.W. Polymer networks formed in liquid crystals. *Liq. Cryst* 1995, *19*, 797–801.
22. Hikmet, R.A.M. Anisotropic gels and plasticized networks formed by liquid crystal molecules. *Liq. Cryst* 1991, *9*, 405–416.
23. Hikmet, R.A.M. Anisotropic gels in liquid crystal devices. *Adv. Mater* 1992, *4*, 679–683.
24. Fung, Y.K.; Borstnik, A.; Zumer, S.; Yang, D.-K.; Doane, J.W. Pretransitional nematic ordering in liquid crystals with dispersed polymer networks. *Phys. Rev. E* 1997, *55*, 1637–1645.
25. Yang, D.-K.; Chien, L.-C.; Doane, J.W. Cholesteric liquid crystal/polymer dispersion for haze-free light shutters. *Appl. Phys. Lett* 1992, *60*, 3102.
26. Dierking, I. Polymer network—Stabilized liquid crystals. *Adv. Mater* 2000, *12*, 167–181.
27. Rajaram, C.V.; Hudson, S.D.; Chien, L.-C. Morphology of polymer-stabilized liquid crystals.*Chem. Mater* 1995, *7*, 2300–2308.
28. Dierking, I.; Kosbar, L.L.; Afazali-Ardakani, A.; Lowe, A.C.; Held, G.A. Network morphology of polymer stabilized liquid crystals. *Appl. Phys. Lett* 1997, *71*, 2454–2456.
29. Dierking, I.; Kosbar, L.L.; Lowe, A.C.; Held, G.A. Polymer network structure and electro-optic performance of polymer stabilized cholesteric textures. I. The influence of curing temperature. *Liq. Cryst* 1998, *24*, 387–395.
30. Held, G.A.; Kosbar, L.L.; Dierking, I.; Lowe, A.C.; Grinstein, G.; Lee, V.; Miller, R.D. Confocal microscopy study of texture transitions in a polymer stabilized cholesteric liquid crystal.*Phys. Rev. Lett* 1997, *79*, 3443–3446.
31. Dierking, I.; Kosbar, L.L.; Lowe, A.C.; Held, G.A. Polymer network structure and electro-optic performance of polymer stabilized cholesteric textures II. The effect of UV curing conditions. *Liq. Cryst* 1998, *24*, 397–406.
32. Fuh, A.Y.-G.; Tsai, M.-S.; Huang, C.-Y. Polymer network formed in liquid crystals: Polymer-network-induced birefringence in liquid crystals. *Jpn. J. Appl. Phys* 1996, *35*, 3960–3963.
33. Dierking, I.; Kosbar, L.L.; Afzali-Ardakani, A.; Lowe, A.C.; Held, G.A. Two-stage switching behavior of polymer stabilized cholesteric textures. *J. Appl. Phys* 1997, *81*, 3007–3014.
34. Mitov, M.; Dessaud, N. Going beyond the reflectance limit of cholesteric liquid crystals. *Nat. Mater* 2006, *5*, 361–364.
35. Dierking, I. Recent developments in polymer stabilised liquid crystals. *Polym. Chem* 2010, *1*, 1153–1159.
36. Archer, P.; Dierking, I. Polymer stabilisation of twisted smectic liquid crystal defect states.*Soft Matter* 2009, *5*, 835–841.
37. Dierking, I.; Archer, P. Imaging liquid crystal defects. *RSC Advances* 2013, *3*, 26433.
38. Hikmet, R.A.M.; Lub, J. Anisotropic networks with stable dipole orientation obtained by photopolymerization in the ferroelectric state. *J. Appl. Phys* 1995, *77*, 6234–6238.
39. Hikmet, R.A.M.; Michielsen, M. Anisotropic network stabilized ferroelectric gels. *Adv. Mater*1995, *7*, 300–304.
40. Hikmet, R.A.M.; Boots, H.M.J.; Michielsen, M. Ferroelectric liquid crystal gels Network stabilized ferroelectric display. *Liq. Cryst* 1995, *19*, 65–76.
41. Guymon, C.A.; Hoggan, E.N.; Walba, D.M.; Clark, N.A.; Bowman, C.N. Phase behaviour and electro-optic characteristics of a polymer stabilized ferroelectric liquid crystal. *Liq. Cryst*1995, *19*, 719–727.
42. Archer, P.; Dierking, I. Elastic coupling in polymer stabilized ferroelectric liquid crystals. *J. Phys. D* 2008, *41*.
43. Petit, M.; Daoudi, A.; Ismaili, M.; Buisine, J.M.; Da Costa, A. Effect of the polymer network density formed in short pitch ferroelectric liquid crystal on the electroclinic effect. *Mol. Cryst. Liq. Cryst* 2008, *487*, 61–73.
44. Archer, P.; Dierking, I. Electro-optic properties of polymer-stabilized ferroelectric liquid crystals before, during and after photo-polymerization. *J. Opt. A* 2009, *11*.
45. Zeks, B. Landau free energy expansion for chiral ferroelectric smectic liquid crystals. *Mol. Cryst. Liq. Cryst* 1984, *114*, 259–270.
46. Carlsson, T.; Zeks, B.; Levstik, A.; Filipic, C.; Levstik, I.; Blinc, R. Generalized Landau model of ferroelectric liquid crystals. *Phys. Rev. A* 1987, *36*, 1484–1487.
47. Li, J.; Zhu, X.; Xuan, L.; Huan, X. "V-Shaped" electro-optic characteristics in FLC gels. *Ferroelectrics* 2002, *277*, 85–105.
48. Gieβelmann, F.; Zugenmaier, P. Mean-field coefficients and the electroclinic effect of a ferroelectric liquid crystal. *Phys. Rev. E* 1995, *52*, 1762–1772.
49. Gieβelmann, F.; Heimann, A.; Zugenmaier, P. Experimental determination of Landau-expansion coefficients in ferroelectric liquid crystals. *Ferroelectrics* 1997, *200*, 237–256.
50. Archer, P.; Dierking, I.; Osipov, M.A. Landau model for polymer-stabilized ferroelectric liquid crystals: Experiment and theory. *Phys. Rev. E* 2008, *78*.

51. Petit, M.; Daoudi, A.; Ismaili, M.; Buisine, J.M. Distortion and unwinding of the helical structure in polymer-stabilized short-pitch ferroelectric liquid crystal. *Eur. Phys. J. E* 2006, *20*, 327–333.

52. Pirs, J.; Bos, P.; Petkovšek, R.; Kralj, S.; Pirš, S.; Zumer, S.; Matuszczyk, T. Influence of the polymer network on the switching behaviors in the polymer stabilized ferroelectric liquid crystals. *Jpn. J. Appl. Phys* 2002, *41*, 6011–6015.

53. Manohar, R.; Yadav, S.P.; Pandey, K.K.; Srivastava, A.K.; Misra, A.K. Comparative study of dielectric and electro-optical properties of pure and polymer ferroelectric liquid crystal composites. *J. Polym. Res* 2011, *18*, 435–441.

54. Zheng, W.J. Article title. *Mol. Cryst. Liq. Cryst* 2007, *475*, 173–181.

55. Beckel, E.R.; Cramer, N.B.; Harant, A.W.; Bowman, C.N. Electro-optic properties of thiol-ene polymer stabilized ferroelectric liquid crystals. *Liq. Cryst* 2003, *30*, 1343–1350.

56. Guymon, C.A.; Dougan, L.A.; Martens, P.J.; Clark, N.A.; Walba, D.M.; Bowman, C.N. Polymerization conditions and electrooptic properties of polymer-stabilized ferroelectric liquid crystals. *Chem. Mater* 1998, *10*, 2378–2388.

57. Petit, M.; Hemine, J.; Daoudi, A.; Ismaili, M.; Buisine, J.M.; Da Costa, A. Effect of the network density on dynamics of the soft and the Goldstone modes in short-pitch ferroelectric liquid crystals stabilized by an anisotropic polymer network. *Phys. Rev. E* 2009, *79*.

58. Kaur, S.; Dierking, I.; Gleeson, H.F. Dielectric spectroscopy of polymer stabilised ferroelectric liquid crystals. *Eur. Phys. J. E* 2009, *30*, 265–274.

59. Cherfi, Y.; Hemine, J.; Douali, R.; Beldjoudi, N.; Ismaili, M.; Leblond, J.; Legrand, C.; Daoudi, A. Linear and non-linear dielectric properties of a short-pitch ferroelectric liquid crystal stabilized by a polymer network. *Eur. Phys. J. E* 2010, *33*, 335–342.

60. Kundu, S.; Nayek, P.; Ray, T.; Majumder, T.P.; Roy, S.K.; Kobayashi, S. Influence of network stabilization on the dielectric and electrooptical properties of ferroelectric liquid crystal FELIX-M4851/100. *Jpn. J. Appl. Phys* 2009, *48*.

61. Rudquist, P.; Lagerwall, J.P.F.; Buivydas, M.; Gouda, F.; Lagerwall, S.T.; Clark, N.A.; Maclennan, J.E.; Shao, R.; Coleman, D.A.; Bardon, S.; *et al.* The case of thresholdless antiferroelectricity: Polarization-stabilized twisted SmC* liquid crystals give V-shaped electro-optic response. *J. Mater. Chem* 1999, *9*, 1257–1261.

62. Blinov, L.M.; Pozhidaev, E.P.; Podgornov, F.V.; Pikin, S.A.; Palto, S.P.; Sinha, A.; Yasuda, A.; Hashimoto, S.; Haase, W. Thresholdless hysteresis-free switching as an apparent phenomenon of surface stabilized ferroelectric liquid crystal cells. *Phys. Rev. E* 2002, *66*.

63. Li, J.; Wang, Z.; Cai, Y.; Huang, X. Study of EO properties of polymer network stabilized ferro64. electric liquid crystal in smectic C* phase. *Ferroelectrics* 1998, *213*, 91–99.

64. Shikada, M.; Tanaka, Y.; Xu, J.; Furuichi, K.; Hasebe, H.; Takatsu, H.; Kobayashi, S. Novel mesogenic polymer stabilized ferroelectric liquid crystal display device exhibiting V-shaped electrooptic characteristics. *Jpn. J. Appl. Phys* 2001, *40*, 5008–5010.

65. Furue, H.; Koizumi, Y.; Hatano, J.; Yokoyama, H. Molecular alignment structure and switching of a ferroelectric liquid crystal stabilized by a polymer network created in the SmA phase. *Mol. Cryst. Liq. Cryst* 2005, *437*, 1439–1446.

66. Pozhidaev, E.; Chgrinov, V.; Hegde, G.; Xu, P.Z. Multistable electro-optical modes in ferroelectric liquid crystals. *J. Soc. Inf. Disp* 2009, *17*, 53–59.

67. Beresnev, L.A.; Blinov, L.M.; Baikalov, V.A.; Pozhidayev, E.P.; Purvanetskas, G.V.; Pavluchenko, A.I. Ferroelectricity in tilted smectics doped with optically active additives.*Mol. Cryst. Liq. Cryst* 1982, *89*, 327–338.

68. Levelut, A.M.; Germain, C.; Keller, P.; Liebert, L.; Billard, J. Two new mesophases in a chiral compound. *J. Phys. Fr* 1983, *44*, 623–629.

69. Galerne, Y.; Liebert, L. Antiferroelectric chiral smectic-O* liquid crystal. *Phys. Rev. Lett* 1991,*66*, 2891–2895.

70. Chandani, A.D.L.; Hagiwara, T.; Suzuki, Y.; Ouchi, Y.; Takezoe, H.; Fukuda, A. Tristable switching in surface stabilized ferroelectric liquid crystals with a large spontaneous polarization. Article title. *Jpn. J. Appl. Phys. Lett* 1988, *27*, L729–L732.

71. Strauss, J.; Kitzerow, H.-S. Gray-scale in polymer-stabilized antiferroelectric liquid crystal displays. *Appl. Phys. Lett* 1996, *69*.

72. Artal, M.C.; Ros, M.B.; Serrano, J.L. Antiferroelectric liquid-crystal gels. *Chem. Mater* 2001, *13*, 2056–2067.

73. Lagerwall, S.T.; Dahlgren, A.; Jagemalm, P.; Rudquist, P.; D'have, K.; Pauwels, H.; Dabrowski, R.; Drzewinski, W. Unique electro-optical properties of liquid crystals designed for molecular optics. *Adv. Funct. Mater* 2001, *11*, 87–94.

74. Rudquist, P. Orthoconic antiferroelectric liquid crystals. *Liq. Cryst* 2013, *40*, 1678–1697.

75. Rudquist, P.; Elfstrom, D.; Lagerwall, S.T.; Dabrovski, R. Polymer-Stabilized orthoconic anti-ferroelectric liquid crystals. *Ferroelectrics* 2006, *344*, 177–188.

76. Link, D.R.; Natale, G.; Shao, R.; Maclennan, J.E.; Clark, N.A.; Korblova, E.; Walba, D.M. Spontaneous formation of macroscopic chiral domains in a fluid smectic phase of achiral molecules. *Science* 1997, *278*, 1924–1927.

77. Gorecka, E.; Cepic, M.; Mieczkowski, J.; Nakata, M.; Takezoe, H.; Zeks, B. Enhanced chirality by adding achiral molecules into the chiral system. *Phys. Rev. E* 2003, *67*.

78. Nakata, M.; Takanishi, Y.; Watanabe, J.; Takezoe, H. Blue phases induced by doping chiral nematic liquid crystals with nonchiral molecules. *Phys. Rev. E* 2003, *68*.

79. Dierking, I.; Blenkhorn, W.; Credland, E.; Drake, W.; Kociuruba, R.; Kayser, B.; Michael, T. Stabilising liquid crystalline blue phases. *Soft Matter* 2012, *8*, 4355–4362.

80. Archer, P.; Dierking, I. Experimental determination of

the full Landau potential of bent-core doped ferroelectric liquid crystals. *Phys. Rev. E* 2005, *72*.

81. Archer, P.; Dierking, I. Chirality enhancement through addition of achiral molecules. *Chem. Commun* 2010, *46*, 1467–1469.

82. Atorf, B.; Hoischen, A.; Ros, M.B.; Gimeno, N.; Tschierske, C.; Dantlgraber, G.; Kitzerow, H. Switching performance of a polymer-stabilized antiferroelectric liquid crystal based on bent-core molecules. *Appl. Phys. Lett* 2012, *100*, 223301:1–223301:4.

CHAPTER 10 FERROELECTRIC POLYMER THIN FILMS FOR ORGANIC ELECTRONICS

Manfang Mai, Shanming Ke, Peng Lin, and Xierong Zeng

INTRODUCTION

In recent years, driven by the rapidly developing miniaturized electronics, new organic devices based on polyvinylidene fluoride (PVDF) and its copolymer with trifluoroethylene (TrFE) thin films have attracted intensive research interest. Generally, ferroelectrics exhibit a striking finite size effect [1]. The switching time and the coercive field increase with decreasing film thickness while reversely the remanent polarization and the phase transition temperature decrease with decreasing thickness. For applications which require low-voltage operation such as nonvolatile memories, the ferroelectric films should be thin enough for low operation voltage at the same time they should have sufficient and stable ferroelectric properties to maintain the memory functionality.

PVDF and its copolymer with TrFE are the representative and the most important organic ferroelectric materials. The ferroelectricity in PVDF and P(VDF-TrFE) originates from molecular dipoles associated with electropositive hydrogen atoms and electronegative fluoride atoms, which are switchable upon an external field [2]. They are promising candidates for the next generation nonvolatile high-density memory applications which can replace perovskite ceramics currently used in the commercial ferroelectric random access memories (FRAMs) [3–5]. Their attractive advantages include low temperature processing, low-cost solution processing, outstanding chemical stability, and nontoxicity [6]. In the past few years, there have been considerable research activities to combine P(VDF-TrFE) films as nonvolatile memory cells such as metal-ferroelectric-insulator-semiconductor (MFIS) diode [7–11] and ferroelectric-field-effect-transistor (FeFET) [12–15]. The key issues for optimizing the memory performance including switching time, switching voltage, retention, and endurance still need to be overcome. Recently, PVDF and its copolymer find their new way for the emerging applications in organic electronics, such as negative capacitance devices [16], organic photovoltaic cells [17], and multiferroic tunnel junctions [18].

The discovery of Langmuir-Blodgett (LB) ferroelectric polymer films in 1995 [19] led to investigation of ferroelectric properties at the nanoscale. For P(VDF-TrFE) films, the ferroelectricity and polarization switching have been found for two monolayers (l=1 nm) [20]. It opened the way for investigation of finite size effect at the nanoscale. Since then, it has been a subject of intensive study not only in theoretical aspects but also in technical applications. The results on P(VDF-TrFE) LB films provide many new insights into the fundamental physics: the existence of two-dimensional ferroelectrics with nearly absence of finite size effects on the bulk transition (a first order phase transition with a transition temperature nearly equal to the bulk value) [20], the transition from extrinsic (switching with nucleation and domain wall motion) to intrinsic (switching without domain) ferroelectric switching in the context of Landau-Ginzburg mean-field theory below a thickness of 15 nm [21], and so forth. However, there have been strong debates of the theoretical models used to explain the experiments. Scott pointed out that the presence of two apparent phase transition temperatures interpreted as true two-dimensional surface ferroelectricity by the Moscow-Nebraska group, instead, can be explained by the seminal Tilly-Zeks theory and its extension by Duiker to first order phase transition [22]. Several groups spoke against the intrinsic nature of the switching as the experiments lack several key features of intrinsic switching [23–27]. Lately, the Moscow-Nebraska group demonstrated some new experiment evidences to support their claim of domain-free Landau switching in P(VDF-TrFE) LB films [28, 29] as well as in ultrathin $BaTiO_3$ films [30],

Citation: Manfang Mai, Shanming Ke, Peng Lin, and Xierong Zeng, "Ferroelectric Polymer Thin Films for Organic Electronics," Journal of Nanomaterials, vol. 2015, Article ID 812538, 14 pages, 2015. doi:10.1155/2015/812538.

suggesting that homogeneous switching in ultrathin films is a common phenomenon for all ferroelectric materials. Currently, the explanations for the switching behavior in P(VDF-TrFE) thin films are still controversial.

PVDF is a rich system for both practical applications and academic investigations including phase transition and switching mechanism. The main objective of this paper is to review the current research status of ferroelectric polymer films with emphases on switching mechanism and ferroelectric devices based on P(VDF-TrFE) films particularly with the thickness below one hundred nanometers. The development of ferroelectric polymer thin film devices will have a significant impact on organic electronics.

FERROELECTRIC POLYMERS

The piezoelectricity in poled PVDF was discovered in 1969 [31] followed by the discovery of the pyroelectricity in PVDF in 1971 [32]. However, it was not until the late 1970s that the ferroelectricity in PVDF was confirmed [33, 34]. The distinguishing characteristics of PVDF and its copolymers include highly compact structure, high chemical stability, large permanent dipole moment, ease of fabrication, and low annealing temperature which can be integrated with other materials and processes, for example, silicon microfabrication.

PVDF exhibit four polymorphic crystalline forms: α, β, γ, and δ phases, whose formations depend on crystallization conditions, electrical poling, and mechanical drawing. The β phase has an orthorhombic structure with an all-trans (TTTT) molecule conformation possessing a large spontaneous polarization and is responsible for the ferroelectric and piezoelectric properties. The dipoles in the β-PVDF extend from the electronegative fluoride atoms to the slightly electropositive hydrogen atoms perpendicular to the polymer chain direction, producing a dipole moment of 7×10^{-30} C·m (2 Debyes), as shown in Figure 1. Recently -PVDF is successfully fabricated as an overlooked polymorph of PVDF proposed 30 years ago [35]. The remanent polarization and coercive field are comparable to those of -PVDF and of P(VDF-TrFE) while the thermal stability is enhanced due to its higher Curie temperature.

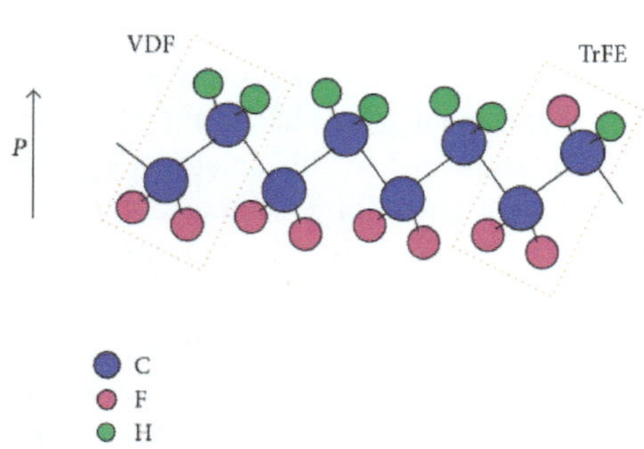

Figure 1: Schematic of molecular structure of P(VDF-TrFE) in the β phase. The polarization direction points from fluoride to hydrogen atoms.

The introduction of TrFE, which has three fluoride atoms per monomer, to copolymerize with PVDF, can enhance the all-trans conformation associated with the β phase. This is because the fluoride atom is larger than hydrogen atom and therefore it can induce a stronger steric hindrance [2]. In addition, the copolymer with TrFE can be annealed to much higher crystallinity than pure PVDF [36]. The most studied copolymers have a composition of P(VDF-TrFE 70/30), which has a maximum spontaneous polarization around 10 μC/cm^2 [37]. Another feature for the introduction of TrFE is that it can lower the phase transition temperature T_c in the copolymers. P(VDF-TrFE) with molar ratios of 50–80% of PVDF undergoes a clear ferroelectric-paraelectric phase transition with a measurable Curie point. The T_c increases with increasing PVDF content from 70°C at 50 mol% to 140°C at 80 mol%. The transition is an order-disorder type while the disorder is a random mixture of trans- and gauche combinations in the conformation of the molecules. Moreover, the transition tends to change from a first order to a second order as the PVDF content decreases to a 52/48 composition with evidence of the disappearance of thermal hysteresis [38]. In the bulk PVDF the phase transition is not accessible with experiments because the melting temperature of this ferroelectric polymer (~170°C) is lower than the point of the possible phase transition T_c. Until very recently, a first order ferroelectric phase transition in the Langmuir-Blodgett ultrathin PVDF films was observed for the first time [39]. We have found the thickness interval where ferroelectric phase transition disappears and transition from ferroelectric to pyroelectric state takes place in PVDF Langmuir-Blodgett films and explain the shift of T_c by the finite size effect at the nanoscale using Landau-Ginzburg-Devonshire theory and the Weiss mean field model [40].

THIN FILMS FABRICATION

Spin Coating

Conventionally, ferroelectric thin films are prepared by solvent spinning method which is very attractive for researchers due to its ease of processing and good quality of film surface. PVDF and its copolymers can be dissolved in a variety of organic solvents for spin coating, such as 2-butanone, dimethylsulfoxide (DMSO), dimethylformamide (DMF), and methylethylketone (MEK). The sample thickness can be controlled by varying the solution concentrations and rotating speeds. P(VDF-TrFE) films prepared by spin coating consist of both crystalline and noncrystalline regions [2]. Uniform and compact ferroelectric films thinner than 60 nm are usually difficult to achieve by spin coating. Lately, with the improvement of experimental conditions, some groups can produce ferroelectric copolymer films even down to 10 nm by spin coating [35, 44, 45]. Li et al. [46] reported that homogeneous and smooth thin PVDF films can be prepared either at low relative humidity or at high substrate temperature. Figure 2 shows the surface morphology evolution of PVDF thin films as a function of both relative humidity and substrate temperature. At high relative humidity the film surface is porous and loose. With decreasing humidity, the film becomes homogeneous and dense (Figure 2(a)). On the other hand (Figure 2(b)), fixing the relative humidity at 25% the surface morphology of PVDF films is denser and smoother as the substrate temperature is increased from 20°C to 100°C. The microstructure of PVDF thin films is determined by the phase separation mechanism.

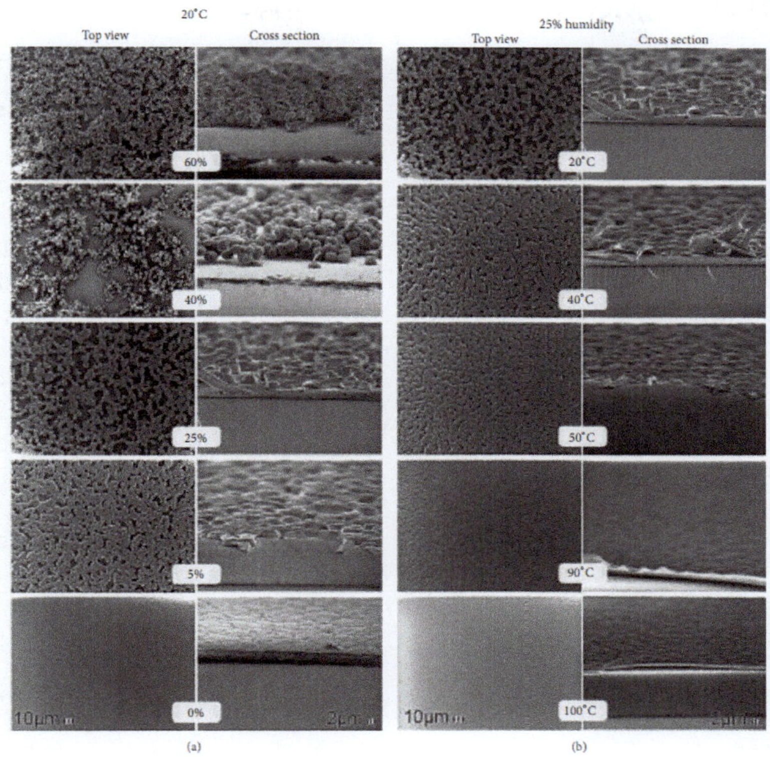

Figure 2: Surface morphology evolution of PVDF thin films. (a) Top view and cross section SEM micrographs of PVDF thin films presented as a function of the relative humidity between 0% and 60%. The films were deposited at room temperature, 20°C. The scale bars of 10 μm and 2 μm are presented in the bottom micrographs. (b) Top view and cross section SEM micrographs of PVDF thin films presented as a function of the deposition temperature between 20°C and 100°C. The relative humidity was fixed at 25% [46].

Langmuir-Blodgett Technique

Two-dimensional ferroelectric polymer films can be fabricated by Langmuir-Blodgett (LB) technique which allows precise control of film thickness down to one monolayer with high crystallinity. The principle of LB technique is the amphiphilic property of molecules which are composed of a hydrophilic part and a hydrophobic part [47]. Amphiphilic molecules are trapped at the interface with one type of bonding being attracted to polar media such as water and the other type of bonding being much less water soluble. The LB films are achieved through subsequent transfer of monolayers from gas-liquid interface onto a solid substrate. The transfer can be done by either vertical or horizontal dipping with the latter called horizontal Langmuir-Schaefer method, as shown in Figure 3(a) [37]. Although P(VDF-TrFE) copolymer is not amphiphilic, the large molecular weight of polymers with low solubility in water can be dispersed to form a sufficiently stable layer on water with a repeatable pressure area isotherm (Figure 3(b)), which identifies the conditions forming closest packing of a film of monolayer thickness [37]. Good quality P(VDF-TrFE 70/30) LB films are deposited from a copolymer solution in dimethylsulfoxide (DMSO) with concentration of 0.01 wt.% and target pressure of 3 mN/m [48, 49].

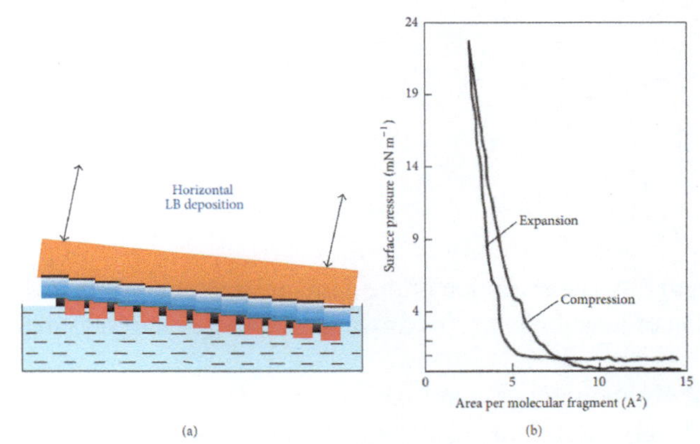

Figure 3: (a) Diagram of the Langmuir-Schaefer monolayer deposition method. (b) Pressure-area isotherm of P(VDF-TrFE 70/30) [37].

Generally, the as-prepared films need to be annealed close to the crystallization temperature to increase the crystallinity. It is reported that the thermal annealing increases the film surface roughness [46]. Alternately, smooth PVDF thin films can be prepared using elevated substrate temperatures and applying a short electrical pulse [35]. Different materials have served as electrodes for P(VDF-TrFE) films including aluminum [20, 50,51], platinum [52, 53], gold [44, 54, 55], silver [54, 56], copper [52], nickel [52], indium [45], sodium [57], and conducting PEDOT/PSS copolymer [58]. The most commonly used electrode is aluminum. The advantages of aluminum electrode are its low deposition temperature, chemical passivation, and suppressing of charge injection from the metal electrode. However, the formation of the oxide layer (Al_2O_3) between the electrode and the ferroelectric film inhibits fast switching of the films.

POLARIZATION SWITCHING MECHANISM

The switching behavior of ferroelectric polymer thin films has been extensively investigated to understand the kinetics of polarization switching, which is very critical for intentional design and preparation of materials for various purposes. Domain nucleation and domain wall motion mechanism for polarization reversal was established early by Merz [59] and Little [60] and later by Miller [61, 62] and Fatuzzo [63]. The studied ferroelectric material was barium titanate ($BaTiO_3$). Numerous models have been proposed to quantify domain phenomena and their effects on ferroelectric switching. The classical model to describe ferroelectric switching kinetics is the Kolmogorov-Avrami-Ishibashi (KAI) model [1, 64, 65], which is based on the study of statistical behavior and the probability of nucleation and growth of domains. An alternative model is the nucleation-limited switching (NLS) model for thin films developed by Tagantsev et al. [66, 67]. For P(VDF-TrFE) ultrathin films, ferroelectric switching may not include the usual nucleation and domain propagation processes. Here, the existing theoretical models for ferroelectric switching are summarized.

Kolmogorov-Avrami-Ishibashi (KAI) Model

Polarization reversal in ferroelectrics proceeds by the nucleation of reversed nuclei followed by the forward and sideways growth of the reversed areas, as shown in Figure 4 [63]. When a field is applied with the opposite polarity, small nuclei of domain with reverse polarization direction appear, usually at the interfaces or grain boundaries (Figure 4(a)). Each nucleus grows parallel or antiparallel to the applied field until it becomes a domain and reaches the opposite side (Figure 4(b)). The domain formed in this way starts expanding sideways (Figure 4(c)). After some time these domains are large enough to join each other and become coalescence (Figure 4(d)). The above successive processes will repeat until the polarization is completely reversed in the whole sample.

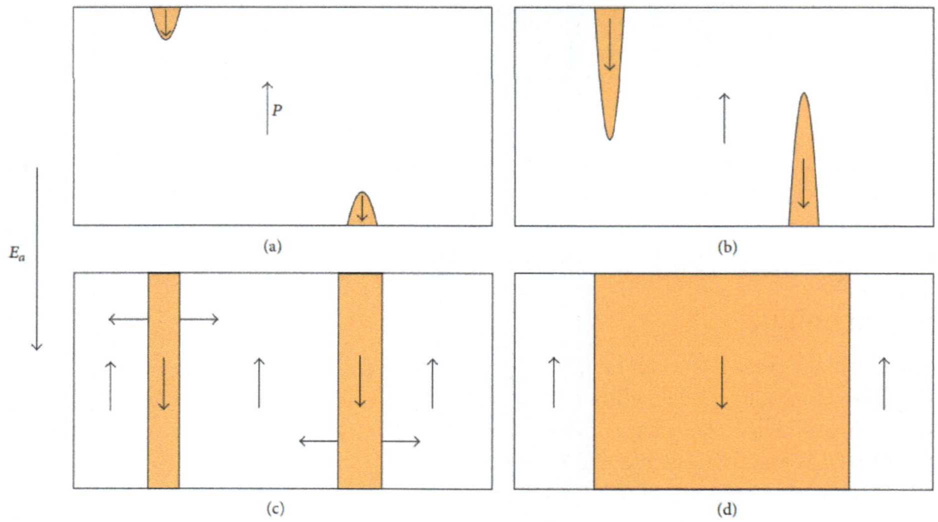

Figure 4: The evolution of the domain structure in ferroelectric during switching under an application of electric field (a) nucleation of new domains, (b) forward growth, (c) sideways expansion, and (d) coalescence, after [63].

The Kolmogorov-Avrami-Ishibashi (KAI) model is applicable and very useful for analyzing the above switching process in many materials including single crystals [66, 68] as well as epitaxial thin film ferroelectrics [69]. In the scope of KAI model, the ferroelectric media where the polarization reversal takes place are infinite. The reversed polarization P(t) and the switching current I(t) can be written as

$$P(t) = 2P_s\left\{1 - \exp\left[-\left(\frac{t}{t_0}\right)^n\right]\right\}, \qquad (1)$$

$$I(t) = 2P_s S_0 \frac{n}{t_0}\left(\frac{t}{t_0}\right)^{n-1} \exp\left[-\left(\frac{t}{t_0}\right)^n\right], \qquad (2)$$

where P_s is spontaneous polarization, S_0 is electrode area, t_0 is switching time, and n can be an integer (α-model or β-model) or non-integer value (mixture of α-model and β-model) between 1 and 3 depending on the dimension of the system. A plot of the switching current profiles for different values of n according to (2) is depicted in Figure 5. The tick labels in the axes would be different for a variation of parameters P_s, S_0 , and t_0 in different samples. Figure 5 shows qualitatively different shapes of the switching currents. For $n > 1$, there is a peak in the current transients. For $n \leq 1$, it is not possible to identify a current maximum. However, there is no physical meaning in the framework of KAI model for an n value less than 1 since the growth dimensionality could never be less than 1. Therefore a current maximum should always exist in the switching transient when the sample is suppressed by a field above the coercive field in the framework of KAI model. Moreover, KAI model describes a thermally activated switching.

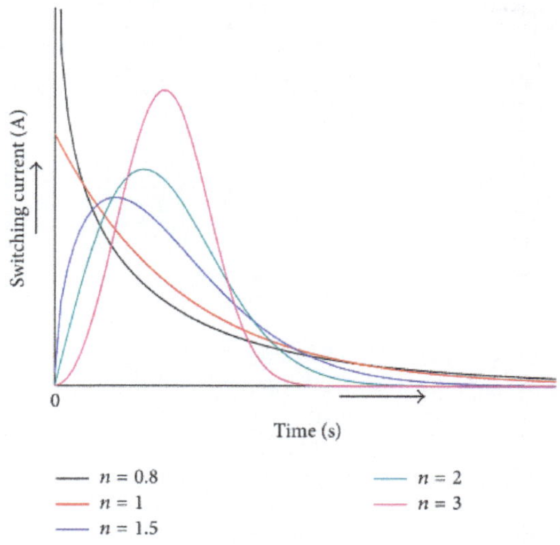

Figure 5: Switching current profiles for n with different values according to (2).

The KAI model assumes unrestricted domain growth in an infinite media. Due to the simplified assumptions, it encounters problems when it comes to precisely describe the switching behavior in real finite ferroelectrics especially for polycrystalline films and ceramics. Therefore, later on new models are constructed to give a more realistic and appropriate description of polarization reversal.

Nucleation-Limited Switching (NLS) Model

Tagantsev et al. [67] developed an alternative scenario of a nucleation-limited switching (NLS) model which is successful to describe the switching kinetics in $Pb(Zr,Ti)O_3$ thin films. This model assumes that the film consists of many areas, which switch independently. The switching in an area is considered to be triggered by an act of the reverse domain nucleation. They used a distribution function of the switching times to describe the ensemble of the elementary regions. In the framework of the NLS model, the nucleation time is assumed to be much longer than the time of domain wall motion and a distribution of switching time is modified in the expression of the time dependence of polarization, written as [67]

$$P(t) = 2P_s \int_{-\infty}^{\infty} \left\{1 - \exp\left[-\left(\frac{t}{t_0}\right)^n\right]\right\} F(\log t_0)\, d(\log t_0), \quad (3)$$

where $F(\log t_0)$ is the distribution function for $\log t_0$. It meets the normalizing condition:

$$\int_{-\infty}^{\infty} F(\log t_0)\, d(\log t_0) = 1. \quad (4)$$

The value of n is assumed to be 2 for thin film ferroelectrics and 3 for bulk ferroelectrics [67]. The switching dynamic extended in a wide time range can be described in terms of this model with a simple shape of the spectrum $F(\log t_0)$. Jo et al. [70] suggested to use the Lorentzian distribution of $\log t_0$, which can account for variations in the local electrical field related to the randomly distributed dipole defects. A Gaussian distribution can also be well adjusted to the random electric field distribution suggested by Zhukov et al. [71]. Mao et al. used the NLS model based on region-by-region switching to explain the switching kinetics of P(VDF-TrFE) thin films, which is successful to describe the polarization reversal behavior in the time domain for samples as thick as 100 nm [72].

Both KAI model and NLS model neglect the size of nuclei and domains. Fatuzzo [63] reported that the switching current maximum in time could be as broad as that of a single relaxation process regardless of the domain growth dimensions, if the size of nuclei is taken into consideration. Numerical simulations in small systems using a local field method combined with dynamic Monte Carlo steps result in a monotonic decrease in the switching current [73].

The Weiss Mean Field Model

One hundred years ago Weiss [74, 75] proposed a model to describe phase transition in ferromagnetic materials. This model is actually a positive feedback model. The magnetic dipoles mutually evoke a field at the location of other dipoles which results in an alignment of all dipoles into the same direction. This local field which is also called Weiss field is only defined at the location of the dipoles. This model for magnetism was later applied to ferroelectrics [76].

In dielectrics Lorentz [77] developed a similar local field model. The local field at induced point dipoles for an infinite cubic lattice can be expressed as

$$E_{loc} = E_a + \frac{P}{3\varepsilon_0}, \quad (5)$$

with E_a being the applied field and P the polarization of the matter. Equation (5) illustrates that the local field vice versa produces the dipole moments at the lattice points. It is the superposition of the applied field and the dipolar field which is proportional to the polarization itself. There are two types of dipoles: permanent dipoles fluctuating thermally activated in double well potentials and induced dipoles. The interaction between all these dipoles is considered with mean fields averaged over all particular local fields. The double well potentials, as shown in Figure 6, are characterized by their activation energy φ_0 and the distance between the wells l [78]. When no local field is present, the double wells are symmetrical and the dipoles are equally occupied in both directions. The net polarization of the whole system is zero. With a local field E_{loc} acting at the permanent dipoles the occupation changes and a net polarization results.

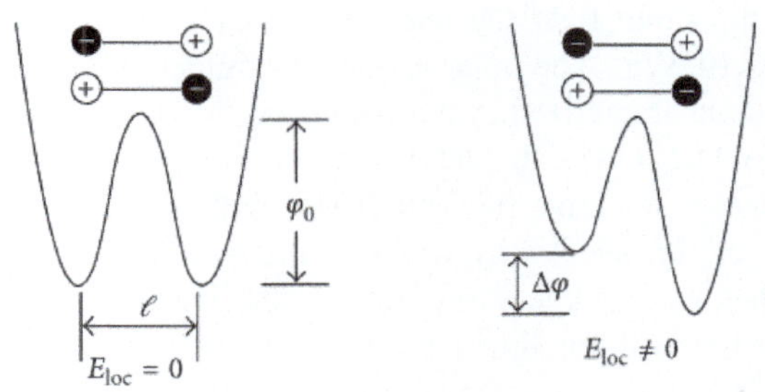

Figure 6: Permanent dipoles are assumed to fluctuate thermally activated in double well potentials. A local field shifts one well against the other redistributing the dipoles [78].

If only permanent dipoles are present, the local field within the Weiss model is assumed to be similar to (5):

$$E_{loc} = E_a + \alpha P, \quad (6)$$

where P is the macroscopic polarization which can be spontaneous below T_c or paraelectric above T_c. α is a coupling factor describing the strength of the interaction between the dipoles. The model is a mean field approach. It also contains microscopic elements including the properties of the double wells with the magnitude of the fluctuating dipole moments, the dipole densities, and the coupling constant α . Therefore the model can be considered as a link between phenomenological descriptions and molecular modeling [78].

With the Weiss model, several ferroelectric properties can be described, that is, ferroelectric hysteresis loop, the temperature dependence of the coercive field, the Curie-Weiss law for the electrical susceptibility, the buttery curves of the susceptibility in an external field, and a second order ferroelectric to paraelectric transition with its typical protraction in an applied electric field [78–80]. It has been shown that if additionally a strain in the material caused by the piezoelectric effect, that is, by the polarization itself, is taken into account, a second order phase transition changes to a first order [81]. In addition, the appearance of a double hysteresis close to the Curie temperature can be explained [82, 83].

Lately, Kliem and Kuehn [84] presented an analytical approach using the Weiss molecular mean-field model combined with the fluctuation of dipoles in double-well potentials to simulate the switching process with a maximum in the switching current as observed by Merz. The time-dependent switching process and the static polarization hysteresis loop are simulated which are in good qualitative agreement with experiments using the material parameters of P(VDF-TrFE). This theory is exact in the limit of infinite dimensions with homogeneous state and single domain.

The Homogenous Non-Domain Switching

Recent studies using P(VDF-TrFE) copolymer Langmuir-Blodgett ultrathin films suggested that the switching kinetics in these films is different from the rules of nucleation and domain wall motion (extrinsic switching). This type of switching is called intrinsic switching which is defined by the absence of nucleation center and domain. The distinction between domain nucleated and continuous intrinsic switching mechanisms can be illustrated in Figure 7 [85]. For intrinsic switching, it has a threshold field the intrinsic coercive field which is as high as 10^9 V/m. The polarization switching exhibits a critical behavior characterized by a pronounced slowing just above the threshold field [86]. In 2000 Ducharme et al. [21] reported that the intrinsic coercive field predicted by the Landau-Ginzburg theory of ferroelectricity was observed in P(VDF-TrFE) LB films below a thickness of 15 nm for the first time. They found that with decreasing thickness the coercive field first increases and then saturates for samples thinner than 15 nm, to a value of 5 MV/cm, which is in good agreement with the theoretical intrinsic value.

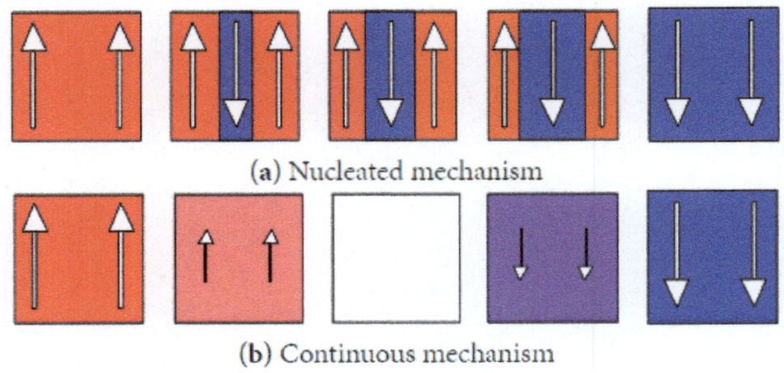

Figure 7: Schematic of polarization evolution during switching in a uniaxial ferroelectric. (a) Discontinuous nucleation and growth of inverted domains. (b) Continuous, uniform decrease of the polarization magnitude through zero without domain formation. Arrows indicate the orientation and magnitude of polarization [85].

In intrinsic switching, the dipoles in the systems are highly correlated and tend to switch coherently or not at all. It can be described by Landau-Khalatnikov equations in the context of mean-field theory:

$$\xi\frac{dP}{dt} = -\frac{\partial G}{\partial P}, \tag{7}$$

$$G = \alpha(T - T_0)P^2 + \beta P^4 + \gamma P^6 - EP, \tag{8}$$

where P, G, E, and T are the polarization, the free energy, the electric field, and the temperature, respectively. ξ is a damping coefficient and $T0$ is the Curie-Weiss temperature. α, β, and γ are constants assumed to be independent of temperature. By minimizing the free energy G in (8), a steady state hysteresis loop (E) can be derived, as shown in Figure 8 [86]. The solid lines in Figure 8(a) denote stable (AB and A' B') or metastable (BC and B' C') minima in the free energy. The dotted line (COC') denotes unstable minima. Starting from a stable state B, if an opposite electric field which has an amplitude lower than the coercive field is applied, the polarization will remain in the metastable state (BC) and then return to its initial state B after the field is removed. On the contrary, if a field larger than the coercive field is applied, the polarization will switch to the other state (A' B') and remain in B' when the field is suppressed. The vertical dashed lines mark the limits of the quasi steady state hysteresis loop. The evolution of the polarization from a negative polarization state to a positive polarization state in Figure 8(b) is obtained by numerical integration of (7). There is switching only for $E>Ec$. Any field larger than Ec results in the same polarization value. The switching time τin in intrinsic homogenous switching frame can be defined as the time needed by the polarization from a stable state (e.g., $+Pr$) to cross the zero axis of the polarization (e.g., B → O in Figure 8(a)) [23]. From the relationship between the derivative of the polarization P and the free energy G in (7) and (8) the reciprocal switching time τin is found to have a square-root critical dependence of the form [86]:

$$\frac{1}{\tau_{\text{in}}} \approx \frac{1}{\tau_0}\left(\frac{E}{E_c} - 1\right)^{1/2}\left(1 - \frac{T - T_0}{T_1 - T_0}\right)^{1/2}, \tag{9}$$

where $\tau_0 \approx 6.3\gamma\xi/\beta^2$ and $T = T_0 + 3\beta^2/4\gamma\alpha$. From this equation it can be seen that the switching time τin will go to infinity as the applied field E approaches Ec or the temperature increases to T_1. There is a true coercive field that for fields below this value no switching should occur. In addition, the temperature dependence of switching process is different from a thermally activated behavior which has an exponential dependence

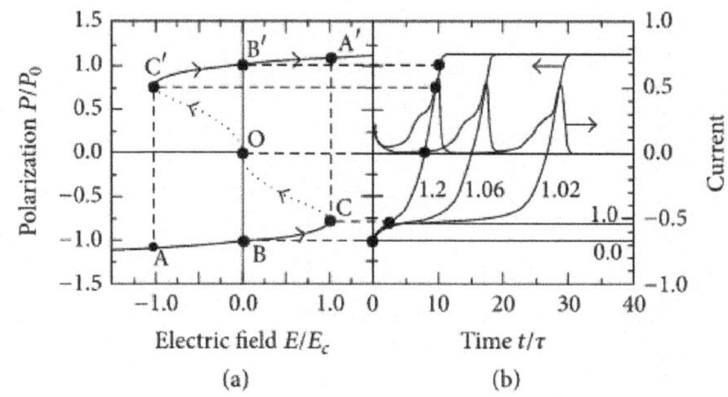

Figure 8: (a) The theoretical hysteresis loop P(E) calculated from (7) and (8) for $T \approx T_0$. The solid lines represent stable or metastable states and the dotted line denotes unstable states. (b) Time evolution of the normalized polarization and current during switching at $T = T_0$ for several values of the normalized electric field. The horizontal dashed lines connect corresponding points in (a) to states in (b) [86].

Followed by the observation of intrinsic switching in ultrathin P(VDF-TrFE) LB films, several groups argued against the intrinsic nature of the switching [23–27]. In [23] the authors found that the experiments lack several key features of intrinsic switching: the coercive field does not saturate with decreasing film thickness; the electric field and temperature dependence of the switching time are not well described by the intrinsic switching dynamics in (9); the films have switching below the coercive field. Naber et al. [27] presented measurements using gold electrode for spin coated P(VDF-TrFE) films down to 60 nm. A thickness independence of coercive field is found which leads to the conclusion that the thickness dependence of coercive field in LB films is due to the influence of the electrode interfaces. Very recently, the same group in [21] demonstrated new experiment evidences to confirm the existence of intrinsic switching kinetics in ultrathin ferroelectric copolymer films [28, 29]. They used piezoresponse force microscopy (PFM) to switch the sample at the nanoscale. The dependence of the switching rate on voltage for a 54 nm thick film exhibits extrinsic nucleation and domain growth kinetics without true threshold coercive field and is qualitatively different from the behavior of an 18 nm thick film, which exhibits intrinsic switching kinetics with a true threshold coercive field. The absence of top electrode on the sample can exclude the effect of electrode interface. Intrinsic homogeneous switching is also found in ultrathin ferroelectric BaTiO3 films [30] by PFM and ultrathin epitaxial $PbTiO_3$ films [85] by in situ synchrotron X-ray scattering measurements. Consequently, the authors claimed that homogeneous switching in ultrathin films is a common phenomenon for all ferroelectric materials.

The applicability of these models for PVDF and its copolymer depends on the sample thickness, morphology, external field, and temperature. Up to date, the switching process in P(VDF-TrFE) thin films is not fully understood. Recently, thermal Barkhausen effect is observed in P(VDFTrFE) copolymer films for the first time [87]. The Barkhausen pulse is a microscopic phenomenon and can be a direct way to prove the existence of domains. The investigation of Barkhausen effect should be a new tool to distinguish different switching mechanisms in ferroelectric materials in the future.

APPLICATIONS

Nonvolatile Memories

One of the most important applications of P(VDF-TrFE) thin films is ferroelectric nonvolatile memory. The principle of nonvolatile ferroelectric random access memories (FRAMs) is based on the polarization reversal by an external applied electric field. The binary logic states "1" and "0" are represented by the nonvolatile storage of the positive or negative remanent polarization states. The nonvolatile property is due to the fact that the sample can hold the polarization state when the external field is removed. Compared to other nonvolatile memories, for example, flash, electrically erasable and programmable read-only memories (EEPROM) FRAMs have faster write and read times, lower power consumption, and high write and read endurance. A summary of the performances for different devices is given in Table 1. FRAMs can be applied in a variety of consumer products, such as smart cards, power meters, printers, video games, and radio-frequency identification (RFID) tags

Table 1: Comparison of the performances for different types of nonvolatile memories [41–43]

Type	Area/cell (normalized)	Read access time	Write access time	Energy per 32 b write	Energy per 32 b read	Endurance
EEPROM	2	50 ns	10 μs	1 μJ	150 pJ	$>10^6$
Flash	1	50 ns	100 ns	2 μJ	150 pJ	10^4
FRAMs	2–5	100 ns	100 ns	1 nJ	1 nJ	$>10^{15}$

The available commercial FRAMs are based on perovskite-type ferroelectrics such as lead zirconium titanate (PZT) and strontium barium titanate (SBT). Researchers have overcome many difficulties including degradation and retention. Besides, perovskites generally require high annealing temperature (>400°C), which is harmful to other components on the chip. Therefore, the perovskite memory chips require buffer layers and complex mask sets [3, 88]. P(VDF-TrFE) copolymer is the promising material to replace perovskites which has attractive advantages: the device can be achieved by low-temperature processing (<200°C) and solution-processing techniques, which would enable its use in ultra-low-cost applications; the copolymer has outstanding chemical stability which barely reacts to other components and it is nontoxic.

There are three main memory elements based on ferroelectric films: metal-ferroelectric-metal (MFM) capacitor, metal-ferroelectric-insulator-semiconductor (MFIS) diode, and ferroelectric-field-effect-transistor (FeFET). The MFM capacitors can be connected to the transistors to integrate into 1T1C (one-transistor-one-capacitor) or 2T2C (two-transistor-two-capacitor) memory cells [89]. The devices based on MFM capacitors are destructive which require restoration by a rewriting cycle after a readout operation. The MFIS and FeFET devices are nondestructive with the ferroelectric polarization state modulating the electrical conductance of the semiconductor channel and thus distinguishing two logic states.

MFIS devices based on P(VDF-TrFE) films have been investigated by several groups [7–11]. Earlier relatively thick P(VDF-TrFE) films (>100 nm) with SiO_2 buffer layer were used resulting in a higher operation voltage (>10 V). More recently, using sub-100 nm P(VDF-TrFE) films, a ±3 V CV loop with a memory window of 1 V was achieved in a stack consisting of 10 nm SiO_2 layer and 36 nm P(VDF-TrFE) LB film [9]. A ±4 V loop with a large memory window of 2.9 V was obtained using a 3 nm Ta_2O_5 buffer layer and a 100 nm P(VDF-TrFE) spin coated film [10]. In [11], the authors fabricated MFM and MFIS devices using epitaxially grown P(VDF-TrFE) thin films (20–30 nm). A memory window of 2.4 V with a ±5 V loop was obtained in MFIS and their printed micropattern application was demonstrated. All-organic FeFET devices based on P(VDF-TrFE) have been reported in [5, 12–15]. High-performance solution-processed FeFETs were first reported in 2005 by Naber et al. [13] using P(VDF-TrFE) film as the gate insulator and MEH-PPV (poly (2-methoxy-5-(2-ethylhexyloxy)-p-phenylenevinylene)) as a semiconductor. The FeFETs have a long retention time (>one week) with a stable I_{on}/I_{off} of 10^4, a high programming cycle endurance (>1000 cycles), and a short programming time of 0.3 ms. However, with 60 V operation voltages, the devices need to be scaled down. To develop a storage medium that enables an independent tuning of the conductive and ferroelectric properties, Asadi et al. used phase separated films of P(VDF-TrFE) and the polymer semiconductor rir-P3HT (region-irregular poly(3-hexylthiophene)) to realize a nonvolatile resistive memory [90]. The charge injection of the electrode into the semiconducting phase is modified by the poling of the ferroelectric phase. A rectifying diode with a LiF/Al top electrode and a silver bottom electrode can be switched and read out nondestructively. Retention, switching time, and cycle endurance of the blend devices are comparable to those of the state-of-the-art polymer FeFETs. Lately, Gerber et al. [15] fabricated FeFET with very thin P(VDF-TrFE) films prepared by Langmuir-Blodgett technique. The memory element consists of a 10 nm thick SiO_2 layer and a 35 nm thick P(VDF-TrFE) film, which can be operated at 4 V. The memory window can also be measured using a Pt/PVDF/SiO_2/p-Si gate MFIS stack diode with device capacitance C versus gate voltage V_g, as shown in Figure 9. The high capacitance at negative gate bias is due to accumulation while the low capacitance at positive gate bias arises from depletion. The memory window increases linearly from 0.15 to 1.3 V as the amplitude of the gate voltage cycle increases from ±1 V to ±6 V. In contrast, the memory window decreases with increasing temperature toward the phase transition temperature at a fixed gate voltage cycle (inset in Figure 9). The short retention time (<10 s) is likely limited by incomplete polarization saturation, which can be improved by further reducing the thickness of the oxide layer so that its capacitance is at least ten times that of the ferroelectric film.

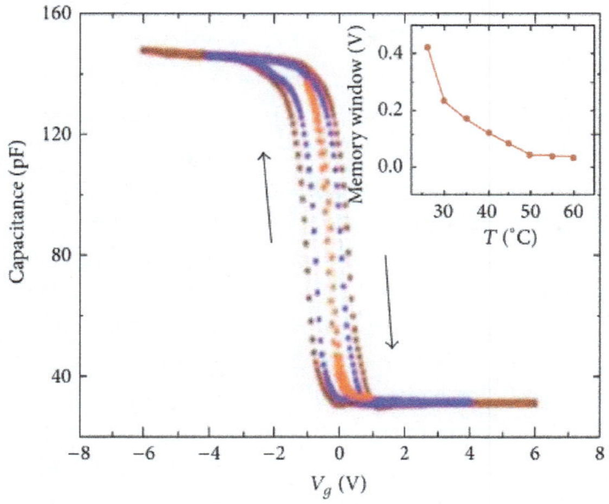

Figure 9: Capacitance-voltage hysteresis loops of a Pt/PVDF/SiO_2/p-Si gate MFIS stack diode measured with gate voltage sweeps over ±1, ±4 and ±6 V at a rate of 0.05 V/s. The measurement frequency was 100 kHz. The inset shows the width of the memory window as a function of temperature for gate voltage sweeps over ±5 V [15].

Negative Capacitance Devices

Conventional field-effect transistors (FETs) require a change in the channel potential of at least 60 mV at room temperature to induce a change in the current by a factor of 10, which is determined by the Boltzmann limit. Recently the concept of coupling the ferroelectric layer to the channel of a field effect transistor for lowering the subthreshold swing due to the negative capacitance effect is actively studied. This concept is proposed by Salahuddin and Datta in 2008 [91] which opens a new route for the realization of transistors with steeper subthreshold characteristics and thus enabling low power dissipation. The experimental validity of Fe-FET with a sub-60 mV/decade switching behavior is firstly demonstrated incorporating a thin P(VDF-TrFE) film into a gate stack [16]. Later, capacitance enhancement is manifested in an epitaxially grown single crystalline bilayer consisting of strontium titanate ($SrTiO_3$) as the dielectric and lead zirconate titanate ($Pb_xZr_{1-x}TiO_3$) as the ferroelectric at higher temperature [92] and in an epitaxially grown single crystalline bilayer of $SrTiO_3$ and barium titanate ($BaTiO_3$) at room temperature [93].

Figure 10 shows the device configuration of the negative capacitance FET [94]. The gate insulator consists of a layer of 40 nm P(VDF-TrFe) and a layer of a 10 nm SiO_2. The ferroelectric negative capacitance can be stabilized by the in-series positive oxide capacitance. The layer thicknesses and material properties should be properly engineered so as to suppress the hysteretic operation. The gate structure acts as an internal voltage transformer. Thus a small voltage swing on the gate electrode causes a larger change in the internal potential barrier which gates the channel current. The subthreshold swing value depends on the P(VDF-TrFE) thickness and the size of the channel. In this device (channel size of 50 μm × 50 μm) a subthreshold swing as low as 13 mV/decade is measured at room temperature which offers the possibility of lower voltage operation in electronic devices. The challenges include successful integration of ferroelectric/dielectric gate stacks onto FET structures especially for oxides and efficient designs to ensure hysteresis free operation [95]. More works need to be done including temperature and interface effect on the device performance.

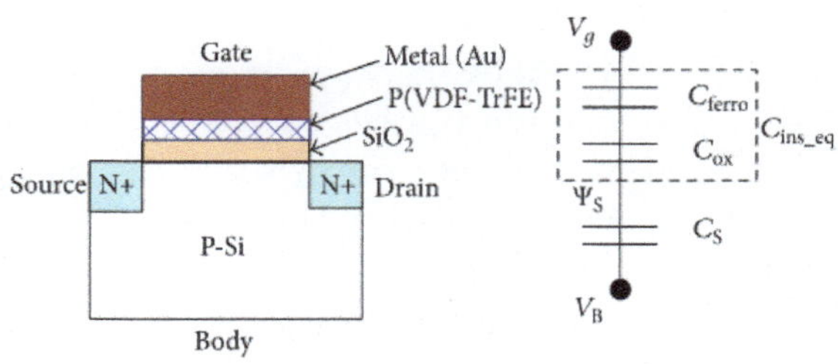

Figure 10: Investigated ferroelectric transistor Fe-FET and equivalent capacitive divider of gate potentials [94].

Photovoltaic Devices

Recently it is reported that, by using a permanent electric field of an ultrathin layer of ferroelectric P(VDF-TrFE) introduced at the interface between the electrode and a semiconductor layer in an organic photovoltaic (OPV) device, the charge pair separation and charge extraction efficiency can be increased and thus the power conversion efficiency (PCE) is enhanced by up to 200% [17, 96]. OPV devices have been intensively investigated during the last few years due to their promising application for future low cost and high performance energy sources. The energy loss in OPV devices is mainly caused by the recombination of electrons and holes in semiconducting polymer-fullerene blends. To separate the electrons and holes and prevent their recombination by an external field is essential to increase the OPV efficiency. A large internal electric field can be induced by incorporating a ferroelectric layer thus eliminating the need for an external field. The device structure and working principle of the OPV based on a P(VDF-TrFE) thin film are illustrated in Figure 11. The ultrathin ferroelectric layer was prepared by Langmuir-Blodgett technique and inserted at the interface of the organic semiconducting layer and metal electrode, which has a large polarization of the order of 10 μC/cm^2. The ferroelectric layer can induce a high density of charges at the FE-semiconductor interface which cannot be compensated by the low-concentration free charges in organic semiconductors. The built-in electric field introduced by the P(VDF-TrFE) layer is derived as $E = d\sigma_P/\varepsilon_0\varepsilon_{FE}L$, with σ_P being the polarization charge density, d the thickness of the P(VDF-TrFE) layer, L the thickness of the polymer semiconductor layer, and ε_{FE} the relative dielectric constant of the P(VDF-TrFE) layer. A three monolayer (5 nm) P(VDF-TrFE) film can induce an internal built-in field as high as 50 V/μm in a 150 nm P3HT:PC_{70}BM layer and thus suppresses the recombination in the device. According to the above derived equation, the higher the built-in electric field is, the thicker the P(VDF-TrFE) layer is. However, the increase of the P(VDF-TrFE) layer thickness also increases the contact resistance, which would reduce device efficiency [17]. The authors found that one or two monolayers of P(VDF-TrFE) were the optimum thickness range for P3HT:PC_{70}BM to achieve the highest efficiency.

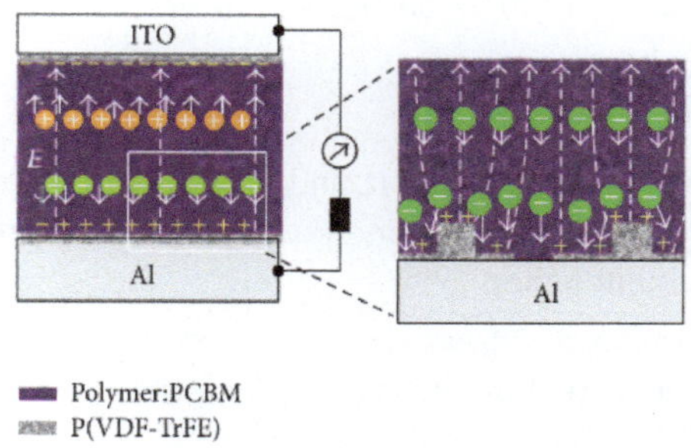

Figure 11: The structure of polymer photovoltaic devices with FE interfacial layers and a schematic diagram of electric field induced by the polarized FE layer and the field-assisted charge extraction. The right-hand panel illustrates the electric-field distribution and electron conduction through the P(VDF-TrFE) on the Al side [17].

Multiferroic Tunnel Junctions

Incorporating P(VDF-TrFE) thin films in multiferroic tunnel junctions is also an emerging application. Ferroelectric polymer has a high electric polarization and weakly binds to the metal ferromagnets, which are favorable for use as barriers in multiferroic tunnel junctions. Experimentally, Mardana et al. have demonstrated that the use of P(VDF-TrFE) LB films can efficiently tailor the interface magnetocrystalline anisotropy (MCA) of an adjacent ferromagnetic Co layer [18]. In a wedge-shaped Co film of varying thickness overlaid with a P(VDF-TrFE) film the magnetic anisotropy of the Co films changes as much as 50% when the ferroelectric polarization of the copolymer is switched from up to down. The large mismatch in the stiffness coefficients between the copolymer and the metal ferromagnet manifests that the magnetic changes are caused by the large electric field at the ferroelectric/ferromagnet interface. First-principles density functional calculations [97, 98] show that PVDF based multiferroic tunnel junctions have very high tunability of the tunneling magnetoresistance and electroresistance effects, which are promising candidates for logic and memory applications. More experiment works need to be done.

SUMMARY AND OUTLOOK

Significant advances have been made on the development of ferroelectric devices based on PVDF and P(VDF-TrFE) thin films in the last ten years. Still, we are in the early stages of commercialization of thin ferroelectric polymer devices in organic electronics. There remain some questions about the microscopic mechanisms of polarization reversal in ferroelectric polymer films. Understanding of these issues is critical. The material quality in terms of processing and properties could be evaluated and improved. Future works to establish a clear relationship between ferroelectric properties and device performance are required. The prospects of ferroelectric polymer devices are promising for microelectronic industry.

ACKNOWLEDGMENTS

This work was supported by the National Natural Science Foundations of China (nos. 51302172, 21405106, and 51272161) and the Science and Technology Research Items of Shenzhen (no. JCYJ20140509172719306).

REFERENCES

1. C. P. de Araujo, J. F. Scott, and G. W. Taylor, Ferroelectric Thin Films: Synthesis and Basic Properties, Gordon and Breach Publicher, 1996.
2. T. Furukawa, "Ferroelectric properties of vinylidene fluoride copolymers," Phase Transitions, vol. 18, pp. 143–211, 1989. ·
3. S. Ducharme, T. J. Reece, C. M. Othon, and R. K. Rannow, "Ferroelectric polymer Langmuir-Blodgett films for a nonvolatile memory applications," IEEE Transactions on Device and Materials Reliability, vol. 5, no. 4, pp. 720–735, 2005. · ·
4. T. Furukawa, Y. Takahashi, and T. Nakajima, "Recent advances in ferroelectric polymer thin films for memory applications," Current Applied Physics, vol. 10, no. 1, pp. e62–e67, 2010. · ·
5. R. C. G. Naber, K. Asadi, P. W. M. Blom, D. M. de Leeuw, and B. de Boer, "Organic nonvolatile memory devices based on ferroelectricity," Advanced Materials, vol. 22, no. 9, pp. 933–945, 2010. · ·
6. M. Poulsen and S. Ducharme, "Why ferroelectric polyvinylidene fluoride is special," IEEE Transactions on Dielectrics and Electrical Insulation, vol. 17, no. 4, pp. 1028–1035, 2010. · ·
7. T. J. Reece, S. Ducharme, A. V. Sorokin, and M. Poulsen, "Non-volatile memory element based on a ferroelectric polymer Langmuir-Blodgett film," Applied Physics Letters, vol. 82, no. 1, pp. 142–144, 2003. · ·
8. S. H. Lim, A. C. Rastogi, and S. B. Desu, "Electrical properties of metal-ferroelectric-insulator-semiconductor structures based on ferroelectric polyvinylidene fluoride copolymer film gate for nonvolatile random access memory application," Journal of Applied Physics, vol. 96, no. 10, pp. 5673–5682, 2004. · ·
9. A. Gerber, H. Kohlstedt, M. Fitsilis et al., "Low-voltage operation of metal-ferroelectric-insulator-semiconductor diodes incorporating a ferroelectric polyvinylidene fluoride copolymer Langmuir-Blodgett film," Journal of Applied Physics, vol. 100, no. 2, Article ID 024110, 2006. · ·
10. S. Fujisaki, H. Ishiwara, and Y. Fujisaki, "Low-voltage operation of ferroelectric poly(vinylidene fluoridetrifluoroethylene) copolymer capacitors and metal-ferroelectric- insulator-semiconductor diodes," Applied Physics Letters, vol. 90, no. 16, Article ID 162902, 2007. · ·
11. Y. J. Park, S. J. Kang, Y. Shin, R. H. Kim, I. Bae, and C. Park, "Non-volatile memory characteristics of epitaxially grown PVDF-TrFE thin films and their printed micropattern application," Current Applied Physics, vol. 11, no. 2, pp. e30–e34, 2011. · ·
12. R. Schroeder, L. A. Majewski, and M. Grell, "All-organic permanent memory transistor using an amorphous, spin-cast ferroelectric-like gate insulator," Advanced Materials, vol. 16, no. 7, pp. 633–576, 2004. · ·
13. R. C. G. Naber, C. Tanase, P. W. M. Blom et al., "High-performance solution-processed polymer ferroelectric field-effect transistors," Nature Materials, vol. 4, no. 3, pp. 243–248, 2005. · ·
14. K. N. Narayanan Unni, S. Dabos-Seignon, and J. M. Nunzi, "Improved performance of pentacene field-effect transistors using a polyimide gate dielectric layer," Journal of Physics D: Applied Physics, vol. 38, no. 8, pp. 1148–1151, 2005. · ·
15. A. Gerber, M. Fitsilis, R. Waser et al., "Ferroelectric field effect transistors using very thin ferroelectric polyvinylidene fluoride copolymer films as gate dielectrics," Journal of Applied Physics, vol. 107, no. 12, Article ID 124119, 2010. · ·
16. A. Rusu, G. A. Salvatore, D. Jiménez, and A. M. Ionescu, "Metal-ferroelectric-metal-oxide-semiconductor field effect transistor with sub-60mV/decade subthreshold swing and internal voltage amplification," in Proceedings of the IEEE International Electron Devices Meeting (IEDM '10), pp. 395–398, December 2010. · ·
17. Y. Yuan, T. J. Reece, P. Sharma et al., "Efficiency enhancement in organic solar cells with ferroelectric polymers," Nature Materials, vol. 10, no. 4, pp. 296–302, 2011. · ·
18. A. Mardana, S. Ducharme, and S. Adenwalla, "Ferroelectric control of magnetic anisotropy," Nano Letters, vol. 11, no. 9, pp. 3862–3867, 2011. · ·
19. S. Palto, L. Blinov, A. Bune et al., "Ferroelectric Langmuir-Blodgett films," Ferroelectrics, Letters Section, vol. 19, no. 3-4, pp. 65–68, 1995. · ·
20. A. V. Bune, V. M. Fridkin, S. Ducharme et al., "Two-dimensional ferroelectric films," Nature, vol. 391, no. 6670, pp. 874–877, 1998. · ·
21. S. Ducharme, V. M. Fridkin, A. V. Bune et al., "Intrinsic ferroelectric coercive field," Physical Review Letters, vol. 84, no. 1, pp. 175–177, 2000. · ·
22. J. F. Scott, "Domain wall kinetics: nano-domain nucleation in lead germanate and Tilley-Zeks theory for PVDF," Ferroelectrics, vol. 291, no. 1, pp. 205–215, 2003. ·
23. H. Kliem and R. Tadros-Morgane, "Extrinsic versus intrinsic ferroelectric switching: experimental investigations using ultra-thin PVDF Langmuir–Blodgett films," Journal of Physics D: Applied Physics, vol. 38, no. 12, pp. 1860–1868, 2005. · ·
24. A. M. Bratkovsky and A. P. Levanyuk, "Comment on:

'intrinsic ferroelectric coercive field'," Physical Review Letters, vol. 87, no. 1, Article ID 019701, 2001. ·

25. R. L. Moreira, "Comment on 'intrinsic ferroelectric coercive field'," Physical Review Letters, vol. 88, no. 17, Article ID 179701, 2002. ·
26. J. F. Scott, M. Dawber, A. Q. Jiang, and F. D. Morrison, "New developments in ferroelectric thin films,"Ferroelectrics, vol. 286, no. 1, pp. 223–235, 2003. ·
27. R. C. G. Naber, P. W. M. Blom, and D. M. De Leeuw, "Comment on 'extrinsic versus intrinsic ferroelectric switching: experimental investigations using ultra-thin PVDF Langmuir-Blodgett films',"Journal of Physics D: Applied Physics, vol. 39, no. 9, pp. 1984–1986, 2006. · ·
28. R. V. Gaynutdinov, S. Mitko, S. G. Yudin, V. M. Fridkin, and S. Ducharme, "Polarization switching at the nanoscale in ferroelectric copolymer thin films," Applied Physics Letters, vol. 99, no. 14, Article ID 142904, 2011. · ·
29. R. Gaynutdinov, S. Yudin, S. Ducharme, and V. Fridkin, "Homogeneous switching in ultrathin ferroelectric films," Journal of Physics Condensed Matter, vol. 24, no. 1, Article ID 015902, 2012. · ·
30. S. Ducharme, V. Fridkin, R. Gaynutdinov, M. Minnekaev, A. Tolstikhina, and A. Zenkevich, "Homogeneous switching in ultrathin ferroelectric $BaTiO_3$ films," http://arxiv.org/abs/1204.4792.
31. H. Kawai, "The piezoelectricity of poly(vinylidene fluoride)," Japanese Journal of Applied Physics, vol. 8, no. 7, pp. 975–976, 1969. ·
32. J. G. Bergman Jr., J. H. McFee, and G. R. Crane, "Pyroelectricity and optical second harmonic generation in polyvinylidene fluoride films," Applied Physics Letters, vol. 18, no. 5, pp. 203–205, 1971. · ·
33. R. G. Kepler and R. A. Anderson, "Ferroelectricity in polyvinylidene fluoride," Journal of Applied Physics, vol. 49, no. 3, p. 1232, 1978. · ·
34. T. Furukawa, M. Date, and E. Fukada, "Hysteresis phenomena in polyvinylidene fluoride under high electric field," Journal of Applied Physics, vol. 51, no. 2, pp. 1135–1141, 1980. · ·
35. M. Li, H. J. Wondergem, M.-J. Spijkman et al., "Revisiting the δ-phase of poly(vinylidene fluoride) for solution-processed ferroelectric thin films," Nature Materials, vol. 12, no. 5, pp. 433–438, 2013. · ·
36. T. Furukawa, G. E. Johnson, H. E. Bair, and Y. Tajitsu, "Ferroelectric phase transition in a copolymer of vinylidene fluoride and trifluoroethylene," Ferroelectrics, vol. 32, no. 1, pp. 61–67, 1981. ·
37. L. M. Blinov, V. M. Fridkin, S. P. Palto, A. V. Bune, P. A. Dowben, and S. Ducharme, "Two-dimensional ferroelectrics," Physics-Uspekhi, vol. 43, no. 3, pp. 243–257, 2000. · ·
38. V. V. Kochervinskii, "Ferroelectricity of polymers based on vinylidene fluoride," Russian Chemical Reviews, vol. 68, no. 10, pp. 821–857, 1999. · ·
39. A. R. Geivandov, S. G. Yudin, V. M. Fridkin, and S. Ducharme, "Manifestation of a ferroelectric phase transition in ultrathin films of polyvinylidene fluoride," Physics of the Solid State, vol. 47, no. 8, pp. 1590–1594, 2005. · ·
40. M. Mai, V. Fridkin, B. Martin, A. Leschhorn, and H. Kliem, "The thickness dependence of the phase transition temperature in PVDF," Physica B: Condensed Matter, vol. 421, pp. 23–27, 2013. · ·
41. A. Sheikholeslami and P. G. Gulak, "A survey of circuit innovations in ferroelectric random-access memories," Proceedings of the IEEE, vol. 88, no. 5, pp. 667–689, 2000. · ·
42. G. W. Burr, B. N. Kurdi, J. C. Scott, C. H. Lam, K. Gopalakrishnan, and R. S. Shenoy, "Overview of candidate device technologies for storage-class memory," IBM Journal of Research and Development, vol. 52, no. 4-5, pp. 449–464, 2008. · ·
43. M. H. Kryder and C. S. Kim, "After hard drives—what comes next?" IEEE Transactions on Magnetics, vol. 45, no. 10, pp. 3406–3413, 2009. · ·
44. T. Nakajima, Y. Takahashi, S. Okamura, and T. Furukawa, "Nanosecond switching characteristics of ferroelectric ultrathin vinylidene fluoride/trifluoroethylene copolymer films under extremely high electric field," Japanese Journal of Applied Physics, vol. 48, no. 9, Article ID 09KE04, 2009. · ·
45. K. Müller, D. Mandal, K. Henkel, I. Paloumpa, and D. Schmeisser, "Ferroelectric properties of spin-coated ultrathin (down to 10 nm) copolymer films," Applied Physics Letters, vol. 93, no. 11, Article ID 112901, 2008. · ·
46. M. Li, I. Katsouras, C. Piliego et al., "Controlling the microstructure of poly(vinylidene-fluoride) (PVDF) thin films for microelectronics," Journal of Materials Chemistry C, vol. 1, no. 46, pp. 7695–7702, 2013. · ·
47. P. Martin and M. Szablewski, Langmuir-Blodgett Troughs Operating Manual, Nima Technology, 2002.
48. M. Mai, B. Martin, and H. Kliem, "Ferroelectric switching in Langmuir-Blodgett and spin-coated thin films of poly(vinylidene fluoride/trifluoroethylene) copolymers," Journal of Applied Physics, vol. 110, no. 6, Article ID 064101, 2011. · ·
49. M. Mai, B. Martin, and H. Kliem, "Polarization relaxation and charge injection in thin films of poly(vinylidene fluoride/trifluoroethylene) copolymer," Journal of Applied Physics, vol. 114, no. 5, Article ID 054104, 2013. · ·
50. B. Xu, C. N. Borca, S. Ducharme et al., "Aluminum doping of poly(vinylidene fluoride with trifluoroethylene) copolymer," Journal of Chemical Physics, vol. 114, no. 4, pp. 1866–1869, 2001. · ·
51. B. Martin, M. Mai, and H. Kliem, "Broadband dielectric dispersion in ferroelectric P(VDF-TrFE) copolymer films," Physica B: Condensed Matter, vol. 407, no. 21, pp. 4398–4404, 2012. · ·
52. F. Xia and Q. M. Zhang, "Influence of metal electrodes on the ferroelectric responses of poly(vinylidene fluoride-trifluoroethylene) copolymer thin films," MRS Proceedings, vol. 734, 2003. ·

53. G. Vizdrik and E. Rije, "Fast polarization switching and influence of platinum electrodes in ultrathin langmuir films of poly(vinylidene fluoride-trifluoroethylene)," Ferroelectrics, vol. 370, no. 1, pp. 74–84, 2008. · ·

54. P. K. Wu, G.-R. Yang, X. F. Ma, and T.-M. Lu, "Interaction of amorphous fluoropolymer with metal,"Applied Physics Letters, vol. 65, no. 4, pp. 508–510, 1994. · ·

55. G.-D. Zhu, X.-Y. Luo, J.-H. Zhang, Y. Gu, and Y.-L. Jiang, "Electrical fatigue in ferroelectric P(VDF-TrFE) copolymer films," IEEE Transactions on Dielectrics and Electrical Insulation, vol. 17, no. 4, pp. 1172–1177, 2010. · ·

56. R. C. G. Naber, B. de Boer, P. W. M. Blom, and D. M. de Leeuw, "Low-voltage polymer field-effect transistors for nonvolatile memories," Applied Physics Letters, vol. 87, no. 20, Article ID 203509, 2005. · ·

57. B. Xu, J. Choi, C. N. Borca et al., "Comparison of aluminum and sodium doped poly(vinylidene fluoride-trifluoroethylene) copolymers by x-ray photoemission spectroscopy," Applied Physics Letters, vol. 78, no. 4, pp. 448–450, 2001. · ·

58. R. C. G. Naber, P. W. M. Blom, A. W. Marsman, and D. M. De Leeuw, "Low voltage switching of a spin cast ferroelectric polymer," Applied Physics Letters, vol. 85, no. 11, pp. 2032–2034, 2004. · ·

59. W. J. Merz, "Domain formation and domain wall motions in ferroelectric $BaTiO_3$ single crystals,"Physical Review, vol. 95, no. 3, pp. 690–698, 1954. · ·

60. E. A. Little, "Dynamic behavior of domain walls in barium titanate," Physical Review, vol. 98, no. 4, pp. 978–984, 1955. · ·

61. R. C. Miller, "Some experiments on the motion of 180° domain walls in $BaTiO_3$," Physical Review, vol. 111, no. 3, pp. 736–739, 1958. · ·

62. R. C. Miller and G. Weinreich, "Mechanism for the sidewise motion of 180° domain walls in barium titanate," Physical Review, vol. 117, no. 6, pp. 1460–1466, 1960. · ·

63. E. Fatuzzo, "Theoretical considerations on the switching transient in ferroelectrics," Physical Review, vol. 127, no. 6, pp. 1999–2005, 1962. · ·

64. A. N. Kolmogorov, "A statistical theory for the recrystallization of metals," Izvestia Akademia Nauk USSR Serie Mathematica, vol. 3, pp. 355–359, 1937.

65. M. Avrami, "Kinetics of phase change. I. General theory," The Journal of Chemical Physics, vol. 7, no. 12, p. 1103, 1939. · ·

66. A. K. Tagantsev, L. E. Cross, and J. Fousek, Domains in Ferroic Crystals and Thin Films, Springer, New York, NY, USA, 2010.

67. A. K. Tagantsev, I. Stolichnov, and N. Setter, "Non-Kolmogorov-Avrami switching kinetics in ferroelectric thin films," Physical Review B—Condensed Matter and Materials Physics, vol. 66, no. 21, Article ID 214109, 2002. ·

68. S. Hashimoto, H. Orihara, and Y. Ishibashi, "Study on D-E hysteresis loop of TGS based on the Avrami-type model," Journal of the Physical Society of Japan, vol. 63, no. 4, pp. 1601–1610, 1994. · ·

69. W. Li and M. Alexe, "Investigation on switching kinetics in epitaxial Pb $(Zr_{0.2}Ti_{0.8})O_3$ ferroelectric thin films: role of the 90° domain walls," Applied Physics Letters, vol. 91, no. 26, Article ID 262903, 2007. · ·

70. J. Y. Jo, H. S. Han, J.-G. Yoon, T. K. Song, S.-H. Kim, and T. W. Noh, "Domain switching kinetics in disordered ferroelectric thin films," Physical Review Letters, vol. 99, no. 26, Article ID 267602, 2007. · ·

71. S. Zhukov, Y. A. Genenko, and H. von Seggern, "Experimental and theoretical investigation on polarization reversal in unfatigued lead-zirconate-titanate ceramic," Journal of Applied Physics, vol. 108, no. 1, Article ID 014106, 2010. · ·

72. D. Mao, I. Mejia, H. Stiegler, B. E. Gnade, and M. A. Quevedo-Lopez, "Polarization behavior of poly(vinylidene fluoride-trifluoroethylene) copolymer ferroelectric thin film capacitors for nonvolatile memory application in flexible electronics," Journal of Applied Physics, vol. 108, no. 9, Article ID 094102, 2010. · ·

73. M. Kuehn and H. Kliem, "The method of local fields: a bridge between molecular modelling and dielectric theory," Journal of Electrostatics, vol. 67, no. 2-3, pp. 203–208, 2009. · ·

74. P. Weiss, "Sur la Rationalite des rapports des moments magnetiques moleculaires et le Magneton,"Archives des Sciences Physiques et Naturelles, vol. 4, p. 401, 1911.

75. P. Weiss, "Sur la nature du champ moleculaire," Archives des Sciences Physiques et Naturelles, vol. 37, p. 201, 1914.

76. J. C. Burfoot and G. W. Taylor, Polar Dielectrics and Their Applications, Macmillan, London, UK, 1979.

77. H. A. Lorentz, The Theory of Dielectrics, Teubner, 1909.

78. H. Kliem, M. Kuehn, and B. Martin, "The weiss field revisited," Ferroelectrics, vol. 400, no. 1, pp. 41–51, 2010. · ·

79. M. Kuehn, V. M. Fridkin, and H. Kliem, "Influence of an external field on the first order ferroelectric phase transition," Ferroelectrics, Letters Section, vol. 37, no. 3, pp. 55–59, 2010. · ·

80. M. Kuehn and H. Kliem, "Phase transitions in the modified weiss model," Ferroelectrics, vol. 400, no. 1, pp. 52–62, 2010. · ·

81. H. Kliem, "Strain induced change from second order to first order ferroelectric phase transition,"Ferroelectrics, vol. 425, no. 1, pp. 54–62, 2011. · ·

82. H. Kliem, A. Leschhorn, and T. Edtbauer, "A model for the double loop of ferroelectric polarization close to T_C," Journal of Applied Physics, vol. 113, no. 10, Article ID 104102, 2013. · ·

83. M. Mai, A. Leschhorn, and H. Kliem, "The field and temperature dependence of hysteresis loops in P(VDF–TrFE) copolymer films," Physica B: Condensed Matter, vol. 456, pp. 306–311, 2015. ·

84. H. Kliem and M. Kuehn, "Modeling the switching kinetics in ferroelectrics," Journal of Applied Physics, vol. 110, no. 11, Article ID 114106, 2011. · ·

85. M. J. Highland, T. T. Fister, M.-I. Richard et al., "Polarization switching without domain formation at the intrinsic coercive field in ultrathin ferroelectric $PbTiO_3$," Physical Review Letters, vol. 105, no. 16, Article ID 167601, 2010. · ·

86. G. Vizdrik, S. Ducharme, V. M. Fridkin, and S. G. Yudin, "Kinetics of ferroelectric switching in ultrathin films," Physical Review B, vol. 68, no. 9, Article ID 094113, pp. 1–6, 2003. ·

87. M. Mai and H. Kliem, "Observation of thermal Barkhausen effect in ferroelectric films of poly(vinylidene fluoride/trifluoroethylene) copolymer," Journal of Applied Physics, vol. 114, no. 22, Article ID 224104, 2013. · ·

88. M. Mort, G. Schindler, W. Hartner et al., "Low temperature process and thin SBT films for ferroelectric memory devices," Integrated Ferroelectrics, vol. 30, no. 1–4, pp. 235–244, 2000. ·

89. H. Ishiwara, M. Okuyama, and Y. Arimoto, Eds., Ferroelectric Random Access Memories: Fundamentals and Applications, Springer, Berlin, Germany, 2004.

90. K. Asadi, D. M. de Leeuw, B. de Boer, and P. W. M. Blom, "Organic non-volatile memories from ferroelectric phase-separated blends," Nature Materials, vol. 7, no. 7, p. 547, 2008. ··

91. S. Salahuddin and S. Datta, "Use of negative capacitance to provide voltage amplification for low power nanoscale devices," Nano Letters, vol. 8, no. 2, pp. 405–410, 2008. · ·

92. A. Islam Khan, D. Bhowmik, P. Yu et al., "Experimental evidence of ferroelectric negative capacitance in nanoscale heterostructures," Applied Physics Letters, vol. 99, no. 11, Article ID 113501, 2011. ·

93. D. J. R. Appleby, N. K. Ponon, K. S. K. Kwa et al., "Experimental observation of negative capacitance in ferroelectrics at room temperature," Nano Letters, vol. 14, no. 7, pp. 3864–3868, 2014. ·

94. G. A. Salvatore, D. Bouvet, and A. M. Ionescu, "Demonstration of subthrehold swing smaller than 60mV/decade in Fe-FET with P(VDF-TrFE)/SiO_2 gate stack," in Proceedings of the IEEE International Electron Devices Meeting (IEDM '08), pp. 1–4, San Francisco, Calif, USA, December 2008. ·

95. S. Salahuddin, "Negative capacitance in a ferroelectric-dielectric heterostructure for ultra low-power computing," in Spintronics V, vol. 8461 of Proceeding of SPIE, p. 846111, 2012.

96. Y. Yuan, P. Sharma, Z. Xiao et al., "Understanding the effect of ferroelectric polarization on power conversion efficiency of organic photovoltaic devices," Energy and Environmental Science, vol. 5, no. 9, pp. 8558–8563, 2012. · ·

97. P. V. Lukashev, T. R. Paudel, J. M. López-Encarnación, S. Adenwalla, E. Y. Tsymbal, and J. P. Velev, "Ferroelectric control of magnetocrystalline anisotropy at cobalt/poly(vinylidene fluoride) interfaces,"ACS Nano, vol. 6, no. 11, pp. 9745–9750, 2012. · ·

98. J. P. Velev, J. M. López-Encarnación, J. D. Burton, and E. Y. Tsymbal, "Multiferroic tunnel junctions with poly(vinylidene fluoride)," Physical Review B—Condensed Matter and Materials Physics, vol. 85, no. 12, 6 pages, 2012.

CHAPTER 11

MODEL FOR CHARGE TRANSPORT IN FERROELECTRIC NANOCOMPOSITE FILM

Meng H. Lean and Wei-Ping L. Chu

INTRODUCTION

Nanocomposite films comprising ceramic nanofillers in amorphous polymer matrices are of interest in capacitive energy storage for rapid power cycling applications. Advantages include high energy density, low cost, lightweight, and small footprint. Nanocomposite films combine the processability and high breakdown field strength of the polymer with the high dielectric constant of the fillers, engineered to an acceptable level of dielectric loss. Most of the increase in the effective dielectric constant comes from an increase in the average field in the polymer matrix with very little of the energy being stored in the high permittivity phase. Large contrast in permittivity between the two phases gives rise to highly inhomogeneous electric fields (E) in the "interaction zone," defined as the interfacial region that surrounds the nanofillers and interspaces. Within this region, mobility is field-dependent and may also depend on the physicochemical bonds on the common surfaces. Morphology plays an important role in film layers with very high surface-to-volume aspect ratios. Highly inhomogeneous fields and structural inhomogeneity generally lead to a significant reduction in the effective breakdown field strength of the composite, limiting the increase in the energy storage capacity and energy density. So a careful compromise needs to be engineered at the polymer-ceramic interfaces to attain a viable solution [1]. Interface engineering attempts to improve the dispersion of filler in polymer matrix. In a well-dispersed composite, interfaces between the ceramic nanoparticle and polymer phases create effective electron scattering and transport centers, thus reducing the breakdown probability. Moreover, well-dispersed ceramic nanoparticles block degradation tree growth and thus increase the long-term breakdown strength.

The polymers currently used as matrices in the dielectric nanocomposites, including polyethylenes (PE), poly(methyl methacrylate)s (PMMA), epoxy resins, and polyimides (PI), usually possess dielectric permittivities of ~2–5. significantly lower than their inorganic counterparts. Poly(vinylidene fluoride) (PVDF) based ferroelectric polymers exhibit large spontaneous polarization and high dielectric constants (~10 at 1 kHz) because of the presence of highly electronegative fluorine on the polymer chains and the spontaneous alignment of C–F dipoles in the crystalline phases. PVDF copolymers, such as poly(vinylidene fluoride-co-trifluoroethylene), P(VDF-TrFE), and poly(vinylidene fluoride-co-hexafluoropropylene), P(VDF-HFP), have also been utilized in the formation of the dielectric nanocomposites. However, dispersion of inorganic fillers in these fluorinated polymers is always problematic because of the low surface energy of the polymers. The agglomeration of the ceramic dopants gives rise to electron conduction for a high dielectric loss and undesirable porosity for dielectric failure at much lower fields. PVDF is a semicrystalline polymer with varying degrees of crystallinity. The ferroelectric properties of crystalline PVDF stem from the net molecular dipole moment, oriented in the plane of the carbon backbone and along the polymer chain [2].

Considerable progress has been made over the past several years in the enhancement of the energy densities of the polymer nanocomposites by tuning the chemical structures of ceramic fillers and polymer matrices and engineering the polymer-ceramic interfaces. Ferroelectric ceramics, such as $BaTiO_3$ (BT), $Pb(Mg_{1/3}Nb_{2/3})O_3$-$PbTiO_3$ (PMN-PT), or other ferroelectrics or relaxor ferroelectrics, possess a very large

Citation: Meng H. Lean and Wei-Ping L. Chu, "Model for Charge Transport in Ferroelectric Nanocomposite Film,"Journal of Polymers, vol. 2015, Article ID 745056, 17 pages, 2015. doi:10.1155/2015/745056.

dielectric permittivity. Large enhancements in the electric energy density and the electric displacement in the nanocomposites of P(VDF-TrFE-CFE) terpolymer/ZrO_2 nanoparticles have been demonstrated through the interface effect with the presence of 1.6 vol% of ZrO_2 nanoparticles [3]. Surface modifications include grafting, functionalization, and bonding. Interface engineering attempts to improve the dispersion of filler in polymer matrices include the grafting of polystyrene (PS) directly onto TiO_2 surfaces via atom transfer radical polymerization (ATRP) to result in a dielectric constant enhancement of over three times that of bulk polystyrene. The high permittivity core lends high capacitance, while the flexible polystyrene shell endows the materials with good dispersability and film-forming properties. A second method is the functionalization or Phosphonic acid surface-modification of BT nanoparticles to yield well-dispersed P(VDF-HFP) composites [4]. Chemical functionalization of the BT nanoparticles with ethylene diamine moieties on the surface also rendered the particles as homogeneous distributions in the poly(vinylidene fluoride-co-chlorotrifluoroethylene) [P(VDF-CTFE)] and poly(vinylidene fluoride-ter-trifluoroethylene-ter-chlorotrifluoroethylene) [P(VDF-TrFE-CTFE)] matrices [5]. Another method is chain-end functionalization of the ferroelectric polymer and subsequent covalent-bonding of the polymer nanocomposites to form uniform nanoparticle dispersion. Ferroelectric polymers have also been terminated with Phosphonic acids together with subsequent utilization of the reactive end-groups of the polymer for directly coupling with oxide nanoparticles to yield the covalent-bonded nanocomposites. It was demonstrated that covalent assembly of polymer matrix and inorganic filler not only led to highly dispersed nanofillers without additional surface modification but also provided great stability and offered enhanced interfacial interactions for high polarization responses under the applied fields [6]. High-energy-density polymer nanocomposites based on surface-functionalized TiO_2 nanocrystals used as dopants in a ferroelectric P(VDF-TrFE-CTFE) have been shown to possess comparable dielectric permittivities of 42 and 47, respectively, measured at room temperature and 1 kHz. High dielectric performance in the nanocomposites is realized via the large enhancement in polarization response at high electric fields and changes in polymer microstructure induced by the nanofillers. The energy density of the P(VDF-TrFE-CTFE) nanocomposites on the TiO_2 concentration was reported to peak at around 10 vol% TiO_2 content. This feature is likely associated with the interface effect that is proportional to the interfacial area. The thickness of the interface region is generally estimated to be ~20 nm when the volume fraction of the polymer chains residing in the interfacial area reaches a maximum at 10 vol%. The large interface area in the nanocomposites would produce the Maxwell-Wagner-Sillars (MWS) interfacial polarization at low frequencies contributing to the creation of the interaction zone with the Gouy-Chapman diffuse layer, thereby greatly affecting polarization and the dielectric response of the polymer matrix. The incorporation of the TiO_2 nanoparticles into the polymer induces an improved electric displacement, which accounts for high energy densities observed in the nanocomposites [7]. Finally, surface hydroxylation treatment using hydrogen peroxide to modify the surface of BT nanofillers dispersed in a ferroelectric copolymer host has resulted in up to two orders of magnitude reduction in the leakage current of nanocomposite thin-film capacitors. This reduction is observed concurrently with the enhancement of the effective permittivity and breakdown strength of the thin-film nanocomposites [8].

Characterization studies on PVDF have involved dynamic pyroelectrical measurements and thermally stimulated depolarization (TSD) experiments, from which the permanent polarization of the dipoles in the crystalline phase, the frozen-in polarization of the dipoles in the amorphous phase, and the effect of the excess charges of the MWS interface polarization can be identified [9]. The influence of uniaxial stretching on the dielectric properties and remanent polarization of α-PVDF into α-phase oriented films were investigated by differential scanning calorimetry (DSC) and infrared spectroscopy (FTIR) and verified the increase in crystallinity by the reduction of the amorphous-crystalline interface, stable remanent polarization, and the permittivity at 1 kHz [10]. Experimental studies on the dependence of polarization fatigue on crystallinity of P(VDF-TrFE) copolymer films indicated that high crystallinity of the film correlated with slower fatigue rate of the film. A possible explanation was the space charge trapped at the boundaries of crystallites and/or captured by defects in both the amorphous and crystalline phases as the major contributor to polarization fatigue [11].

In matching nanofillers with the polymer matrix, the challenge is to understand the role of the interaction zone where the very large area to volume ratio of the interfaces in nanocomposites would have significant impact on the electrical and dielectric properties of the polymers.

MODELING RATIONALE

Nanocomposite Models

The physics of the interfaces is still not well understood. Key questions remain on how the interfaces are related to the structures of the components and how the interfacial properties change with the filler size, orientation, loading, and the applied electric field. Detailed modeling guided by theoretical insight and in-depth dielectric characterization is necessary to derive understanding before rational designs can be used to guide the synthesis of hybrid structures with optimally engineered interfaces. Creation of such a model begins with an appropriate representation for the nanofiller/nanocrystallite and its charge transport behavior.

Several models of nanoparticles are discussed in the literature [12]. The Tanaka Multicore 3-layer model consists of (1) a first layer on the nanoparticle surface only a few nm thick forming a region of chemical bonding that is ionic, covalent, hydrogen, or van der Waals' force; (2) a second layer corresponding to a region of some ordered polymer chains, which is some 10 nm in thickness; and (3) a third layer which is dominated by the far-distance force originating from the electric double layer (EDL) which is several 100 nm thick, corresponding to the Debye shielding length. Another is the Lewis 2-layer model comprising an inner bound Stern layer and a diffuse Gouy-Chapman (Debye-Hückel) outer layer created around each nanoparticle to determine the dielectric properties. The classical EDL is similar to the Lewis model. It assumes a monopole net charge for the core particle with a Stern-Helmholtz first layer of counter-charge tightly attached to the particle surface forming a double-layer charge structure.

The second Gouy-Chapman layer of diffuse charge is loosely attached by Coulomb force, electrically screening the first layer. Charge transport is enabled by the increase in nanoparticle loading within the dielectric composite eventually leading to overlap of the diffuse double layers forming a conducting path and thereby greatly affecting polarization and dielectric response of the polymer matrix. Bulk charge accumulation is reduced due to this internal conductivity, and the dielectric breakdown strength of the nanocomposite is improved.

In this paper, we propose to extend the EDL to model ferroelectric nanofillers and nanocrystallites by substituting a dipole for the core. This eEDL model would allow initial charge attachment on the opposite-signed end of the dipole to form the Stern-Helmholtz layer. Subsequent arriving charge may be repelled by the initial attached charge forming the diffuse transport layer. Charge transport will occur by meandering of the trajectories due to Coulomb repulsion and use of a field-dependent mobility, $\mu(\mathbf{E})$. The MWS effect due to charge build-up could lead to polarization inversion. It is well known that the polarization and dielectric properties of the PVDF based ferroelectric polymers originate from their crystalline domains. We propose to simulate semicrystalline PVDF as nanocrystallites in an amorphous polymer matrix where the ferroelectric nanoregions are treated with the eEDL model. Ceramic nanofillers such as the Perovskite BT in an amorphous polymer matrix are also represented with the eEDL model. The charge particle transport simulation will determine electrode field modification and allow the passage of charge particles, tracking of trajectories, and calculation of charge fractions.

Current Simulation Efforts

Current simulation efforts on nanocomposites are based on effective medium theory (EMT) for charge percolation to calculate complex permittivity using Maxwell-Garnett, Bruggeman, and Loyenga models [13]. For continuum calculations of energy density, effective permittivity may be estimated by any of the preceding methods, including the Lichtenecker logarithmic rule. Another uses Matlab to create phenomenological 2-layer models to estimate relative permittivity incorporating water uptake and matrix reorganization although actual model details are lacking [14]. Percolation methods [15, 16] are applied when conductive or semiconductive fillers are used in the insulator matrix to improve effective dielectric constant, k. As the volume fraction f of the fillers increases to the vicinity of the percolation threshold f_c, f of the composites can be described by the power law:

$$\frac{k}{k_m} \propto |f_c - f|^{-s}, \quad (1)$$

where k_m is the dielectric constant of the polymer matrix and s is an exponent of near unity. Using this strategy, the dielectric constant of composites can be increased one to two orders of magnitude over the polymer matrix. The end result is the qualitative prediction of possible conductive paths through the nanocomposite medium. This picture of the electrical behavior of particulate nanocomposite materials is attributed to the formation of a connected network of conductive inclusions within the insulating polymer matrix. The implicit assumption is that charge is able to transfer between localized states through hopping conduction and/or electron tunneling.

Bipolar charge transport models capable of handling leakage current up to prebreakdown levels have been successfully applied to layered polymer films [17–19]. Continuum charge transport models are not suited to simulate material with morphology, especially at the nanometer length scale. Probing the role of interfaces in nanocomposites for charge transport requires quantitative details of charge migration through the interspaces, a requirement beyond the capability of continuum models. Particle simulation or specifically a class of "particle-in-cell" (PIC) methods is needed to represent the discrete charge entities and track their migration as particles through a nanocomposite rheology consisting of random nanofillers and/or nanocrystallites in an amorphous matrix. In order to study the role of interfaces, it is necessary to facilitate charge particle transport through the interspaces with a field-dependent mobility under the composite E field while allowing for particle-particle and particle-nanofiller/nanocrystallite interactions. Nanoregions with ferroelectric properties may be modeled as spherical dipoles with remanent dipole moment and polarization orientation that are random, in-plane, parallel, and antiparallel. This paper will study the effects of nanofiller loading, degree of crystallinity, and polarization orientation on charge conduction, field modification, and distribution of charge fractions.

The Proposed eEDL Model

The eEDL model for ferroelectric nanofillers and nanocrystallites is shown in Figure 1 where the core is replaced by a dipole of the appropriate physical diameter and dipole moment. In Figure 1(a), incoming positive charge is repelled by the positive end and attracted towards the negative end of the dipole. Charge is allowed to attach on impact forming the bound Stern-Helmholtz layer.

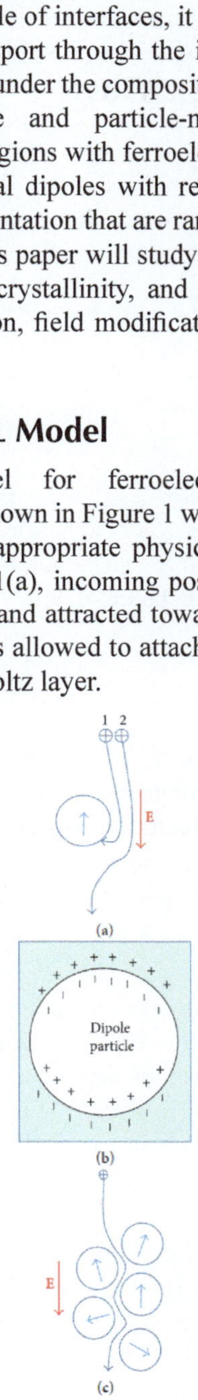

Figure 1: Extended EDL (eEDL) model for ferroelectric nanofillers and nanocrystallites: (a) Initial charge "1" attached to opposite-signed end of dipole particle repels subsequent particle "2" after charge build-up; (b) Maxwell-Wagner-Sillas (MWS) polarization effect; and (c) charge particle trajectory meandering through interspaces.

Subsequent incoming charge may be repelled to form the diffuse outer Gouy-Chapman transport layer. The cumulative charge build up on opposing ends of the dipole leads to MWS polarization as shown in Figure 1(b). The gradual charge deposition and formation of the diffuse layers as charge percolates through the nanocomposite film create the interaction zone which consists of the interspaces formed from overlapping interfacial regions. Trajectories for charge that make it through the film to the counter-electrode are curvilinear paths that meander through the interspaces, as shown in Figure1(c). Charges that arrive at the counter-electrode contribute to the conduction of the film. These charges are considered as neutralized and therefore do not contribute to the field.

PROBLEM FORMULATION

Electrostatic Fields

The composite field due to the bias, space charge, and polarization charge is given by

$$\mathbf{E} = \mathbf{E}_{\text{bias}} + \mathbf{E}_q + \mathbf{E}_{\text{dipole}}. \tag{2}$$

The electrostatic field, E, is the gradient of a scalar potential, ϕ, resulting in the Poisson equation for charge conservation where ρ is the space charge:

$$\nabla \bullet \mathbf{E} = -\nabla^2 \phi = \frac{\rho}{\varepsilon_0}. \tag{3}$$

The potential for a point charge, q, is

$$\phi = \frac{q}{4\pi\varepsilon_0 r}. \tag{4}$$

When the particle count, n, is small to moderate, fields from the space charge are computed from superposing or summing the point source solutions for all particles considered:

$$\mathbf{E}_q = -\int_{V'} \nabla G \frac{1}{\varepsilon_0} \rho\left(v'\right) dV' \approx \sum_{i=1}^{n} \nabla G \frac{1}{\varepsilon_0} q_i, \tag{5}$$

with G as the Green function fundamental solution to a point charge in the volume V'. The polarization, P, from sphere-shaped ferroelectric materials may be written as

$$\mathbf{P} = 3\frac{\varepsilon - 1}{\varepsilon + 2}\mathbf{E} = \chi \mathbf{E}, \tag{6}$$

where ε is the ratio of the permittivity of the sphere to the polymer matrix, $(\varepsilon-1)/(\varepsilon+2)$ is the Clausius-Mossotti factor, and χ is the susceptibility. The dipole moment, potential, and field are given by

$$\mathbf{p} = \frac{4}{3}\pi a^3 \mathbf{P} = 4\pi \frac{\varepsilon - 1}{\varepsilon + 2} a^3 \mathbf{E},$$

$$\phi(r) = \frac{1}{4\pi\varepsilon_0} \frac{\mathbf{p} \bullet \mathbf{r}}{r^3},$$

$$\mathbf{E}_{\text{dipole}} = -\nabla\phi = \frac{1}{4\pi\varepsilon_0}\left(\frac{3(\mathbf{p} \bullet \mathbf{r})\mathbf{r} - \mathbf{p}}{r^3}\right), \tag{7}$$

where p is the intrinsic dipole moment, r is the distance of the evaluation, and a is the nanocrystallite radius. The total E field is obtained by summing both dipole contributions in (5).

Solution of Poisson Equation with the BIEM

The solution of the Poisson equation is obtained using an integral equation method derived from the integral form of the divergence theorem. Volume integrals for homogeneous space may be reduced to surface integrals thus reducing the dimensionality of the problem region. The boundary integral equation method (BIEM) field solver is based on the source distribution technique (SDT) where mobile space charge and bulk trapped charge are treated as volume sources and conducting boundaries and material interfaces are replaced with appropriate distributions of free and bound interfacial polarization and trapped charges to satisfy the specified boundary and interface conditions. Once the charge distributions are ascertained, field parameters are evaluated by superposition of the integral contributions from all sources.

The solution to the Laplacian given by the following equation:

$$-\nabla^2 \phi = 0, \tag{8}$$

is a Fredholm integral of the first kind for potential:

$$\phi(\mathbf{r}) = \int_{S'} G\left[\mathbf{r} \mid s'\right] \frac{\sigma\left(s'\right)}{\varepsilon_0} dS', \tag{9}$$

where $\sigma(s')$ is a charge distribution on surface S' and G, the Green function, is given by

$$G\left[\mathbf{r} \mid \mathbf{r}'\right] = \frac{1}{4\pi\left|\mathbf{r} - \mathbf{r}'\right|}, \tag{10}$$

which naturally satisfies the vanishing far-field condition. The E field is related to the potential by

$$\mathbf{E} = -\nabla\phi = -\int_{S'} \nabla G \frac{\sigma\left(s'\right)}{\varepsilon_0} dS', \tag{11}$$

and it can be computed by integration of analytically differentiable kernels where, for example, the flux which is the normal E field on the surface S' is given by

$$\frac{\partial\phi(s)}{\partial n} = \int_{S'} \frac{\partial G\left[s \mid s'\right]}{\partial n} \frac{\sigma\left(s'\right)}{\varepsilon_0} dS' - \frac{\sigma(s)}{2\varepsilon_0}, \tag{12}$$

where the coefficient in the second term on the right is the Cauchy principal value (CPV) integral with a reentrant angle of π. Enforcement of boundary conditions for potential and flux and interface conditions at material interfaces for continuity of tangential E and normal D (= εE):

$$\mathbf{n} \times (\mathbf{E}_1 - \mathbf{E}_2) = 0,$$

$$\mathbf{n} \bullet (\varepsilon_1 \mathbf{E}_1 - \varepsilon_2 \mathbf{E}_2) = \frac{\lambda}{\varepsilon_0}, \tag{13}$$

for the Poisson equation results in

$$\int_{S'} G \frac{\sigma(s')}{\varepsilon_0} dS' + \int_{C'} G \frac{\gamma(c')}{\varepsilon_0} dC' = \phi(s) - \int_{V'} G \frac{\rho(v')}{\varepsilon_0} dV',$$

$$\left\{ \int_{S'} \frac{\partial G}{\partial n} \frac{\sigma(s')}{\varepsilon_0} dS' - \frac{\sigma(s)}{2\varepsilon_0} \right\} + \int_{C'} \frac{\partial G}{\partial n} \frac{\gamma(c')}{\varepsilon_0} dC'$$

$$= \frac{\partial \phi(s)}{\partial n} - \int_{V'} \frac{\partial G}{\partial n} \frac{\rho(v')}{\varepsilon_0} dV',$$

$$\left\{ \int_{C'} \frac{\partial G}{\partial n} \frac{\gamma(c')}{\varepsilon_0} dC' - \frac{\varepsilon_1 + \varepsilon_2}{\varepsilon_1 - \varepsilon_2} \frac{\gamma(c)}{2\varepsilon_0} \right\} + \int_{S'} \frac{\partial G}{\partial n} \frac{\sigma(s')}{\varepsilon_0} dS'$$

$$= -\frac{\lambda(c)}{\varepsilon_0(\varepsilon_1 - \varepsilon_2)} - \int_{V'} \frac{\partial G}{\partial n} \frac{\rho(v')}{\varepsilon_0} dV', \tag{14}$$

for Dirichlet, Neumann, and dielectric interface conditions, respectively, where $\phi(s)$ and $\partial\phi(s)/\partial n$ are the potential and normal derivative on the surface. Kernel functions, including the Green function and its derivatives (G, $\partial G/\partial n$, and ∇G), are integrated numerically using Gauss-Legendre quadrature. Singular kernels are accurately computed using minimumorder sampling by tying the quadrature weight function to the singularity. Details on the computation of these integrals are discussed in detail for axisymmetric and 2D geometries [19, 20]. Discretized forms of these three equations are solved simultaneously for (s) on electrodes and (c) on interfaces. Equations (14) invoke superposition of sources that include free charge, σ, on electrodes (S'); polarization charge, γ, and trapped charge, λ, on material and physical interfaces (C'); and mobile and trapped space charge, ρ in the volume (V'). With A as electrode area and J as the current density, finite numbers of positive and negative charge particles, Δn, are injected at each time step, Δt, from the cathode and the anode into the respective adjacent polymer according to

$$\Delta n = \frac{JA}{q} \Delta t, \tag{15}$$

where the charges are injected from randomized locations on the surface of the corresponding electrodes. Free charge at the two electrode-polymer interfaces and polarization and trapped charge at the polymer-polymer interface are related to the electric fields on sides 1 and 2 by

$$\sigma_{\text{electrode}} = \varepsilon_0 \varepsilon_r \mathbf{n} \bullet \mathbf{E},$$

$$\sigma_{\text{polarization}} = \varepsilon_0 (\mathbf{n} \bullet \mathbf{E}_2 - \mathbf{n} \bullet \mathbf{E}_1),$$

$$\sigma_{\text{trapped}} = \varepsilon_0 (\varepsilon_2 \mathbf{n} \bullet \mathbf{E}_2 - \varepsilon_1 \mathbf{n} \bullet \mathbf{E}_1), \tag{16}$$

therefore allowing direct inference of the E fields.

Matrix Algebra for Rapid Computation and Reuse

Elements of the discretized matrices on the left hand sides of (14) are calculated from spatial integrals and therefore need not be recomputed for every time step. Large portions of the matrices may be saved and reused in matrix algebra for rapid computation of the sources or independent variables, σ and γ. The forcing functions on the right hand side may be approximated as summations of contributions from point charges as shown in (5). More details are available in the literature [20].

For large numbers of particles, n, and spatially varying distributions a "scatter-gather" method may be used to scatter the space charge, ρ, to qi on the vertices of the inscribed volume mesh using, for example, trilinear interpolation:

$$\rho = \frac{\sum_{i=1}^{8} \alpha_i q_i}{V}, \tag{17}$$

where αi are the trilinear interpolation functions and V is the mesh volume. This particle-mesh scheme facilitates integration by gathering the contributions from the vertices, i, of all volume meshes, j:

$$\mathbf{E} = \int_{V'} \nabla G \frac{1}{\varepsilon_0} \rho(v') \, dV' \approx \sum_{j=1}^{k} \sum_{i=1}^{8} \nabla G \frac{1}{\varepsilon_0} \alpha_i q_i, \tag{18}$$

to compute the E field at the ρ position. This method requires only nk ($\ll n^2$) calculations for a mesh of size $k \ll n$. Following (17), the current density, J, can be similarly expressed as

$$\mathbf{J} = \frac{\sum_i \alpha_i q_i \mathbf{v}_i}{V}, \tag{19}$$

with v_i (= μE) as the drift velocity. Contributions of these volume sources to the E field are evaluated as the algebraic summations of the field from each volume mesh as shown in (18).

Equations of Motion

The electrodynamic simulation of space charge migration through the nanocomposite film involves the time-dependent integration of the two equations of motion:

$$m \frac{d\mathbf{v}}{dt} = q\mathbf{E} + \nabla (\mathbf{p} \bullet \mathbf{E}), \tag{20}$$

$$\frac{d\mathbf{x}}{dt} = \mathbf{v}, \tag{21}$$

where m and q are the mass and the charge of the particle, v is its velocity, and E is the electric field at the particle location. The equations above are followed in time as they evolve in velocity and position space (phase space) to determine the trajectory of each particle. The driving terms in (20) are the Coulomb and dipole forces. Due to the small acceleration time and short mean free path, the charge is assumed to migrate with a drift velocity, v = μE, where μ is the mobility in the polymer.

SIMULATION ALGORITHM

Particle Simulation

Particle simulation is achieved within a computational cell with the solution of the Poisson equation for charge conservation and integration of the equations of motion performed at each time interval as a two-step field solve-particle push algorithm [21]. Nanofillers and nanocrystallites are randomly distributed within the cell to the prescribed vol% loading using "hard sphere" logic that is, allowing contact but no overlap. High loading will require some ordering, like hexagonal close-packed (HCP) to achieve a volume fraction of 0.9069. The 3D PIC model simulates the dynamics of charge particle injection, particleparticle interactions, and particle attachment to nanofillers and nanocrystallites during migration through the nanocomposite layers between the top and bottom electrodes. In accordance with the proposed eEDL model, particles initially attach to nanofillers and nanocrystallites on contact to form the bound Stern-Helmholtz layer on the opposite charged end of the dipole. Subsequent particles form the diffuse GouyChapman transport layer due to Coulomb and dipole force repulsion. The self-consistent E field solution is obtained using the BIEM with Dirichlet boundary conditions for the electrodes, homogeneous Neumann boundary conditions for the four vertical side-walls, and includes contributions from the mobile, attached, and polarization charge. Charge particles are injected from the electrodes over an E field range using Schottky emission at low-to-moderate fields and Fowler-Nordheim tunneling at high fields. Charge particles migrate with Poole-Frenkel field-dependent mobility subject to velocity saturation at high field. Recombination is based on Monte Carlo statistics. Particle tracking involves a predictorcorrector algorithm to integrate the equations of motion subject to Coulomb, dipole, and bias field forces. A selfconsistent particle-particle, particle-mesh (P^3M) scheme may be implemented for very large numbers of particles [22–24].

Computational Cell

A computational cell is used to cater to the nanoscale simulations of discrete, randomly distributed nanofiller and nanocrystallite particles in an amorphous polymer matrix. The computational cell is a cuboid of edge dimension 1 μm, volume of 1 μm^3 bounded by 4 vertical side walls with zero flux condition ($\partial\phi/\partial n = 0$), and top (anode) and bottom (cathode) electrodes at constant potential to maintain the bias E field. Discrete numbers of charge particles, commensurating with the current density and time step, are continuously injected from the electrodes as a function of the averaged electrode E field. Injection locations are randomly dispersed over the electrode surface. Charge particles migrate through the nanofiller and nanocrystallite distribution under the composite E field, initially attaching on contact to form the bound layer. Subsequent charge particles are repelled by the bound layer and meander through the interspaces formed by the interaction zone, outlining curvilinear trajectories. Charge particles that arrive at the counterelectrode are considered to be neutralized and therefore have no further contribution to the composite E field

To minimize local error, displacement, Δh, is required to be smaller than the Debye length; that is, $LD = (\varepsilon kBT/q^2\ Ni)^{1/2}$, where Ni is the largest charge number density. The time step, Δt, needs to be shorter than the dielectric relaxation time, $\tau = \varepsilon/qNi\mu$, characteristic of charge fluctuations to decay. The stability criterion of the explicit algorithm is given by the Courant-Friedrichs-Levy (CFL) limit, c:

$$c = \left|\frac{v\Delta t}{\Delta h}\right| \leq 1, \tag{22}$$

which represents the ratio of mobile charge velocity, V, to trace velocity, $\Delta h/\Delta t$. In accordance with (22), for stability, the trace velocity cannot be faster than the speed of the charge [25]. Simulation proceeds with the following steps where setup consists of tasks (1) and (2) and time-dependent solution involves iteration of tasks (3) and (4) as follows:

- Generate geometric model of computation cell with surface meshes for numerical discretization and integration.
- Generate random distribution of nanoregions and assigned polarization orientation to given vol% loading.
- Solve for surface sources on electrodes and interfaces.
- Integrate surface sources to compute E field on charge particles for use in dynamic tracking through nanoregions using the PIC algorithm.

Time Integration Algorithm

The two differential equations given by (20) and (21) are integrated numerically using a predictor-corrector method of order Δt^2 or secondorder accurate in time. The dependent variables, v and x, are iteratively improved in two steps at each time level. To start, we assume that, at time level n, the sets (x^{n-1}, x^n) and (v^{n-1}, v^n) are known. Then, the predictor-corrector equations are as follows.

Predictor:

$$\begin{aligned}
\mathbf{v}^{n+1} &= \mathbf{v}^{n-1} + 2\Delta t\mathbf{A}^n, \\
\mathbf{x}^{n+1} &= \mathbf{x}^{n-1} + 2\Delta t\mathbf{v}^n, \\
\mathbf{v}^{n-1'} &= \mathbf{v}^n + \frac{1}{4}\left(\mathbf{v}^{n+1} - \mathbf{v}^{n-1}\right), \\
\mathbf{x}^{n-1'} &= \mathbf{x}^n + \frac{1}{4}\left(\mathbf{x}^{n+1} - \mathbf{x}^{n-1}\right).
\end{aligned} \tag{23}$$

Corrector:

$$\begin{aligned}
\mathbf{v}^{n+1} &= \mathbf{v}^{n-1'} + \frac{\Delta t}{2}\mathbf{A}^{n+1}, \\
\mathbf{x}^{n+1} &= \mathbf{x}^{n-1'} + \frac{\Delta t}{2}\mathbf{v}^{n+1}.
\end{aligned} \tag{24}$$

Update the following:

$$\begin{aligned}
&\mathbf{v}^{n-1} = \mathbf{v}^n \qquad \mathbf{x}^{n-1} = \mathbf{x}^n, \\
&\mathbf{v}^n = \mathbf{v}^{n+1} \qquad \mathbf{x}^n = \mathbf{x}^{n+1}, \\
&\mathbf{A}^n = \frac{q}{m}\mathbf{E} + \frac{\nabla(\mathbf{p}\bullet\mathbf{E})}{m}.
\end{aligned} \tag{25}$$

In the predictor stage, the initial values for the $(n-1)$ iteration level are improved. These values are then used in the corrector stage to obtain the updated values for the next time level. The time-dependent algorithm loops through the following steps.

(1) Inject a number of particles according to Schottky or Fowler-Nordheim injection.

(2) For each particle, we have the following.

Predictor

Compute **E** field at x^n.
Compute $\mathbf{v}^{n+1}$ and $\mathbf{x}^{n+1}$.
Compute new $\mathbf{v}^{n-1'}$ and $\mathbf{x}^{n-1'}$.

Corrector

Compute **E** field at x^{n+1}.
Compute new $\mathbf{v}^{n+1}$, $\mathbf{x}^{n+1}$.
Particle push is obtained.

(3) Update iteration arrays.

(4) Check particle location: (a) remove if deposited on counter-electrode or (b) reintroduce from opposite side-wall if exited from side-wall.

(5) If no mobile charge left in polymer, exit. Else return to step (1).

The preceding tacitly assumes that the field is "frozen-in" relative to the time step of charge migration and the transient time to attain terminal velocity is much shorter than the mean free path or the time between collision events between oppositely charged particles

CHARGE INJECTION AND TRANSPORT

Charge Injection

Charge injection from a metal like Al into vacuum requires a work function of ~4.1 eV. For injection into a polymer such as LDPE, the work function is ~1.2 eV. Charge injection from the electrode into the polymer is further facilitated by lowering of the potential barrier through a combination of the image force and the applied field. Assuming Boltzmann statistics for charge carriers in the organic semiconductor, electrons need to migrate from the highest occupied molecular orbital (HOMO) to the lowest unoccupied molecular orbital (LUMO) through a barrier defined as the difference between the Fermi level in the injecting electrode and the center of the density of states (DOS) distribution at the LUMO-level. On the other hand, holes are injected to the HOMO through a barrier defined as the difference between the Fermi level in the injecting electrode and the center of the DOS distribution at the HOMO-level. At moderate applied fields, electrons or holes within the electrode arrive at the metal-polymer interface with enough energy to leave the metal surface.

Charge injection from a metal electrode into the LUMO band of the polymer by Schottky barrier thermionic emission are given by

$$J_n = AT^2 e^{-(W_n-\Delta w)/(k_B T/q)},$$
$$J_p = AT^2 e^{-(W_p-\Delta w)/(k_B T/q)},\qquad (26)$$

where A is the Richardson constant (= 1.2 × 10^6 A/m^2 ·K^2) and W_n and W_p are the energy barriers to injection in eV. The combined effect of the image force and the applied field results in a lowering of the barrier potential given by

$$\Delta w = \sqrt{\frac{qE}{4\pi\varepsilon}}.\qquad (27)$$

where $C = (q^3/16\pi^2 h\Phi_e)$, $\beta = (4\sqrt{2m_e}/3hq)\Phi_e^{3/2}$, m_e, me is the electron effective mass, h is Planck's constant

(4.1356 × 10^{-15} eV·s), and Φ_e is the effective potential barrier. Φ_e is equal to qW_p or qW_n for positive and negative charge, respectively. This occurs when the applied field is high enough that the potential barrier seen by the emitted electrons is sufficiently thin to allow tunneling through the barrier. The transition from Schottky emission to Fowler Nordheim tunneling occurs at about 166 V/μm [18].

Field-Dependent Mobility

At low-fields and low bulk samples, carriers are almost in equilibrium with the lattice vibrations so the low-field mobility is mainly affected by phonon and Coulomb scattering. The mobility increases until the velocity approaches the random thermal velocity. In a moderately large electric field, less thermal fluctuation is required to free charge allowing for higher conduction via the Poole-Frenkel mobility:

$$\mu = \mu_0 e^{\gamma\sqrt{E}} \approx \mu_0 e^{(\Delta w/(k_B T/q))},\qquad (28)$$

where γ is a constant and Δw is as defined in (27). At higher electric fields, mobility decreases with increasing electric field due to increased lattice scattering at higher carrier energies, and the velocity of carriers begins to saturate. The following expression is used to implement a field-dependent mobility that provides a smooth transition between low-field and high-field behavior:

$$\mu = \frac{\mu_0}{\left[1 + (\mu_0 E/v_{sat})^{\beta}\right]^{1/\beta}},\qquad (29)$$

where μ_0 is the low-field mobility at a field of E_0, Vsat is the saturation velocity, and β=1 is commonly used [26].

Charge Attachment/Detachment

Charge trapping takes place at a hopping site that requires energy substantially greater than the average energy to release charge carriers. Trapping and detrapping of space charge in polymeric materials are related to the microstructure and morphology of the materials. Trapping mechanisms include physical defects such as dangling bonds which lead to shallow traps; "selftraps" due to field modification which alters the length of the polymer chain and their potential well; and chemical defects or impurities which result in deep traps. Detrapping mechanisms may be photon-assisted by illumination or phonon-assisted through lattice vibration, impact ionization, and tunneling, with the latter two occurring at high fields. Charge trapping in the polymer bulk is not considered for now. In accordance with the eEDL model, for particle simulation, charge trapping in/attachment to nanocrystallites are assumed to be deterministic, that is, attachment on impact or probability $Pt(t) = 1$ initially, with subsequent taper-off due to Coulomb force and dipole field repulsion from buildup of the attached charge in accordance with the MWS effect, that is, $Pt(t) \rightarrow 0$ as the limiting behavior.

First principles molecular dynamics (MD) calculations of trap depths and metal-polymer barrier potentials may be used together with AC-biased *I-V* curve measurements of leakage current to extract practical transport and attachment/detachment coefficients.

The probability and coefficient for charge detachment/detrapping from nanocrystallites are given by

$$P_d(t) = 1 - e^{-k_d t},$$

$$k_d = n_0 v \sigma e^{-E_t/k_B T}, \qquad (31)$$

where $\sigma(E)$ is the detachment cross-section (nm^2), n_0 is the nanocrystallite density (μm^{-3}), Et is the trap depth (eV), and V is the charge velocity (cm/s). Monte Carlo selection [23] uses random numbers to determine the outcomes of charge detachment events and proceeds as follows.

(1) Given $\sigma = (E)$.

(2) Calculate (t).

(3) Generate random number: $0<R<1$.

(4) If $R>P \rightarrow$ detachment event.

(5) If $R<P \rightarrow$ no detachment.

Detachment is not considered for now due to a lack of published data. However, this model is capable of incorporating future parameters to be abstracted from empirical fits to MD calculations.

Recombination

Neglecting recombination between mobile charge and trapped bulk charge of the opposite polarity, the Monte Carlo collision model may also be used to describe particle-particle events between oppositely charged entities resulting in recombination. The probability of a collision of the ith charge particle with a charge particle of the opposite polarity in a time step Δt is given by

$$P_i = 1 - e^{-n_0 \sigma(E_i) v_i \Delta t}, \qquad (32)$$

where n_0 is the number density of the opposite polarity mobile charge, $\sigma(Ei)$ is the total collision cross-section which in general depends on the kinetic energy of the ith particle, and Vi is the velocity of this particle relative to the velocity of the opposite polarity particle. Therefore, $n_0\sigma$ is a measure of the number of collisions per unit length, $1/n_0\sigma$ is the mean free path between collisions, and $n0\sigma Vi$ is the collision frequency. From (32) it is clear that, for $\Delta t = 0$, $P_i = 0$ and for $\Delta t = \infty$, $P_i = 1$ means the particle will eventually collide. For a finite Δt, the probability of a collision is determined by comparing P_i with R_i, representing a uniform random number between 0 and 1. Following the same logic as in charge detachment, for $P_i > R_i$, the particle i is assumed to have sustained a collision within the time step Δt resulting in recombination and neutralization of the two charge particles.

RESULTS AND DISCUSSION

The model for the nanocomposite film is a random distribution of nanofillers and/or nanocrystallites in a 500 nm layer sandwiched on the top and bottom by 250 nm amorphous polymer layers. Nanofillers and/or nanocrystallites are represented as spherical inclusions with ferroelectric properties and much higher permittivity in the case of ceramic nanofillers. The linear dimension of the computational cell is 1 μm. Top and bottom electrodes form the anode and cathode, respectively, between which a bias E field is imposed. Only unipolar injection from the anode is considered here. The four vertical side walls have zero flux and periodic boundary conditions where, in the latter, charge particles exiting a side-wall reenter from the opposite side-wall. Charge transport studies are performed for several nanoparticle polarization orientations with respect to the bias *E* field including random, in-plane, parallel, and antiparallel (Figure 2). Table 1 summarizes the simulation parameters, including the dipole moment of BT nanofillers and PVDF nanocrystallites [27].

Table 1: Simulation parameters

Parameter	Value	Description
μ_p	9×10^{-11} cm^2/V·s	Normal mobility of positive charge
μ_n	9×10^{-11} cm^2/V·s	Normal mobility of negative charge
W_p	1.2 eV	Metal-polymer work function
W_n	1.2 eV	Metal-polymer work function
ε_{PVDF}	12.4	Dielectric constant of PVDF
ε_{BT}	100	Dielectric constant of $BaTiO_3$
d_{PVDF}	70 nm	Diameter of PVDF nanocrystallites
m_{PVDF}	1.183×10^{-30} C·m	Dipole moment of PVDF nanocrystallites
d_{BT}	150 nm	Diameter of $BaTiO_3$ nanofillers
m_{BT}	4.25×10^{-30} C·m	Dipole moment of $BaTiO_3$ nanofillers

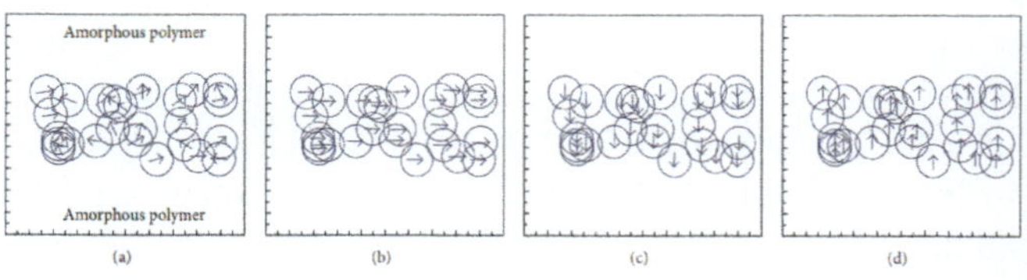

Figure 2: Nanocomposite film with nanofiller polarizations including (a) random; (b) in-plane; (c) parallel; and (d) antiparallel with respect to the bias *E* field.

Sample trajectories are shown in Figure 3 for randomly distributed 150 nm nanofillers (large circles) at 10 vol% loading with dipole moment of 4.25×10^{-30} C·m and (a) in-plane; (b) parallel; and (c) antiparallel polarization. Figure 3(a) for in-plane polarization shows the rear view with 7 attachments and 3 trajectories passing through to the cathode, one of which suffers multiple deflections enroute. The side view in Figure 3(b) shows 8 trajectories terminating on upper half-spheres due to the bias field and the added attraction of the dipole field and 2 passing through to the cathode. Finally, Figure 3(c) contains a perspective view showing trajectories terminating on the tail-end of the spheres and 1 trajectory that is repelled by the front-end of the dipole, deflected laterally, and subsequently moved to the counter-electrode.

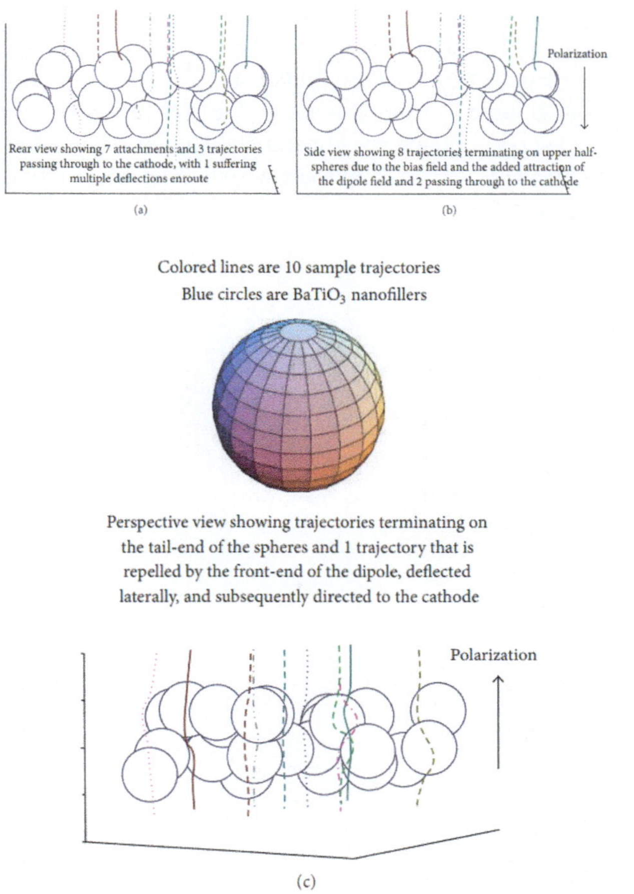

Figure 3: Charge trajectories for randomly distributed 150 nm nanofillers (large circles) at 10 vol% loading with dipole moment of 4.25×10^{-30} C·m and (a) in-plane; (b) parallel; and (c) antiparallel polarization.

Figures 4(a) and 4(b) shows the charge fractions, normalized to total charge, of those attached to nanofillers and those that arrived at the counter-electrode for the 4 polarization cases. Lower attachment results in higher conduction or higher fractions of arrivals at the counter-electrode and vice versa. In all runs, the tracking time is adjusted so that the same amount of charge is injected. The random and in-plane cases have similar results with the latter case exhibiting a slightly higher attachment due to all dipoles being aligned in the plane to provide equal access to the opposite-signed ends. The parallel case presents the negative end of the dipoles directly to the positive charge and results in the highest attachment. The antiparallel case is just the opposite, making it more difficult for incoming positive charge to be initially repelled before they can be drawn to the underside of the dipoles. Depending on the proximity of the neighboring nanoparticles, a dipole polariz ation cancellation effect may cause the charge particle to continue migration therefore increasing the probability of propagating through to the counter-electrode. In particular, for the antiparallel case, very little attachment is observed to the nanofillers but there is >90% arrival at the counter-electrode.

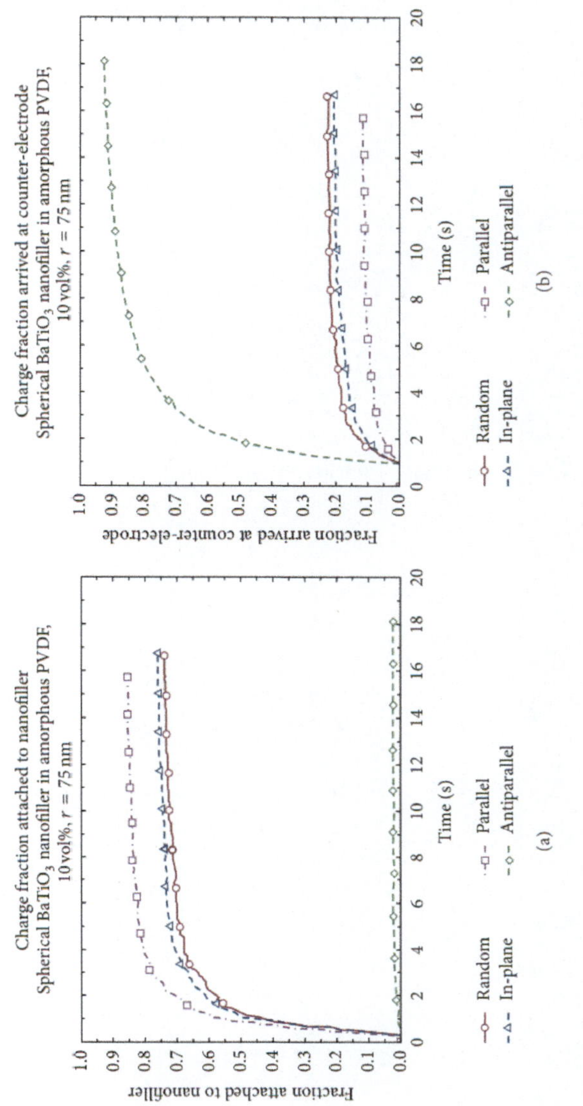

Figure 4: Charge particle fractions: (a) attached to nanofillers and (b) arrived at counter-electrode. Very little attachment to nanofillers and >90% arrival at the counter-electrode for the antiparallel case.

Shown in Figure 5 are the fractions of positive charge captured on nanofillers, in transit, and arrived at the counter-electrode. The curves asymptote with time to 2.24%, 5.05%, and 92.70%, respectively. Table 2summarizes the asymptotic fractions for all the polarization cases.

Table 2: Charge fractions for BT in amorphous PVDF binder

Polarization	% mobile	% attached to nanofiller	% arrived at counter-electrode
Random	3.41	73.95	22.64
In-plane	3.30	76.05	20.65
Parallel	3.13	85.40	11.46
Antiparallel	5.41	2.23	92.36

10 v% $BaTiO_3$; dipole moment = 4.25×10^{-30} C·m; d = 150 nm.

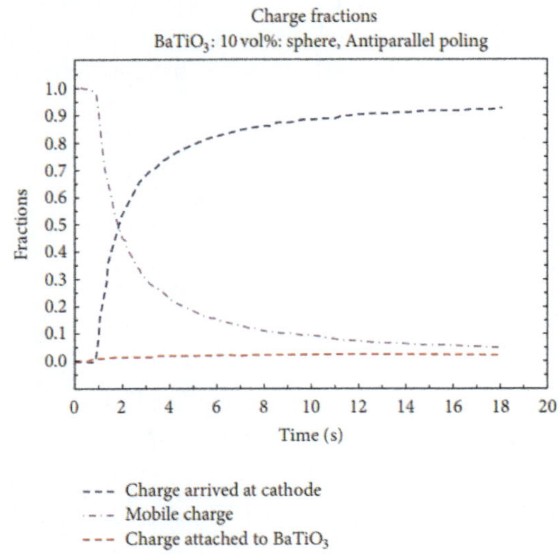

Figure 5: Fractions of positive charge captured on nanofillers, in transit, and arrived at the counter-electrode asymptote with time for any given volume fraction. The curves asymptote to 2.24%, 5.05%, and 92.70%, respectively.

Figure 6 shows the top views of scatter plots for the randomly distributed 150 nm nanofillers (large circles) at 10 vol% loading with dipole moment of 4.25×10^{-30} C·m and attached charge (orange), mobile charge (magenta), and charge arrived at the counter-electrode (blue) for (a) random; (b) in-plane; (c) parallel; and (d) antiparallel polarizations. Figure 6(a) shows mobile positive charge attracted to the tail-end of the randomly polarized spheres. The upwards-directed in-plane polarization in Figure 6(b) shows charge attachment to the back end of the spheres. Figure 6(c) shows mobile positive charge attracted to the tail-end of the sphere, forming fully coated upper hemispheres. Finally, in Figure 6(d) the spherical nanofillers are rendered transparent to show the mobile positive charge attached to the underside or lower hemisphere.

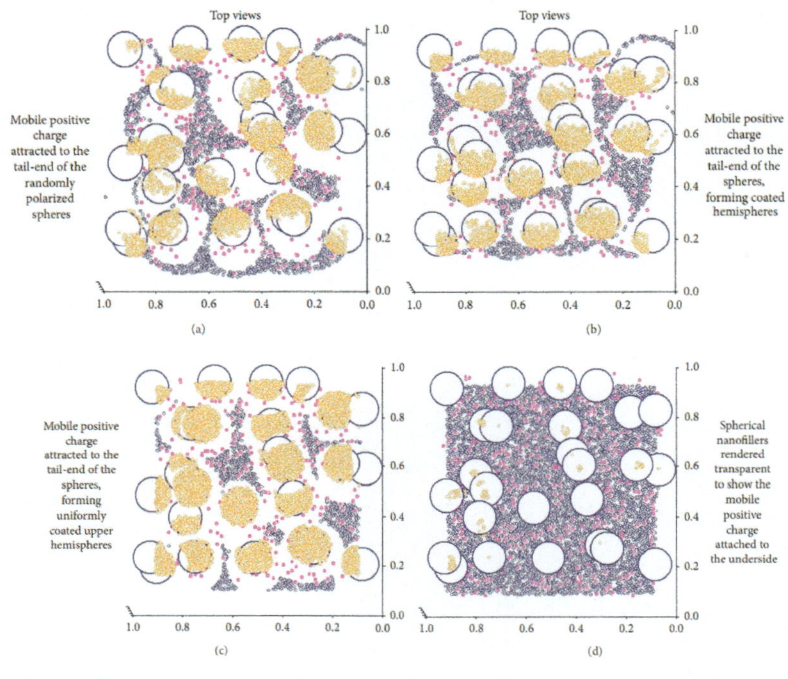

Figure 6: Top views of scatter plots showing randomly distributed 150 nm nanofillers (large circles) at 10 vol% loading with dipole moment of 4.25×10^{-30} C·m and attached charge (orange), mobile charge (magenta), and charge arrived at counter-electrode (blue) for (a) random; (b) in-plane; (c) parallel; and (d) antiparallel polarizations.

The combined effects of 10 vol% BT nanofillers and varying degrees of crystallization (10 to 50 vol%) of the PVDF polymer matrix are studied where, for maximum effect, only antiparallel polarization is considered. Figure 7 shows side views of scatter plots and charge trajectories for randomly distributed 150 nm nanofillers (10 vol%, dipole moment of 4.25×10^{-30} C·m) and 70 nm nanocrystallites (10 to 40 vol% loading, dipole moment of 1.183×10^{-30} C·m). Curvilinear trajectories show charge particles migrating through the interspaces under the combined influence of bias and polarization fields and charge. Red trajectories signify attached charge, black trajectories denote unattached (mobile) charge, green circles are PVDF nanocrystallites, and solid blue circles are BT nanofillers. The incident trajectories at low degree of crystallization are almost normal to the plane of the nanocomposite film. With increasing vol% loading of the nanocrystallites, the incident trajectories bend outwards to reflect the increasing repulsion seen by the upward directed polarization. The proportion of red lines is also increased to show increasing attachment with higher degree of crystallization.

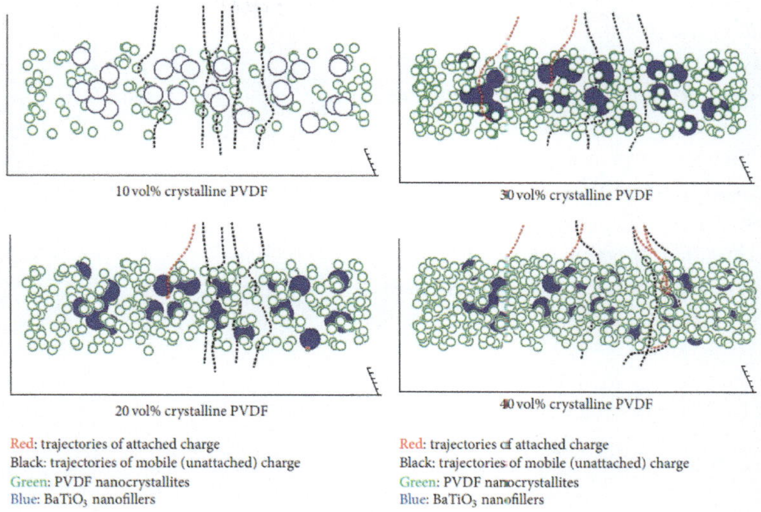

Figure 7: Side views of scatter plots and charge trajectories for randomly distributed 150 nm nanofillers (10 vol%, dipole moment of 4.25×10^{-30} C·m) and 70 nm nanocrystallites (10 to 50 vol% loading, dipole moment of 1.183×10^{-30} C·m) and antiparallel polarization. Curvilinear trajectories show charge particles migrating through interspaces under combined influence of bias and polarization fields and charge.

Figure 8 shows side views of scatter plots of randomly distributed, antiparallel polarized, 150 nm nanofillers (10 vol%, dipole moment of 4.25×10^{-30} C·m) and 70 nm nanocrystallites (10 to 50 vol% loading, dipole moment of 1.183×10^{-30} C·m) and attached charge (orange), mobile charge (magenta), and charge arrived at counter-electrode (blue). Charge attaches mostly to the tail (negative end) of the dipolar PVDF nanocrystallites and nanofillers. Solid blue circles denote BT nanofillers and green circles are the PVDF nanocrystallites.

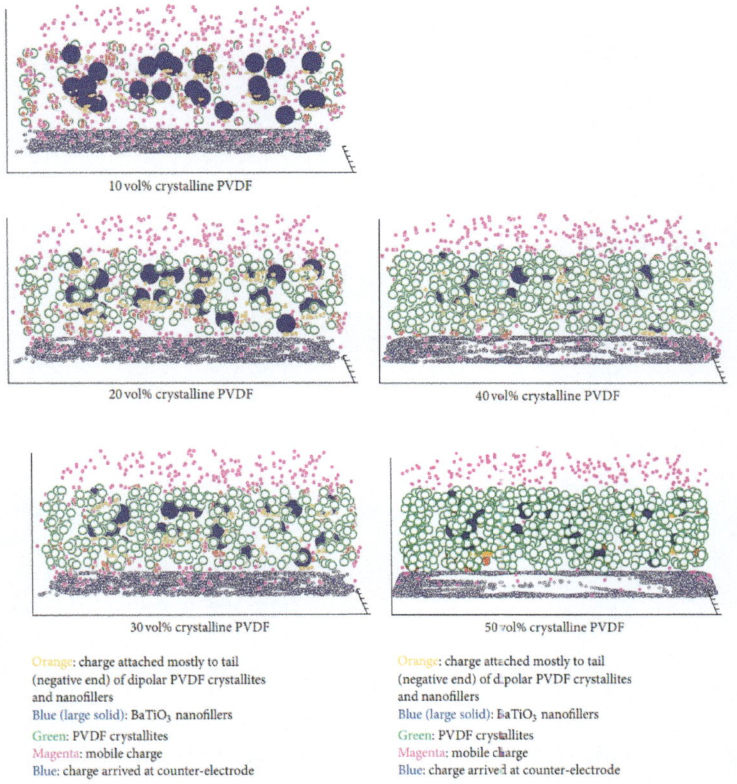

Figure 8: Side views of scatter plots showing randomly distributed 150 nm nanofillers (10 vol%, dipole moment of 4.25×10^{-30} C·m) and 70 nm nanocrystallites (10 to 50 vol% loading, dipole moment of 1.183×10^{-30} C·m), antiparallel polarized, and attached charge (orange), mobile charge (magenta), and charge arrived at counter-electrode (blue).

Asymptotes of charge fractions over time are tabulated in Table 3 for the range of PVDF nanocrystallite loadings and cases with and without 10 vol% of BT nanofillers. It is observed that the combined effect of nanocrystallites and nanofillers contributes additively to the polarization field resulting in increased attachment up to about 40 vol% of nanocrystallites. For the current set of parameters, leakage conduction appears to asymptote to approximately 55%, which seems to be an inflection point beyond which the fraction arriving at the counter-electrode reincreases. This observation may be nonphysical as the higher predicted values may be attributed to charge leakage to the counter-electrode along the side-walls as seen in the 40 and 50 vol% cases in Figure 8. Therefore, this may be due to the periodic boundary condition that reintroduces exiting particles into the opposite side-wall thereby localizing their trajectories to the edges of the cell. From the 40 and 50 vol% trajectories in Figure 7, higher vol% results in increased repulsion for the antiparallel polarization case, as evidenced by both the deflection of the incident trajectories and the larger proportion of attachments. A clear need is a self-consistent set of empirical measurements to pin down the polarization limit and relate that to the degree of crystallization.

Table 3: Charge fractions for BT in semicrystalline PVDF

vol% $BaTiO_3$	vol% PVDF degree of crystallinity	% mobile charge	% on PVDF nanocrystallites	% attached to $BaTiO_3$	% arrived at cathode
0	10	5.19	1.56	0.0	93.26
	20	4.81	22.43	0.0	72.76
	30	4.43	39.62	0.0	55.95
	40	4.30	40.38	0.0	55.33
	50	4.20	25.51	0.0	70.29
10	10	5.04	10.95	5.43	78.58
	20	4.60	31.48	5.60	58.33
	30	4.34	34.86	7.90	52.90
	40	4.35	24.60	5.12	65.92
	50	4.13	8.60	2.22	85.05

Reduction of the average anode E field due to the antiparallel polarized nanoparticles reduces the amount of injected current following the Schottky emission model used here. E field reduction also reduces transport through the nanocomposite film layer. Because antiparallel polarization leads to increased leakage current conduction, it is anticipated that there will be reduced accumulation of space charge in the interaction zone and decreased probability for breakdown. Table 4 summarizes the average E field reduction on the anodes. The field reduction is approximately 10 V/μm for every 10 vol% increase in degree of crystallinity for this set of parameters.

Table 4: E field reduction due to antiparallel polarization

vol% $BaTiO_3$	vol% PVDF degree of crystallinity	E (anode) V/um	E (cathode) V/um	Average ΔE^* V/um
0	0	−100	−100	0.0
0	10	−88	−92	10.0
	20	−78	−81	20.5
	30	−67	−70	31.5
	40	−56	−59	42.5
	50	−46	−48	53.0
10	10	−81	−85	17.0
	20	−71	−75	27.0
	30	−60	−64	38.0
	40	−50	−53	48.5
	50	−39	−42	59.5

$^*\Delta E$: field reduction due to antiparallel poling.

The charge fractions for 10 vol% BT nanofiller and 30 vol% degree of crystallinity of PVDF is shown in Figure 9where charge captured on nanocrystallites, nanofillers, in transit, and arrived at the counter-electrode asymptote with time to 34.86%, 7.90%, 4.34%, and 52.70%, respectively.

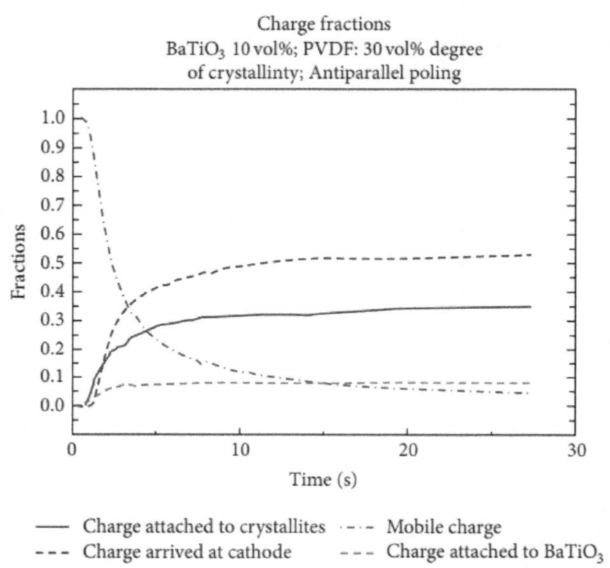

Figure 9: Charge particle fractions for antiparallel polarization. Fractions of positive charge captured on nanocrystallites, nanofillers, in transit, and arrived at the counter-electrode asymptote with time for any given volume fraction. The curves asymptote to 34.86%, 7.90%, 4.34%, and 52.70%, respectively, for 10 vol% $BaTiO_3$ nanofiller and 30 vol% PVDF nanocrystallites.

Figure 10(a) for 10 vol% BT nanofiller and 30 vol% PVDF nanocrystallites shows that average anode and cathode E fields are reduced due to the antiparallel polarization. The fields take up to 1 transit time to reach steady-state, with the anode E field lowered by the proximity of the injected positive charge. In Figure 10(b), injected current density at the anode stabilizes after an interval of 1 transit time.

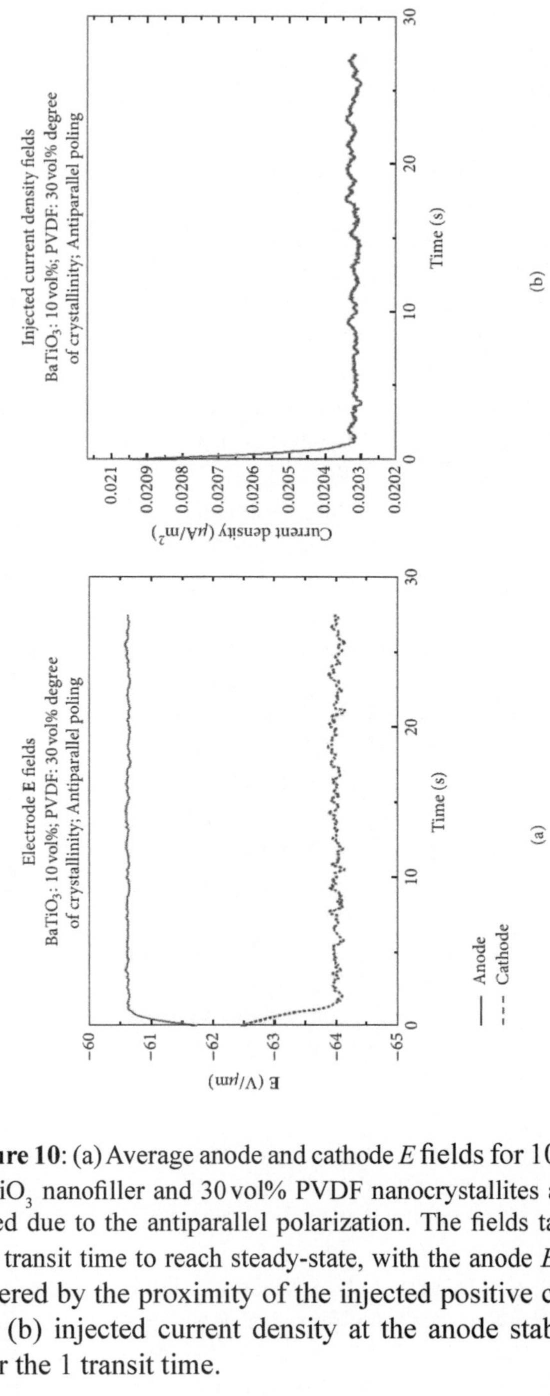

Figure 10: (a) Average anode and cathode E fields for 10 vol% $BaTiO_3$ nanofiller and 30 vol% PVDF nanocrystallites are reduced due to the antiparallel polarization. The fields take up to 1 transit time to reach steady-state, with the anode E field lowered by the proximity of the injected positive charge and (b) injected current density at the anode stabilizes after the 1 transit time.

Shown in Figure 11(a) is the average E field at the anode which drops from the bias value of 100 V/μm due to the field from antiparallel poling of ferroelectric BT nanofillers and PVDF nanocrystallites. Field reduction is seen to increase linearly with increasing degree of crystallinity of PVDF. In Figure 11(b), injected current density assumes Schottky thermionic emission which drops due to decreasing anode E field. As expected, current density reduction goes exponentially with E field reduction. From the preceding results, it is clear that antiparallel polarization has dramatically higher conduction, a property which has immediate implications on reducing residual charge accumulation and extending the lifetime and reliability of nanocomposite films. The mechanism acts to lower E field, reduce charge injection, and minimize attachment to nanoparticles by orienting the dipole for field repulsion. Suitable ferroelectric materials include Perovskites, namely, BT, TiO_2, and ZrO_2 as nanofillers.

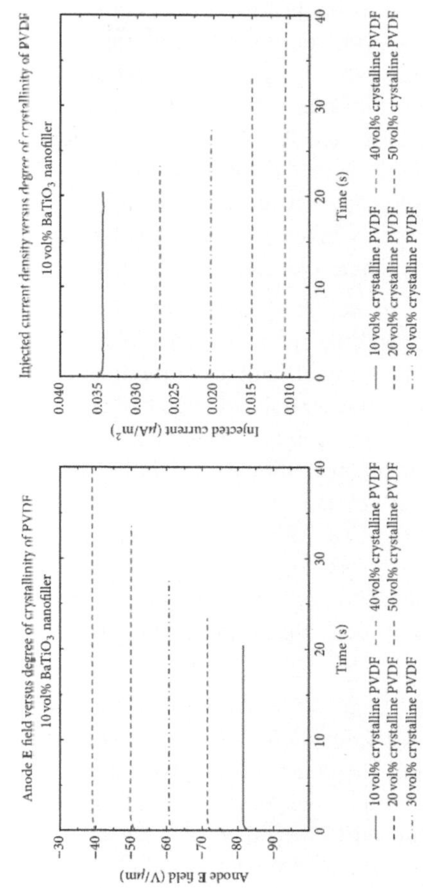

Figure 11: (a) E field at anode drops from bias value of 100 V/μm due to field from antiparallel poling of ferroelectric $BaTiO_3$ nanofillers and PVDF nanocrystallites. Field reduction increases linearly with increasing degree of crystallinity of PVDF and (b) injected current density assumes Schottky thermionic emission and drops due to decreasing anode E field. Current density reduction goes exponentially with E field reduction.

Another beneficial implication may be to increase the power conversion efficiency (PCE) of organic photovoltaic (OPV) devices by inserting an ultrathin film of a ferroelectric copolymer, P(VDF-TrFE), at the metal-organic interface to enhance the charge extraction efficiency. Highly crystalline P(VDF-TrFE) films prepared by the Langmuir-Blodgett method spontaneously polarize and has been shown to be responsible for the enhancement of PCE in ferroelectric OPV devices [28].

Too high vol% loading of nanoparticles and/or too strong antiparallel polarization may significantly reduce the electrode E field and severely limit charge injection. Results are derived from parameters drawn from the literature. There is therefore a need for verified simulation parameters derived from empirical or first principles calculations to ensure computed results are comparable to laboratory data.

CONCLUSIONS

This paper has described a rapid 3D particle simulation algorithm which couples the BIEM with robust predictor-corrector time integration of the equations of motion to efficiently simulate charge transport through nanocomposite film comprised of ferroelectric nanofillers and/or nanocrystallites in amorphous polymer matrices. An extended EDL model was implemented which substituted a dipole for the monopole in the classical EDL model. The rationale was to allow for the initial charge particles to attach onto the nanofillers and/or nanocrystallites on impact to form the bound Stern-Helmholtz layer. Subsequent charge particles are repelled due to charge build-up and the MWS effect, leading to the formation of the diffuse Gouy-Chapman transport layer. Careful trajectory tracking revealed the curvilinear paths taken by charge particles that meander through the interspaces that form the interaction zone.

The eEDL model is used to simulate the ferroelectric properties of crystalline nanoparticles and ceramic nanofillers such as BT as statistical ensembles of randomly oriented polar nanoregions with the associated dipole moments embedded in a continuous amorphous matrix. Preliminary simulation results corroborate the notion of modeling the ferroelectric nanoparticles as remanent polarized nanoregions in an amorphous matrix. Extension of the EDL model to ferroelectric simulations shows promise as the predicted rate of charge migrations around both crystalline nanoregions and ceramic nanofillers for different polarization orientations is seen to be field-dependent.

Antiparallel polarization has shown much higher charge conduction. The explanation is the preferential attachment of charge to the tail (negative end) of the dipole due to charge polarity. In the antiparallel case, the mobile charge is repelled before it is attracted to the other end; a situation which may repeat as almost colinear dipole fields of aligned nanoparticles cancel, thus facilitating the charge to move through the interspaces. Charge captured on the oppositely charged end of the dipole may ultimately lead to polarization reversal.

One area of future work is to study the shape-dependence and effects of physical barriers on charge transport. The barrier properties of polymers can be significantly altered by inclusion of inorganic platelets with sufficient aspect ratio to alter the diffusion path of penetrant molecules. These models are generally based on random, parallel filler platelets perpendicular to the permeation direction (random in only two directions), introducing an orientation factor. At high aspect ratio in nanocomposites, significant decreases in permeability are predicted and observed in practice [29]. High aspect ratio nanolayers provide properties that are not possible for larger-scaled composites. The impermeable layers mandate a tortuous pathway to transverse the nanocomposite. The enhanced barrier characteristics and lower charge migration are benefits from the hindered diffusion pathways through the nanocomposite [30].

Another area of future work will involve derivation of transport and attachment/detachment coefficients from empirical data on time-dependent *I-V* curve measurements of leakage current and matching with model predictions to extract and refine estimates for key parameters. The model will then be used to predict results that can be compared against experiment. Ultimately the validated model may be used for parameter exploration and optimization.

Yet another effort would be to seek corroboration with estimates from first principles quantum mechanical (QM) and molecular dynamics (MD) calculations of trap depths and metal-polymer barrier potentials.

ACKNOWLEDGMENT

Support by the Department of the Navy, Office of Naval Research, Grant no. N000141310064, is gratefully acknowledged.

REFERENCES

1. P. Barber, S. Balasubramanian, Y. Anguchamy et al., "Polymer composite and nanocomposite dielectric materials for pulse power energy storage," Materials, vol. 2, no. 4, pp. 1697–1733, 2009.
2. M. Poulsen and S. Ducharme, "Why ferroelectric polyvinylidene fluoride is special," IEEE Transactions on Dielectrics and Electrical Insulation, vol. 17, no. 4, pp. 1028–1035, 2010.
3. Q. Wang and L. Zhu, "Polymer nanocomposites for electrical energy storage," Journal of Polymer Science, Part B: Polymer Physics, vol. 49, no. 20, pp. 1421–1429, 2011.
4. P. Kim, N. M. Doss, J. P. Tillotson et al., "High energy density nanocomposites based on surface-modified $BaTiO_3$ and a ferroelectric polymer," ACS Nano, vol. 3, no. 9, pp. 2581–2592, 2009.
5. J. Li, J. Claude, L. E. Norena-Franco, S. I. Seok, and Q. Wang, "Electrical energy storage in ferroelectric polymer nanocomposites containing surface-functionalized $BaTiO_3$ nanoparticles," Chemistry of Materials, vol. 20, no. 20, pp. 6304–6306, 2008.
6. J. Li, P. Khanchaitit, K. Han, and Q. Wang, "New route toward high-energy-density nanocomposites based on chain-end functionalized ferroelectric polymers," Chemistry of Materials, vol. 22, no. 18, pp. 5350–5357, 2010.
7. J. Li, S. I. Seok, B. Chu, F. Dogan, Q. Zhang, and Q. Wang, "Nanocomposites of ferroelectric polymers with TiO_2 nanoparticles exhibiting significantly enhanced electrical energy density," Advanced Materials, vol. 21, no. 2, pp. 217–221, 2009.
8. M. N. Almadhoun, U. S. Bhansali, and H. N. Alshareef, "Nanocomposites of ferroelectric polymers with surface-hydroxylated $BaTiO_3$ nanoparticles for energy storage applications," Journal of Materials Chemistry, vol. 22, no. 22, pp. 11196–11200, 2012.
9. D. Rollik, S. Bauer, and R. Gerhard-Multhaupt, "Separate contributions to the pyroelectricity in poly(vinylidene fluoride) from the amorphous and crystalline phases, as well as from their interface,"Journal of Applied Physics, vol. 85, no. 6, pp. 3282–3288, 1999.
10. A. B. Silva, R. Gregorio, J. V. Esteves, and C. Wisniewski, "Influence of the amorphous-crystalline interface on the dielectric and ferroelectric polarization of α-PVDF," in Proceedings of the 11th International Conference on Advanced Materials, Rio de Janeiro, Brazil, 2009.
11. Z.-G. Zeng, G.-D. Zhu, L. Zhang, and X.-J. Yan, "Effect of crystallinity on polarization fatigue of ferroelectric P(VDF-TrFE) copolymer films," Chinese Journal of Polymer Science (English Edition), vol. 27, no. 4, pp. 479–485, 2009.
12. D. Pitsa and M. G. Danikas, "Interfaces features

in polymer nanocomposites: a review of proposed models," Nano, vol. 6, no. 6, pp. 497–508, 2011.

13. R. Mekala and N. Badi, "Modeling and simulation of high permittivity core-shell ferroelectric polymers for energy storage solutions," in Proceedings of the COMSOL Conference, Boston, Mass, USA, 2013.
14. I. A. Tsekmes, R. Kochetov, P. H. F. Morshuis, J. J. Smit, and T. Andritsch, "Modeling the dielectric response of epoxy based nanocomposites," in Proceedings of the IEEE Electrical Insulation Conference (EIC ‹14), pp. 47–50, 2014.
15. Y. P. Mamunya, V. V. Davydenko, P. Pissis, and E. V. Lebedev, "Electrical and thermal conductivity of polymers filled with metal powders," European Polymer Journal, vol. 38, no. 9, pp. 1887–1897, 2002.
16. J. Hicks, A. Behnam, and A. Ural, "A computational study of tunneling-percolation electrical transport in graphene-based nanocomposites," Applied Physics Letters, vol. 95, no. 21, Article ID 213103, 2009.
17. M. H. Lean and W.-P. L. Chu, "Effect of gaseous void on bipolar charge transport in layered polymer film," Journal of Physics D: Applied Physics, vol. 47, no. 7, Article ID 075303, 2014.
18. M. H. Lean and W.-P. L. Chu, "Dynamic charge mapping in layered polymer films," IEEE Transactions on Dielectrics and Electrical Insulation, vol. 21, no. 3, pp. 1319–1329, 2014.
19. M. H. Lean and W.-P. L. Chu, "Simulation of charge packet formation in layered polymer film,"COMPEL: The International Journal for Computation and Mathematics in Electrical and Electronic Engineering, vol. 33, no. 4, pp. 1396–1415, 2014.
20. M. H. Lean and W.-P. L. Chu, "Dielectric breakdown model for polymer films," IEEE Transactions on Dielectrics and Electrical Insulation, vol. 21, no. 5, pp. 2259–2266, 2014.
21. M. H. Lean, "Particle simulations of ion cloud in a magnetic field," IEEE Transactions on Magnetics, vol. 34, no. 5, pp. 3118–3121, 1998.
22. R. W. Hockney and J. W. Eastwood, Computer Simulation Using Particles, McGraw-Hill, New York, NY, USA, 1981.
23. J. M. Dawson, "Particle simulation of plasmas," Reviews of Modern Physics, vol. 55, no. 2, pp. 403–447, 1983.
24. C. K. Birdsall, "Particle-in-cell charged-particle simulations, plus Monte Carlo collisions with neutral atoms, PIC-MCC," IEEE Transactions on Plasma Science, vol. 19, no. 2, pp. 65–85, 1991.
25. D. M. Caughey and R. E. Thomas, "Carrier mobilities in silicon empirically related to doping and field,"Proceedings of the IEEE, vol. 55, no. 12, pp. 2192–2193, 1967.
26. S. Gottlieb, C.-W. Shu, and E. Tadmor, "Strong stability-preserving high-order time discretization methods," SIAM Review, vol. 43, no. 1, pp. 89–112, 2001.
27. C.-G. Duan, W. N. Mei, W.-G. Yin et al., "Simulations of ferroelectric polymer film polarization: the role of dipole interactions," Physical Review B: Condensed Matter and Materials Physics, vol. 69, no. 23, Article ID 235106, 2004. Y. Yuan, P. Sharma, Z. Xiao et al., "Understanding the effect of ferroelectric polarization on power conversion efficiency of organic photovoltaic devices," Energy and Environmental Science, vol. 5, no. 9, pp. 8558–8563, 2012.
28. D. R. Paul and L. M. Robeson, "Polymer nanotechnology: nanocomposites," Polymer, vol. 49, no. 15, pp. 3187–3204, 2008.
29. P. C. Lebaron, Z. Wang, and T. J. Pinnavaia, "Polymer-layered silicate nanocomposites: an overview,"Applied Clay Science, vol. 15, no. 1-2, pp. 11–29, 1999.

CHAPTER 12

DIELECTRIC PROPERTIES OF $PBNB_2O_6$ UP TO 700°C FROM IMPEDANCE SPECTROSCOPY

Kriti Ranjan Sahu and Udayan De

INTRODUCTION

A generator of pressure or movement or ultrasonic wave and their sensor/detector can be fabricated utilizing inverse and direct piezoelectric effect, respectively. These possibilities opened up a huge array [1] of medical, industrial, and other applications of piezoelectric materials with significant commercial implications. Commercial piezoelectric materials have mostly been barium titanate (BT), lead zirconate titanate (PZT), or materials based on BT or PZT. But Curie temperature (T_c), the upper limit for piezoelectricity, is at best 130°C [2] for BT and 365°C for a modified PZT. So, higher T_c materials are being developed worldwide for high temperature (HT) applications. Lead metaniobate ($PbNb_2O_6$), shortened here as PNO, with a higher Curie temperature (517 to 570°C in different papers), is one of the present candidate materials [2–14]. It was discovered [3] in 1953. However, synthesis of its piezoelectric phase, the metastable orthorhombic structure, in a pure state is difficult [4]. During the preparation of the orthorhombic PNO by quenching from a temperature above 1250°C, a few competing compounds and phases (like the rhombohedral PNO phase) tend to form. The rhombohedral PNO is not piezoelectric. In fact, the difficulty [4] of forming single-phase and high-density orthorhombic PNO, together with the success of BT, PZT, and related piezomaterials for near room temperature applications, somewhat halted R & D on PNO after the initial years of pioneering work and publications [5–9]. In the meantime, a need has developed for high temperature (HT) piezoelectric materials (i) for HT ultrasonic imaging of nuclear fuel rods through molten metal coolants [15] in Fast Breeder Reactors and (ii) for HT applications like sensors in car exhaust systems. So, since the late nineties, work on PNO has been revived by a few groups (Table 1) in search of higher Curie temperature materials. The stable forms of $PbNb_2O_6$ are rhombohedral (at low temperature) and tetragonal (at high temperature). Quenching from the tetragonal form, as mentioned earlier, leads to metastable orthorhombic form at room temperature. Roth showed by X-ray diffraction [5, 6] that the orthorhombic PNO (ferroelectric) changes to a tetragonal structure (paraelectric) above ~570°C.

Table 1: Comparison of recently reported Curie temperatures of orthorhombic $PbNb_2O_6$ samples

Sample	Sintering Temp. (°C)	Sintering Time (h)	Curie temperature $T_{(Curie)}$ in °C
Present	1270	4.5	$T_{(Curie)}$ > 576 for 4.0 MHz, and >574 for 0.5 kHz.
[17]	1270	4.5	517
[18]	1270	—	534 and 540 (both at 10 kHz)
[19]	—	—	540 (presumably at 80 kHz)
[20]	1240–1260	4	530 (at 10 kHz)
[13]	1270	4.5	531 (at 1 kHz), 528 (at 1 MHz) and 533 (at 10 MHz)

Goodman [3] had earlier concluded the $T_{(Curie)}$ of quenched PNO to be 570°C, from their dilatometric and dielectric characterizations. Although this temperature is regarded even in recent papers [10] as the Curie temperature of $PbNb_2O_6$, almost all the recent electrical measurements have obtained much lower values of its Curie temperature (Table 1). Reference [10] quoted $T_{(Curie)}$ of 500°C for Ferroperm (Denmark) PNO samples (1995) and 495°C for Morgan Matroc (UK) PNO samples (1997). Lower $T_{(Curie)}$ in these commercial samples may or may not be due to efforts to optimize other piezoelectric parameters. Table 1 has been compiled from recent publications. It shows $T_{(Curie)}$ to be in the range 517°C to 540°C only, except in present samples. So,

Citation: Kriti Ranjan Sahu and Udayan De, "Dielectric Properties of $PbNb_2O_6$ up to 700°C from Impedance Spectroscopy," Journal of Materials, vol. 2013, Article ID 702946, 15 pages, 2013. doi:10.1155/2013/702946.

there is a need to optimize preparation conditions of quenched $PbNb_2O_6$ to get highest possible Curie temperature, high purity and high density. A pointer in this direction is our observation on quenched PNO [11] of an endothermic signal at 570.6°C in Differential Scanning Calorimetry [12].

Here, our preparation of high quality orthorhombic PNO and results of our impedance spectroscopy up to 700°C and over a wide frequency range from 20 Hz to 5.5 MHz, with a detailed analysis, have been presented. Intrinsic cationic defects in niobates like $LiNbO_3$ and their role are well known [16]. Likely lattice defects in a quench-prepared $PbNb_2O_6$ sample should similarly be regarded as an intrinsic part of the sample. The structure of the samples has, therefore, been characterized fully before other studies. Although some work on PNO (Table 1) and a few on various targeted chemical substitutions in PNO have been reported in the literature, not all possible dielectric and electrical parameters for PNO itself have yet been derived from IS data. This has been attempted here. The dielectric response in such an electroceramic is known to be complex with many contributions that need to be known and understood. So, desirable observations like the highest $T_{(Curie)}$ as well as all uncommon observations of the present work are reproduced. All possible data analysis has been carried out, yielding important parameters like bulk resistance, relaxation times, activation energies, AC and DC conductivity, and phase angle.

MATERIALS AND METHODS

Aldrich lead(II) oxide (PbO, 99.999%, yellowish orange and Melting Point = 886°C) and niobium pentoxide (99.9% GR) of Loba Chemie, weighed in proper proportion to give Pb : Nb = 1 : 2, plus 2% of excess PbO, have been well ground under isopropyl alcohol, dried, and pelletized. Additions of excess PbO were introduced by some earlier workers to compensate possible evaporative loss of PbO. To get best $PbNb_2O_6$, present firings have been done in stages at 1050°C, 1290°C and 1270°C in a box furnace with PID temperature controller. Extra PbO (3%) has been added before the 3rd firing. All our firings have been done in palletized form to provide intimate contact and reduce Pb loss. But grinding and repalletizing at each step have also been undertaken, to improve mixing and the sample quality. After the last firing, the samples have been quenched. The samples have been characterized by XRD with Rietveld analysis [11] and Transmission Electron Microscopy (TEM).

The impedance spectroscopy has been carried out using Solatron "SI 1260 Impedance Gain Phase Analyzer" with a high temperature attachment. Our PNO pellets (diameter ~ 12 mm and 2.2 to 2.4 mm thick) have not been poled before measurement. Silver paint coat on the flat faces of the pellet under study served as the electrodes for IS measurements. This measurement of ε' and ε'' (the real and imaginary components, respectively, of the relative dielectric constant i.e., permittivity) has been done for different measuring frequencies at different values the sample temperature (T) while heating from room temperature to 700°C.

RESULTS AND DISCUSSION

XRD pattern and Rietveld analysis [11], Figure 1(a), confirmed the orthorhombic structure and absence of any other phase. The smallness of the difference of the fitted graph and raw data proved our Rietveld fit to be good. It yields a = 17.6468(7) A° b = 17.9512(7)A, and °c = 3.8704(1)A°. The samples showed about 80% of the theoretical density on average. TEM diffractograms [6] with the electron beam on individual grains of the ceramic sample always gave diffraction patterns that matched these lattice parameters from XRD result. One TEM diffraction pattern is shown in Figure 1(b). This way, all of the nearly forty grains examined have been found to be in orthorhombic phase.

Figures 2–4 show ε'versus frequency, ε'' versus frequency, and tan $\delta = \varepsilon''/\varepsilon'$ versus frequency results (over 20 Hz to 5.5 MHz range) for 12 temperatures from 50∘ C to 700∘ C. In Figure 2, real part (ε') of dielectric constant decreases with increase of measuring frequency up to about ln f = 12 to 14 region, implying ~162 kHz to ~1203 kHz region, and then increases sharply with increase of frequency as reported earlier [13] for all temperatures except at low temperatures like 200°C. Moreover, ε' is low and less dependent on frequency at low temperatures up to 300°C, leading to overlap of several graphs. It increases with increase of temperature, up to 573° C. Graph of ε' shows significantly increased values in the 573°C graphs with respect to the 570° C graph. This temperature is the region of ferroelectric to paraelectric phase transition, as is clearer in graph on temperature variation (Figure 5). One must note the interesting decrease of ε' with further increase of temperature above 573° C. This, in fact, shows up in ε' versus T plot as peaking at around 573°C. Here, it was studied up to 700° C. The values at different temperatures of dielectric dispersion with respect to

frequency, which will be proportional to dε' /d(ln f), will be discussed from the low frequency, specifically ln f (Hz) < 7, part of Figure 2. At a low temperature of 100° C, dε' /d(ln f) = 2.483 ± 0.739 only, increasing to 41.158 ± 7.805 at 300° C, whereas at the high temperature of 573° C, dε' /d(ln f) = −4121.578 ± 226.784, a high value.

In Figure 3, frequency dependence of ε'' or the imaginary part of relative permittivity is more at low frequencies and higher temperatures, the ε'' versus frequency graph for 700° C being at top in the lower frequency region. Here, ε'' decreases steeply as frequency increases from 20 Hz to about ln f=8 implying f = 2.98 kHz. A weak peaking of ε'' at ln f = 13.8, implying f = 984.6 kHz, is visible for graphs for low temperatures like 50° C, 300° C, 500° C, and 550° C in the enlarged view. Our higher temperature graphs show no peak in this temperature region but only a slow increase of ε'' to allow peaking at higher frequencies. This fall and peaking of ε'' has been shown and discussed by earlier workers [13], although the details differ. Dielectric loss (tan δ) in Figure 4 approximately follows ε'' with respect to frequency and temperature dependence.

Figures 5 and 6 are plots with respect to temperature, of ε' and tan $\delta = \varepsilon''/\varepsilon'$, respectively. The peak value of ε' is denoted by ε' max and its position on the temperature axis by T_m. Data for eight frequencies, selected out of 20 Hz to 5.5 MHz data in Figures 2–4, have been plotted in Figures 5 and 6 to avoid overlaps and provide a clearer view of the graphs

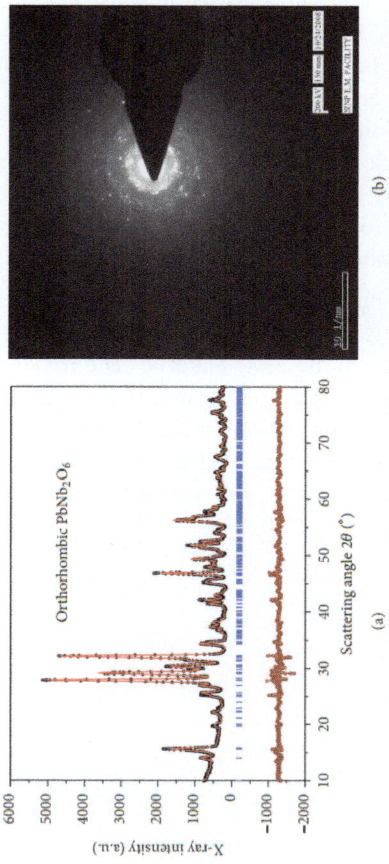

Figure 1: (a) Rietveld fit (red graph) to XRD data points (black circles) for orthorhombic PbNb$_2$O$_6$ with their difference shown by the continuous curve in the smaller-sized graph. The vertical blue bars below the main graph are the calculated peak positions. (b) Electron diffraction pattern for HR Transmission Electron Microscopy on a selected grain (microcrystal) of quenched lead metaniobate (PNO).

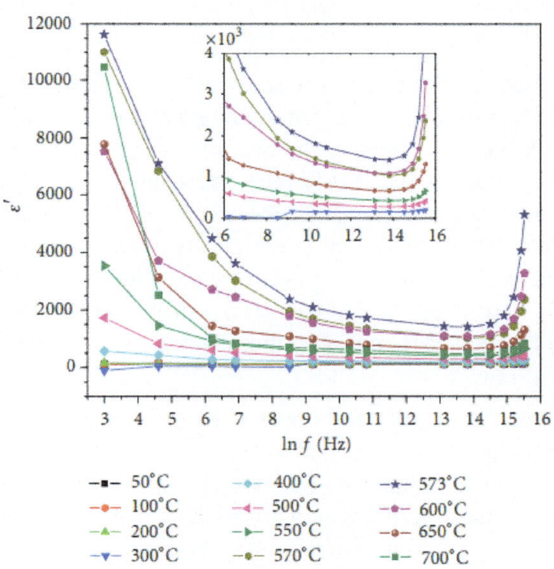

Figure 2: Frequency dependence of the real part of dielectric constant (ε') in the frequency range 20 Hz (ln f = 2.995) to 5.5 MHz (ln f = 15.52), compiled from heating runs at different temperatures for quenched PNO sample, the orthorhombic phase of PbNb$_2$O$_6$. The inset is an enlarged view of the variation of ε' in the higher frequency range.

Fall of ε' and, hence, of the dielectric susceptibility, χ' = (ε' − 1), with increasing T, a characteristic of the paraelectric phase and exhibited here for $T>T_m$ region, is due to opposition of the increasing thermal agitation, to dipole alignment. The rise of ε' with T in the T be discussed first. A wide minimum (Figure 5), appearing at 300° C, for frequencies 0.5 kHz, 1 kHz, and 5 kHz in ε' versus T, as well as in tan δ versus T graph (Figure 6), cannot be readily explained. However, the more important feature of this work is the large permittivity peak at T_m with T_m ranging from 574° C to 576° C as the measurement frequency is increased from 0.5 kHz to 4 MHz, as mentioned earlier. In Figure 7, the peak value, ε'_{max}, of ε' shows a sharp fall initially and then a slow rise with increase of measuring frequency. The value of ε'_{max} decreases from 11623 to 1450 as the frequency (f) increases from 20 Hz to 1 MHz. Then, for f>1 MHz, the value of ε'_{max} increases slowly, being about 5465 at 5.5 MHz. The prominent peak of the imaginary part ε' in the ε' versus T graph can be seen to be around T_m as expected. The dielectric loss (tan δ) also peaks at around T_m.

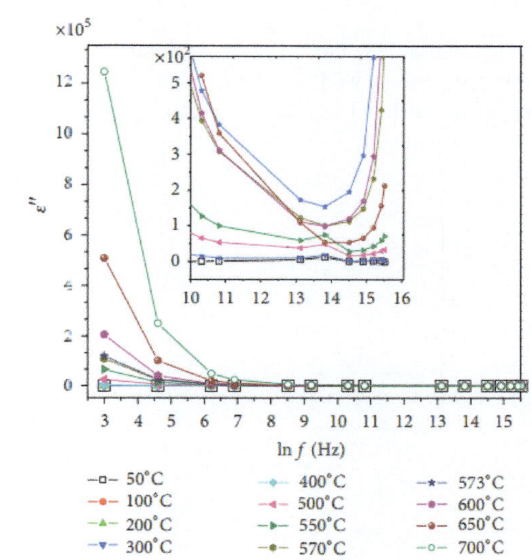

Figure 3: Frequency dependence of the imaginary part (ε'') of dielectric constant in the frequency range 20 Hz to 5.5 MHz, compiled from heating runs at different temperatures, for quenched PNO sample, the orthorhombic phase of PbNb$_2$O$_6$. The inset is an enlarged view of the variation of ε'' in the higher frequency range

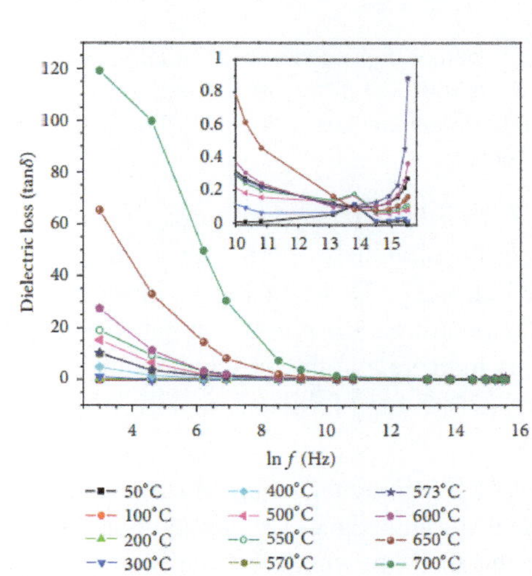

Figure 4: Frequency dependence of the dielectric loss (tan δ) in the frequency range 20 Hz to 5.5 MHz, compiled from heating runs at

different temperatures, for quenched PNO sample, the orthorhombic phase of $PbNb_2O_6$.

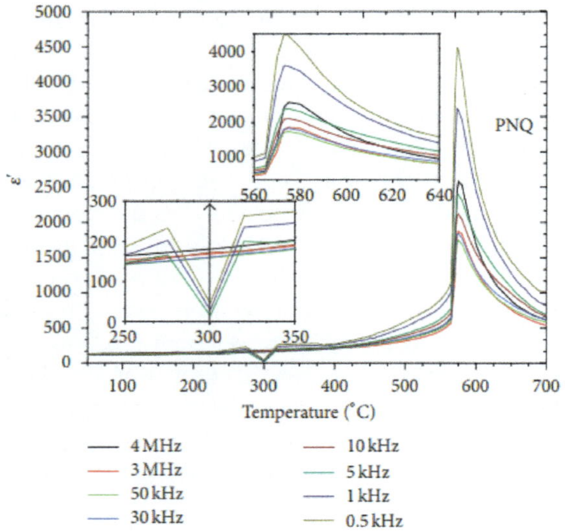

Figure 5: Temperature dependence of the real part (ε') of dielectric constant in the temperature range of 50° C to 700° C, at different frequencies, for quenched PNO sample, the orthorhombic phase of Pb-Nb_2O_6.

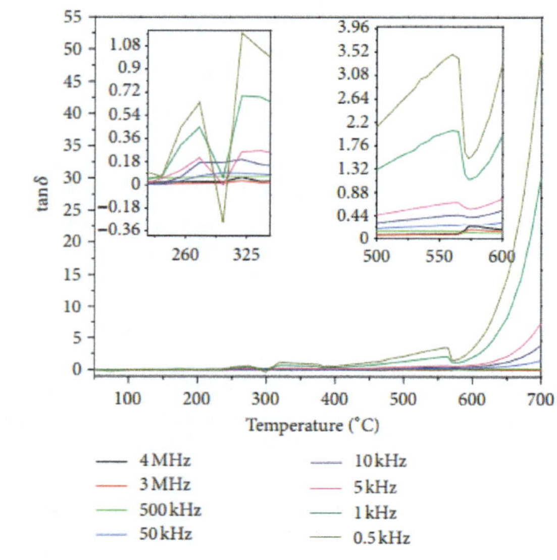

Figure 6: Temperature dependence of the dielectric loss (tan δ) in the range 50° C to 700° C at different frequencies for quenched PNO sample, the orthorhombic phase of $PbNb_2O_6$. Two insets magnify two important features.

Above T_m, ε' decreases with increase of T, and, hence, the material is obviously paraelectric above T_m. It is ferroelectric below T_m. This transition temperature gives [22] the Curie temperature. So, present impedance spectroscopy measurements indicate a high Curie temperature that depends on frequency, 576° C for 4.0 MHz and 580° C for 5.5 MHz, for example.

The plot of Z'' (imaginary part of the impedance) versus Z' (real part of the impedance), called Nyquist diagram [23], is based on the well-known Cole-Cole plot of ε'' versus ε'. The Z'' versus Z' plot for our orthorhombic PNO is shown in Figures 8(a) and 8(b). The straight line segments joining of the data points in Figures 8(a) and 8(b) are drawn only as a guide to the eye. Single semicircular nature of the plots is clearer in the higher temperature plots, as the lower temperature plots, like that for 200° C in the insert of Figure 8(a), show a near linear rise, possible for a large semicircle. An important observation is that the Nyquist diagram here shows only one semicircular arc [23] over the studied frequency range 20 Hz to 5.5 MHz. One semicircle implies only one contribution, here from the sample grains and not from any second source (e.g., grain boundaries or electrode effect). A sample with one semicircle Nyquist diagram or Cole-Cole plot is equivalent

[23] to a bulk resistance R and a capacitance C in parallel. These elements give rise to a time constant $\tau = CR$, the dielectric relaxation time of the basic material. The frequency ω_m of the peak of the Nyquist diagram can find the relaxation time τ as $(\omega_m)(\tau) = 1$ for this peak point [23]. At this point, Z'' has its maximum value. So, many authors [24, 25] use the frequency of peak (i.e., maximum) of Z'' versus frequency plot to find the relaxation time τ.

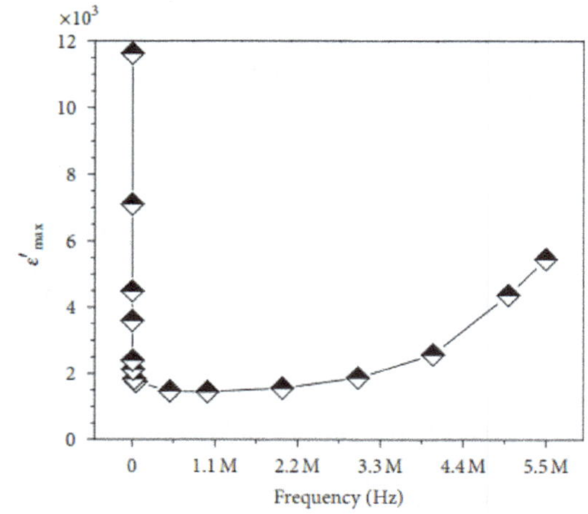

Figure 7: The maximum value (ε' max) of the real part of dielectric constant in the in the temperature range of 50°C to 700∘ C, as a function of frequency (Hz), for quenched PNO sample, the orthorhombic phase of $PbNb_2O_6$.

Present work additionally suggests an alternative method (based [23] on $Z'' = R/2 = Z'$ criterion) of determining relaxation time (τ). We later compare the two methods, with the finding that they match (Figure 14) fairly well, match being better at higher temperatures. Often the centre of the semicircular arc of the Nyquist diagram is not on the Z' -axis, so that the semicircle is seen to be depressed by an angle (θ) below the Z' -axis. This is the case, if there is a distribution of relaxation times. This θ is determined by the width of the distribution of relaxation time.

If the semicircle starts near the origin at $Z' = R_\infty$ and intercepts the Z' -axis again at R_0, $R = (R_0 - R_\infty)$ gives [23] the bulk resistance. Fitting the impedance spectroscopy (Z'' versus Z') data, usually forming an arch, to the equation of a circle, its radius r and θ and, hence, $R0$ can be determined. Bulk resistance $R = (R_0 - R_\infty) = 2r \cdot \cos\theta$ has thus been calculated from our data in Figure 8 and plotted in Figure 9. It is found that the semicircle of data points for a particular temperature in Figures 8(a) and 8(b) reduces in size with increase of temperature, implying smaller bulk resistance of the sample at higher temperatures as documented in Figure 9. Real part ε' of permittivity in present samples in the paraelectric region has been fitted to the modified CurieWeiss law [26], $(1/\varepsilon' - 1/\varepsilon'_{max})$

$= (T - T_m)/C_w$, where C_w is Curie-Weiss constant and γ is the diffusivity parameter with $1<\gamma$.

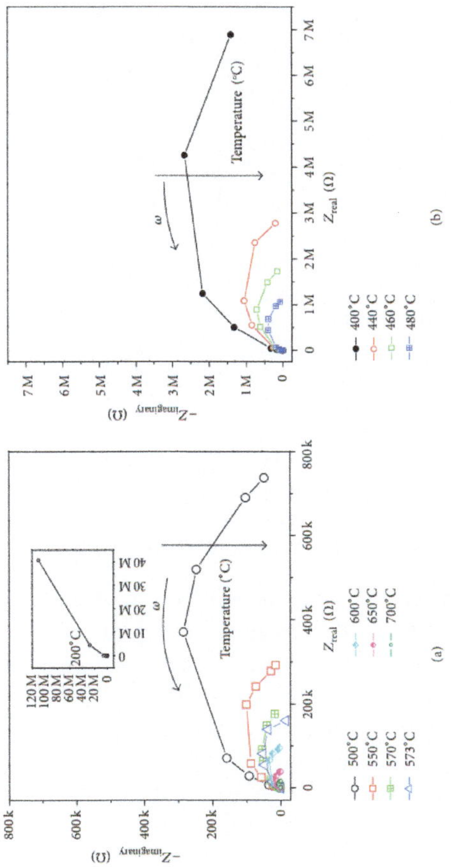

Figure 8: (a) Plot of the imaginary part, Zimaginary in ohm, of impedance versus its real part, Zreal in ohm, for quenched PNO sample, $PbNb_2O_6$ in orthorhombic phase, at temperatures: 500∘ C, 550° C, 570∘ C, 573° C, 600° C, 650° C, and 700° C. Since the complex impedance $Z* = Z_{real} - jZ_{imaginary}$, quantity $(-Z_{imaginary})$ has actually been plotted in the figure. (b) Plot of the imaginary part, Zimaginary in ohm, of impedance versus its real part, Zreal in ohm, for the quenched PNO sample, PbNb2O6 in orthorhombic phase, at temperatures: 400° C, 440° C, 460° C, and 480° C. Since the complex impedance $Z* = Z_{real} - jZ_{imaginary}$, quantity $(-Z_{imaginary})$ has actually been plotted in the figure.

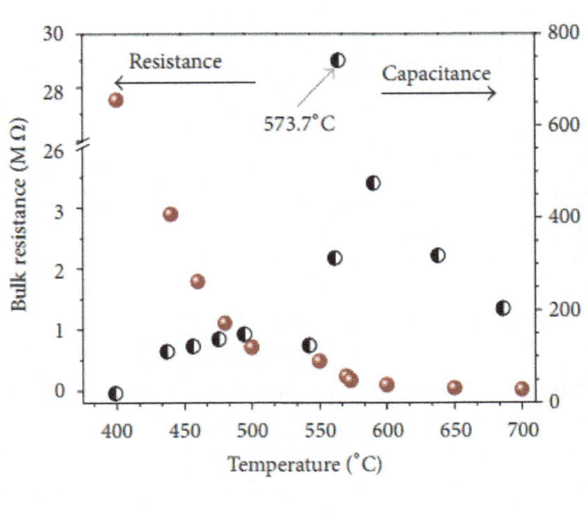

Figure 9: Bulk resistance R in MΩ and capacitance C in nF of the equivalent RC circuit of quenched PbNb2O6 sample, plotted against temperature, T. Here, R and C have been calculated from Nyquist diagram of Z' and Z'' in 20 Hz to 5.5 MHz range and the corresponding time constant $\tau = CR$, related to the dielectric relaxation time.

The previous equation reduces to Curie-Weiss law for $\gamma = 1$. The full equation can be rewritten as $\ln(1/\varepsilon' - 1/\varepsilon'_{max}) = \gamma \ln(T - T_m) - \ln C_w$. So, plotting of $\ln(1/\varepsilon' - 1/\varepsilon'_{max})$ versus $\ln(T-T_m)$ for different frequencies, as shown in Figures 10(a) and 10(b), gives the fit parameters (Table 2). Present finding that γ is nearly 1 in all cases proves that the relaxor property is minimum for these samples. This γ=1 leads to the observed sharpness of the peak at T_m. Relaxor piezoelectric materials, on the other hand, show a broadened peak.

Next, AC conductivity (σac) has been calculated from the dielectric loss (tan δ) data using the empirical formula, $\sigma_{ac} = \omega\varepsilon_0\varepsilon' \tan \delta = 2\pi f \varepsilon_0\varepsilon'$, where $\varepsilon_0 = 107/4\pi c^2 = 8.845 \times 10^{-12}$ $F \cdot m^{-1}$ = vacuum permittivity and c = speed of light in vacuum. Figure 11 presents this result on AC conductivity as a function of $1000/T$. Now the maxima in different graphs of Figure 11 (prominent only in high frequency graphs) are at $1000/T$ ~ 1.18K−1, that is, at T ~ 573∘ C that corresponds to the ferroelectric-to-paraelectric transition. AC activation energy (Eac) was calculated using the empirical Arrhenius relation $\sigma_{ac} = \sigma_0 \exp(-E_{ac}/kT)$.

Table 2: Result of fitting ε versus T data to the modified Curie-Weiss equation, as detailed in the text

Frequency (Hz)	T_m (°C)	$C_w \times 10^5$ (°C)	γ	ε'_{max}
5.5 M	579.5 ± 4	2.187	1.141	5472.6
3.0 M	575.2 ± 2	1.739	1.136	1876.1
2.0 M	574.8 ± 1	1.471	1.110	1574.2
50 k	574.9 ± 1	1.236	1.035	1752.2
1.0 k	573.8 ± 1	5.237	1.283	3613.9
0.5 k	573.0 ± 1	2.635	1.117	3632.5
0.1 k	572.5 ± 1	0.877	0.744	7103.5

The aforementioned maximum, known to occur in the AC conductivity graph at around the ferroelectricparaelectric phase transition temperature (T_c), is due to domain reorientation, domain wall motion, and the dipolar behavior [27]. In our data in Figure 11, there are three distinct temperature regions of Arrhenius type conduction, (a) ~50 to ~300° C section, corresponding to [28] electron or hole hopping, (b) ~400 to ~550° C part, due to conduction by small polarons and oxygen vacancies, and (c) ~580 to ~700° C high temperature region due to intrinsic ionic conduction [29]. Data in these three regions have been separately fitted to the Arrhenius relation as shown in Figure 12. The slope of $\ln(\sigma_{ac})$ versus $1/T$ plot gives AC activation energy Eac.

The previous fitting avoided the region between 300 and 400° C, as it showed an unexplained peak. However, the rise in some graphs at ~300° C to ~322° C region of Figure 11 reminds one of the anomaly at ~300° C in ε" versus T as well as in tan δ versus T graphs in Figures 5 and 6. It is noteworthy that temperature dependence of σac is radically different below and above this temperature zone, indicating different AC activation energies for AC conduction. However, somewhat wavy nature of the data at low frequencies and low temperatures, in this and other work [30], and suggestion of the rise by only one point in high frequency graphs prevent definite conclusions on the rise. These data at low frequencies (here 0.5, 1, and 10 kHz in Figure 11) and low temperature (T < 401° C) have not been considered for present AC activation energy calculation in next paragraph. In the high temperature graph in Figure 12, Eac falls rapidly with increase of measuring frequency up to 500 kHz and then becomes weakly dependent on frequency at higher frequencies. The low temperature graph of Eac is practically independent of frequency over the range (500 kHz to 5 MHz) studied here. As seen in the graphs, the AC activation energy is higher at lower frequencies, 0.5 kHz data giving 0.954 eV (with $\sigma 0 = 57.56\ \Omega\cdot m^{-1}$) for 400 to 550° C, and 1.17 eV (with $\sigma_0 = 1.508 \times 103\ \Omega\cdot m^{-1}$) for 580 to 700° C. These values fall to 0.523 eV (with $\sigma_0 = 24.62\ \Omega\cdot m^{-1}$) for 400 to 550° C and −1.994 eV (with $\sigma_0 = 6.253 \times 10^{-13}\ \Omega\cdot m^{-1}$) for 580 to 700° C, that is, in paraelectric region, respectively. Figure 13 gives the variation of $Z_{real} = Z'$ and $Z_{imaginary} = Z''$ against the measuring frequency. Here, Z' and Z'' graphs are seen to merge in the high frequency region approaching zero value. The single semicircular Cole-Cole plot suggests an equivalent circuit [23] of resistance R and capacitance C in parallel giving the well-known complex impedance $Z* = R/(1 + j\omega_{RC}) = Z' - jZ''$. Applying the relation $\tau = CR$ and the earlier-mentioned condition: $\omega_m\tau=1$, one gets $Z'' = R/2$ and $Z' = R/2$. So, relaxation time related frequency is the frequency at which Z' versus frequency and Z'' versus frequency graphs cross each other making $Z' = Z''$. We calculate this relaxation time and present it as the relaxation time τ_{cross} in Figure 14.

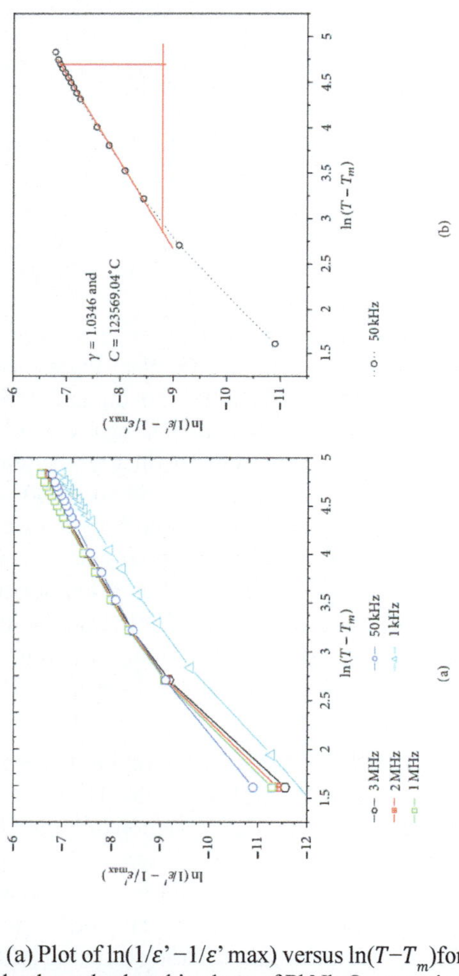

Figure 10: (a) Plot of $\ln(1/\varepsilon' - 1/\varepsilon'\max)$ versus $\ln(T-T_m)$for quenched PNO sample, the orthorhombic phase of $PbNb_2O_6$, at various frequencies (3 MHz, 2 MHz, 1 MHz, 50 kHz, and 1 kHz). It examines the relationship, expected for the paraelectric phase of a normal ferroelectric. (b) A specific example of finding the parameters ($\gamma = 1.0346$ and $C = 123569.04$° C) of the modified Curie-Weiss law from the 50 kHz plot of $\ln(1/\varepsilon' - 1/\varepsilon'_{max})$ versus $\ln(T - T_m)$ for the paraelectric phase of quenched $PbNb_2O_6$, which is orthorhombic.

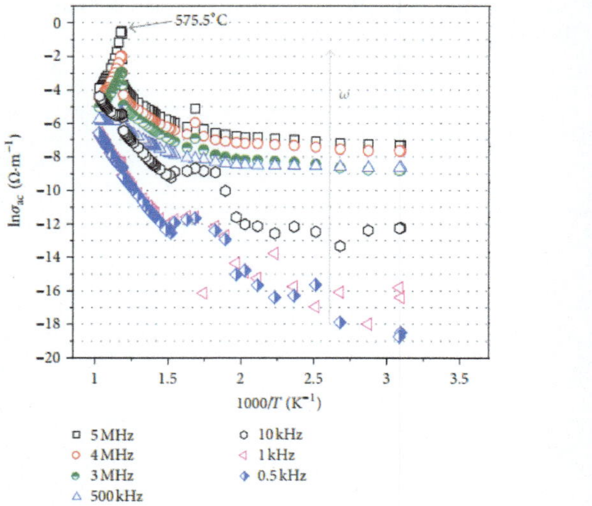

Figure 11: Frequency dependence of AC conductivity in the temperature range 50° C to 700° C, at different frequencies, for quenched PNO sample, the orthorhombic phase of $PbNb_2O_6$.

We determine relaxation time also from the frequency at the Z'' peak of Z'' versus frequency graphs (Figure 13) as done by others [30], calling it τpeak to compare it in Figure 14 with τcross. Figure 14 shows that τcross and τpeak behave similarly proving correctness of each approach. But these two graphs show different values with more relative scatter in the τpeak graph. This point shows up again in DC conductivity discussion in next paragraphs.

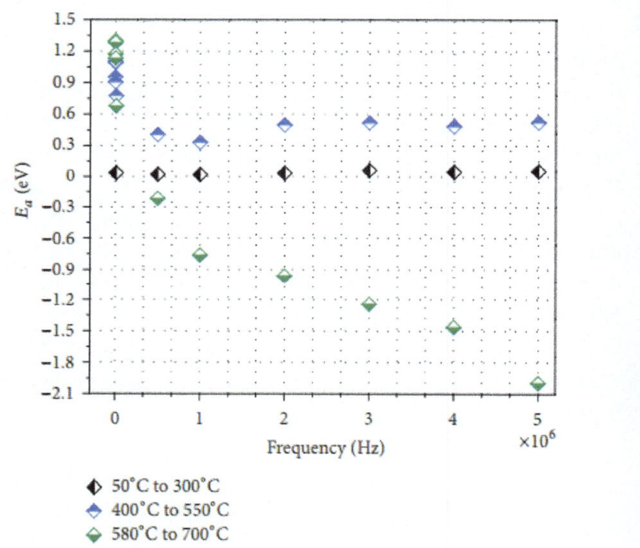

Figure 12: AC conductivity activation energy, E_a, as calculated from the slope of the Arrhenius plot of $\ln(\sigma_{ac})$ versus $1/T$, resulting in three Ea values for three temperature ranges. This figure depicts the variation of Ea with measuring frequency

The decrease of relaxation time with temperature has been adequately discussed in the literature [31]. Relaxation time for this material is seen to fall with increasing temperature, except for a prominent peaking at the ferroelectric to paraelectric transition temperature. $R = (R_0 - R_\infty) \sim R_0$, since $R_\infty \ll R_0$. This R has already been calculated from Figure 8. So, presently determined $CR = \tau$ value gives us C. These two equivalent circuit parameters C and R have been presented in Figure 9. Here, there is an impressive peaking of capacitance C at the Curie temperature. Another related parameter, the angle (θ), has been plotted against temperature in Figure 15. Clearly, it may be positive or negative, depending on the position of the centre of the semicircular arc. Variation of the impedance phase angle, $\delta = \tan{-1} - (Z' / Z'')$, with frequency, plotted on a logarithmic scale, is called Bode plot [23, 26]. Bode plots at different temperatures have been shown in Figures 16(a) and 16(b), graphs for 570∘ C, 573°C, and 600°C being given in Figure 16(b)to avoid overlap. At low frequencies and high temperatures (e.g., the 700°C graph), the constant phase part of the phase angle can be seen to be predominating. For a pure capacitor, $\delta = -90°$, while for a resistor it is zero. For a capacitor, the voltage, Vc, lags behind the response current by 90°. In the phase angle plot, an approach to pure capacitive behavior at low frequency values can usually be identified by $\delta \rightarrow -90°$. For an RC circuit, it will have an intermediate value and depends on ω as seen in the graph. Figure 16(b) contains data sets at Tm region temperatures and shows a prominent increase of the phase angle between 1 MHz and 5.5 MHz. The frequency ranges for near constancy of the phase angle (e.g., over 20 Hz to ~1 kHz for 700°C) can be observed from the figures. This constant phase region appears at low frequencies for the high temperature runs and at high frequencies for low temperature runs.

DC conductivity has been calculated using the relation [32], $\sigma_{dc} = (\varepsilon_0\varepsilon_\infty)/\tau$ from experimentally found τ and ε_∞(= ε' value as $f \rightarrow \infty$). Using τcross and τpeak data, as found in Figure 14, temperature dependence of σdc(cross) and σdc(peak), and, hence, that of two sets of DC activation energy have been calculated for the present PNO samples. Result from each technique shows expected increase of σ_{dc} with temperature (T), as shown in Figure 17. DC activation energy (E_{dc}) due to DC conductivity [14] was calculated using the empirical Arrhenius relation $\sigma dc = \sigma 0 \exp(-E_{dc}/kT)$, where k is Boltzmann constant, σ_0 is preexponential factor and E_{dc} is the DC activation energy for the conduction process [33–36], by plotting $\ln(\sigma_{dc})$ versus $1/T$. Figure 19 gives this DC activation energy for different temperature ranges, indicated by the horizontal bars, as found independently from τcross and τpeak data, for two- and three-region fits. Almost all earlier authors determined DC activation energy, E_{dc}, from relaxation time via DC conductivity. The 2-segment fit has been utilized in the literature for the usual τpeak-based evaluation of E_{dc} below and above (Curie).

Gonzalez et al [36] calculated relaxation times ($\tau = \tau_{CC}$, τ_T, τ^*) and other parameters for PNO sample from their 450°C to 590°C measurement of the temperature dependence of the imaginary dielectric modulus, separately by frequency domain analysis and by time domain analysis. Their τ^*, relaxation time due to thermal effect, was calculated in time domain using the Kohlrausch-Williams-Watts function, $(t) = e{-}(t/\tau^*)\ \beta$. Finally $\langle\tau_T\rangle$, the average relaxation time, was calculated using the relation, $\langle\tau_T\rangle = (\Gamma(1/\beta)/\beta)\tau^*$, where Γ is the Euler gamma function. Next, σdc was calculated using the relation $\sigma_{dc} = \varepsilon_\infty / \langle\tau_T\rangle$. They made separate liner fits to their $\ln(\sigma_{dc})$ versus $1/T$ data for data below T(Curie) and for data above T(Curie), finding, however, similar values E_{dc} = 1.08 eV below T(Curie) and E_{dc} = 1.11 eV above T(Curie).

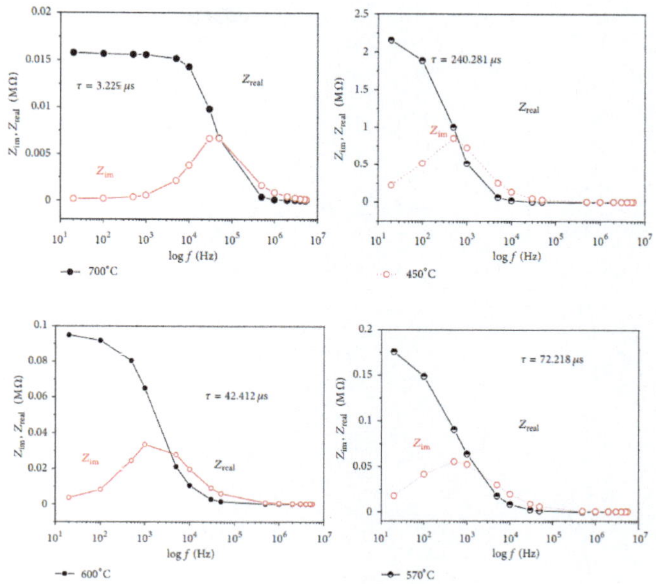

Figure 13: Variation of Z_{real} (real part of impedance) and Z_{im} (imaginary part of impedance) with frequency (plotted in log scale) at different temperatures. Frequency at the intersection of the two graphs gives a value of the relaxation time of the system, to be called τcross. Relaxation time estimated from the frequency) at the peaking in Z'' is to be called τ_{peak}.

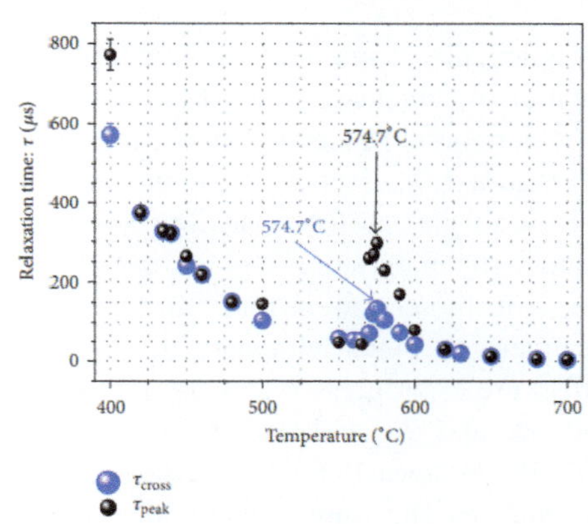

Figure 14: Variation of the relaxation time (τpeak and τcross) with temperature (T) has been shown in one graph, relaxation times τpeak and τcross being estimated by two different techniques as defined in Figure 13.

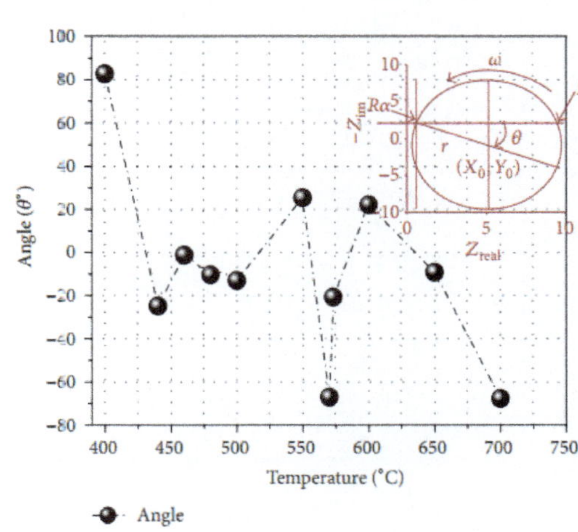

Figure 15: Diagrammatic definition of the angle (θ) of Nyquist Plot (in the inset), as detailed in the text, and its variation with temperature.

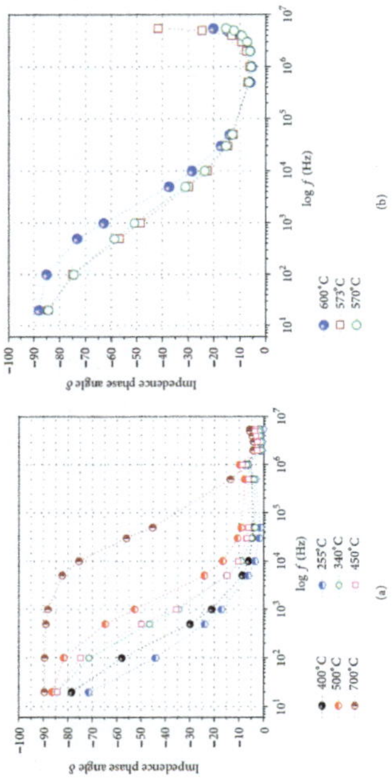

Figure 16: (a) Variation of the impedance phase angle, $\delta = \tan^{-1} - (Z_{real}/Z_{im})$, with frequency at different temperatures (255° C, 340° C, 400° C, 450° C, 500° C, and 700° C). (b) Variation of the impedance phase angle, $\delta = \tan^{-1} - (Z_{real}/Z_{im})$, with frequency at a few important temperatures (570°C, 573°C, and 600°C).

Finding practically same values apparently implies a single conduction mechanism [36]. However, the fact that more than one conduction mechanism is expected [36] from the observed (Figure 2) "departure of the real permittivity data from a low frequency plateau-like behaviour towards the lowest frequencies and high temperature" has also been pointed out by these authors, with the comment that "further measurements and data treatment to better elucidate this additional low-frequency effect are in progress." This motivated us (i) to use τcross to find σdc and E_{dc}, and also (ii) to try three segment fitting (Figure 18) as needed for the data resulting from the τcross-based analysis.The data actually demands a middle segment (here 570° C to 600° C) fit.

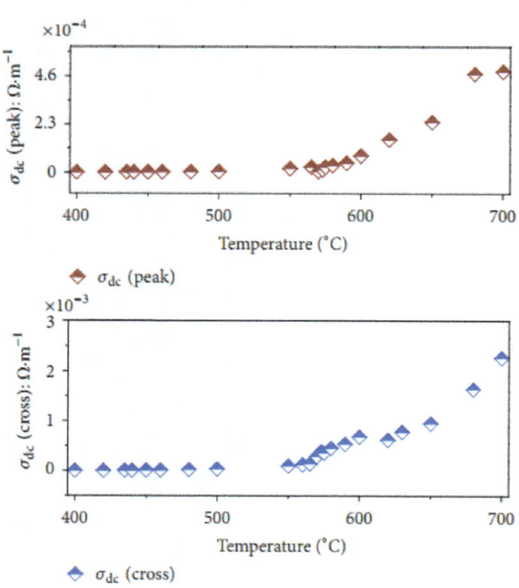

Figure 17: Variation of DC conductivity, σ_{dc} (cross/peak), with temperature in 400°C to 700°C range, for quenched PNO sample. Here, two values of DC conductivity, σ_{dc}(cross) = $(\varepsilon_0\varepsilon_\infty)/\tau$cross in the lower graph and σ_{dc}(peak) = $(\varepsilon_0\varepsilon_\infty)/\tau_{peak}$ in the upper graph have been calculated from τ_{cross} and τ_{peak} given in Figure 14.

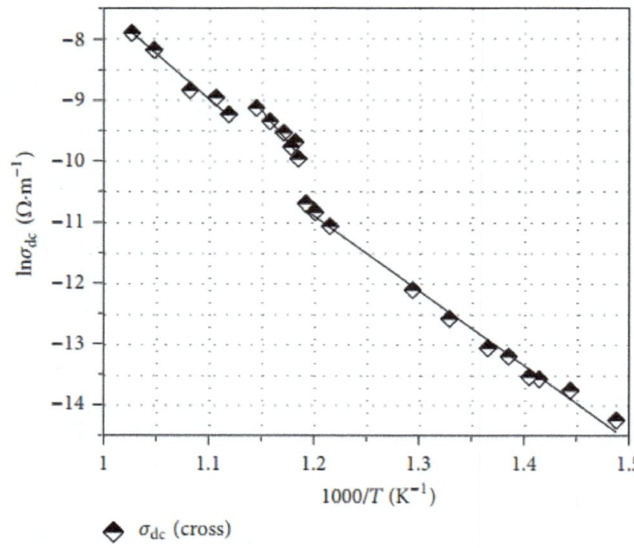

Figure 18: Variation of ln σ_{dc}(cross) with $1/T(K^{-1})$ in temperature 400°C to 700°C range, and least square fitting of the liner equation in three different temperature regions that yield three different values of DC activation energy

Here, one notes in Figure 17 that σ_{dc}(cross) versus T data clearly fall into three linear segments:

400°C to 565°C giving (Figure 18) E_a = 1.0724 ± 0.043 eV,

570°C to 600°C giving (Figure 18) E_a = 1.595 ± 0.120 eV,

610°C to 700°C giving (Figure 18) E_a = 1.263 ± 0.156 eV,

which implies a peaking of the activation energy $E_a = E_{dc}$ (cross) in the Curie temperature (T_c) region and practically the same value for low and high temperature regions ("C3" data points in Figure 19). This discovers the expected [36] existence of different E_{dc} and, hence, different conduction mechanisms, retaining same or similar values of E_{dc} below and above the T_c region.

A 2-segment fit to σ_{dc}(cross) versus T data of Figure 17 yields ("C2" points in Figure 19)

E_a = 1.072 ± 0.043 eV over 400°C to 565°C,

E_a = 1.026 ± 0.086 eV over 570°C to 700°C,

missing the significant rise and fall within a narrow Tc region (570 to 600°C) to wrongly conclude no significant change across T_c (the Curie temperature). This is in line with the limited knowledge, currently available in the literature as already discussed.

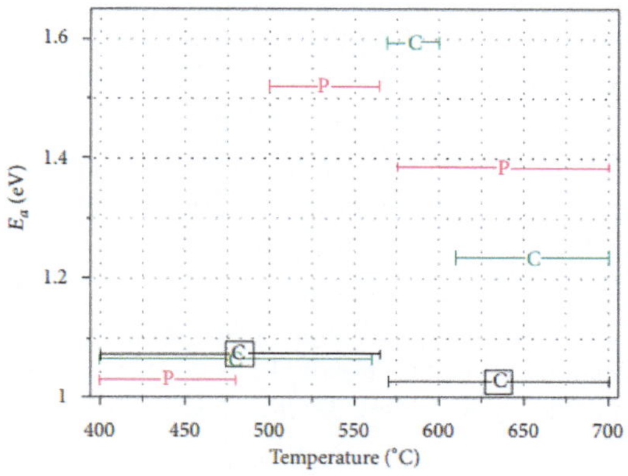

Figure 19: Activation energy (E_{dc}) for DC conduction in different temperature regions. *P* in symbol name indicates the use of peak technique of finding relaxation time, and *C* indicates the cross technique case. Here, 2 stands for fit to 2 temperature regions and 3 for 3 temperature regions fit, the chosen temperature region of fit being indicated by corresponding horizontal bar through the symbol.

σ_{dc}(peak) versus *T* data of Figure 17 can be seen to have more scatter than σ_{dc}(cross) versus *T* data with the former not falling well either on a set of straight lines or even on a smooth graph, as should be the case from already discussed higher scatter of τ_{peak} data. We feel that τcross should be used as relaxation time if there is a choice to choose from τpeak and τ_{cross} data. Still we tried linear fits to σ_{dc}(peak) versus *T* data. The 3-segment fit could tackle this scattered data set relatively better as follows:

400° C to 480° C $E_a = 1.028 \pm 0.101$ eV,

500° C to 590° C $E_a = 1.523 \pm 0.130$ eV,

600° C to 700° C $E_a = 1.387 \pm 0.168$ eV,

as shown in Figure 19 as "P3" data points.

So, the three-region fits of τ_{cross} and τ_{peak} suggest a peaking in E_{dc} versus *T* variation, at or below (resp.) the Curie temperature. This better match of the E_{dc} peak with the Curie temperature for τ_{cross} data further supports the better reliability of τ_{cross} data. Near the transition temperature, DC activation energy is highest (1.595 ± 0.120 eV), a new observation that needs further verification and explanation.

CONCLUSION

Presently prepared $PbNb_2O_6$ samples revealed pure orthorhombic phase by X-ray Rietveld analysis, with confirmation down to granular level, by High Resolution Transmission Electron Microscopy. These structural characterizations as well as impedance spectroscopy over 20 Hz to 5.5 MHz and up to 700°C established the high quality of these samples.

Present evaluations of DC and AC conductivities, capacitance (*C*) and resistance (*R*) of the equivalent *CR* circuit, relaxation times and related activation energies from the analysis of our wide-ranging impedance spectroscopy data on orthorhombic PNO, and discussions have been either more extensive than in the literature or done for the first time. This is the first reporting of the Nyquist diagram, the plot of Z'' versus Z', for orthorhombic lead metaniobate. An impressive peak (~760 nF) of the equivalent *CR* circuit capacitance (*C*) in the *C* versus *T* graph has been obtained at T_m as shown. AC conductivity (σ_{ac}) is seen to peak at T_m, more prominently in the higher frequency plots. Relaxation time registers a clear peak at T_m, while falling slowly with increasing temperature in other temperature regions.

Measured DC electrical conductivity appears to show three regions of temperature dependence (Figure 17) rather than the usually tried ferroelectric and paraelectric regions. A suitably chosen transition region in the plot of $\ln(\sigma_{dc})$ versus $1/T$ in Figure 18 forms the third region, interestingly giving the highest activation energy (Figure 19). A 3-region fit should be carefully examined for different systems, since the third region may be missed due to data points, inadequate with respect to either the temperature range or the density of data points or both. Moreover, a forced two-region fit (C2 points in Figure 19) to data needing a three-region fit will mask the large up and down (C3 points) of the activation energy value in the Curie temperature region and indicate a small change of activation energy across the Curie temperature, similar to the result of 2-segment fit to time domain analysis [35] result. It appears that the DC activation energy can be estimated better from τcross and 3-segment fit (Figure 19), unless the result in a particular experiment is against 3-segment fit. Presently observed peaking of the activation energy E_{dc}(cross) in the Curie temperature (T_c) region is a new result.

The peak at T_m in the ε' versus *T* graph is sharp, leading to a value of ~1 for the diffusivity parameter and implying near absence of any relaxor property. Impedance spectroscopy showed (Table 1) highest record of ferroelectric-paraelectric transition temperature (T_m), T_m > 580○ C for 5.5 MHz, and Tm > 574○ C for 0.5 kHz. This implies potential use of this material for piezoelectric sensors and actuators up to temperatures much higher than the limits of BT and PZT based piezomaterials.

ACKNOWLEDGMENTS

The dielectric measurement set-up at the Deptartment of Ceramic Engineering, NIT, Rourkela 769008, India, has been used with the active encouragement of Professor Swadesh K Pratihar. Authors also thank Shri Pulak Ray and his team at SINP for help in TEM operation. U. De thanks Alexander von Humboldt Foundation (Germany) for partial support that led to preliminary presentation of a part of this work in international conferences and discussions.

REFERENCES

1. I. Ihara, "Ultrasonic sensing: fundamentals and its applications to nondestructive evaluation," Lecture Notes Electrical Engineering, vol. 21, pp. 287–305, 2008. ·
2. R. Kazys, A. Voleisis, and B. Voleisiene, "High temperature ultrasonic transducers: review," Ultragarsas, vol. 63, no. 2, pp. 7–17, 2008.
3. G. Goodman, "Ferroelectric properties of lead metaniobate," Journal of the American Ceramic Society, vol 36, no. 11, pp. 368–372, 1953. ·
4. A. A. Nesterov, E. V. Karyukov, and K. S. Masurenkov, "Synthesis of phases of composition $PbNb_2O_6$ and

$BaNb_2O_6$ with the use of active precursors," Russian Journal of Applied Chemistry, vol. 82, no. 3, pp. 370–373, 2009. · ·

5. R. S. Rath, "Unit-cell data of the lead niobate $PbNb_20_6$," Acta Crystallographica, vol. 10, part 6, pp. 437–437, 1957. ·

6. U. De, "TEM, HT XRD, electrical, optical and thermal characterizations of materials for high temperature piezoelectric applications," in Proceedings of the IUMRS-ICA, Abstract no. 21_1078, BEXCO, Busan, Korea, 2012, http://www.iumrs-ica2012.org/program.php.

7. M. H. Francombe and B. Lewis, "Structural, dielectric and optical properties of ferroelectric leadmetaniobate," Acta Crystallographica, vol. 11, part 10, pp. 696–6703, 1958. ·

8. E. C. Subbarao and G. Shirane, "Nonstoichiometry and ferroelectric properties of $PbNb_2O_6$-type compounds," The Journal of Chemical Physics, vol. 32, no. 6, pp. 1846–1851, 1960. ·

9. E. C. Subbarao, "X-ray study of phase transitions in ferroelectric $PbNb_2O_6$ and related materials," Journal of the American Ceramic Society, vol. 43, no. 9, pp. 439–442, 1960. ·

10. S. Ray, E. Günter, and H. J. Ritzhaupt-Kleissl, "Manufacturing and characterization of piezoceramic lead metaniobate $PbNb_2O_6$," Journal of Materials Science, vol. 35, no. 24, pp. 6221–6224, 2000.

11. K. R. Chakraborty, K. R. Sahu, A. De, and U. De, "Structural characterization of orthorhombic and rhombohedral lead meta-niobate samples," Integrated Ferroelectrics, vol. 120, no. 1, pp. 102–113, 2010. · ·

12. K. R. Sahu and U. De, "Thermal characterization of piezoelectric and non-piezoelectric lead meta-niobate," Thermochimica Acta, vol. 490, no. 1-2, pp. 75–77, 2009. ·

13. F. Guerrero, Y. Leyet, M. Venet, J. de Los, and J. A. Eiras, "Dielectric behavior of the $PbNb_2O_6$ferroelectric ceramic in the frequency range of 20 Hz to 2 GHz," Journal of the European Ceramic Society, vol. 27, no. 13–15, pp. 4041–4044, 2007. · ·

14. S. K. Barbar and M. Roy, "Synthesis, structural, electrical, and thermal studies of $Pb_{1-x}Ba_xNb_2O_6$ (x = 0.0 and 0.4) ferroelectric ceramics," ISRN Ceramics, vol. 2012, Article ID 710173, 5 pages, 2012. ·

15. J. W. Griffin, T. J. Peters, G. J. Posakony, H. T. Chien, and L. J. Bond, Under-Sodium Viewing: A Review of Ultrasonic Imaging Technology for Liquid Metal Fast Reactors, PNNL-18292, U.S. Department of Energy, Pacific Northwest National Laboratory, Richland, Wash, USA, 2009.

16. I. V. Kityk, M. Makowska-Janusik, M. D. Fontana, M. Aillerie, and A. Fahmi, "Influence of non-stoichiometric defects on optical properties in $LiNbO_3$," Crystal Research and Technology, vol. 36, no. 6, pp. 577–589, 2001.

17. M. Venet, A. Vendramini, F. L. Zabotto, F. D. Garcia, and J. A. Eiras, "Piezoelectric properties of undoped and titanium or barium-doped lead metaniobate ceramics," Journal of the European Ceramic Society, vol. 25, no. 12, pp. 2443–2446, 2005. ·

18. H. S. Lee and T. Kimura, "Effects of microstructure on the dielectric and piezoelectric properties of lead metaniobate," Journal of the American Ceramic Society, vol. 81, no. 12, pp. 3228–3236, 1998.

19. J. Soejima, K. Sato, and K. Nagata, "Preparation and characteristics of ultrasonic transducers for high temperature using $PbNb_2O_6$," Japanese Journal of Applied Physics 1, vol. 39, no. 5, pp. 3083–3085, 2000.

20. Y. M. Li, L. Cheng, X. Y. Gu, Y. P. Zhang, and R. H. Liao, "Piezoelectric and dielectric properties of $PbNb_2O_6$-based piezoelectric ceramics with high Curie temperature," Journal of Materials Processing Technology, vol. 197, no. 1–3, pp. 170–173, 2008.

21. P. Q. Mantas, "Dielectric response of materials: extension to the Debye model," Journal of the European Ceramic Society, vol. 19, no. 12, pp. 2079–2086, 1999. ·

22. S. Zhang, C. A. Randall, and T. R. Shrout, "High Curie temperature piezocrystals in the $BiScO_3$-$PbTiO_3$perovskite system," Applied Physics Letters, vol. 83, no. 15, pp. 3150–3152, 2003. · ·

23. J. R. Macdonald and W. B. Johnson, in Impedance Spectroscopy: Theory, Experiment, and Applications, E. Barsoukov and J. R. Macdonald, Eds., John Wiley & Sons, Hoboken, NJ, USA, 2nd edition, 2005.

24. S. Sen and R. N. P. Choudhary, "Impedance studies of Sr modified $BaZr_{0.5}Ti_{0.95}O_3$ ceramics," Materials Chemistry and Physics, vol. 87, no. 2-3, pp. 256–263, 2004. · ·

25. A. Shukla, R. N. P. Choudhary, and A. K. Thakur, "Thermal, structural and complex impedance analysis of Mn^{4+} modified $BaTiO_3$ electroceramic," Journal of Physics and Chemistry of Solids, vol. 70, no. 11, pp. 1401–1407, 2009. ·

26. K. Uchino and S. Nomura, "Critical exponents of the dielectric constants in diffused phase transition crystals," Ferroelectrics Letters Section, vol. 44, no. 11, pp. 55–61, 1982. ·

27. L. E. Cross, "Relaxor ferroelectrics," Ferrooelectrics, vol. 76, no. 1, pp. 241–267, 1987. ·

28. O. Raymond, R. Font, N. Suárez-Almodovar, J. Portelles, and J. M. Siqueiros, "Frequency-temperature response of ferroelectromagnetic Pb $(Fe_{1/2}Nb_{1/2})O_3$ ceramics obtained by different precursors—part I: structural and thermo-electrical characterization," Journal of Applied Physics, vol. 97, no. 8, Article ID 084107, 8 pages, 2005. · ·

29. B. Behera, E. B. Arajo, R. N. Reis, and J. D. S. Guerra, "AC conductivity and impedance properties of $0.65Pb(Mg_{1/3}Nb_{2/3})O_3$-0.35 $PbTiO_3$ ceramics," Advances in Condensed Matter Physics, vol. 2009, Article ID 361080, 6 pages, 2009. · ·

30. R. N. P. Choudhary, D. K. Pradhan, G. E. Bonilla, and R. S. Katiyar, "Effect of La-substitution on structural and dielectric properties of $Bi(Sc_{1/2}Fe_{1/2})O_3$ ceramics," Journal of Alloys and Compounds, vol. 437, no. 1-2, pp. 220–224, 2007. · ·

31. K. S. Rao, D. M. Prasad, P. M. Krishna, and J. H.

Lee, "Synthesis, electrical and electromechanical properties of tungsten-bronze ceramic oxide: $Pb_{0.68}K_{0.64}Nb_2O_6$," Physica B, vol. 403, pp. 2079–2087, 2008.

32. Z. Lu, J. P. Bonnet, J. Ravez, and P. Hagenmuller, "Correlation between low frequency dielectric dispersion (LFDD) and impedance relaxation in ferroelectric ceramic $Pb_2KNb_4TaO_{15}$," Solid State Ionics, vol. 57, no. 3-4, pp. 235–244, 1992. ·

33. P. S. Sahoo, A. Panigrahi, S. K. Patri, and R. N. P. Choudhary, "Impedance spectroscopy of $Ba_3Sr_2DyTi_3V_7O_{30}$ ceramic," Bulletin of Materials Science, vol. 33, no. 2, pp. 129–134, 2010. · ·

34. S. Rachna, S. M. Gupta, and S. Bhattacharyya, "Impedance analysis of $Bi_{3.25}La_{0.75}Ti_3O_{12}$ ferroelectric ceramic," Pramana, vol. 71, no. 3, pp. 599–610, 2008. · ·

35. S. Devi and A. K. Jha, "Dielectric and complex impedance studies of $BaTi_{0.85}W_{0.15}O_{3+\delta}$ ferroelectric ceramics," Bulletin of Materials Science, vol. 33, no. 6, pp. 683–690, 2010.

36. R. L. Gonzalez, Y. Leyet, F. Guerrero, et al., "Relaxation dynamics of the conductive process for $PbNb_2O_6$ ferroelectric ceramics in the frequency and time domain," Journal of Physics: Condensed Matter, vol. 19, no. 13, Article ID 136218, 2007

CHAPTER 13

$ZNO_{1-X}S_X$ NANOSPHERE IN FERROELECTRIC LIQUID CRYSTAL MATRIX: THE EFFECT OF AGGREGATION AND DEFECTS ON THE DIELECTRIC AND ELECTRO-OPTICAL PROPERTIES

Dharmendra Pratap Singh, Swadesh Kumar Gupta, and Rajiv Manohar

INTRODUCTION

The scope of nanotechnology, in the present age, is not only limited within the field of material science but also extended to all the fields of research and application. The fields like electronics, condense matter, and medical sciences are benefited from the application point of view when used with nanoscience [1–3]. The use of nanoscience in fields of liquid crystal has created some interesting and useful properties [4].

Ferroelectric liquid crystals (FLCs) have salient features over the trivial liquid crystals (LCs) because of their tilted smectic phase and the chirality of the molecules [5]. They have proved their applicability of showing faster response time and the instruments based on less threshold voltage [6]. In addition to these properties, they have also shown their potential for the field effect transistors (FETs), spatial light modulators (SLMs), switchable devices, and nonvolatile memory effects [7, 8]. The prospective of tuning the properties of liquid crystals due to the addition of nanomaterial is currently a hot topic in liquid crystal research [9–11]. The composite system of the pure FLC and nanomaterial has already shown its utility in the FLC-based devices with greater efficiency [12].

In the present paper, we report a comprehensive study of the dielectric and electro-optical (E-O) properties of a ferroelectric liquid crystal highly doped with $ZnO_{1-x}S_x$ nanospheres (NS). The present work has a motive to analyze the effect of high concentration of the nanomaterial on the FLC matrix and also on the E-O and dielectric parameters. The $ZnO_{1-x}S_x$ NS were added in 5% wt/wt concentration to the pure FLC 16/100. The results show that the doping of NS alters the physical properties of the pure FLC. The investigated results are discussed and explained on the basis of the interactions taking place between the NS and FLC molecules in the composite geometry, effect of high doping of NS on the pure FLC, and molecular dynamics of NS and FLC molecules with their dipolar contributions.

Experimental Details

A commercially available ferroelectric liquid crystal (FLC) 16/100, purchased from Clariant Chemicals Co. Ltd., Germany, was used as host material in the present study. The phase sequence of the FLC material is Cr, SmC*, SmA, N*, and Iso at −20°C, 72°C, 82°C, and 90°–94°C. For the pure FLC 16/100, the standard values (as provided by the manufacturer) of spontaneous polarization (P_s), cone angle (2θ), and rotational viscosity (γG)) at 25°C were −10.5 nC/cm², 54.30°, and 60 m Pa.s, respectively.

The $ZnO_{1-x}S_x$ nanospheres (NS) were added to the pure FLC as a guest entity. A typical coprecipitation method was used to synthesize the present NS (nanocrystals of spherical shape) taking CH_3OH as solvent. The prepared NS were characterized by X-ray diffraction (XRD) and transmission electron microscopy (TEM). The diameter of the nanospheres was found to be ~10–15 nm. The detailed information about the synthesis of the nanosphere (spherical shape nanoparticle) and the effect of small concentration of the NS on FLC

Citation: Dharmendra Pratap Singh, Swadesh Kumar Gupta, and Rajiv Manohar, $ZnO_{1-x}S_x$ Nanosphere in Ferroelectric Liquid Crystal Matrix: The Effect of Aggregation and Defects on the Dielectric and Electro-Optical Properties; doi.org/10.1155/2013/250301.

parameters has already been published by Singh et al. [13]. Highly conducting (~10–15 nm) indium tin oxide (ITO) coated glass plates were used to make sample cells. The desired electrode pattern on the ITO substrates was achieved using a photolithographic technique. The active electrode area was 7 mm × 7 mm. The uniform thickness (6 μm in our case) of both the sample cells was maintained by means of Mylar spacer. The planar alignment was obtained by coating polymer (Nylon 6/6) on ITO glass plate and unidirectional rubbing. The empty sample cells were calibrated using analytical reagent (AR) grade carbon tetrachloride (CCl_4) and benzene (C_6H_6).

The nanosphere-ferroelectric liquid crystal (NS-FLC) composite system was prepared by the mixing of NS in the 5% wt/wt ratio with pure FLC 16/100. To prepare the composite system, an appropriate amount (in the weight ratio, i.e., 5.0%) of nanospheres was mixed into the pure FLC and then homogenized with an ultrasonic mixer at 90°C for one hour, and uniform dispersion of nanospheres was ensured but the high concentration of NS aggregates in the pure FLC matrix which was clearly observed in the polarizing optical microscope (POM) images.

The pure FLC and the NS-doped FLC were filled in the sample cells at a temperature higher than the isotropic temperature by means of capillary action and then cooled gradually up to room temperature with the application of a small a.c. electric field. The alignment of the samples was checked by the polarizing microscope under the crossed polarizer-analyzer arrangement.

The dielectric spectroscopy of the samples was performed by a computer-controlled impedance/gain analyzer (Solartron SL1260) attached with a temperature controller for the frequency interval 100 Hz–32 MHz. The dielectric measurements were carried out as a function of temperature by placing the sample on a computer-controlled hot plate instec (HCS 302). The temperature stability was better than ±0.1°C. The low- and high-frequency corrections were also made to the experimental data.

The FTIR spectrum of the pure and the NS-doped FLC was recorded by the Fourier transform infrared spectrophotometer (IR Affinity-1, SHIMADZU) between 1500 to 3000 cm^{-1}. The photoluminescence (PL) spectra of pure and NS-doped FLC were recorded using Perkin Elmer LS 55 luminescence spectrophotometer. During the PL measurement, all the measuring parameters like slit width (4 nm in our case), excitation step, and so forth, were same for the entire samples. The two excitation wavelengths, that is, 290 and 315 nm, were used to record the PL spectra of the pure and the NS-doped FLC in the present investigation.

The spontaneous polarization (P_s) measurement was performed using a triangular wave method at a voltage of 20 $V_{P\text{-}P}$ and 10 Hz frequency. P_s was determined by using the following relation [14]:

$$P_s = \frac{1}{2A}\int i(t)\,dt, \tag{1}$$

where $\int i(t)dt$ is the area under the current bump and A is the active area of the sample in sample holder. For the measurement of response time (τ_{res}), a square wave signal of 10 Hz and 20 $V_{P\text{-}P}$ was used. The rotational viscosity (γG) of the Goldstone mode in SmC* phase was calculated by the relation

$$\gamma_G = P_s \times \tau_{res} \times E = P_s \times \frac{t_{10} - t_{90}}{1.8} \times E, \tag{2}$$

where E is the applied switching field and $t10$ and $t90$ are the corresponding times for 10% and 90% of the maximum level of response time while other symbols have their usual meaning [15]. The dielectric relaxation phenomena of the pure and the NS-doped FLC were analyzed using the Cole-Cole relation [16]:

$$\varepsilon^* = \varepsilon'(\infty) + \frac{\delta\varepsilon}{1 + (i\omega\tau)^{1-\alpha}}, \tag{3a}$$

where, $\delta\varepsilon$ is the relaxation strength and ε' (∞) is the highfrequency limit of the dielectric permittivity data, $\omega = 2\pi f$, where f is the frequency, τ is the relaxation time, and α is the distribution parameter. The experimental data suffers low- and high-frequency deviations in dielectric data and exceedingly require correction for low- and high-frequency values. On separating the real and imaginary parts of (3a), one may get

$$\varepsilon' = \varepsilon'(dc)\, f^{-n} + \varepsilon'(\infty) + \frac{\delta\varepsilon\left[1 + (2\pi f\tau)^{(1-\alpha)}\sin(\alpha\pi/2)\right]}{1 + (2\pi f\tau)^{2(1-\alpha)} + 2(2\pi f\tau)^{(1-\alpha)}\sin(\alpha\pi/2)}, \tag{3b}$$

$$\varepsilon'' = \frac{\sigma(dc)}{\varepsilon_0 2\pi f^k} + \frac{\delta\varepsilon(2\pi f\tau)^{(1-\alpha)}\cos(\alpha\pi/2)}{1 + (2\pi f\tau)^{2(1-\alpha)} + 2(2\pi f\tau)^{(1-\alpha)}\sin(\alpha\pi/2)} + Af^m. \tag{3c}$$

Here, (dc) is the dc ionic conductance, ε_o is the free space permittivity, and f is the frequency while n, m, and k are the fitting parameters. The terms ε' $(dc)^{-n}$ and $\sigma(dc)/\varepsilon_o 2\pi f^k$ are added in (3b) and (3c) for correcting the low-frequency effect due to the electrode polarization capacitance and ionic conductance. The Afm term is added in (3c) for correcting the high-frequency effect due to the ITO resistance and lead inductance [9]. The other abbreviations are same as given for (3a). The experimental data have been fitted in these equations and corrected for low- and high-frequency values [13].

Results and Discussions

Figures 1(a) and 1(b) represent the SEM image of the $ZnO_{1-x}S_x$ (NS) and aggregation of the NS, respectively. A small agglomeration has been observed in the SEM image of the NS which is the basic problem of most of the nanomaterial. Due to the agglomeration property of NS, it aggregates in the FLC matrix as dispersed in high concentration. The aggregation and topological defects are the main factors which alter the physical properties of the pure FLC material.

Fourier transform infrared (FTIR) spectroscopy is a powerful modern technique for the characterization of samples. This technique provides the information about the concentration of the impurities, their bonding with the host material, and different types of vibrations occurring in the samples.

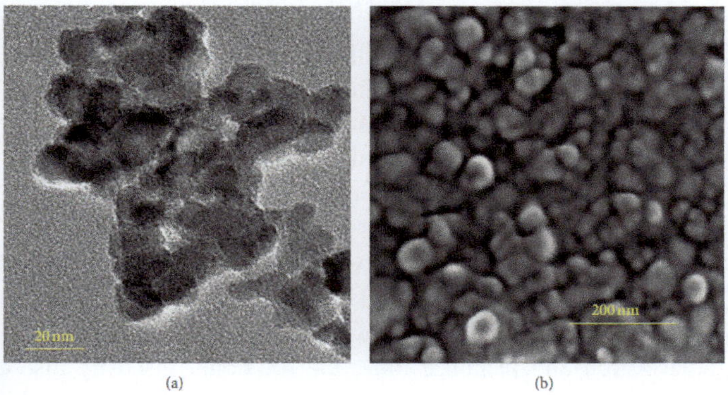

Figure 1: (a) The scanning electron microscopic (SEM) image of the ZnO$_{1-x}$S$_x$ nanosphere (NS) which represents a small agglomeration of the NS. (b) SEM image of the aggregation of NS.

The FTIR analysis, for the wave number intervals of 1500–3000 cm^{-1}, has been performed for the pure and the NS-doped FLC material in Figures 2(a) and 2(b), respectively. The change in FTIR spectra for the NS-doped FLC system is mainly due to the aggregation of the NS in FLC matrix. As, NS behaves like the ionic additive to the FLC matrix, the large concentration of the NS (or aggregation of NS) reflects in the FTIR spectra in the form of impurity. The inset of Figure 2(b) shows the enlarged image of FTIR spectra from 1500 to 1700 cm^{-1}. The transmission peaks in this region show the vibrations corresponding to impurity ions, that is, NS. The change in percentage Transmission is due to the topological defects and scattering of light from the aggregation of NS. Therefore, the change in FTIR spectra clearly indicates the presence of a guest entity in a large concentration in the FLC matrix.

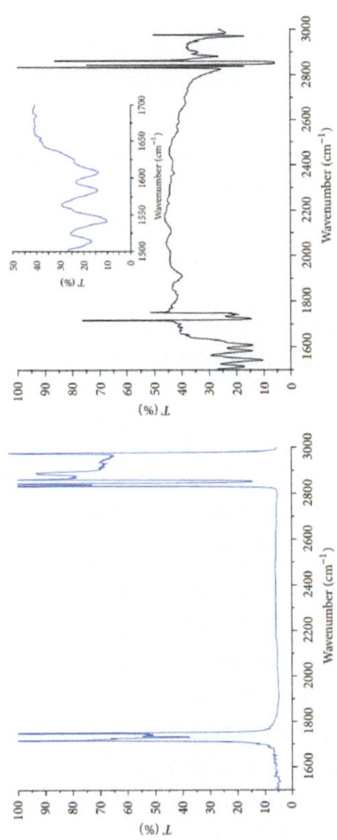

Figure 2: FTIR spectra of the (a) pure FLC and (b) NS-doped FLC

material. The inset of Figure 2(b) shows the enlarged FTIR spectra of the NS-doped FLC material between the interval of 1500–1700 cm^{-1}.

The PL spectra of the pure and the NS-doped FLC using two different excitation wavelengths have been shown in Figure 3. The PL spectra of the pure FLC show a broad emission peak centred at 388 nm corresponding to violet emission. The PL measurement was performed on two different excitation wavelengths, that is, 290 nm and 315 nm. The change in the excitation wavelength only alters the intensity of emission peak, but it does not cause any shifting in the emission peak. Due to the addition of NS, the PL of the pure FLC materials has been enhanced. This enhancement in the PL spectra of the pure FLC materials is attributed to the coupling of localized surface plasmon resonance from NS with FLC molecules [17]. In addition to this, some other feeble PL emission peaks centred at 409, 432, and 460 nm have also been observed. These emission peaks lie in the spectral region due to the self-activated PL centres. These centres have often been attributed to the crystal vacancies. Thus, feeble PL emission peaks centred at 409, 432, and 460 nm arise from the crystal line incorporation of trace amount of Zn^{2+} in the crystal lattice creating the oxygen and sulphur deficiency [18]. The nonsmoothening of the PL spectra also suggests the size distribution of the NS in FLC matrix. The change in excitation wavelength does not produce any change in the magnitude of the PL emission peak of the NS-doped FLC system. This shows that the aggregation of the NS in the pure FLC matrix due to large amount of NS dominate over the intras FLC interactions.

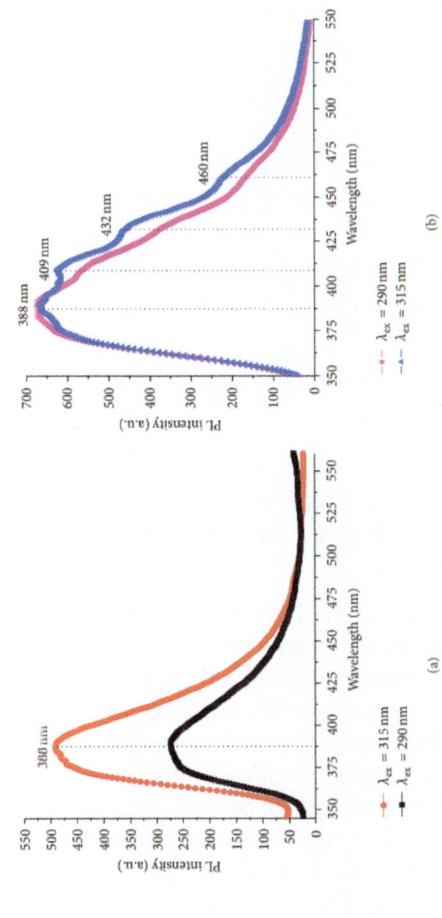

Figure 3: The photoluminescence (PL) spectra of the (a) pure FLC and the (b) NS-doped FLC at two different excitation wavelengths, that is, 290 nm and 315 nm.

With the help of triangular wave electrical response curve (also known as polarization reversal current method), the spontaneous polarization of the pure and the NS-doped FLC has been calculated. The behaviour of spontaneous polarization (P_s) with the change in temperature is shown in Figure 4(a). The value of P_s decreases with increasing temperature for both the pure and the doped FLC, but the P_s value for the NS doped FLC system is higher than that of the pure FLC. When a high concentration of the NS is dispersed in the FLC material, it causes an aggregation of the NS. The NS behaves like an ionic additive in the FLC system. At the place of aggregation, these ionic charges accumulate and form a local internal electric field. This local internal electric field increases the separation between polarization charges on the LC molecules which results in the increased dipole moment of the doped system. The increased dipole moments result the increased spontaneous polarization of the NS-doped FLC system [19].

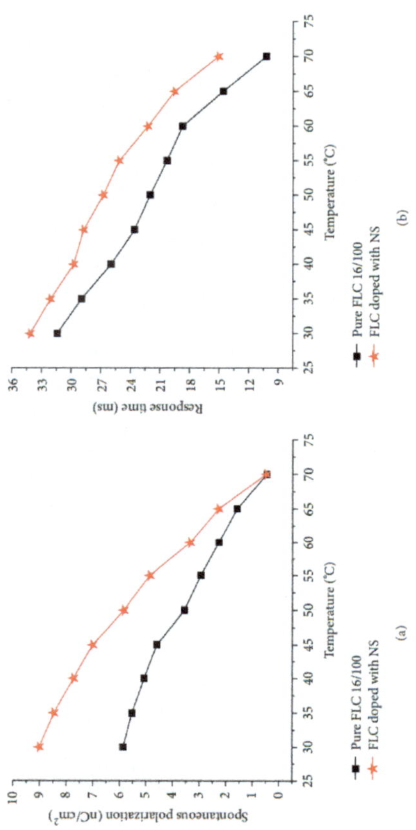

Figure 4: The variation of (a) spontaneous polarization and (b) response time on the temperature scale for the pure and the NS-doped FLC samples.

The response time (τ_{res}) of the pure and the doped FLC system is plotted against the temperature scale in Figure 4(b). The response time of the pure and the doped system decreases with increasing temperature. The response for the doped system is slower as compared to the pure FLC due to the presence of high concentration of NS. The presence of high quantity of NS creates additional constraints in the molecular motion of the FLC molecules under the influence of applied field. These constraints reduce the energy of the doped FLC system; therefore, FLC molecules do not respond quickly with the applied field as compared to the pure FLC molecules which results in a slower response for the NS-FLC system. No noticeable shift in the SmC*-SmA phase transition temperature has been observed for the NS-FLC system.

The rotational viscosity with the variation of temperature is shown in Figure 5. Almost linear decrement in the value of rotational viscosity on increasing the temperature has been observed for both the pure and the doped systems. But the value of rotational viscosity for the NS-doped FLC system is greater than that of the pure FLC system. As rotational viscosity $(\gamma = P_s \cdot \tau_{res} \cdot E)$ is the product of spontaneous polarization, response time, and effective electric field applied on the sample cell, therefore due to increase in the spontaneous polarization and response time, the rotational viscosity also increases as compared to the pure FLC material. As discussed previously, the NS aggregates in the FLC matrix and forms a local internal electric field near the aggregation. This field influences the smectic layer ordering which alters the smectic layer spacing as shown in the inset of Figure 5. This combined effect of disordering in smectic layers and aggregation produces the topological defects in FLC matrix. These two factors affect the rotational motion of FLC molecules which results in the enhanced rotational viscosity of the doped FLC system. The alteration in the smectic layer separation due to spherical gold nanoparticle has already been reported by Pratibha et al. [20].

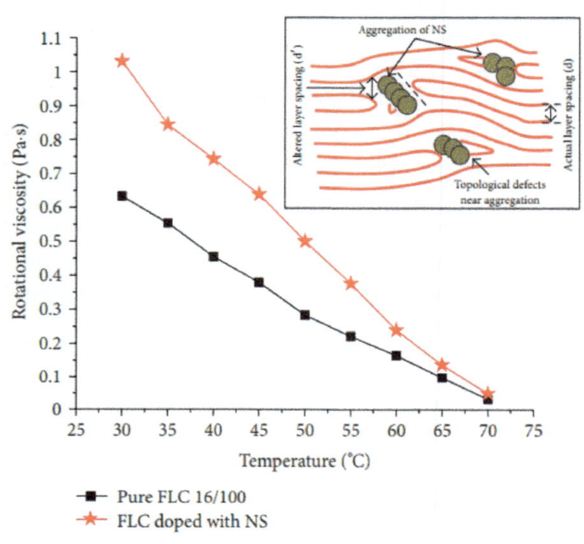

Figure 5: The change in rotational viscosity with the variation of temperature for the pure and the NS-doped FLC whereas the inset of figure shows the paradigm of the warping of the smectic layers due to the presence of aggregation of NS.

In the molecular motion of the composite system, four types of interactions take place, namely,(i)FLC molecule-molecule interaction;(ii)FLC molecule-NS interaction;(iii)NS-NS interaction;(iv)FLC molecule-alignment layer interaction.

In the presence of any guest entity (nanoparticles, CNTs, dyes, polymers, etc.), the interaction of FLC molecules with that of the guest material (NS in our case) takes place. The probability of weak or strong NS-NS interaction depends upon the concentration of NS in the composite system. In the present work, 5% wt/wt concentration of NS causes strong NS-NS interactions in the composite system. This strong NS-NS interaction causes the aggregation of NS in FLC matrix. One should notice that the size of the nanomaterial also plays a crucial role in determining the properties of the composite systems [21]. The bigger size of NS and strong NS-NS interaction are mainly responsible for the modifications in the properties of the pure FLC. From the application point of view, FLC-NS interaction in the composite is accountable depending on the nature, characteristics, size, shape, and quantity of NS.

Mainly, three types of events exist in the NS-FLC composite system; that is, the presence of NS wraps the smectic layers around it, NS produces hindrance in the motion of the FLC molecule, and NS strongly interacts with the FLC molecules. The presence of NS warps the smectic layers and alters the spacing between them. These warped smectic layers radically alter the neighbouring smectic layers. In the cone motion of the FLC molecules, the factor $d\varphi/dt$ in viscous torque ($-\Gamma z = (d\varphi/dt)$, where Γz is the z component of viscous torque and φ is the phase angle) is redistributed because of the warping of the smectic layers. This results in a higher rotational viscosity for the NS-doped FLC system.

The polarizing optical microscopic (POM) images in dark and bright states have been shown in Figure 6. Figures 6(a) and 6(b) represent the POM images of the pure FLC in dark and bright states, respectively, whereas Figures 6(c) and 6(d) represent the NS-doped FLC in the dark and bright states. The dark state POM image corresponds to undercrossed polarizer-analyzer condition whereas the bright state was achieved by rotating the sample cell by 45° from the initial condition. The high concentration of NS causes aggregation and defects in the FLC medium which is clear from Figure 6(c). Due to this aggregation of NS, the molecular ordering of the FLC molecules also impairs. The increase in the relaxation strength of the NS-doped FLC is compatible with the present result. The study of the relaxation strength will be discussed later in the paper. The high doping of the NS in the pure FLC also reduces the anchoring of the FLC sample which is also verified by the optical textures. The occurrence of the intense chevron geometry can be seen in the case of the pure FLC (Figures 6(a)and 6(b)) whereas in the doped FLC system, the chevron geometry was not well observed. The presence of high concentration of the NS causes other topological defects in the FLC matrix causing the scattering of transmitted light through NS which lowers the contrast of the NS-doped FLC system.

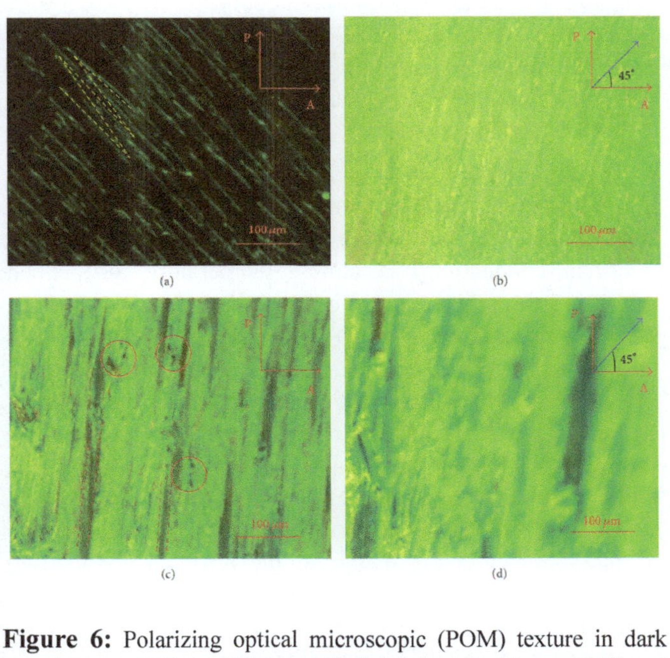

Figure 6: Polarizing optical microscopic (POM) texture in dark state (a) pure FLC, (c) NS-doped FLC, bright state (b) pure FLC, and (d) NS-doped FLC. The dark state was achieved under the crossed polarizer-analyzer condition whereas the bright state was achieved by rotating the sample cell by 45° from the initial condition. The aggregation of NS has been shown in the circle and topological defects are shown between two dotted parallel lines. Chevron or zig-zag defects are shown in Figure 6(a) in V-shaped dotted lines.

From application point of view, the controlled addition of nanomaterial in liquid crystal enhances the contrast of the LC material; therefore, the determination of the suitable concentration of any nanomaterial in host LC is vital factor.

The change in response time can also be understood by this observation. For weak anchoring, the relation can be given as [22]

$$\tau_{\text{Res}} = \frac{4\eta d}{W\pi^2}, \tag{4}$$

where d is the cell thickness and W is the anchoring energy coefficient. The lower anchoring energy coefficient of the NSdoped FLC results in a higher response time.

The temperature dependence of the dielectric permittivity at 100 Hz for both the pure and the NS-doped FLC is depicted in Figure 7. The dielectric permittivity of the pure FLC increases slightly with increase in the temperature. The maximum value of dielectric permittivity for the pure FLC has been noticed at 45°C. Further increment in temperature causes a small decrease in the dielectric permittivity. A sharp decrement in the value of dielectric permittivity has been observed at the vicinity of SmC*-SmA phase transition temperature. This sharp decrease in the dielectric permittivity is due to the change in geometry at the SmC*-SmA phase transition temperature. The dielectric permittivity measurement for the NS-doped FLC shows an increased value of this parameter as compared to the pure FLC. This increase in the value of dielectric permittivity is due to the increased polarization (i.e., net dipole moment) of the NS-FLC system [23]. The variation of dielectric permittivity on temperature scale for the NS-doped FLC is not similar to that of the pure FLC. For the NS doped FLC system, the value of dielectric permittivity increases almost linearly with increase in the temperature for the temperature interval of 30°C–55°C. Beyond this temperature interval, the dielectric permittivity shows a reducing trend on the temperature scale due to the change in geometry of the LC phase.

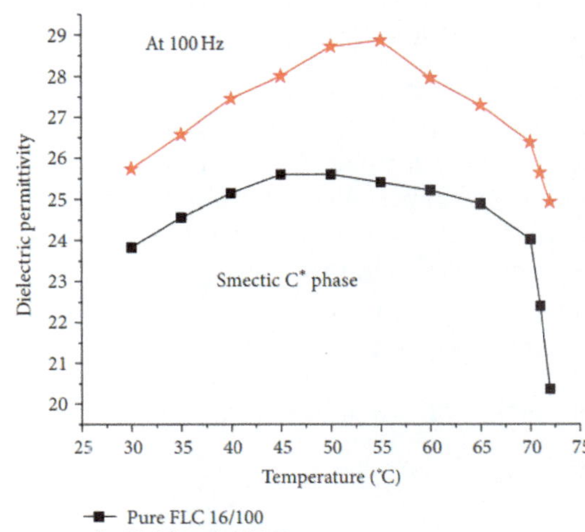

Figure 7: Temperature dependence of the dielectric permittivity at 100 Hz for the pure and the NS-doped FLC

The relaxation strength ($\Delta\varepsilon$) is plotted against the temperature in Figure 8. The relaxation strength for both the pure and the NS-doped FLC increases with increase in the temperature. The relaxation strength shows its maximum value near the temperature 55∘ C, and further increment in temperature causes a decrease in the value of $\Delta\varepsilon$ because FLC sample approaches from SmC∗ to SmA phase. The relaxation strength for the doped FLC system is greater than that of the pure FLC. This greater value of $\Delta\varepsilon$ for the doped system can be explained by the following relation [24]:

$$\Delta\varepsilon = \frac{1}{2K\varepsilon_0} \cdot \frac{P_s^2}{q^2\theta^2}, \tag{5}$$

where $\Delta\varepsilon$ is the relaxation strength, Ps the spontaneous polarization, θ the tilt angle, q the wave vector of helix, and K is the elastic constant. The quantity P_s/θ is a constant in the first approximation and the variation of K with temperature is also very small; therefore, the variation of relaxation strength mainly depends upon the wave vector of helix (q). The wave vector of helix ($q = 2\pi/p$, where p is pitch) is decreased; thus, p increases due to aggregation of NS and defects formation in the FLC matrix.

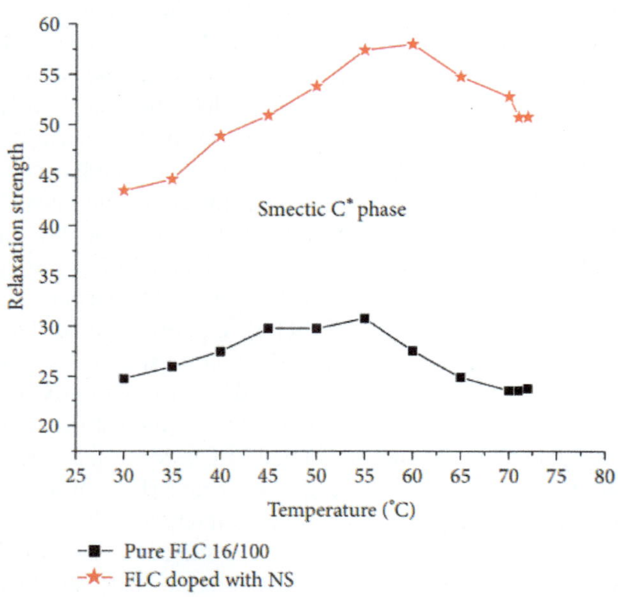

Figure 8: The variation of the relaxation strength with change in temperature for the pure and the NS-doped FLC.

A nonlinear increase in relaxation frequency (f_R) has been shown in Figure 9. In the whole SmC∗ phase, relaxation frequency increases nonlinearly with increasing temperature and attains its maximum value at SmC∗-SmA phase transition temperature. Further increment in the temperature causes a sharp decrease in the relaxation frequency value due to change in phase geometry and the removal of helical structure of the FLC molecules. The value of relaxation frequency for the doped FLC system is less as compared to the pure FLC system. This behaviour of fR can be explained by the relation [25]

$$f_R = \frac{Kq^2}{2\pi\gamma}, \tag{6}$$

where γ is the rotational viscosity and other symbols have their usual meaning.

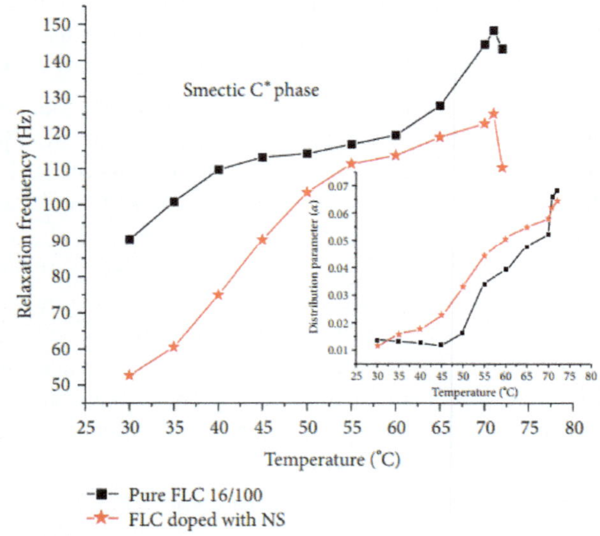

Figure 9: The variation of the relaxation frequency with change in temperature for the pure and the NS-doped FLC while the inset of figure shows the variation of the distribution parameter with change in temperature.

The above relation shows that relaxation frequency is inversely proportional to the rotational viscosity. As the rotational viscosity has greater value for the doped system, a reduced value of the relaxation frequency is found for it. In addition to this, the relaxation frequency also depends upon the wave vector of helix. It is assumed that the wave vector of helix has also decreased due to the presence of NS which results in the decrease in relaxation frequency. The reduction in the relaxation frequency has also verified by the distribution parameter α shown in the inset of Figure 9. The value of α is higher for the NS-doped FLC sample; therefore, the relaxation frequency of the NS-doped FLC get reduced. The distribution parameter of the pure FLC is almost constant for the first four consecutive observations on temperature scale (i.e. 30°C to 45°C). Beyond these observations, distribution parameter increases with increase in the temperature. For the NS doped FLC system, distribution parameter also increases with increase in the temperature. In both cases, the small value of distribution parameter (0.01–0.07) indicates that Goldstone mode predominates over the soft mode in the SmC* phase [26], and the doped FLC system completely follows the Cole-Cole-type relaxation process.

Figure 10 represents the change in electrical conductivity [27] with the variation of temperature at 133 Hz for both the pure and the NS-doped FLC. For the pure FLC system, the electrical conductivity is almost temperature independent whereas it strongly depends on temperature for the NS-doped FLC system. Electrical conductivity sharply increases with increase in the temperature and shows its maximum value at 50°C. Further increment in the temperature causes a reduction in the electrical conductivity. As NS behaves like the ionic additives to the FLC materials, it increases the electrical conductivity in the doped FLC system. In the temperature range of 35°C–50°C, the conductivity is mainly driven by the semiconducting nature of the NS, and conductivity increases with increase in the temperature. Beyond 50°C, the FLC molecules get sufficient energy and move with more kinetic energy which increases the number of collisions of FLC molecules with NS. Thus, the resistive forces increase in the NS-FLC system which results in the decrement in the conductivity.

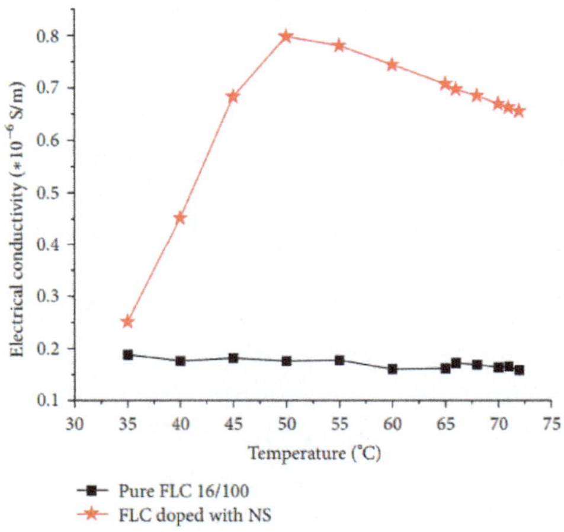

Figure 10: The change in the electrical conductivity with the variation of temperature for the pure and the NS-doped FLC.

CONCLUSIONS

The salient features of the present investigation can be concluded as follows: the addition of nanospheres (NS) in the pure FLC material changes the dielectric and E-O parameters of the pure FLC. The bigger size of NS occupies greater volume as compared to the FLC molecule; therefore, they produce hindrance in the motion of the FLC molecules and wrap the smectic layer. The high quantity of NS produces strong NS-NS interaction (which causes the aggregation of NS in the FLC matrix) which dominates other interactions taking place in the composite system. The presence of high concentration of NS was confirmed by the FTIR analysis. The change in FTIR spectra clearly indicates the presence of a guest entity. The change in percentage transmission in the NS-doped FLC system is due to the topological defects and scattering of the light from the aggregation of NS in FLC matrix. The enhanced PL in case of the NS-doped FLC has been attributed to the surface Plasmon resonance between the NS and FLC molecules. Some other secondary PL emission peak has also been noticed due to the result of the oxygen and sulphur vacancies in the crystal structure. The aggregation of NS supports the FLC molecular dipole moment which increases the net ferroelectricity (or net dipole moment) in the composite. The increased ferroelectricity increases the value of spontaneous polarization and dielectric permittivity. The variation of relaxation strength and relaxation frequency depends upon the wave vector of helix and rotational viscosity. The distribution parameter strongly recommends for the existence of Cole-Cole relaxation phenomena in the composite system. The topological defects, produced by NS, create a slower response for the composite system, but its order is still important in case of constant rate video processing. High dopant concentration of the NS (i.e., 5%) in FLC plays a role as ionic additive which results in a higher conductivity for the NS-doped FLC system. The better electrical response in the NS doped FLC system indicates its applicability in electronic device fabrication. The intense chevron geometry in case of the pure FLC has not been observed in the NS-doped FLC system, but the high concentration of NS produces the topological defects in the composite system.

ACKNOWLEDGMENTS

The authors are thankful to the Department of Science and Technology, Government of India, for the financial assistance for present work in the form of project. The authors (Swadesh Kumar Gupta and Dharmendra Pratap Singh) are thankful to CSIR for SRF Grant. Authors are also very thankful to Professor A. C. Pandey, Nano Application Centre, University of Allahabad, for providing the nanospheres for present study.

REFERENCES

1. L. J. Yu and M. M. Labes, "Fluorescent liquid-crystal display utilizing an electric-field-induced cholesteric-nematic transition," Applied Physics Letters, vol. 31, no. 11, pp. 719–720, 1977.
2. R. C. Y. King and F. Roussel, "Transparent carbon nanotube-based driving electrodes for liquid crystal dispersion display devices," Applied Physics A, vol. 86, no. 2, pp. 159–163, 2007.
3. L. Calucci, G. Ciofani, D. de Marchi et al., "Boron nitride nanotubes as T_2-weighted MRI contrast agents," Journal of Physical Chemistry Letters, vol. 1, no. 17, pp. 2561–2565, 2010.
4. D. P. Singh, S. K. Gupta, A. Srivastava, and R. Manohar, "The phenomenon of induced photoluminescence in ferroelectric mesophase," Journal of Luminescence, vol. 139, pp. 60–63, 2013.
5. R. B. Meyer, L. Liebert, L. Strzelecki, and P. Keller, "Ferroelectric liquid crystals," Journal de Physique Lettres, vol. 36, no. 3, pp. 69–71, 1975.
6. T. Joshi, A. Kumar, J. Prakash, and A. M. Biradar, "Low power operation of ferroelectric liquid crystal system dispersed with zinc oxide nanoparticles," Applied Physics Letters, vol. 96, no. 25, Article ID 253109, 3 pages, 2010.
7. R. Martel, V. Derycke, C. Lavoie et al., "Ambipolar electrical transport in semiconducting single-wall carbon nanotubes," Physical Review Letters, vol. 87, no. 25, Article ID 256805, 4 pages, 2001.
8. J. Prakash, A. Choudhary, A. Kumar, D. S. Mehta, and A. M. Biradar, "Nonvolatile memory effect based on gold nanoparticles doped ferroelectric liquid crystal," Applied Physics Letters, vol. 93, no. 11, Article ID 112904, 3 pages, 2008.
9. D. P. Singh, S. K. Gupta, K. K. Pandey, S. P. Yadav, M. C. Varia, and R. Manohar, "Ferroelectric liquid crystal matrix dispersed with Cu doped ZnO nanoparticles," Journal of Non-Crystalline Solids, vol. 363, pp. 178–186, 2013.
10. V. N. Vijayakumar and M. L. N. M. Mohan, "Enhancement of the display parameters of 4′-pentyl-4-cyanobiphenyl due to the dispersion of functionalised gold nano particles," Journal of Dispersion Science and Technology, vol. 33, no. 1, pp. 111–116, 2012.
11. A. Kumar, G. Singh, T. Joshi, G. K. Rao, A. K. Singh, and A. M. Biradar, "Tailoring of electro-optical properties of ferroelectric liquid crystals by doping Pd nanoparticles," Applied Physics Letters, vol. 100, no. 5, Article ID 054102, 4 pages, 2012.
12. A. K. Srivastava, E. P. Pozhidaev, V. G. Chigrinov,

and R. Manohar, “Single walled carbon nano-tube, ferroelectric liquid crystal composites: excellent diffractive tool,” Applied Physics Letters, vol. 99, no. 20, Article ID 201106, 3 pages, 2011.

13. D. P. Singh, S. P. Yadav, P. K. Tripathi et al., “Concentration dependent physical parameters of ferroelectric liquid crystal and ZnOS nano material composite system,” Soft Materials, vol. 11, no. 3, pp. 305–314, 2013.
14. R. Manohar, A. K. Srivastava, P. K. Tripathi, and D. P. Singh, “Dielectric and electro-optical study of ZnO nano rods doped ferroelectric liquid crystals,” Journal of Materials Science, vol. 46, no. 18, pp. 5969–5976, 2011.
15. K. Skarp and S. T. Lagerwall, “Rotational viscosities in ferroelectric smectic liquid crystals,” Ferroelectrics, vol. 84, no. 1, pp. 119–142, 1988.
16. K. S. Cole and R. H. Cole, “Dispersion and absorption in dielectrics I. Alternating current characteristics,” The Journal of Chemical Physics, vol. 9, no. 4, pp. 341–351, 1941.
17. A. Kumar, J. Prakash, D. S. Mehta, A. M. Biradar, and W. Haase, “Enhanced photoluminescence in gold nanoparticles doped ferroelectric liquid crystals,” Applied Physics Letters, vol. 95, no. 2, Article ID 023117, 3 pages, 2009.
18. A. Kumar, J. Prakash, A. D. Deshmukh, D. Haranath, P. Silotia, and A. M. Biradar, “Enhancing the photoluminescence of ferroelectric liquid crystal by doping with ZnS quantum dots,” Applied Physics Letters, vol. 100, no. 13, Article ID 134101, 4 pages, 2012.
19. P. Malik, A. Chaudhary, R. Mehra, and K. K. Raina, “Electrooptic and dielectric studies in cadmium sulphide nanorods/ferroelectric liquid crystal mixtures,” Advances in Condensed Matter Physics, vol. 2012, Article ID 853160, 8 pages, 2012.
20. R. Pratibha, W. Park, and I. I. Smalyukh, “Colloidal gold nanosphere dispersions in smectic liquid crystals and thin nanoparticle-decorated smectic films,” Journal of Applied Physics, vol. 107, no. 6, Article ID 063511, 5 pages, 2010.
21. F. V. Podgornov, A. V. Ryzhkova, and W. Haase, “Influence of gold nanorods size on electro-optical and dielectric properties of ferroelectric liquid crystals,” Applied Physics Letters, vol. 97, no. 21, Article ID 212903, 3 pages, 2010.
22. X. Nie, R. Lu, H. Xianyu, T. X. Wu, and S. Wu, “Anchoring energy and cell gap effects on liquid crystal response time,” Journal of Applied Physics, vol. 101, no. 10, Article ID 103110, 5 pages, 2007.
23. S. K. Gupta, D. P. Singh, and R. Manohar, “SWCNT doped ferroelectric liquid crystal: the electro-optical properties with enhanced dipolar contribution,” Current Applied Physics, vol. 13, no. 4, pp. 684–687, 2013.
24. T. P. Majumder, M. Mitra, and S. K. Roy, “Dielectric relaxation and rotational viscosity of a ferroelectric liquid crystal mixture,” Physical Review E, vol. 50, no. 6, pp. 4796–4800, 1994.
25. H. S. Kitzerow and C. Bahr, Chirality in Liquid Crystals, Springer, Berlin, Germany, 2001.
26. Y. P. Panarin, Y. P. Kalmykov, S. T. M. Lughadha, H. Xu, and J. K. Vij, “Dielectric response of surface stabilized ferroelectric liquid crystal cells,” Physical Review E, vol. 50, no. 6, pp. 4763–4772, 1994.
27. F. Kremer and A. Schonhals, Broadband Dielectric Spectroscopy, Springer, Berlin, Germany, 2003.

CHAPTER 14

VIBRATIONAL SPECTRAL STUDIES OF PURE AND DOPED TGSP CRYSTALS

N. Kartheeswari and K. Viswanathan

INTRODUCTION

Triglycine sulfate is a ferroelectric material and belongs to the pyroelectric family. It plays a major role in FT-IR instrumentation and infrared detector. It exhibits order-disorder phase transition at the Curie point 49°C. The crystal structure of TGS was reported by Hoshino et al. [1]. The unit cell of TGS contains three types of glycines GI, GII, and GIII. Glycine I is in the form of zwitterion. GII and GIII are two planar glycines. They have protonated carboxyl groups which have taken protons from the sulfuric acid. They form chain-like system with SO_4^{2-} group. The neighbouring chains are connected by SO_4^{2-} groups of adjacent chains and GI group. Such configuration of TGS is regarded as particularly important for the ferroelectric behavior of TGS crystal. The orientation of positively charged NH_3 groups determines the direction of spontaneous polarization in TGS. The spontaneous polarization reversal in TGS is due to the proton transfer between glycine and glycinium ions [2]. TGS exhibits order-disorder phase transition at the Curie point. Above Curie temperature, it is in the Paraelectric phase with space group symmetry $P2_1/m$ and below Curie it is in the ferroelectric phase with space group symmetry $P2_1$ with two formula units per unit cell [3]. The unit cell parameters of TGS are .15 Å, Å, Å, and $a = 9.15$ Å, $b = 12.6445$ Å, $c = 5.725$ Å, and $\beta = 105.53°$ [4]. In TGS, majority of domains were found in the form of rod shapped with lenticular cross-sections elongated in the direction perpendicular to c-axis [5]. TGS provides rich lattice vibration spectrum below 300 cm^{-1}[6]. The main disadvantage of TGS is its lower Curie temperature and easy depolarization by time, electrical, mechanical, and thermal means. Because glycine has no asymmetric carbon and it is optically inactive, it is believed that doping of TGS with an optically active molecule will keep permanent polarization in TGS lattice. There are a large number of researches going on TGS to minimize the depolarization effect and to keep permanent polarization in TGS lattice. There is no change in crystal structure for Nd-doped TGS and it is reported that Nd is coordinated with glycine [2]. But in the case of ADP, L-Asparagine- and L-tryptophan-doped samples' change in crystal morphology is observed [7–9]. T_c value is higher for l-threonine-, dl-threonine-, l-methionine-doped crystals and TGSP crystals [10, 11]. Spectral investigation of TGS doped with ADP, L-tryptophan, phosphoric acid, L-lysine, L-cystine reveals that doped samples have lower spontaneous polarization values *Ps* and higher coercive field values compared to pure TGS. Higher coercive field value implies that the crystal is in the monodomain state [7, 9, 11–13]. Pyroelectric coefficient is increased for thiourea-, L-alanine- and DL-alanine-, L-lysine-doped samples. [14–16]. Incorporation of nitric acid and EDTA into TGS crystal increases the dielectric constant value [17, 18]. The substitution of amino acids l-threonine, dl-threonine, l-methionine, and L-cystine results in the decrease of dielectric constant value compared to pure TGS crystal [10, 13]. Zinc chloride- and urea-doped samples exhibit nonlinear optical property and have SHG efficiency very much greater than standard KDP crystal [19, 20]. In case of LGLM, La-, Ce-, Nd-, L-lysine-doped TGS crystal strong internal bias field is created, indicating that the dopant reduces the depolarizing effects in TGS crystals. The internal bias field fixes the polarization in a preferential direction with minimum possibility of polarization reversal [16, 21, 22]. Doping of TGS with metal ions Fe^{3+}, Cr^{3+}, and Co^{2+} decreased the indirect band gap [23]. Investigation of TGS samples previously influenced by electric field E perpendicular to ferroelectric b-axis reveals that there are rigid stripped domains parallel to c-axis [24]. Vickers' microhardness study reveals that hardness value is increased for TGS when doping with L-lysine and imino diacetic acid [25, 26]. But hardness value of crystal is decreased for doping TGS with l-threonine, dl-threonine and

Citation: N. Kartheeswari and K. Viswanathan, Vibrational Spectral Studies of Pure and Doped TGSP Crystals, doi.org/10.1155/2013/272069.

l-methionine, L-cystine, thiourea, and EDTA [10, 13, 14, 18]. In present work, pure and ADP-, KDP- (0.2 mol) doped TGSP crystals are grown from solution.

Crystal Growth and Characterization

Pure and ADP-, KDP- (0.2 mol) doped TGSP crystals were grown by solution growth method [4]. Pure TGSP crystal is grown by taking 3 moles of Analar grade Glycine, 0.5 mol of sulphuric acid, and 0.5 mol of phosphoric acid 100 mL of distilled water. The solution is stirred well for 3 hrs. Then the solution is filtered and left to slow evaporation. ADP- and KDP-(0.2 mol) doped crystals are grown by the same method. Good-quality crystals are obtained within the growth period of 1–3 months. Grown crystals are shown in Figure 1. UV-Vis absorption spectra of grown crystals are recorded in the wavelength range from 200 to 800 nm using UV-Vis 2450 Make Shimadzu model spectrophotometer. FT-IR spectra are recorded in the frequency range from 400 to 4000 cm^{-1} using IR Affinity Make Shimadzu model spectrophotometer. FT-Raman spectra were recorded in the frequency range from 0 to 3500 cm^{-1} using the excitation radiation of 5145 A° using Lab Ram HR 800 model spectrophotometer. Ferroelectric hysteresis study was carried out using homemade Sawyer-Tower circuit [27].

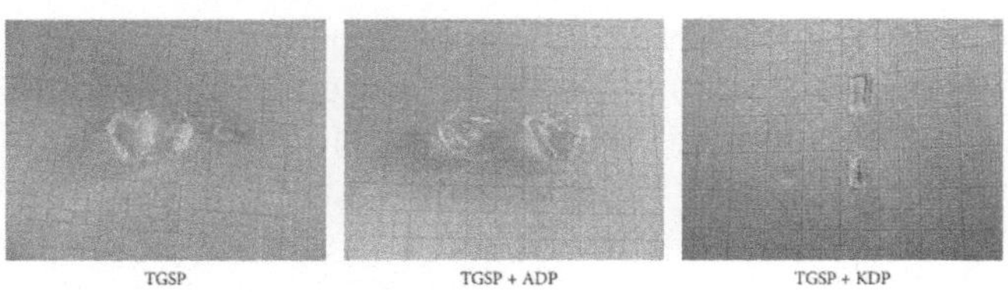

Figure 1: Grown crystals of pure and ADP-, KDP- (0.2 mol) doped TGSP.

RESULTS AND DISCUSSION

UV-Vis Spectral Investigation of Grown Crystals

Figure 2 shows UV-Vis absorption spectra of pure and ADP-, KDP-(0.2 mol) doped TGSP crystals. For TGSP crystal, λ_{max} = 232 nm. For TGSP + ADP, λ_{max} = 233 nm and for TGSP + KDP, λ_{max} = 233.5 nm. For doped crystals, there is slight shift in wavelength of maximum absorption λ_{max}. This is referred to as bathochromic shift. From this, it is observed that there is change in energy levels to effect transition. For doped samples, the energy required to effect the electron promotion is lesser than that of pure TGSP so that the wavelength that provides this energy is increased for doped crystals. Absorption at lower wavelength reveals that there must be higher energy transition corresponding to C=C– NO_2 group. It is observed that all these crystals have transmission percentage of above 90%. Energy band gap values were found out using the relation $E = 1240/\lambda_{max}$ eV [28]. Also energy band gap values are obtained from Urbach plot and results are shown in Table 1. Figure 3 shows Urbach plots for grown crystals.

Table 1: Energy band gap values of grown crystals.

Crystal	Eg (calculated from Eg = $1240/\lambda_{max}$ eV) eV	Eg (obtained from Urbach Plot) eV
TGSP	5.34	5.1496
TGSP + ADP	5.32	5.1606
TGSP + KDP	5.3	5.1496

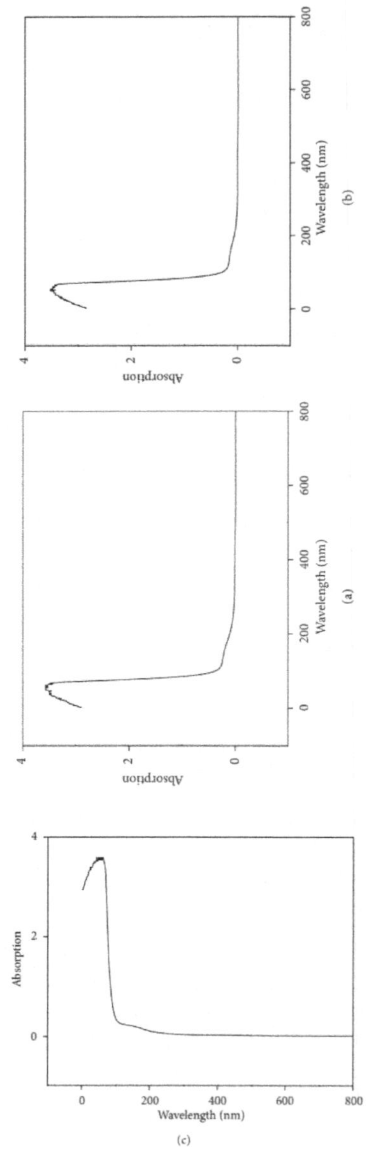

Figure 2: UV-Vis absorption spectra of (a) TGSP, (b) TGSP + ADP, and (c) TGSP + KDP crystals.

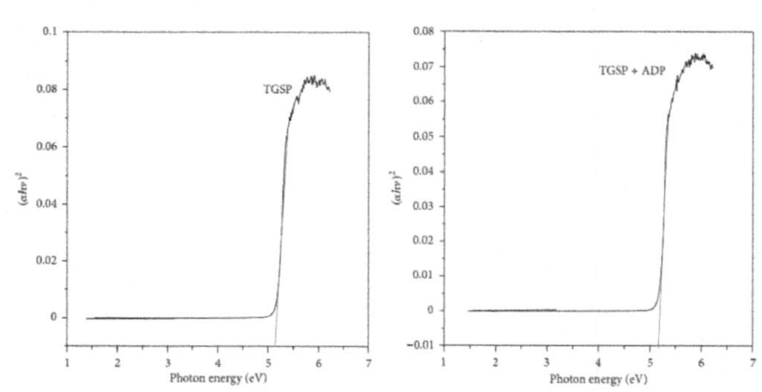

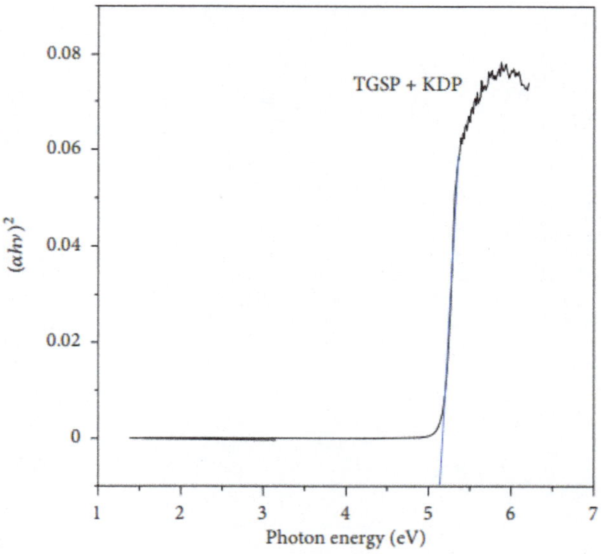

(c)

Figure 3: Urbach plot of (a) TGSP, (b) TGSP + ADP, and (c) TGSP + KDP crystals.

FT-IR Spectral Analysis

Figure 4 shows FT-IR spectra of pure and ADP-, KDP-doped TGSP crystals. FT-IR spectrum of pure TGSP crystal matches very well with the earlier reported values of pure TGS crystal [17]. All expected characteristic vibrations are observed and assignments were tabulated in Table 2. IR bands at 1425 cm^{-1} and 1622 cm^{-1}corresponding to symmetric and asymmetric stretching vibrations of COO^- indicate the zwitterion configuration of glycine [17]. IR bands observed in the region between 1683 cm^{-1} to 1869 cm^{-1} corresponding to stretching vibration of C=O indicate the presence of glycinium ion configuration [17]. Degenerate modes of NH_3 bending and C=O, NH_4, C–H, and O–H stretching vibrations are observed. FT-IR spectra of ADP-, KDP doped samples provide very similar features as those of pure TGSP. More bands were located at the same positions as those of pure TGSP. There is a very slight shift observed in band positions compared to pure TGSP. But doped samples provide less resolution of bands. Some bands are broadened and some are narrowed. Degeneracy is more for doped samples than that of pure TGSP.

Table 2: FT-IR analysis

Pure TGSP	TGSP + ADP	TGSP + KDP	Assignment
401 w–472 w (8 lines)	412 w–455 w (5 lines)	403 w–472 w (7 lines)	$\nu(PO_4)$, $\nu(SO_4)$
503 w	501 w–536 w (4 lines)	501 w	τ(C–N)
572 w	571 w	569 w	$\nu(SO_4^-)$
615 m 646 w	615 m–646 w (3 lines)	615 m 646 w	δ(C–N), NH_3^+ oscillation
669 m 678 w	667 m 680 w	669 m 678 w	ν(S–H)
866 m 898 m	866 m	866 m	ν(C–C)
908 w	908 w	906 w	ν(P–O–H), $\rho(NH_3)$
979 s	979 s	978 s	ν(C–C–N), $\nu(SO_4^-)$
1018 w	—	1018 w	$\omega(CH_2)$, δ(C–C), ν(C–N)
1051 w	1051 w	1051 w	ν(P–O–H)
1083 w	1083 w	1082 w	ν(P–O–H), ν(C=O)
1111 m 1124 s	1109 m	1109 m 1130 s	$\rho(CH_2)$, ν(C–C), $\nu(SO_4)$
1309 m	1309 m	1309 m	$\delta(CH_2)$ of glycine, ν(C–C)
1338 w	1338 w	—	$\nu(NO_2)$
1363 w	1363 w	—	ν(C–N)
1375 m	1375 m	1377 m	$\delta(CH_2)$
1386 w	1386 w	1388 w	$\nu(NH_4)$
1398 w	1396 w	—	ν(C=O), $\nu(NH_4)$
1404 w	1404	—	$\delta(NH_4)$
1409 w	—	—	$\nu_{as}(COO^-)$, $\delta(NH_4)$
1417 m 1425 m	1417 m 1423 m	1421 m 1427 m	$\nu(COO^-)$, $\delta(CH_2)$
1435 w 1448 w	1435 w	1435 w	$\delta(NH_3)$
1456 m–1498 w (5 lines)	1456 m–1496 w (5 lines)	1458 m–1490 w (3 lines)	$\nu(NH_4)$, $\nu(COO^-)$
1506 m–1558 w (5 lines)	1506 m–1558 w (7 lines)	1506 m–1558 w (4 lines)	$\delta(NH_3)$
1570 m–1595 w (3 lines)	1568 m 1575 w	1570 m 1575 w	$\nu(COO^-)$
1616 w 1622 w	1616 w 1622 w	1622 w	$\nu_{as}(COO^-)$, $\delta(NH_3)$
1635 w 1647 w 1653 w	1633 w 1645 w 1653 w	1635 w 1647 w 1653 w	ν(O=P–OH), ν(C=C)
1662 w 1670 w	1662 w–1674 w (3 lines)	1662 w 1672 w	$\delta(H_2O)$, ν(C=C)
1683 w–1869 w (12 lines)	1683 w–1867 w (19 lines)	1685 w–1869 w (12 lines)	ν(C=O), $\nu(C{=}NH_4)$
1942 w–2360 w (5 lines)	1942 w–2358 w (4 lines)	2017 w 2357 w	Overtone and combination bands
3502 w–3547 w (5 lines)	3502 w–3545 w (5 lines)	3502 w–3545 w (4 lines)	ν(O–H), ν(N–H)
3566 w–3701 w (11 lines)	3550 w–3701 w (13 lines)	3566 w–3701 w (9 lines)	ν(O–H), ν(C–H), ν(C=O), δ(P–O–H)
3711 w–3947 w (24 lines)	3711 w–3961 w (24 lines)	3711 w–3963 w (25 lines)	ν(O–H) of water

ν_s: symmetric stretching, ν_{as}: asymmetric stretching, δ: bending, ρ: rocking, ω: wagging, and τ: torsional vibrations.

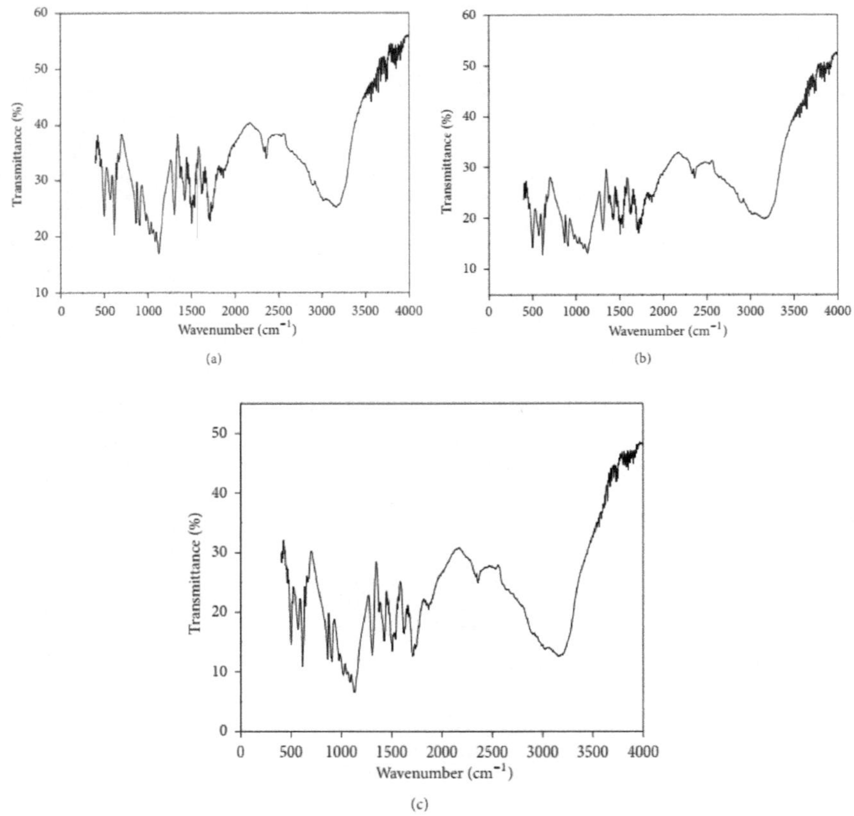

Figure 4: FT-IR spectra of (a) TGSP, (b) TGSP + ADP, and (c) TGSP + KDP crystals.

FT-Raman Spectral Analysis

Figure 5 shows FT-Raman spectra of pure and doped TGSP crystals. FT-Raman spectrum of pure TGSP matches very well with the earlier reported values of pure TGS [2, 17]. Obtained Raman bands and their assignments are tabulated in Table 3. Raman band observed at 1414 cm^{-1} and 1605 cm^{-1} corresponding to symmetric and asymmetric stretching vibration of COO^- group confirms the zwitterion configuration of glycine [2, 17].

Table 3: FT-RAMAN analysis

Pure TGSP	TGSP + ADP	TGSP + KDP	Assignment
74 m 103 m	74 m–340 m (4 lines)	73 m–234 m (5 lines)	Lattice-mode vibration of glycine
—	466 m	451 m	Lattice-mode vibration of SO_4, δ(C–N) out of plane, $\nu(PO_4)$
—	506 w–590 w (3 lines)	506 w 590 w	τ(C–N), $\nu(PO_4)$, δ(C–CO), $\nu(SO_4^-)$
632 w 633 w	614 w	628 w	$\nu(SO_4)$, δ(C–N) in plane, NH_3^+ oscillation
—	678 m	677 m	ν(C–C)
869 m	871 m	—	ν(C–C)
889 w	890 w	890 w	$\rho(CH_2)$, $\rho(NH_3)$, ν(C–C), ν(P–O–H)
978 s	978 s	978 s	ν(C–C–N), $\nu(SO_4^-)$
1051 w	1042 w	1045 w	$\omega(CH_2)$, δ(C–C), ν(C–N), ν(C–N–H), ν(P–O–H)
1114 w 1122 w	1123 w 1128 w	1124 w 1157 w	$\rho(CH_2)$, ν(C–C), $\nu(SO_4)$
1304 w 1314 w	1306 w	1309 w	δ(C–H), ν(C–C)
1414 w	1416 w	1413 w	$\nu_s(COO^-)$, $\delta(CH_2)$
1439	1441 1496	1443	$\nu(NH_4)$, $\delta(CH_2)$
1605	1607	1606	$\nu(COO^-)$, $\nu(NH_4)$, ν(P–O–H), ν(C=C)
1683 w	1678	1683 1762	ν(C=O), ν(O–H–O)
2648	—	—	$\delta(H_2O)$, ν(COOH)
2982	2958 2980	2958 2982	ν(C–H)
3014 w 3057 w	3019 w 3140 w	3017 w 3143 w	$\nu(NH_4)$

ν_s: symmetric stretching, ν_{as}: asymmetric stretching, δ: bending, ρ: rocking, ω: wagging, and τ: torsional vibrations.

The band at 1683 cm^{-1} corresponding to symmetric vibration of C=O group confirms the presence of glycinium ion configuration [17]. In FT-Raman spectra of ADP- and KDP-doped samples, some peaks were shifted to a considerable range compared to pure TGS. A change in intensity of all peaks was observed. The amount of polarisability change will determine the Raman scattering intensity. So, it can be concluded that the change in intensity of peaks may be due to incorporation of dopants.

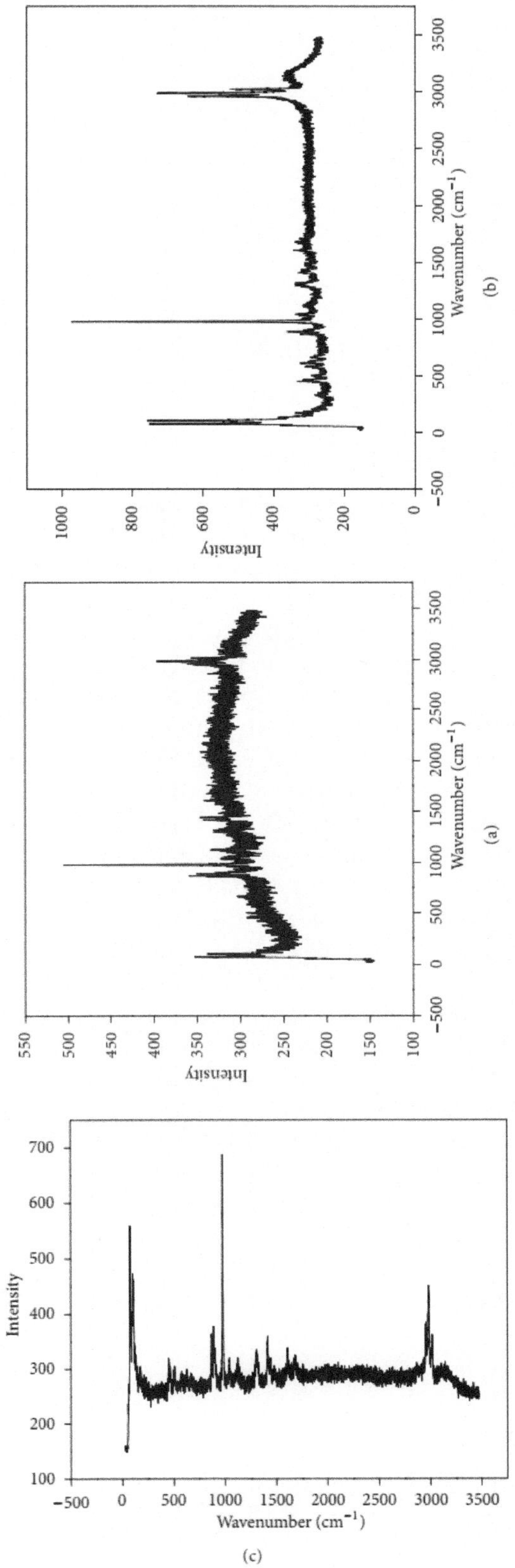

Figure 5: FT-Raman spectra of (a) TGSP (b) TGSP + ADP (c) TGSP + KDP crystals.

Powder XRD Study

Figure 6 shows powder XRD pattern of pure and doped TGSP crystals. All crystals belong to monoclinic structure. XRD pattern of grown crystals differ from each other in intensity of reflection. Sharp and high intensity of reflection planes revealed that all crystals have good crystalline nature. Changes in Intensity of peaks for doped samples compared to pure TGSP revealed that there must be some changes in electron density in reflection planes due to the incorporation of dopants. Small change in 2θ value is observed for doped samples. So, pure and doped samples have different morphologies. Lattice parameter values are shown in Table4. Obtained lattice parameter values for pure and doped crystals slightly differ from each other. Because of doping, there may be some defects or strains in grown crystals.

Table 4: Lattice parameter values of grown crystals

Crystal	*a* (nm)	*b* (nm)	*c* (nm)	*β* (deg)
TGSP	9.4173	12.6449	5.7734	110.36
TGSP + ADP	9.5786	12.58434	5.8198	110.36
TGSP + KDP	9.4566	12.67908	5.7222	110.36

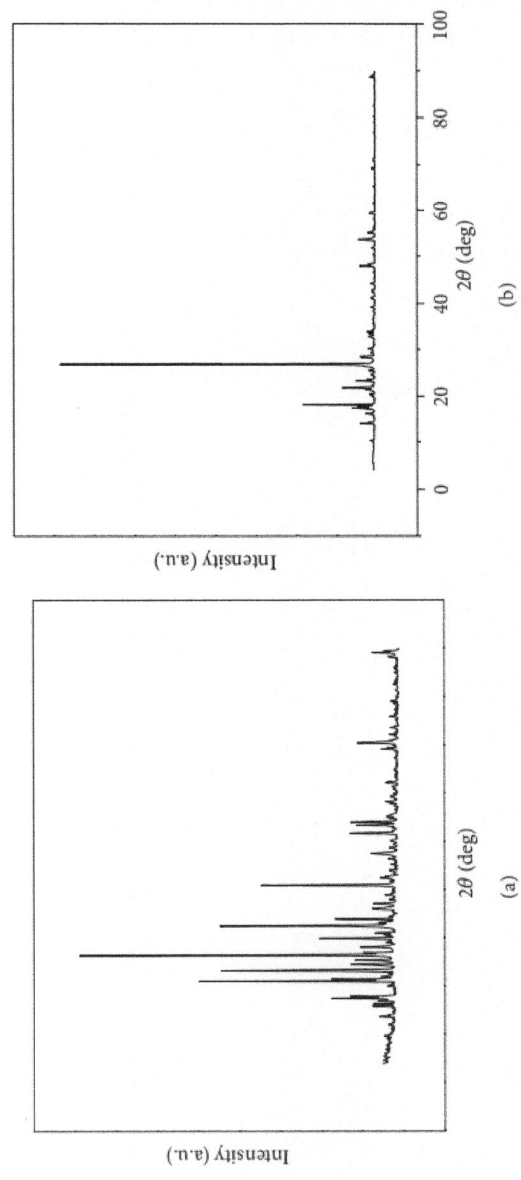

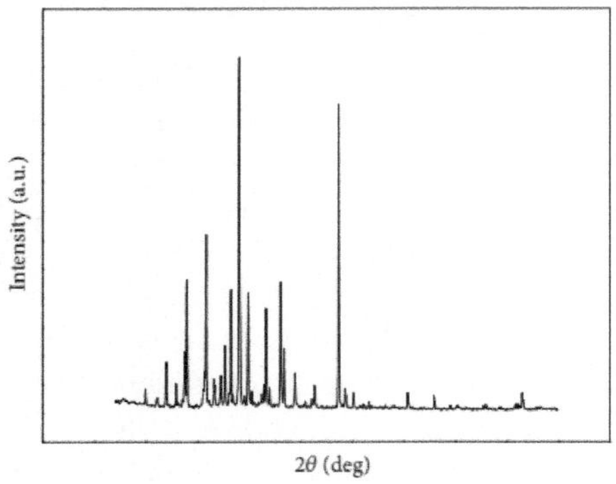

(c)

Figure 6: Powder XRD pattern of (a), TGSP (b), TGSP + ADP (c), and TGSP + KDP crystals.

Ferroelectric Hysteresis Study

Homemade Sawyer-Tower circuit is constructed [27]. Sample capacitors were prepared by using aluminium foils and the pure and doped TGSP samples as dielectric in between the aluminium foils [29]. Spontaneous polarization values (*Ps*) for all samples were obtained by using the equation $P = Q/A$ micro coulomb/cm^2. Here, Q is charge measured on sample capacitor (coulomb) and A is area of capacitor plate (cm^2) [29]. Results are shown in Table 5. Figure 7 shows obtained ferroelectric hysteresis loops of grown crystals.

Table 5: Ferroelectric hysteresis loop measurement values

Crystal	Spontaneous polarization Ps (μC/cm^2)	Remnant polarization Pr (μC/cm^2)	Coercive field value (kV/cm)
TGSP	1.712	1.04	6
TGSP + ADP	3.1388	1.5023	6.01
TGSP + KDP	3.6804	1.8731	6.02

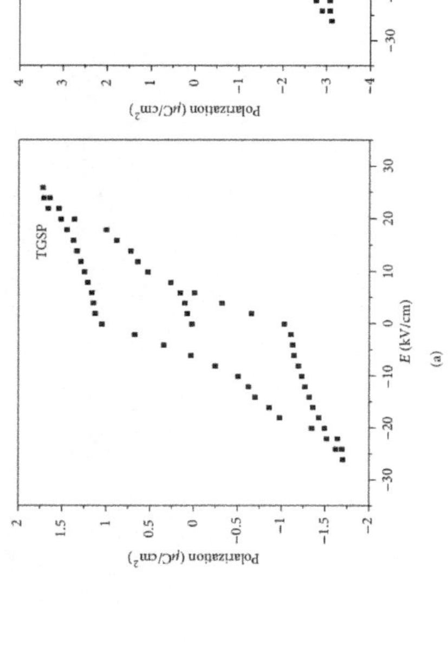

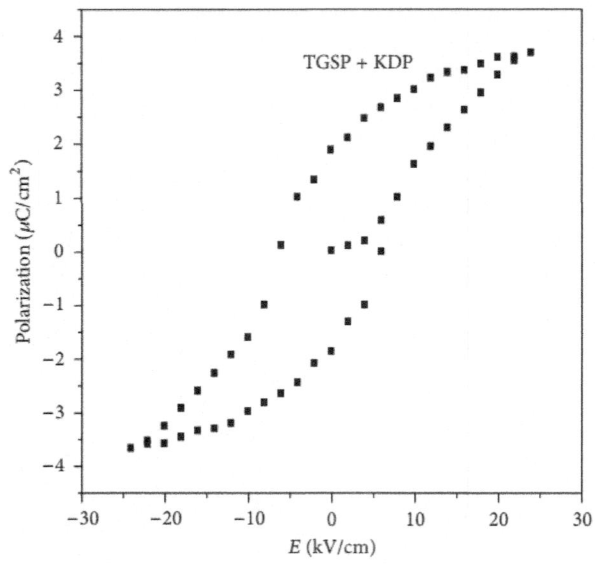

(c)

Figure 7: Ferroelectric hysteresis loops of (a) TGSP, (b) TGSP + ADP, and (c) TGSP + KDP crystals.

Electrical Measurement

Grown crystals are subjected to electrical characterization using IMPEDENCE ANALYSER IM3570. All crystals conduct electricity linearly. Doped crystals have higher electrical conductivity than pure TGSP crystal. Results are shown in Table 6. Figure 8 shows electrical conductivity graphs of grown crystals. Electrical conductivity graphs of pure and doped TGSP crystals contain two regions: frequency-independent Ac conductivity region in low-frequency range and frequency-dependent dc conductivity region in high-frequency range. Conductivity graphs obey Arrhenius' relation and Jonscher's power law [30–33]. For all grown crystals, dc conductivity linearly increases with an increase in frequency. This indicates that electrical conductivity of these crystals is due to hopping mechanism. Electrical conductivity σ_{dc} is obtained by nonlinear fitting for the conductivity graphs. Then, hopping frequency ω_p is obtained by using the relation $\omega_p = (\sigma_{dc}/A)^{1/n}$. Here, n is frequency exponent. A is temperature-dependent parameter. Charge carrier concentration is obtained by $N=\sigma_{dc}T/\omega_p$. Mobility of charge carriers is obtained by $\mu=\sigma_{dc}/Ne$. Here, e is charge of electron [32]. From electrical conductivity analysis, TGSP + KDP crystal has maximum dc electrical conductivity. It has higher hopping frequency value. Though TGSP + KDP has less charge carrier concentration compared to pure TGSP, it has higher value of mobility of charge carrier. This may contribute to the increase of dc electrical conductivity for TGSP + KDP crystal.

Table 6: Electrical conductivity Analysis

Crystal	Electrical conductivity Ec (Siemen/cm)	Hopping frequency ω_p (Hz)	Charge carrier concentration N/cm^{-3}	Mobility μ (cm^2/Vs)
TGSP	1.3048e − 6	12690.11	0.3290e − 8	24.787e + 20
TGSP + ADP	2.5054e − 6	27080.365	0.2961e − 8	52.8833e + 20
TGSP + KDP	8.5078e − 6	102880.996	0.2646e − 8	200.95899e + 20

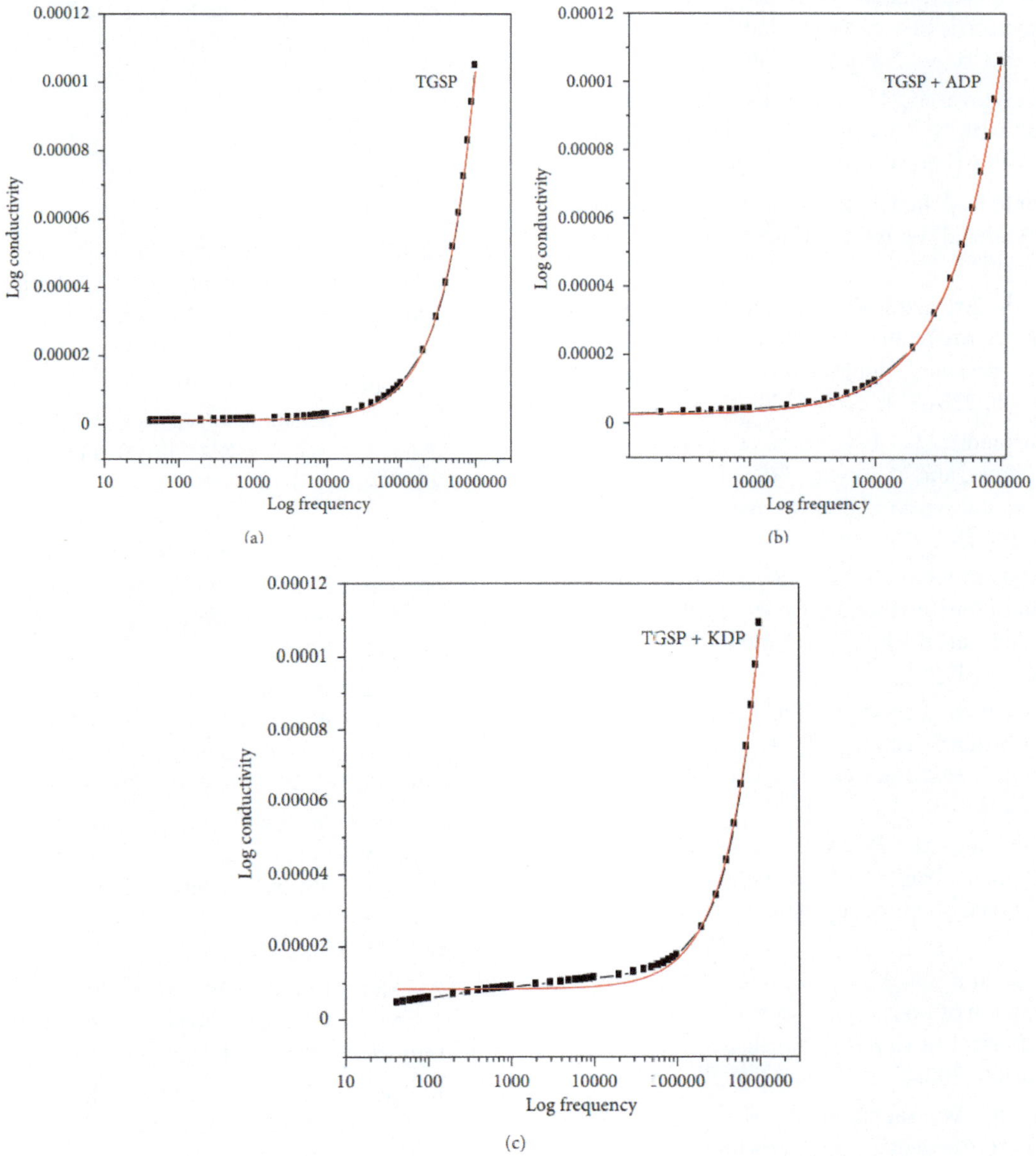

Figure 8: Frequency versus conductivity graphs of (a) TGSP, (b) TGSP + ADP, and (c) TGSP + KDP crystals.

CONCLUSION

From UV-Vis spectra of grown crystals, it is confirmed that all these crystals have excellent optical quality. This property makes these crystals useful for applications in lasers, holographic recording, optical filters, and nonlinear optical applications and electrooptic applications. From FT-IR and FT-Raman spectral investigations, molecular structures of pure and doped TGSP crystals are verified. Less resolution of peaks and change in intensity of peaks for doped samples compared to pure TGSP are due to interaction between parent and dopants. It is concluded that ADP and KDP were well incorporated into the lattice of TGSP crystal. From powder XRD pattern, it is confirmed that all crystals are crystallizes in monoclinic structure. Ferroelectric hysteresis study reveals that for ADP- and KDP-doped samples spontaneous polarization values were slightly increased. So doped crystals have improved ferroelectric behavior than pure TGSP. So it can be concluded that ADP- and KDP-doped TGSP crystals are most suitable for infrared detector applications. From electrical conductivity measurements, it is observed that all crystals conduct electricity linearly and doped crystals have higher electrical conductivity than pure TGSP crystal.

ACKNOWLEDGMENTS

The authors are thankful to the Management of Karpagam University for providing the facilities for the work, and to the Head, Department of Nanoscience and Technology, Bharathiar University, for recording the Raman spectra. Also thanks are due to the faculty and research scholars of the Department of Physics, Karpagam University, Coimbatore, India, for their help during the course of the work.

REFERENCES

1. S. Hoshino, Y. Okaya, and R. Pepsinsky, "Crystal structure of the ferroelectric phase of $(Glycine)_3 \cdot H_2SO_4$," Physical Review, vol. 115, no. 2, pp. 323–330, 1959.
2. E. M. Mihaylova and H. J. Byrne, "Raman studies of TGS doped with N_d," Journal of Physics and Chemistry of Solids, vol. 61, no. 12, pp. 1919–1925, 2000.
3. M. Kay and R. Kliengberg, "The crystal structure of triglycine sulfate," Ferroelectrics, vol. 5, no. 1, pp. 45–52, 1973.
4. K. Meera, S. Aravazhi, P. Santhana Raghavan, and P. Ramasamy, "Growth and characterisation of L-tyrosine-doped TGS crystals," Journal of Crystal Growth, vol. 211, no. 1, pp. 220–224, 2000.
5. H. V. Alexandru, C. Berbecaru, F. Stanculescu, L. Pintile, I. Matei, and M. Lisca, "Doped TGS crystals for IR detection and sensors," Sensors and Actuators A, vol. 113, no. 3, pp. 387–392, 2004.
6. A. M. Malyarevich and M. R. Posledovich, "The assignment of lattice vibrations in triglycine sulfate-type crystals," Journal of Molecular Structure, vol. 375, no. 1-2, pp. 43–51, 1996.
7. G. Arunmozhi, S. Lanceros-Méndez, and E. de Matos Gomes, "Antiferroelectric ADP doping in ferroelectric TGS crystals," Materials Letters, vol. 54, no. 5-6, pp. 329–336, 2002.
8. S. Kalainathan, M. B. Margaret, and T. Irusan, "Morphological changes of L-asparagine doped TGS crystal," Crystal Engineering, vol. 5, no. 1, pp. 71–78, 2002.
9. D. Jayalakshmi and J. Kumar, "Growth and characterization of l-tryptophan-doped ferroelectric TGS crystals," Journal of Crystal Growth, vol. 310, no. 7–9, pp. 1497–1500, 2008.
10. K. Meera, R. Muralidharan, A. K. Tripathi, and P. Ramasamy, "Growth and characterisation of l-threonine, dl-threonine and l-methionine admixtured TGS crystals," Journal of Crystal Growth, vol. 263, no. 1–4, pp. 524–531, 2004.
11. A. Saxena, V. Gupta, and K. Sreenivas, "Characterization of phosphoric acid doped TGS single crystals,"Journal of Crystal Growth, vol. 263, no. 1–4, pp. 192–202, 2004.
12. W. Kulita and M. Trybus, "TGS single crystals doped with lysine: new material for IR detectors," SPIE Proceedings, vol. 5124, 2003.
13. K. Meera, R. Muralidharan, P. Santhanaraghavan, R. Gopalakrishnan, and P. Ramasamy, "Growth and characterisation of L-cystine doped TGS crystals," Journal of Crystal Growth, vol. 226, no. 2-3, pp. 303–312, 2001.
14. K. Meera, R. Muralidharan, A. K. Tripathi, R. Dhanasekaran, and P. Ramasamy, "Growth of thiourea-doped TGS crystals and their characterisation," Journal of Crystal Growth, vol. 260, no. 3-4, pp. 414–421, 2004.
15. C. Berbecaru, H. V. Alexandru, L. Pintilie, A. Dutu, B. Logofatu, and R. C. Radulescu, "Doped versus pure TGS crystals," Materials Science and Engineering B, vol. 118, no. 1–3, pp. 141–146, 2005.
16. G. Su, Y. He, H. Yao, Z. Shi, and Q. Wu, "New pyroelectric crystal L-lysine-doped TGS (LLTGS)," Journal of Crystal Growth, vol. 209, no. 1, pp. 220–222, 2000.
17. R. Parimaladevi, C. Sekar, and V. Krishnakumar, "The effect of nitric acid (HNO_3) on growth, spectral, thermal and dielectric properties of triglycine sulphate (TGS) crystal," Spectrochimica Acta A, vol. 75, no. 2, pp. 617–623, 2010.
18. K. Meera, A. Claude, R. Muralidharan, C. K. Choi, and P. Ramasamy, "Growth and characterisation of EDTA-added TGS crystals," Journal of Crystal Growth, vol. 285, no. 3, pp. 358–364, 2005.
19. A. Wojciechowski, I. V. Kityk, G. Lakshminarayana et al., "Laser-induced optical effects in triglycine-zinc chloride single crystals," Physica B, vol. 405, no. 13, pp. 2827–2830, 2010.
20. V. Krishnakumar, S. Sivakumar, R. Nagalakshmi, S. Bhuvaneswari, and M. Rajaboopathi, "Effect of doping an organic molecule ligand on TGS single crystals," Spectrochimica Acta A, vol. 71, no. 2, pp. 480–485, 2008.
21. T. Bharthasarathi, V. Siva Shankar, R. Jayavel, and P. Murugakoothan, "Growth and characterization of biadmixtured TGS single crystals," Journal of Crystal Growth, vol. 311, no. 4, pp. 1147–1151, 2009.
22. R. Muralidharan, R. Mohankumar, P. M. Ushasree, R. Jayavel, and P. Ramasamy, "Effect of rare-earth dopants on the growth and properties of triglycine sulphate single crystals," Journal of Crystal Growth, vol. 234, no. 2-3, pp. 545–550, 2002.
23. A. Abu El-Fadl, "Optical properties of TGS crystals doped with metal ions in the vicinity of phase transition," Physica B, vol. 269, no. 1, pp. 60–68, 1999.
24. K. Ćwikiel, B. Fugiel, and M. Mierzwa, "Rigid domain structure in TGS ferroelectric," Physica B, vol. 293, no. 1-2, pp. 58–66, 2000.
25. R. Mohan Kumar, R. Muralidharan, D. Rajan Babu et al., "Growth and characterization of L-lysine doped TGS and TGSP single crystals," Journal of Crystal Growth, vol. 229, no. 1, pp. 568–573, 2001.
26. C. Rai, K. Sreenivas, and S. M. Dharmaprakash, "Improved ferroelectric and pyroelectric parameters in iminodiacetic acid doped TGS crystal," Journal of Crystal Growth, vol. 312, no. 2, pp. 273–275, 2010.
27. C. B. Sawyer and C. H. Tower, "Rochelle salt as a dielectric," Physical Review, vol. 35, no. 3, pp. 269–273, 1930.
28. T. Balu, T. R. Rajasekaran, and P. Murugakoothan, "Studies on the growth, structural, optical and mechanical properties of ADP admixtured TGS crystals," Current Applied Physics, vol. 9, no. 2, pp. 435–440, 2009.
29. M. Dawber, I. Farnan, and J. F. Scott, "A classroom experiment to demonstrate ferroelectric hysteresis,"The

American Journal of Physics, vol. 71, no. 8, pp. 819–822, 2003.

30. C. S. Ramya, S. Selvasekarapandian, T. Savitha et al., "Conductivity and thermal behavior of proton conducting polymer electrolyte based on poly (N-vinyl pyrrolidone)," European Polymer Journal, vol. 42, no. 10, pp. 2672–2677, 2006.

31. R. Baskaran, S. Selvasekarapandian, N. Kuwata, J. Kawamura, and T. Hattori, "Conductivity and thermal studies of blend polymer electrolytes based on PVAc-PMMA," Solid State Ionics, vol. 177, no. 26-32, pp. 2679–2682, 2006.

32. V. D. Nithya, R. Jacob Immanuel, S. T. Senthilkumar et al., "Studies on the structural, electrical and magnetic properties of $LaCrO_3$, $LaCr_{0.5}Cu_{0.5}O_3$ and $LaCr_{0.5}Fe_{0.5}O_3$ by sol-gel method," Materials Research Bulletin, vol. 47, no. 8, pp. 1861–1868, 2012.

33. V. D. Nithya and R. Kalai Selvan, "Synthesis, electrical and dielectric properties of $FeVO_4$ nanoparticles," Physica B, vol. 406, no. 1, pp. 24–29, 2011.

CHAPTER 15 INFLUENCE OF BIAS ON THE FRICTION IMAGING OF FERROELECTRIC DOMAINS IN SINGLE CRYSTAL BARIUM TITANATE ENERGY STORAGE MATERIALS

Lan Chen,Di Wu, Xiao-Min Xiong, Hai-Yan Wang, and Jin-Xiu Zhang

INTRODUCTION

Ferroelectric barium titanate ceramic ($BaTiO_3$) is a promising advanced material with the outstanding performance in dielectric properties for energy storage applications. With the high permittivity and volumetric efficiency [1, 2], $BaTiO_3$ was extensively used in electrical energy storage especially in supercapacitor and multilayer ceramic capacitors (MLCCs).

It is worth mentioning that the performance of the $BaTiO_3$-based MLCCs was significantly affected by microstructures such as domain orientation, domain walls, and point defects [3, 4]. The ferroelectric domain structure can be imaged by variety of techniques such as electron microscopy techniques, optical microscopy and X-ray techniques, and scanning probe microscopy including piezoresponse force microscopy (PFM) and atomic force microscope (AFM) [5, 6]. Actually, friction mode of atomic force microscopy plays an important role in characterizing the surface of specimen by applying a friction force between tip and specimen [7–11].

Many previous works have reported the applications of friction imaging on characterizing the ferroelectric domain structure, which provided a clear picture for domain structures due to variant friction coefficients of different domains [12–14]. Recently, owing to the great importance of friction in industry, more and more research works were carried out to investigate the origin of friction in the micro- or even nanometer scale [15,16] and even to explore efficient ways to control the friction force. Bluhm et al. [17] studied the friction imaging of domains with polarization perpendicular to the surface of specimen of guanidinium aluminum sulfate hexahydrate (GASH) and triglycine sulfate (TGA) and proposed that the chemical difference inside domains was the origin for the different friction coefficient, which could not be affected by the static electric force between tip and specimen. Park et al. [9] reported the possibility of making use of electric field to control the friction force between tip and semiconductor, although its exact mechanism is not clear until now. Although extensive attempts have been made in last decade, there is still no convincing explanation on the exact mechanism for different polarization domain with different friction coefficient.

In this report we investigate the friction imaging of newly cleaved surface of single crystal $BaTiO_3$ under different voltage biases. It was found that different domains showed different dependence on voltage. Under different voltage biases, the friction imaging of a fraction of domains displayed a great change in brightness while that of the other domains displayed only a very small change. Possible mechanisms of the interesting phenomena originating from different static force between different charged tip and the periodical array of surface charges inside the inplane domains were proposed.

MATERIALS AND METHODS

The multidomain single crystal $BaTiO_3$ was prepared by the Institute of Physics, Chinese Academy of Sciences. First, the specimen was broken into two parts, one of which was about

Citation: Lan Chen, Di Wu, Xiao-Min Xiong, Hai-Yan Wang, and Jin-Xiu Zhang, "Influence of Bias on the Friction Imaging of Ferroelectric Domains in Single Crystal Barium Titanate Energy Storage Materials," Journal of Nanomaterials, vol. 2014, Article ID 617436, 5 pages, 2014. doi:10.1155/2014/617436

1 millimeter thick. The newly formed cross section with the dimensions of 4 mm × 1 mm was regarded as the newly cleaved surface. The newly cleaved surface was preferred because it was almost not affected by the pollutants and the free charges in air. The relative humidity was controlled to be <40%, and the experiments were carried out in about one hour.

Friction imaging was obtained through a commercial scanning force microscope SPA-300HV (Seiko, Japan). The scanning direction was perpendicular to the long beam of the cantilever Sil-AF01A. The cantilever with a resonant frequency of 12 kHz, a spring constant k of 0.15 N/m, and a beam length of 400 μm was used. The tip of the cantilever was coated by a gold film. During the experiment, the specimen stage was grounded and a voltage bias (+12 V or −12 V) was applied onto the cantilever tip via a direct current voltage source.

The friction imaging was obtained on a same region of the newly cleaved surface under three different experimental conditions: without voltage bias, +12 voltage, and −12 voltage. The imaging region is about 20 μm × 20 μm where there coexist many domains that could be imaged.

RESULTS AND DISCUSSIONS

Figure 1 is the morphologic image and the friction images under different voltage biases. Three regions, respectively, marked A, B, and C in Figure 1, were discussed. From the morphologic image of Figure 1(a), it could be seen that the roughness of the cleaved surface was small. But the normal friction image without voltage bias (Figure 1(b)) has much contrast between three regions. In the normal friction image of Figure 1(b)without bias, Regions A and B showed almost the same brightness while Region C was darker than A and B. In addition, a significant fraction of domain boundaries was observed clearly. When +12 V bias was applied, it was a surprise that the friction image in Figure 1(c) was much different from Figure 1(b). In Figure 1(c), Region A became darker, being a real difference from that in Figure 1(b). However, the bright Region B still remained bright and the dark Region C remained dark in Figure 1(c). When −12 V bias was applied, there was almost no change in the brightness of Regions B and C, whereas Region A changed back to bright (Figure 1(d)). That was to say, the friction images in Figures 1(b) and 1(d) were almost the same.

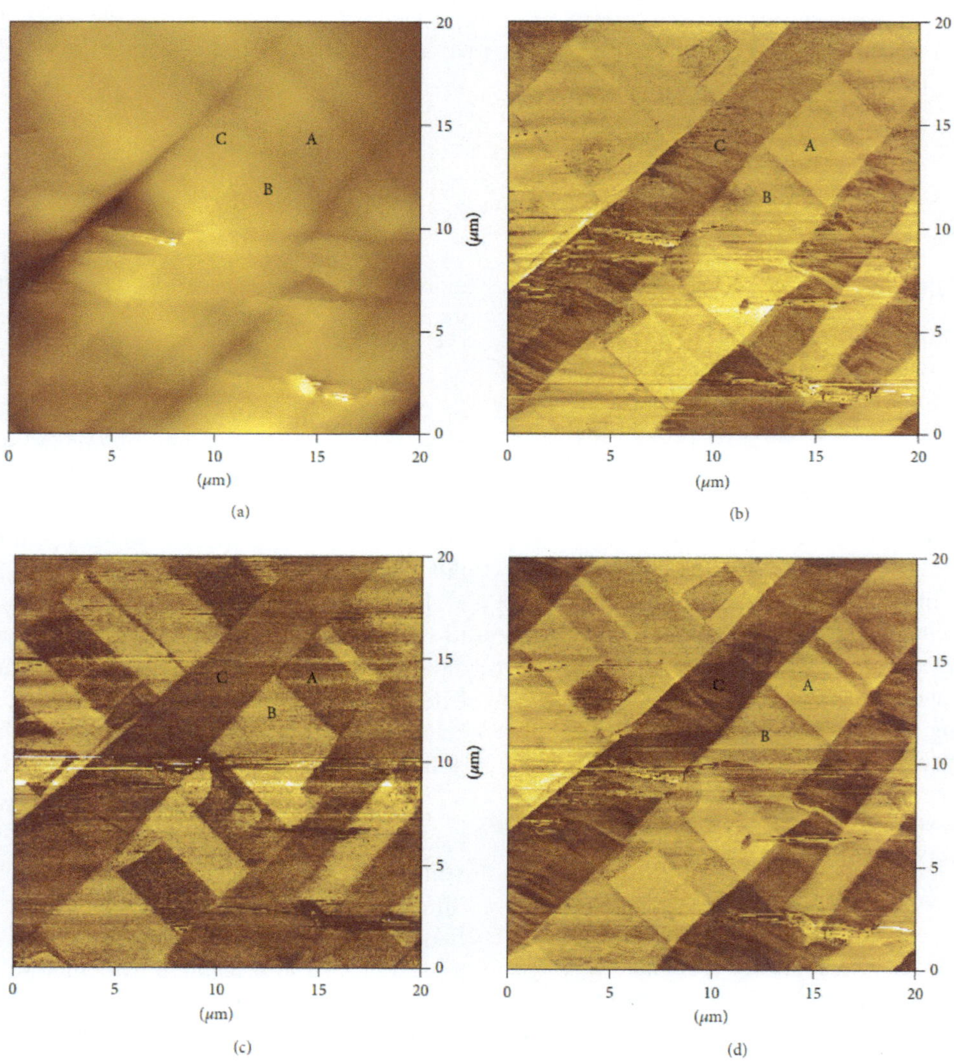

Figure 1: The morphologic image (a) and the friction images on a same region of the newly cleaved surface of single crystal multidomain $BaTiO_3$ under three different experimental conditions: without voltage bias (b), +12 voltage (c), and −12 voltage (d). The specimen stage is grounded.

For single crystal materials, the regions with different brightness in friction image were sometimes considered as ferroelectric domains with different polarization directions [18, 19]. Moreover, the cross section of single crystal multidomain $BaTiO_3$ had many domains with different polarization directions including both perpendicular and parallel to the surface as reported by Soergel [20]. Therefore, it was suggested that the three regions in Figure 1 were associated with different oriented domains, which is similar to the results of PFM in [21]. In our experiment results, when the bias changed from 0 V to +12 V, domain A darkened while domain B remained bright. But while the bias changes into −12 voltage, domain A changed back into bright and domains B and C are kept the same. Three domains exhibiting different lateral force contrast were observed. The voltage bias had no effect on the friction force in domains B and C but had significant effect in domain A. When the tip was applied to positive voltage, the friction force between positive charged tip and domain A was reduced dramatically compared with that for the negative or zero voltage bias. The dramatic change of friction force could be ascribed to the change of static electric interaction induced by the different charged tip.

Correia et al. [14] suggested that the static electric interaction could not change the friction force. Although they observed electric domains of GASH and TGS via scanning force microscope in the friction mode, what they concerned were the domains with the polarization direction perpendicular to the surface. However, in our experiments, we imaged the newly broken cross section of single multidomain $BaTiO_3$ where there were all kinds of electric domains with different polarization directions scattered on the surface. The polarizations included not only the one perpendicular to the surface but also the parallel to the surface. It was the domains with polarization direction parallel to the surface that changed their brightness in the friction imaging under different voltage bias. According to the particular crystal structure of $BaTiO_3$, it was inferred that the charge distribution on the surface of domains with polarization direction parallel to the surface was inhomogeneous, resulting in the friction force change under different voltage bias.

It was well known that $BaTiO_3$ was tetragonal at room temperature, where Ti^{4+} was at the off-center position of the tetrahedral symmetry [1]. Thus the Ti^{4+} and the negative charge center were denoted as O^{2-} forming a permanent electrical dipole. Figure 2 schematically showed the alignment of the permanent electrical dipoles in the domain with the polarization direction parallel to the surface. As seen in Figure 2, the separation between Ti^{4+} and the negative charge center was 1.633 Å, much less than the separation 2.403 Å between electric dipoles, which leads to an inhomogeneous charge distribution of surface. In friction imaging with scanning force microscope, while the tip was voltage biased, the tip was charged and the interaction between the charged tip and the inhomogeneous charge distribution resulted in an inhomogeneous energetic barrier distribution.

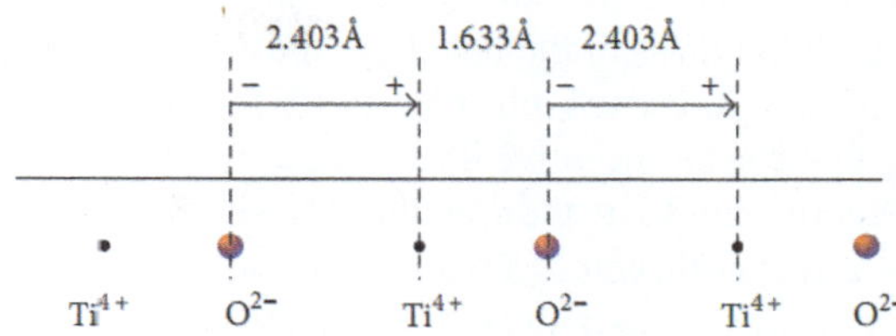

Figure 2: The alignment of the permanent electrical dipoles in the domain of $BaTiO_3$ with the polarization direction parallel to the surface.

Different charged tip corresponds to different inhomogeneous energetic barrier distribution on the domain parallel to the surface, which is sketched as in Figure 3. For the case shown in Figure 3, it was assumed that the tip was scanned from left to right in the friction imaging. When the tip was positive voltage biased, the Coulomb interaction between the positive charged tip and the inhomogeneous charge distribution leaded to a zigzag energetic barrier distribution as in Figure 3(a). In this case, a small friction force was obtained. However, for the negative voltage bias, the situation was inversed. The zigzag energetic barrier was shown in Figure 3(b). A much large friction force was obtained. Additionally, when a certain angle existed between the scan direction and the polarization direction and a bias was applied to the tip, the friction force would change a little. In this case, the qualitative result was the same. At the same time the size of the tip had no effect on it.

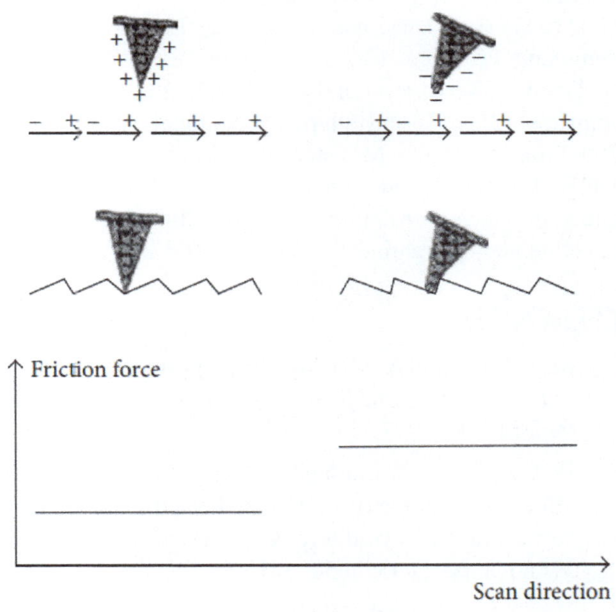

Figure 3: The friction force between the voltage-biased tip and the domain of $BaTiO_3$ with the polarization direction parallel to the surface for the voltage bias is positive (left) and negative (right). It is assumed that the tip is scanned from left to right in the friction imaging. The Coulomb interaction between the charged tip and the inhomogeneous charge distribution of the domain leads to a zigzag energetic barrier distribution, which results in different friction force.

Based on the above discussions and our experimental results, the brightness of the image for the domain A charged a lot: the friction force was much larger for the positive charged tip than that for the negative charged tip. It was thought that region A corresponded to the domain with polarization direction parallel to the surface and is almost the same as the scanning direction of the tip. As to the regions B and C, the contrasts of them were not affected by the biases, and their boundaries well agreed with the morphologic image. So due to the titling of the domains, the polarization directions of domains B and C could not be judged only based on the friction force imaging and morphologic image and further investigations were necessary.

CONCLUSIONS

The newly cleaved surface of a single crystal multidomain $BaTiO_3$ was investigated through the friction imaging with atomic force microscope. It was found that the brightness of some regions changed when the tip was differently voltage biased. It was proposed that the inhomogeneous charge distribution in the domain with polarization direction parallel to surface resulted in this change. The experimental results may provide a new method for the determination of the polarization direction for the domain parallel to the surface, which may be useful in the investigation of ferroelectric energy storage materials.

ACKNOWLEDGMENTS

This investigation was supported by National Natural Science Foundation of China (NSFC) (Grant nos. 10974259, 11274391, and 11104357), Science and Technology Planning Project of Guangdong Province, China (Grant no. 2012B060100003), Fundamental Research Funds for the Central Universities (Grant nos. 121gpy36 and 09lgpy29), National Key Technology R&D Program of the Ministry of Science and Technology (2011BAK15B04), and Science and Technology Planning Project of General Administration of Quality Supervision, Inspection and Quarantine of China (2011QK315).

REFERENCES

1. M. E. Lines and A. M. Glass, Principles and Applications of Ferroelectrics and Related Materials, Clarendon Press, Oxford, UK, 1997.
2. W. Cai, C. Fu, Z. Lin, and X. Deng, "Vanadium doping effects on microstructure and dielectric properties of barium titanate ceramics," Ceramics International, vol. 37, no. 8, pp. 3643–3650, 2011.
3. X. Ren, "Large electric-field-induced strain in ferroelectric crystals by point-defect-mediated reversible domain switching," Nature Materials, vol. 3, no. 2, pp. 91–94, 2004.
4. M. Osada and T. Sasaki, "Two-dimensional dielectric nanosheets: novel nanoelectronics from nanocrystal building blocks," Advanced Materials, vol. 24, no. 2, pp. 210–228, 2012.
5. E. Soergel, "Piezoresponse force microscopy (PFM)," Journal of Physics D: Applied Physics, vol. 44, no. 46, Article ID 464003, 2011.
6. P. R. Potnis, N. T. Tsou, and J. E. Huber, "A review of domain modelling and domain imaging techniques in ferroelectric crystals," Materials, vol. 4, no. 2, pp. 417–447, 2011.
7. K. S. K. Karuppiah, Y. Zhou, L. K. Woo, and S. Sundararajan, "Nanoscale friction switches: friction modulation of monomolecular assemblies using external electric fields," Langmuir, vol. 25, no. 20, pp. 12114–12119, 2009.
8. B. Bhushan, K. J. Kwak, and M. Palacio, "Nanotribology and nanomechanics of AFM probe-based data recording technology," Journal of Physics: Condensed Matter, vol. 20, no. 36, Article ID 365207, 2008.
9. J. Y. Park, D. F. Ogletree, P. A. Thiel, and M. Salmeron, "Electronic control of friction in silicon pn junctions," Science, vol. 313, no. 5784, p. 186, 2006.
10. J. Krim, "Friction at the atomic scale," Scientific American, vol. 275, no. 4, pp. 74–80, 1996.
11. E. Gnecco, R. Bennewitz, T. Gyalog, and E. Meyer, "Friction experiments on the nanometre scale,"Journal of Physics: Condensed Matter, vol. 13, no. 31, pp. R619–R642, 2001.
12. R. Lüthi, H. Haefke, K. P. Meyer, E. Meyer, L. Howald, and H.-J. Güntherodt, "Surface and domain structures of ferroelectric crystals studied with scanning force microscopy," Journal of Applied Physics, vol. 74, no. 12, pp. 7461–7471, 1993.
13. H. Bluhm, U. D. Schwarz, K. P. Meyer, and R. Wiesendanger, "Anisotropy of sliding friction on the triglycine sulfate (010) surface," Applied Physics A: Materials Science & Processing, vol. 61, no. 5, pp. 525–533, 1995.
14. A. Correia, J. Massanell, N. García, A. P. Levanyuk, A. Zlatkin, and J. Przeslawski, "Friction force microscopy study of a cleaved ferroelectric surface: Time and temperature dependence of the contrast, evidence of domain structure branching," Applied Physics Letters, vol. 68, no. 20, pp. 2796–2798, 1996.
15. A. Socoliuc, E. Gnecco, S. Maier et al., "Atomic-scale control of friction by actuation of nanometer-sized contacts," Science, vol. 313, no. 5784, pp. 207–210, 2006.
16. R. W. Carpick, "Controlling friction," Science, vol. 313, no. 5784, pp. 184–185, 2006.
17. H. Bluhm, U. D. Schwarz, and R. Wiesendanger, "Origin of the ferroelectric domain contrast observed in lateral force microscopy," Physical Review B—Condensed Matter and Materials Physics, vol. 57, no. 1, pp. 161–169, 1998.
18. R. W. Carpick, D. Y. Sasaki, and A. R. Burns, "Large friction anisotropy of a polydiacetylene monolayer,"Tribology Letters, vol. 7, no. 2-3, pp. 79–85, 1999.
19. C. Munuera, E. Barrena, and C. Ocal, "Deciphering structural domains of alkanethiol self-assembled configurations by friction force microscopy," Journal of Physical Chemistry A, vol. 111, no. 49, pp. 12721–12726, 2007.
20. E. Soergel, "Visualization of ferroelectric domains in bulk single crystals," Applied Physics B: Lasers and

Optics, vol. 81, no. 6, pp. 729–752, 2005.

21. D. Liu, M. Chelf, and K. W. White, "Indentation plasticity of barium titanate single crystals: dislocation influence on ferroelectric domain walls," Acta Materialia, vol. 54, no. 17, pp. 4525–4531, 2006.

CHAPTER 16

CUSP CATASTROPHE MODEL FOR DESCRIPTION OF BISTABILITY IN FERROELECTRIC LIQUID CRYSTALS

Nail Migranov, Aleksey Kudreyko, and Denis Kondratyev

INTRODUCTION

Ferroelectric liquid crystals (FLCs) are widely used in production of fast displays in electronic equipment due to fast response time (~100 μs) to the electric field. Inherent permanent polarization that interacts with the electric field makes ordering of molecules completely different from usual liquid crystals, and in this connection numerous technical applications arouse great interest in investigations of FLC structures. In 1980 Clark and Lagerwall [1] introduced bistable FLC. These crystals are considered as a potential operating component for production of e-paper due to its fast response time, wide viewing angle, and so forth [2].

However, the progress in the development of optical processing systems is being impeded by the lack of enough knowledge on instabilities of FLCs, which is related with the free energy. One of the main advantages in the application of the methods of bifurcation theory to nonlinear stability problems is that in general it is possible to reduce the dimension of the original system to a low dimensional bifurcation system, mainly in the neighborhood of the critical points.

As known, bifurcation theory and catastrophe theory are two of the best known areas in the field of dynamical systems. Due to its topological character, catastrophe theory has several advantages that allow getting qualitative results. Catastrophe models have been successfully applied in various fields of research [3]. Gibelli and Turzi [4] applied catastrophe theory to derive a criterion for identification of tricritical points in a multiorder parameter system represented by SmC. In [5, 6] authors proposed that the Landau expansion of free energy can be represented as a catastrophe functional that possesses inherent properties for description of instabilities. The lack of this proposition is that the corresponding relation between the behavior parameters is not represented.

In our research we applied asymptotic methods in order to elaborate the relation of the governing parameters in the series of azimuthal angle. Due to the complexity of the torque balance equation, it is difficult to find the values of the governing parameters and calculate the minima of free energy. The applied catastrophe theory to the study of the free energy potential of FLC provided a detailed mathematical description of how the geometry of the energy functional varies with the control parameters. The local geometry around the critical points of the free energy functional is represented by a certain catastrophe function, which is determined by the elastic, surface, spontaneous polarization and dielectric free energy densities. The obtained plots confirm the validity of the catastrophe theory for description of bistability of FLC. The advantage of this qualitative approach is that the catastrophe functional makes it possible to find energy minima and bifurcation points in multiparameter dynamic systems.

This paper is organized as follows: in Section 2 we introduce the relation that models the behavior of the azimuthal angle and free energy of ferroelectric liquid crystal. Then with the aid of the torque balance equation and the approximation theory we establish the relation between the behavior parameters in the proposed relation. In Section 3 we introduce the normalized free energy potential in the form of the cusp catastrophe. The results are discussed in Section 4.

FREE ENERGY OF FERROELECTRIC LIQUID CRYSTAL

Assume that in the absence of the electric field molecules of FLC align in parallel to a substrate (planar alignment). When

Citation: Nail Migranov, Aleksey Kudreyko, and Denis Kondratyev, "Cusp Catastrophe Model for Description of Bistability in Ferroelectric Liquid Crystals," Physics Research International, vol. 2014, Article ID 294723, 5 pages, 2014. doi:10.1155/2014/294723.

electric field is turned on, consider a director n of FLC, which is placed in this field

$$\mathbf{n} = \sin\theta\cos\varphi\ \mathbf{e}_x + \sin\theta\sin\varphi\ \mathbf{e}_y + \cos\theta\ \mathbf{e}_z, \tag{1}$$

where θ is the polar angle between n and the z-axis and φ is the azimuthal angle in the Cartesian coordinate system. As illustrated in Figure 1, the electric field E in the FLC is directed along the y-axis, which is normal to the bounding plates separated by a distance d, and the sample of FLC is uniform along the x- and z-axes.

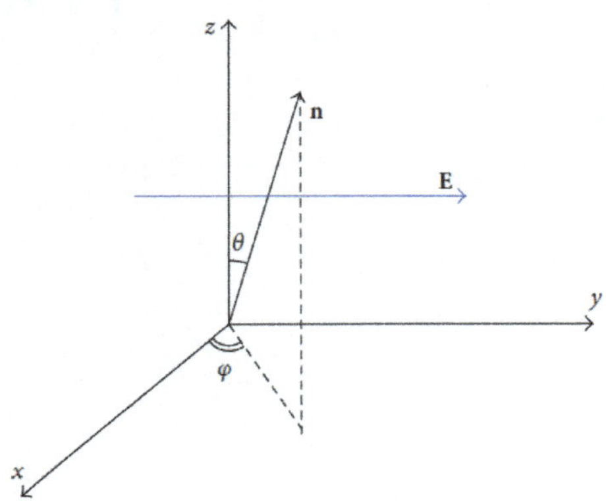

Figure 1: Schematic representation of the director n in the electric field E.

The total free energy density G of FLC, which is confined between two plates, consists of elastic energy density F_{el}, surface energy density F_s, spontaneous polarization density F_p, and the dielectric interaction energy density F_{diel} induced by the applied field. Then the total free energy G of the FLC is given by

$$G(\varphi) = \int_V \left(F_{el} + F_p + F_{diel}\right) dV + \int_S F_s dS, \tag{2}$$

where S, V are area and volume of the sample. The substitution of n into the elastic part of the free energy density gives

$$F_{el} = F_0 + \frac{1}{2}A\left(1 + \nu\sin^2\varphi\right)\left(\varphi_{,y}\right)^2 + A(1+\nu)\,\varphi_{,y}Q_0\sin\varphi, \tag{3}$$

where

$$F_0 = \frac{1}{2}\left(K_2 q_T^2 + K_3 q_B^2\right),$$

$$A = K_1\sin^2\theta, \qquad B = \left(K_2\cos^2\theta + K_3\sin^2\theta\right)\sin^2\theta,$$

$$\nu = \frac{B - A}{A}, \qquad Q_0 = \frac{K_2 q_T\sin\theta\cos\theta + K_3 q_B\sin^2\theta}{A(1+\nu)}. \tag{4}$$

In these expressions q_T and q_B are the inherent twist and bend wavenumbers ($\simeq 1.3\cdot106\ m_{-1}$); K_1, K_2, K_3 are splay, twist, and bend elastic constants. The dielectric interaction energy density F_{diel} is expressed as

$$F_{diel} = -\frac{1}{2}\varepsilon_\perp\varepsilon_0\left(1 + \Delta\varepsilon'\sin^2\varphi\right)E^2, \tag{5}$$

where $\Delta\varepsilon' = (\varepsilon_\parallel' - \varepsilon_\perp)/\varepsilon_\perp$, $\varepsilon_\parallel' = \varepsilon_\perp\cos^2\theta + \varepsilon_\parallel\sin^2\theta$, and ε_0 is the dielectric constant. Next, the polarization density Fp is given by

$$F_p = P_s E\cos\varphi, \tag{6}$$

where P_s is the magnitude of the polarization vector; the physical interpretation of P_s can be found, for example, in [7]. The surface interaction associated with the anchoring is defined as

$$F_s = -g_1\cos^2(\varphi - \varphi_s) - g_2\cos(\varphi - \varphi_s), \tag{7}$$

where g_1, g_2 denote azimuthal and polar anchoring strengths between the molecules of liquid crystal and bounding plates and φ_s is the pretilt angle of the molecules [7, 8]. Suppose that $\varphi \simeq 0$ (e.g., $0 < \varphi \leq \pi/36$ rad); then the following approximations are valid:

$$\sin\varphi \approx \varphi - \frac{1}{6}\varphi^3 + \frac{1}{120}\varphi^5,$$

$$\cos\varphi \approx 1 - \frac{1}{2}\varphi^2 + \frac{1}{24}\varphi^4. \tag{8}$$

It is known that in FLCs the spontaneous polarization vector Ps exhibits inherent rotation within the xy-plane. Assume that the mean of the azimuthal angle φ is defined relative to the y-axis as the following series:

$$\varphi = \sum_{i=0}^{n} a_i\tilde{y}^i, \tag{9}$$

where $\tilde{y}$ is the normalized length $y = y/d^{\sim}$ and ai is a dimensionless parameter. Consider this expansion for n=2; then

$$\varphi = a_0 + a_1\tilde{y} + a_2\tilde{y}^2. \tag{10}$$

In the following, we will assume that y=$\tilde{y}$; then $y \in [0; 1]$. The substitution of (10) and approximations of trigonometric functions into the stationary torque balance equation [7] under the one-constant approximation ($K_1 = K_2 = K_3 = K$)

$$K\sin^2\theta\frac{1}{d^2}\frac{\partial^2\varphi}{\partial y^2} + \frac{\Delta\varepsilon'}{2}\varepsilon_\perp E^2\sin 2\varphi + P_s E\sin\varphi = 0 \tag{11}$$

yields the following equation for the terms up to y^2:

$$\begin{aligned}&\Big(2\Delta\varepsilon'\varepsilon_\perp E^2 a_0 a_1^2 - \frac{1}{2}P_s E a_0 a_1^2 + \frac{1}{24}P_s E a_0^4 a_2 \\ &+ \frac{2}{3}\Delta\varepsilon' E^2 a_0^4 a_2 + 2\Delta\varepsilon'\varepsilon_\perp E^2 a_0^2 a_2 + P_s E a_2 \\ &- \Delta\varepsilon'\varepsilon_\perp E^2 a_2 + \frac{4}{3}\Delta\varepsilon'\varepsilon_\perp E^2 a_0^3 a_1^2 \\ &+ \frac{1}{12}P_s E a_0^3 a_1^2 - \frac{1}{2}P_s E a_0^2 a_2 E\Big)y^2 \\ &+ \Big(-2\Delta\varepsilon'\varepsilon_\perp E^2 a_0^2 a_1 + \frac{2}{3}\Delta\varepsilon'\varepsilon_\perp E^2 a_0^4 a_1 + P_s E a_1 \\ &- \frac{1}{2}P_s E a_0^2 a_1 + \Delta\varepsilon'\varepsilon_\perp E^2 a_1 + \frac{1}{24}P_s E a_0^4 a_1\Big)y \\ &+ \frac{1}{120}P_s E a_0^5 + P_s E a_0 + \Delta\varepsilon'\varepsilon_\perp E^2 a_0 - \frac{2}{3}\Delta\varepsilon'\varepsilon_\perp E^2 a_0^3 \\ &+ \frac{2K\sin^2\theta}{d^2}a_2 + \frac{2}{15}\Delta\varepsilon'\varepsilon_\perp E^2 a_0^5 - \frac{1}{6}P_s E a_0^3 = 0,\end{aligned} \tag{12}$$

where θ is the constant along the y-axis and $] = 0$.

The complexity of (12) makes it unsolvable analytically with respect to coefficients ai. In order to gain some insight into the free energy potential (2), we apply the perturbation methods [9]. Since (12) is an identity for each parameter y, then each factor of y_i ($i = 0, 1, 2$) vanishes independently. Consider the factor of y in the identity (12); then we obtain

$$a_1\left(P_sE - 2\Delta\varepsilon'\varepsilon_\perp E^2a_0^2 + \frac{2}{3}\Delta\varepsilon'\varepsilon_\perp E^2a_0^4 - \frac{1}{2}P_sEa_0^2 + \Delta\varepsilon'\varepsilon_\perp E^2 + \frac{1}{24}P_sEa_0^4\right) = 0. \quad (13)$$

This means that either $a_1 = 0$ or

$$\left(16\Delta\varepsilon'\varepsilon_\perp E^2 + P_sE\right)a_0^4 + \left(12P_sE - 48\Delta\varepsilon'\varepsilon_\perp E^2\right)a_0^2 + 24\left(\Delta\varepsilon'\varepsilon_\perp E^2 + P_sE\right) = 0; \quad (14)$$

then

$$a_0 = \pm\left(\frac{24\Delta\varepsilon'\varepsilon_\perp + 6P_s}{P_s + 16\Delta\varepsilon'\varepsilon_\perp E} \pm \frac{2\left(48\left(\Delta\varepsilon'\varepsilon_\perp E\right)^2 - 30\Delta\varepsilon'\varepsilon_\perp P_s + 3P_s^2\right)^{1/2}}{P_s + 16\Delta\varepsilon'\varepsilon_\perp E}\right)^{1/2}. \quad (15)$$

There exist many FLCs that possess negative and positive dielectric anisotropy [10]. In our example we suppose that $\varepsilon_\parallel = 3.89$, $\varepsilon_\perp = 2.91$, $\theta = 2\pi/15$, $\varphi_s = 7\pi/60$, and $K = 10^{-12}$ N [11]. The substitution of the representative magnitudes of the polarization vector and the electric field in (15) $P_s = 10^{-5}$ C · m^{-2}, $E = 5 \cdot 10^6$ V m^{-1} gives $|a_0| \simeq 1$. This issue violates the condition $\varphi \simeq 0$; therefore $a1 = 0$. From the free term of y in (12) we find that

$$a_2 = -\frac{a_0d^2E}{240K\sin^2\theta} \times \left(120\Delta\varepsilon'\varepsilon_\perp E - 80\Delta\varepsilon'\varepsilon_\perp Ea_0^2 + 16\Delta\varepsilon'\varepsilon_\perp Ea_0^4 + P_sa_0^4 - 20P_sa_0^2 + 120P_s\right); \quad (16)$$

that is, $a_2 \simeq -10^4\, a_0$ for $d = 20$ μm. With the exception of the trivial solution in the factor of y^2, our further analysis assumes that $|a_0| \ll 1$, so we can neglect it with respect to a_2.

Combining functionals (3)–(8) in (2), we find that the total free energy G is a function of y, where $y \in [0; 1]$. Carrying out the integration over y, we obtain

$$G(a_2) = Ra_2^4 + Sa_2^2 + Ta_2 + U, \quad (17)$$

where

$$R = \frac{1}{54}\left(-\frac{1}{8}P_sE + E\Delta\varepsilon'\sin^2\varphi_s\right) - \frac{1}{24}KQ_0\sin^2\varphi_s,$$
$$S = \frac{1}{10}\left(\frac{1}{2}P_sE - \Delta\varepsilon'E^2\sin^2\varphi_s\right) + \frac{1}{2}KQ_0\sin^2\varphi_s + \frac{2}{3}K\sin^2\varphi_s,$$
$$T = -\frac{1}{3}\left(g_2\sin\varphi_s + g_1\sin 2\varphi_s\right),$$
$$U = F_0 - \frac{1}{2}E^2\varepsilon_\perp - \frac{1}{2}P_sE. \quad (18)$$

Note that for the given parameters of FLC, the factor of a^3_2 can be neglected because

$$\frac{1}{21}\left(2g_1\sin 2\varphi_s + \frac{1}{2}g_2\sin\varphi_s\right)a_2^3 \ll |T|\,a_2, \quad (19)$$

where $g_1, g_2 \simeq 10^{-5}$ J· m^{-2} ($\simeq 10^{-2}$ erg · cm^{-2}) [12]; therefore we will not take it into consideration

CUSP CATASTROPHE MODEL

To get free term in higher degree of functional (17), it is convenient to divide it by R. So, we analyze the normalized free energy and have reduced our problem to the famous cusp catastrophe [6]. It is known that two types of variables are distinguished in the catastrophe theory: control and behavior variables. The behavior of the system is described by the relation between these variables. Free energy $\tilde{G}$ written in terms of the cusp catastrophe model is

$$\tilde{G}(a_2) = a_2^4 + \frac{S}{R}a_2^2 + \frac{T}{R}a_2 + \frac{U}{R}, \quad (20)$$

where a_2 plays a role of the behavior parameter. Therefore, functional (20) can be referred to as bifurcation model exhibiting a cusp catastrophe. Depending on the control parameters $a = S/R$ and $b = T/R$, the $\tilde{G}$ has one or two minima. The bifurcation set is given by the condition that a critical point is degenerate. Then the projection of $\tilde{G}$into the (a, b) space can be determined as

$$\frac{\partial\tilde{G}}{\partial a_2} = 0, \qquad \frac{\partial^2\tilde{G}}{\partial a_2^2} = 0 \quad (21)$$

and by eliminating a_2, we obtain [3] the bifurcation set

$$\Delta = 8a^3 + 27b^2 = 0. \quad (22)$$

This set is plotted in Figure 2, where two curves merge according to the power-law behavior.

It is clear that the explicit computation of(a_2)represents a difficulty; therefore we will make a qualitative analysis of (20) and (22). Consider the line *A-B-C-D* in Figure 2 for a fixed mean of the parameter a. In point A only one minimum of free energy exists, and the system is assumed to be in this minimum (solid line in Figure 3). When we increase b, the system shifts to the point B, where a second minimum starts to appear (dashed line in Figure 3).

When the values of E and Ps are such that the state of FLC is between the bifurcation set (line BC), there exist two isolated minima with a potential barrier between them (dash-dotted line in Figure 3). The barrier disappears in the bifurcation point C, and even a small perturbation causes the system to move to the left-hand minimum. Similar behavior is possible when we move from point D to A, but when we change the field polarity and follow from C to B, FLC enters the right-hand minimum (Figure 3). As can be seen, a hysteresis in FLC occurs exhibiting free-energy wells. In other words, the behavior of FLC depends on the direction of the way that the system changes.

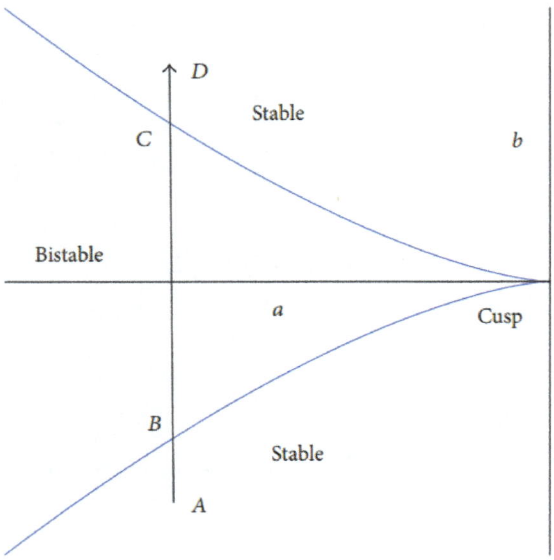

Figure 2: Relation between a and b, where a changeover takes place from one to two minima in the $\tilde{G}$function

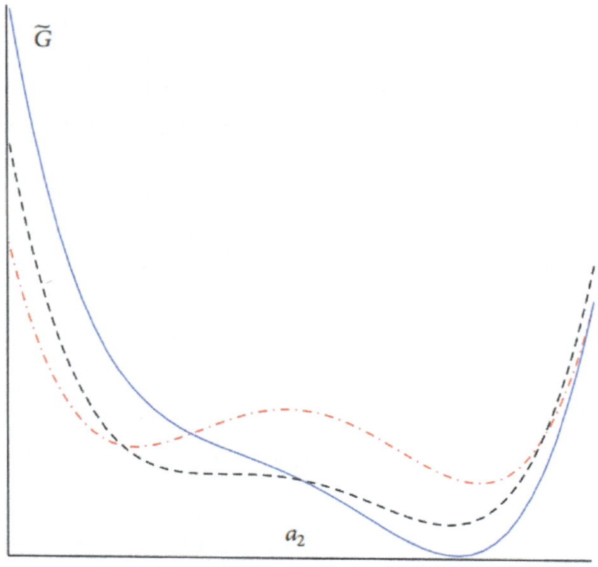

Figure 3: Schematic representation of $\tilde{G}$ versus different means of the control parameters a, b

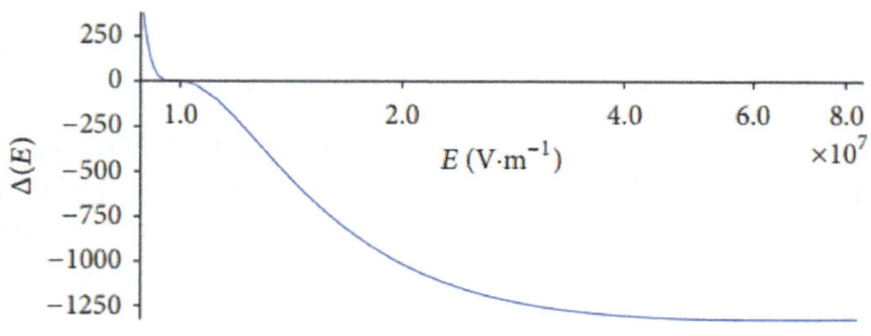

Figure 4: Cross section of $(E)\tilde{}$ for $P_s = 2\ 10^{-5}$ C m^{-2}.

The explanation of this effect is known in literature as the Clark-Lagerwall bistability [1, 10], which occurs when electric field breaks zenithal anchoring. In the surface stabilized FLC geometry, the free-energy wells (Figure 3) at two azimuthal orientations of the director on the tilt cone are created by interactions with the surfaces of the bounding plates in thin cells. This effect occurs if $g1 = g2$ and $\Delta(a, b) \leq 0$. When $\Delta(a, b) > 0$, then expression $dG/d\tilde{a}_2 = 0$ has one root in the a_2-space, and FLC is stable because free energy has one minimum (points A and D in Figure 2). If expression $dG/d\tilde{a}_2 = 0$ has three roots, then bistabilities in FLC occur, and $\Delta(a, b) < 0$ (see Figure 4). Thus, bistable state of FLC is described by using the free energy potential and the cusp catastrophe model. The plot $\Delta(E)$ can help to determine the range of the electric field when bistability in FLC occurs.

CONCLUSIONS

The main result of this paper for theory of FLC is the justification of the relation between behavior parameters in the series expansion of azimuthal angle. For solution of this problem we considered equilibrium state of FLC in the "bookshelf" geometry when an electric field is applied perpendicular to the substrates. The proposed expansion for φ helped us to reach the following numerical results: by using the torque balance equation and the bifurcation theory, one can readily establish the relationship between the behavior parameters ai. The free energy potential given in (2) can be replaced by a functional with a priori known properties. The demo example explicitly shows bistability of FLC from the value of$E$that corresponds to the break of zenithal anchoring.

ACKNOWLEDGMENT

This work was supported by the Russian Foundation for Basic Research (RFBR) under Project no. 14-02-97026.

REFERENCES

1. N. A. Clark and S. T. Lagerwall, "Submicrosecond bistable electro-optic switching in liquid crystals,"Applied Physics Letters, vol. 36, no. 11, pp. 899–901, 1980.
2. C. Wang and P. J. Bos, "Bistable C1 ferroelectric liquid crystal device for e-paper application," Displays, vol. 25, no. 5, pp. 187–194, 2004.
3. T. Poston and I. Stewart, Catastrophe Theory and Its Applications, Dover, New York, NY, USA, 2006.
4. L. Gibelli and S. Turzi, "A catastrophe-theoretic approach to tricritical points with application to liquid crystals," SIAM Journal on Applied Mathematics, vol. 70, no. 1, pp. 63–76, 2009.
5. G. Derfel, "Stationary states of the surface stabilized ferroelectric liquid crystal layers in electric field,"Liquid Crystals, vol. 8, pp. 331–343, 1990.
6. C. R. Ou, C.-M. Ou, C. H. Lin, and S. L. Young, "Catastrophic description of ferroelectric liquid crystal instabilities," Journal of Superconductivity and Novel Magnetism, vol. 23, pp. 917–921, 2010.
7. I.-C. Khoo, Liquid Crystals, John Wiley & Sons, Hoboken, NJ, USA, 2nd edition, 2007.
8. M. Nakagawa, M. Ishikawa, and T. Akahane, "Dynamic responses of ferroelectric liquid crystals in the surface-stabilized geometry I," Japanese Journal of Applied Physics, vol. 27, no. 4, pp. 456–463, 1988.
9. A. H. Nayfeh, Perturbation Methods, Wiley Classics Library, Wiley-Interscience, New York, NY, USA, 2000.
10. P. Archer, I. Dierking, V. Görtz, and J. W. Goodby,

"Probing the material properties and phase transitions of ferroelectric liquid crystals by determination of the Landau potential," The European Physical Journal E, vol. 25, pp. 385–393, 2008.

11. A. Smith and I. W. Stewart, "Asymptotic solutions for an electrically induced Freedericksz transition in a wedge of smectic C liquid crystal," Journal of Physics A: Mathematical and General, vol. 39, pp. 11361–11382, 2006.
12. L. Blinov, Structure and Properties of Liquid Crystals, Springer, 2011.

CHAPTER 17

ASYMPTOTICS OF TIME HARMONIC SOLUTIONS TO A THIN FERROELECTRIC MODEL

Naïma Aïssa and Kamel Hamdache

INTRODUCTION

In this work, we are interested in the model equations of ferroelectric materials introduced in [1] and discussed in [1–3] for example. We consider time harmonic solutions to the model as studied in [4]. We first rewrite the equations of the model given in [1] to precise the boundary conditions we will use. Let (E, H) be the electromagnetic field acting on the ferroelectric material Ω, which is a bounded and a regular domain of R^3. Let P be the electric polarization induced in Ω. The electric displacement is then given by D = (E + P) where > 0 is the electric permittitivity of the vacuum. The Maxwell equations satisfied by the electromagnetic field are

$$\mu\partial_t H - \operatorname{curl} E = 0,$$
$$\epsilon\partial_t(E+P) + \operatorname{curl} H + \sigma E = 0, \tag{1.1}$$

where $\mu > 0$ is the magnetic permeability of the vacuum and $\sigma > 0$ is the conductivity constant of the ferroelectric material. The behavior of the electric polarization P is driven by the nonlinear Maxwell equation.

$$\partial_t^2 P + \lambda^2 \operatorname{curl}^2 P + a\partial_t P = -b(E_{\text{eq}}(P) - E), \tag{1.2}$$

where $\operatorname{curl}^2 P = \operatorname{curl}(\operatorname{curl} P)$, $E_{eq}(P)$ is the nonlinear equilibrium electric field which will be given later, and $\lambda^2 = 1/\mu$. The parameters a and b are some physical positive constants. This model is obtained as follow (see [1]). Denoting by m the internal magnetization and by j the current density which is driven by the difference between the equilibrium field $E_{eq}(P)$ and the electric field E, then with the internal polarization field P they satisfy the set of equations

$$\epsilon(\partial_t P + \delta^{-1}\mathbf{j}) = \operatorname{curl}\mathbf{m},$$
$$\mu(\partial_t \mathbf{m} + \alpha\delta^{-1}\mathbf{m}) = -\operatorname{curl} P,$$
$$(\partial_t \mathbf{j} + \alpha\delta^{-1}\mathbf{j}) = \beta\delta^{-1}(E_{\text{eq}}(P) - E) \tag{1.3}$$

which reduces to the nonlinear Maxwell equation (1.2) satisfied by P. The internal magnetization m satisfies the boundary condition $m \times n = 0$, then the second equation of (1.3) implies that P satisfies

$$\operatorname{curl} P \times n = 0 \text{ on } \partial\Omega. \tag{1.4}$$

In this work, we consider, on $\partial\Omega$, Leontovitch-type boundary conditions for E extending the one used in [1], that is,

$$H \times n + \beta n \times (E \times n) = 0, \qquad \operatorname{curl} P \times n = 0, \tag{1.5}$$

where β is some nonnegative function defined on $\partial\Omega$ and n is the unit outward normal to $\partial\Omega$.

The equilibrium field is assumed to be the gradient of a potential function $\varphi(|P|^2)$. We have $E_{eq}(P) = 2P\varphi(|P|^2)$ where $\varphi : R+ \to R$ is a C^2 convex function satisfying the hypotheses given in [1], more precisely, we assume that there exist $0 < r_1 < r_0$ and $C_2 > 0$ such that

$$\phi(0) = \phi(r_0) = 0, \qquad \phi'(r_1) = 0, \qquad \phi'(0) < 0,$$
$$(s\phi'(s^2))' \le C_2 \quad \forall s \ge 0. \tag{1.6}$$

Hence, for all $s \ge R > r_1$, there exists $C_R > 0$ such that $\varphi(s) \ge C_R$ and for all $s \ge 0$

$$|\phi'(s)| \le C_*, \qquad 0 \le s^2\phi^{(2)}(s^2) \le C_*, \tag{1.7}$$

Citation: Naïma Aïssa and Kamel Hamdache, "Asymptotics of Time Harmonic Solutions to a Thin Ferroelectric Model,"Abstract and Applied Analysis, vol. 2007, Article ID 61629, 14 pages, 2007. doi:10.1155/2007/61629.

where $C_* = \max(|\varphi'(0)|, C_2)$. Examples of such potentials defined on R^+ satisfying the hypotheses are the following: $\varphi_1(s)=b(1+s^2)^{1/2} - as - 1$ with $0<a<b<1$, $\varphi_2(s)= s/2 - \log(1+s)$, $\varphi_3(s) = as+1-(1+s)^\alpha$ with $a>0$ and $a<\alpha<1$.

With these hypotheses, the vector-valued function $E_{eq}(P) = 2P\varphi'(|P|2)$ satisfies the estimate

$$|E_{eq}(P) - E_{eq}(Q)| \le C_* |P-Q| \tag{1.8}$$

for all $P,Q \in R^3$.

Let us mention other interesting models for ferroelectric materials, see [5–7] for example. In the first two papers, the authors consider deformable ferroelectric materials and give the evolution equation for the spontaneous polarization. The model obtained is different from the one given in [1], since it includes the deformation of the bodies. In the second one, a theoretical model is proposed explaining the lamellar morphology of domains of opposite polarization observed in ferroelectric crystals in their polar phases. The jump conditions for the electric field and polarization vector across domain walls play an important role in the characterization of the free energy. Many interesting mathematical problems, as the dimension reduction of domains, are contained in both papers.

In this paper, we are dealing with time harmonic solutions to the model (1.1)-(1.2). We write $H(t,x) = e^{\imath t}H(x)$, $E(t,x) = e^{\imath\omega t}E(x)$, $P(t,x) = e^{\imath\omega t}P(x)$, and $F(t,x) = e^{\imath\omega t}F(x)$ with $\omega > 0$ fixed. The new complex field (E,P) satisfies the set of equations

$$(\zeta_1(\omega) + \lambda^2 \operatorname{curl}^2)E = \omega^2 P + \imath\omega F$$
$$(\zeta_2(\omega) + \lambda^2 \operatorname{curl}^2)P = -b(2P\phi'(|P|^2) - E), \tag{1.9}$$

where

$\zeta_1(\omega) = -\omega^2 + \imath a_1$, $\zeta_2(\omega) = -\omega^2 + \imath\omega a_2$ with $a_1 = \sigma/\epsilon$ and $a_2 = a$. The magnetic field H is recovered from the electric field E by the formula $H = \operatorname{curl}E/\imath\omega\mu$. The boundary conditions on $\partial\Omega$ write

$$\operatorname{curl}E \times n + \imath\omega\mu\beta n \times (E \times n) = 0, \qquad \operatorname{curl}P \times n = 0. \tag{1.10}$$

The main difficulty in this problem is related to the lack of regularity of the polarization field P to prove the stability of the nonlinear equilibrium field $E_{eq}(P)$ with respect to the weak convergence of a sequence Pm. It is easy to prove that a sequence of solutions (E_m, P_m) of (1.9) is such that P_m, curl P_m are bounded in $L^2(\Omega)$. Even if we prove that div P_m is also bounded in $L^2(\Omega)$, the boundary condition curl $P_m \times n = 0$ satisfied by Pm does not allow to deduce compactness in $L^2(\Omega)$ of the sequence Pm. Note that, in [8], the compactness of the sequence (P_m) is obtained in the case of the boundary condition curl $P \times n + \beta n \times (P \times n) = 0$. To avoid this difficulty, we derive new model equations as follows

For a given $P \in L^2(\Omega)$, the Hodge decomposition of P gives the orthogonal decomposition in $L^2(\Omega)$, $P = \nabla\phi + U$ where $\phi \in H^1(\Omega)$ and $U \in L^2(\Omega)$ satisfying $\operatorname{div}U = 0$, $U \cdot n = 0$ on $\partial\Omega$. The scalar potential ϕ is unique up to additive constants (See [9, Corollary 5, page 258]). Hence, curl$P \times n$ = curl$U \times n = 0$. The field U may be decomposed on $\Gamma = \partial\Omega$ as follows: (see [10, page 75]) $U = (U\cdot n)n + U\Gamma$, $U\Gamma = n \times (U \times n)$. Thus, U satisfies the equivalent boundary condition on Γ (see [10, page 77]),

$$\frac{\partial U_\Gamma}{\partial n} + \mathcal{R}(U_\Gamma) = 0, \quad U \cdot n = 0, \tag{1.11}$$

where is the symmetric curvature operator acting in the tangent plane. In what follows, we are interested only in the regular part U of the polarization field P and assume that the potential ϕ is constant in Ω. Hence, we have P = U, then

$$\operatorname{div}P = 0 \quad \text{in } \Omega, \qquad P \cdot n = 0 \quad \text{on } \partial\Omega. \tag{1.12}$$

Next, we assume that the source term F satisfies in Ω the condition

$$\operatorname{div}F = 0 \tag{1.13}$$

By considering the equation satisfied by E in (1.9), we deduce the compatibility condition $\zeta_1(\omega)\operatorname{div}E - \omega^2 \operatorname{div}P = 0$ which implies div$E = 0$. Hence, under the divergence free condition for P, (1.9) shows that (E,P) satisfies in Ω the new problem

$$(\zeta_1(\omega) + \lambda^2 \operatorname{curl}^2)E - \omega^2 P = \imath\omega F,$$
$$(\zeta_2(\omega) - \lambda^2\Delta)P + \nabla\pi = -b(E_{eq}(P) - E),$$
$$\operatorname{div}P = 0, \tag{1.14}$$
$$\operatorname{curl}E \times n + \imath\omega\mu\beta n \times (E \times n) = 0,$$
$$\operatorname{curl}P \times n = 0, \quad P \cdot n = 0 \quad \text{on } \partial\Omega,$$

where π is the Lagrange multiplier associated with the constraint div$P = 0$ and where we used the relation $-\Delta = \operatorname{curl}^2 + \nabla\operatorname{div}$. Combining the condition div$P = 0$ with the compatibility condition div$E = 0$ and using the second equation of (1.14), we see that the equilibrium eelectric field should satisfy the condition $\operatorname{div}E_{eq}(P) = \varphi^{(2)}(|P|2)P \cdot \nabla(|P|^2)$. In the remainder of the paper, we assume that the ferroelectric domain is the cylinder $\Omega\varepsilon = \Omega T \times (0, \varepsilon)$ with thickness $\varepsilon > 0$ and the cross-section ΩT which is an open, bounded, and regular set of R^2. The generic point of $\Omega\varepsilon$ is denoted by $x = (x_T, x_3)$ where $x_T = (x_1, x_2) \in \Omega T$ and $0 \le x_3 \le \varepsilon$. We also assume that the function β appearing in the boundary condition satisfied by the electric field E depends on ε and is given by

$$\beta^\varepsilon(x_3) = \beta \quad \text{on } \partial\Omega_T \times (0,\varepsilon), \qquad \beta^\varepsilon(0) = \varepsilon\beta_0, \qquad \beta^\varepsilon(\varepsilon) = \varepsilon\beta_1 \quad \text{in } \Omega_T, \tag{1.15}$$

where β, β_k are positive constants. The boundary $\partial\Omega^\varepsilon$ writes as $(\Omega_T \times \{x_3 = 0\}) \cup (\Omega_T \times \{x_3 = \varepsilon\}) \cup (\partial\Omega_T \times (0, \varepsilon))$. We denote by $(E\varepsilon, P\varepsilon)$ the solutions satisfying (1.14) in Ω^ε. Let us set some notations. We define the norm of the complex Lebesgue space $L^2(\Omega\varepsilon)$ by setting $|F|_\varepsilon^2 = (1/\varepsilon)\int_0^\varepsilon \int_{\Omega_T} |F(x_T, x_3)|^2 dx_T dx_3$ and its scalar product by $(F;G)_\varepsilon = (1/\varepsilon)\int_0^\varepsilon \int_{\Omega_T} F \cdot G^* dx_T dx_3$ where G^* where $G*$ stands for the complex conjugate of G. We use the same notations for the Lebesgue space $L^2(\partial\Omega_T \times (0,\varepsilon))$. If $\Omega = \Omega T \times (0, 1)$, we write $|\cdot|$ for the norm of $L^2(\Omega)$ and $(\cdot;\cdot)$ for its scalar product. We denote by (u_1, u_2, u_3) the canonical basis of R^3.

The paper is organized as follows. In Section 2, we prove uniform bounds for the solution of the model equations (1.14). In Section 3, we then introduce a change of variable with respect to the vertical variable to transform the thin domain Ω^ε to the cylinder Ω with thickness 1. We deduce the uniform bounds for the scaled solutions satisfying the model equations (3.5). In Section 4, we pass to the limit in the weak formulation of (3.5) and deduce the reduced model. The last section is devoted to some remarks.

UNIFORM BOUNDS

Since Eeq is a Lipschitz perturbation of order 0 of the operator (curl2 ,curl2), then existence and uniqueness of the solution (Eε,Pε) of problem (1.14) can be obtained by introducing, in the Hilbert space = L2(Ωε) × L2(Ωε), the unbounded operator $\mathscr{A}$ with do-

ducing, in the Hilbert space $\mathscr{H} = L^2(\Omega^\varepsilon) \times L^2(\Omega^\varepsilon)$, the unbounded operator $\mathscr{A}$ with do-
main $D(\mathscr{A}) = \{(E,P) \in \mathscr{H},\ (\operatorname{curl}^2 E, \operatorname{curl}^2 P) \in \mathscr{H},\ \operatorname{div} P = 0 \text{ in } \Omega^\varepsilon,\ \operatorname{curl} E \times n + \imath\omega\mu\beta^\varepsilon n \times (E \times n) = 0,\ \operatorname{curl} P \times n = 0,\ P \cdot n = 0 \text{ on } \partial\Omega^\varepsilon\}$ with $\mathscr{A}(E,P) = (\operatorname{curl}^2 E, \operatorname{curl}^2 P)$ for $(E,P) \in D(\mathscr{A})$. Problem (1.14) writes as $(\mathscr{B}(\omega) + \lambda^2 \mathscr{A})(E,P) = \imath\omega\mathscr{F} + b\mathscr{G}(E,P)$ where $\mathscr{B}(\omega)$ is the diagonal block matrix with diagonal $\zeta_1(\omega)I$ and $\zeta_2(\omega)I$, $\mathscr{G}(E,P) = (0, E_{eq}(P) - E)$, and

$\mathscr{F} = (F,0)$. We use classical results (e.g., see [11–13]) to prove existence and uniqueness of the solution for ω > 0 fixed.

In order to obtain uniform estimates, we multiply the first equation of (1.14) by E_ε^* and the second one by P_ε^* and use the Green formula

$$-(\Delta P_\varepsilon; P_\varepsilon)_\varepsilon = |\nabla P_\varepsilon|_\varepsilon^2 + \int_{\Gamma^\varepsilon} \mathscr{R}(P_{\Gamma,\varepsilon}) \cdot P_{\Gamma,\varepsilon}^\star d\sigma. \tag{2.1}$$

We get (notice that $(\nabla\pi_\varepsilon, P_\varepsilon)_\varepsilon = 0$)

$$\zeta_1(\omega)|E_\varepsilon|_\varepsilon^2 + \lambda^2 |\operatorname{curl} E_\varepsilon|_\varepsilon^2 + \imath\omega\lambda^2\mu \int_{\partial\Omega^\varepsilon} \beta^\varepsilon |E_\varepsilon \times n|^2 d\sigma = \omega^2 (P_\varepsilon; E_\varepsilon)_\varepsilon + \imath\omega(F; E_\varepsilon)_\varepsilon,$$
$$\zeta_2(\omega)|P_\varepsilon|_\varepsilon^2 + \lambda^2 |\nabla P_\varepsilon|_\varepsilon^2 + \lambda^2 \int_{\partial\Omega^\varepsilon} \mathscr{R}(P_{\Gamma,\varepsilon}) \cdot P_{\Gamma,\varepsilon}^* d\sigma + b\int_{\Omega^\varepsilon} |P_\varepsilon|^2 \phi'(|P_\varepsilon|^2) dx = b(E_\varepsilon; P_\varepsilon)_\varepsilon. \tag{2.2}$$

The real parts of each equation write as

$$-\omega^2 |E_\varepsilon|_\varepsilon^2 + \lambda^2 |\operatorname{curl} E_\varepsilon|_\varepsilon^2 = \omega^2 \Re(P_\varepsilon; E_\varepsilon)_\varepsilon + \Re(\imath\omega(F; E_\varepsilon)_\varepsilon),$$
$$-\omega^2 |P_\varepsilon|_\varepsilon^2 + \lambda^2 |\nabla P_\varepsilon|_\varepsilon^2 + \int_{\partial\Omega^\varepsilon} \mathscr{R}(P_{\Gamma,\varepsilon}) \cdot P_{\Gamma,\varepsilon}^* d\sigma + b\int_{\Omega^\varepsilon} |P_\varepsilon|^2 \phi'(|P_\varepsilon|^2) dx = b\Re(E_\varepsilon; P_\varepsilon)_\varepsilon \tag{2.3}$$

and the imaginary parts give

$$a_1 |E_\varepsilon|_\varepsilon^2 + \omega\mu\lambda^2 \int_{\partial\Omega^\varepsilon} \beta^\varepsilon |E_\varepsilon \times n|^2 d\sigma = \omega^2 \Im(P_\varepsilon; E_\varepsilon)_\varepsilon + \Im(\imath\omega(F; E_\varepsilon)_\varepsilon),$$
$$\omega a_2 |P_\varepsilon|_\varepsilon^2 = b\Im(E_\varepsilon; P_\varepsilon)_\varepsilon. \tag{2.4}$$

Adding the last equalities and using the property $\Im(P_\varepsilon; E_\varepsilon)_\varepsilon + \Im(E_\varepsilon; P_\varepsilon)_\varepsilon = 0$, we, get

$$a_1 b |E_\varepsilon|_\varepsilon^2 + b\omega\mu\lambda^2 \int_{\partial\Omega^\varepsilon} \beta^\varepsilon |E_\varepsilon \times n|^2 d\sigma + a_2\omega^3 |P_\varepsilon|_\varepsilon^2 = \Im(\imath\omega b(F; E_\varepsilon)_\varepsilon). \tag{2.5}$$

Using the fact that is independent of ε, then there exists c > 0 which is independent of ε such that $|\int_{\Gamma^\varepsilon} \mathscr{R}(P_{\Gamma,\varepsilon}) \cdot P_{\Gamma,\varepsilon}^* d\sigma| \le c|P_{\Gamma,\varepsilon}|_\varepsilon^2 \le c(\eta|\nabla P_\varepsilon|_\varepsilon^2 + C_\eta |P_\varepsilon|_\varepsilon^2)$ for all $\eta > 0$. We obtain, for η small enough, the following result.

Lemma 2.1 There exists C > 0 which is independent of ε (depending on ω and F) such that

$$|E_\varepsilon|_\varepsilon + |\operatorname{curl} E_\varepsilon|_\varepsilon + \left|\sqrt{\beta^\varepsilon} E_\varepsilon \times n\right|_\varepsilon \le C,$$
$$|P_\varepsilon|_\varepsilon + |\nabla P_\varepsilon|_\varepsilon + |\pi^\varepsilon|_\varepsilon \le C. \tag{2.6}$$

Moreover,

$$|\Delta P_\varepsilon|_\varepsilon + |\operatorname{curl}^2 E_\varepsilon|_\varepsilon \le C. \tag{2.7}$$

THE SCALED PROBLEM AND CONVERGENCES

We introduce the change of variable $z = x_3/\varepsilon$ for $x_T \in \Omega_T$ fixed. We define the cylinder $\Omega = \Omega_T \times (0,1)$ with generic point (x_T, z). For a given vector-valued function $G(x_T,x_z)$ defined on Ω^ε we set $G^\varepsilon(x_T, z) = G(x_T, \varepsilon_z)$ which is defined in Ω. We write $\mathbf{G}^\varepsilon = (G_T^\varepsilon, g^\varepsilon)$ where $G^\varepsilon = (\mathbf{G}_1^\varepsilon, \mathbf{G}_2^\varepsilon)$ and $g^\varepsilon = \mathbf{G}_3^\varepsilon$. Denoting ∇_T the gradient with respect to the variable x_T we have $\nabla_T G_T = \nabla_T \mathbf{G}^\varepsilon$ and $\partial_{x_3} G = (1/\varepsilon)\partial_z \mathbf{G}^\varepsilon$ Let g be a scalar function and let $GT = (G_1,G_2)$ be a vector-valued function both defined in Ω. We set

$$\operatorname{curl}_T g = (\partial_2 g, -\partial_1 g), \qquad \Delta_T g = \partial_1^2 g + \partial_2^2 g,$$
$$\operatorname{Curl} G_T = \partial_1 G_2 - \partial_2 G_1, \qquad \operatorname{div}_T G_T = \partial_1 G_1 + \partial_2 G_2. \tag{3.1}$$

With the change of variable, we have $(1/\varepsilon)\int_0^\varepsilon |G(x_3)|^2 dx_3 = \int_0^1 |\mathbf{G}^\varepsilon(z)|^2 dz$ and the differential operators become $G = \operatorname{curl}_\varepsilon \mathbf{G}^\varepsilon$, $\operatorname{div} G = \operatorname{div}_\varepsilon \mathbf{G}^\varepsilon$, $\Delta G = \Delta_\varepsilon \mathbf{G}^\varepsilon$ with

$$\operatorname{curl}_\varepsilon \mathbf{G}^\varepsilon = -\frac{1}{\varepsilon}\partial_z(\mathbf{G}^\varepsilon \times \mathbf{u}_3) + \operatorname{curl}_T g^\varepsilon + (\operatorname{Curl}_T G_T^\varepsilon)\mathbf{u}_3,$$
$$\operatorname{div}_\varepsilon \mathbf{G}^\varepsilon = \operatorname{div}_T G_T^\varepsilon + \frac{1}{\varepsilon}\partial_z g^\varepsilon,$$
$$\Delta_\varepsilon \mathbf{G}^\varepsilon = \Delta_T G_T^\varepsilon + \frac{1}{\varepsilon^2}\partial_z^2 \mathbf{G}^\varepsilon, \qquad \nabla_\varepsilon g^\varepsilon = \left(\nabla_{x_T} g^\varepsilon, \frac{1}{\varepsilon}\partial_z g^\varepsilon\right). \tag{3.2}$$

We rewrite $\operatorname{curl}_\varepsilon \mathbf{G}^\varepsilon$ as follows:

$$\operatorname{curl}_\varepsilon \mathbf{G}^\varepsilon = (\theta^\varepsilon, \operatorname{Curl}_T G_T^\varepsilon), \quad \theta^\varepsilon = \left(\partial_2 g^\varepsilon - \frac{1}{\varepsilon}\partial_z G_2^\varepsilon, \frac{1}{\varepsilon}\partial_z G_1^\varepsilon - \partial_1 g^\varepsilon\right). \tag{3.3}$$

Notice that $\theta^\varepsilon \cdot u_3 = 0$ a.e. Here we have identified the 2D vectors θ^ε and $\operatorname{curl}T\, g^\varepsilon$ with the vectors $(\theta^\varepsilon, 0)$ of R^3 and $(\partial_2 g^\varepsilon, -\partial_1 g^\varepsilon, 0)$, respectively. This identification will be used throughout this manuscript.

Let $(E_\varepsilon, P_\varepsilon)$ be a solution to problem (1.14) associated with the source term F satisfying the hypothesis

$$F = (F_T(x_T), 0), \qquad \operatorname{div}_T F_T = 0. \tag{3.4}$$

Using the previous notations, let $\mathbf{E}^\varepsilon = (E_T^\varepsilon, e^\varepsilon)$ and $\mathbf{P}^\varepsilon = (P_T^\varepsilon, p^\varepsilon)$ be the scaled solution to (1.14) and let Π^ε be the scaled function associated with π^ε. Then $(E^\varepsilon, P^\varepsilon)$ satisfies in Ω the system of equations

$$(\zeta_1(\omega) + \lambda^2 \operatorname{curl}_\varepsilon^2)\mathbf{E}^\varepsilon = \omega^2 \mathbf{P}^\varepsilon + \imath\omega \mathbf{F}(x_T),$$
$$(\zeta_2(\omega) - \lambda^2 \Delta_\varepsilon)\mathbf{P}^\varepsilon + \nabla_\varepsilon \Pi^\varepsilon = -b(E_{eq}(\mathbf{P}^\varepsilon) - \mathbf{E}^\varepsilon),$$
$$\operatorname{div}_\varepsilon \mathbf{P}^\varepsilon = 0, \tag{3.5}$$

$$\operatorname{curl}_\varepsilon \mathbf{E}^\varepsilon \times n + \imath\omega\mu\beta^\varepsilon n \times (\mathbf{E}^\varepsilon \times n) = 0 \quad \text{on } \partial\Omega,$$
$$\operatorname{curl}_\varepsilon \mathbf{P}^\varepsilon \times n = 0, \quad \mathbf{P}^\varepsilon \cdot n = 0 \quad \text{on } \partial\Omega,$$

where $n = (nT, n3)$ is the unit outward normal to Ω. We have $n = u_3$ for $z = 1$, $n = -u_3$ for $z = 0$, and $n = nT = (n_1,n_2)$ on $\partial\Omega_T$ for $0 \leq z \leq 1$.

Let θ^ε be the 2D vector appearing in the definition of $\operatorname{curl}_\varepsilon E^\varepsilon$. We have

$$\theta^\varepsilon = \left(\partial_2 e^\varepsilon - \frac{1}{\varepsilon}\partial_z E_2^\varepsilon, \frac{1}{\varepsilon}\partial_z E_1^\varepsilon - \partial_1 e^\varepsilon\right). \tag{3.6}$$

The boundary conditions satisfied by $(E^\varepsilon, P^\varepsilon)$ are rewritten as follows. On $z = 0$ and $z = 1$, we have

$$(\theta^\varepsilon \times \mathbf{u}_3)(x_T, 1) = -\imath\omega\mu\beta_1\varepsilon E^\varepsilon(x_T, 1), \qquad (\theta^\varepsilon \times \mathbf{u}_3)(x_T, 1) = \imath\omega\mu\beta_0\varepsilon E^\varepsilon(x_T, 0),$$
$$p^\varepsilon(x_T, 1) = p^\varepsilon(x_T, 0) = 0,$$
$$\partial_z P_T^\varepsilon(x_T, 1) + \varepsilon\mathcal{R}(P_T^\varepsilon(x_T, 1)) = -\partial_z P_T^\varepsilon(x_T, 0) + \varepsilon\mathcal{R}(P_T^\varepsilon(x_T, 0)) = 0, \tag{3.7}$$

and on $\partial\Omega_T \times (0,1)$, we have

$$\operatorname{Curl}_T E_T^\varepsilon = \imath\omega\beta E^\varepsilon \times n_T, \qquad \operatorname{Curl}_T P_T^\varepsilon = 0,$$
$$\frac{\partial \mathbf{P}_\Gamma^\varepsilon}{\partial n_T} + \mathcal{R}(\mathbf{P}_\Gamma^\varepsilon) = 0, \qquad P_T^\varepsilon \cdot n_T = 0, \tag{3.8}$$

where $E_T^\varepsilon \times n_T = E_1^\varepsilon n_2 - E_2^\varepsilon n_1$. Recall that $P_T^\varepsilon = n_T \times (P^\varepsilon \times n_T)$. Applying the uniform bounds of Lemma 2.1 to the scaled solution (E^ε,P^ε) and using (3.7), we get the following

Lemma 3.1: There exists C > 0 which is independent of ε such that

$$|\mathbf{E}^\varepsilon| + |\mathbf{P}^\varepsilon| + \left|\Pi^\varepsilon\right| \le C,$$
$$|\theta^\varepsilon| + |\operatorname{Curl}_T E_T^\varepsilon| + |\nabla_T \mathbf{P}^\varepsilon| + \frac{1}{\varepsilon}|\partial_z \mathbf{P}^\varepsilon| \le C,$$
$$|\operatorname{curl}_\varepsilon^2 \mathbf{E}^\varepsilon| + |\Delta_\varepsilon \mathbf{P}^\varepsilon| \le C. \tag{3.9}$$

Moreover, the traces of the solution satisfy the estimates

$$|E^\varepsilon_{T|z=k}| \le C, \quad |\theta^\varepsilon_{|z=k}| \le C\varepsilon, \quad \textit{for } k = 1,2,$$
$$|\mathbf{E}^\varepsilon \times n_{|\partial\Omega_T\times(0,1)}| \le C. \tag{3.10}$$

We will prove the following general result which is useful in the sequel.

Proposition 3.2: Let $\mathbf{U}^\varepsilon = (U_T^\varepsilon, u^\varepsilon)$ be a bounded sequence of $L^2(\Omega)$ such that $\operatorname{curl}_\varepsilon U^\varepsilon = (\theta^\varepsilon, \operatorname{Curl}_T U_T^\varepsilon)$ is bounded in $L^2(\Omega)$ and assume that the tangential trace $U^\varepsilon \times n$ is uniformly bounded in $L^2(\partial\Omega)$. Then, there exists a subsequence, still denoted, U^ε such that $U^\varepsilon = (U_T^\varepsilon, u^\varepsilon) \rightharpoonup \mathbf{U} = (U_T, u)$ weakly in $L^2(\Omega)$, $\operatorname{Curl}_T U_T^\varepsilon \rightharpoonup \operatorname{Curl}_T U_T$ weakly in $L^2(\Omega)$. Moreover, U_T is independent of z and

$$(U_T^\varepsilon \times \mathbf{u}_3)_{|z=k} \rightharpoonup U_T \times \mathbf{u}_3 \quad \textit{weakly in } L^2(\Omega_T) \textit{ for } k = 0,1,$$
$$\int_0^1 U_T^\varepsilon \times n_T dz \rightharpoonup U_T \times n_T \quad \textit{weakly in } L^2(\partial\Omega_T). \tag{3.11}$$

Proof: Let $U = (U_T, u)$ be the weak limit in $L^2(\Omega)$ of a subsequence of U^ε. Let $\varphi \in \mathcal{D}(\overline{\Omega})$ be a test function. The Green formula gives

$$\int_\Omega \operatorname{curl}_\varepsilon \mathbf{U}^\varepsilon \cdot \varphi dx = \int_\Omega \mathbf{U}^\varepsilon \cdot \operatorname{curl}_\varepsilon \varphi dx - \int_{\partial\Omega_T\times(0,1)} \mathbf{U}^\varepsilon \times n \cdot \varphi d\sigma$$
$$-\frac{1}{\varepsilon}\int_{\Omega_T} (\mathbf{U}^\varepsilon \times \mathbf{u}_3)_{|z=1} \cdot \varphi_{|z=1} dx_T + \frac{1}{\varepsilon}\int_{\Omega_T} (\mathbf{U}^\varepsilon \times \mathbf{u}_3)_{|z=0} \cdot \varphi_{|z=0} dx_T. \tag{3.12}$$

Firstly, we choose in the Green formula $\varphi = \varepsilon\phi$ with $\phi = (\phi_1,\phi_2,0) = (\phi_T,0) \in \mathcal{D}(\Omega)$. Since $\operatorname{curl}_\varepsilon \varphi = -\partial_z(\phi \times \mathbf{u}_3) + \varepsilon(\operatorname{Curl}_T \phi_T)\mathbf{u}_3$, then passing to the limit in (3.12), we get

$$\int_\Omega -\mathrm{U}_1\partial_z\phi_2 + \mathrm{U}_2\partial_z\phi_1 dx = 0 \tag{3.13}$$

which implies that $\partial_z U_T = 0$ in the sense of distributions so, U_T is independent of the variable z. Next, let A_j be the weak limit in $L^2(\Omega_T)$ of a subsequence of the traces $(U^\varepsilon \times u^3)_{|z=j}$ for j = 0,1. To identify A_1, we choose in the Green formula $\phi = \varepsilon z\varphi$ with $\varphi = (\phi_1(x_T),\phi_2(x_T),0) \in (\mathcal{D}(\Omega_T))^3$. Passing to the limit in (3.12), we get

$$\int_{\Omega_T} -\mathrm{U}_1\phi_2 + \mathrm{U}_2\phi_1 dx - \int_{\Omega_T} A_1 \cdot \phi dx_T = 0 \tag{3.14}$$

which shows that $A_1 = U_T \times u_3$. Secondly, we use the test function $\phi = \varepsilon(1 - z)\varphi$ with $\phi = (\phi_1(x_T),\phi_2(x_T),0) \in (\mathcal{D}(\Omega_T))^2$ in the Green formula (3.12) and pass to the limit, we get

$$\int_{\Omega_T} \mathrm{U}_1\phi_2 - \mathrm{U}_2\phi_1 dx + \int_{\Omega_T} A_0 \cdot \phi dx_T = 0. \tag{3.15}$$

Thus, we get $A_0 = U_T \times u_3$ and $A_0 = A_1$. Finally, let g be the weak limit in $L^2(\partial\Omega_T \times (0,1))$ of a subsequence of the traces $U^\varepsilon \times n_{T|\partial\Omega_T\times(0,1)}$. To characterize g, we consider the test function $\varphi = (0,0,\phi_3(x_T))$ with $\phi_3 \in \mathcal{D}(\overline{\Omega_T})$. Observing that $\operatorname{curl}_\varepsilon \varphi = \operatorname{curl}_T \phi_3 = (\partial_2\phi_3, -\partial_1\phi_3, 0)$ and passing to the limit in (3.12), since U is independent of the variable z, we deduce that

$$\int_{\Omega_T} \phi_3 \operatorname{Curl}_T \mathrm{U}_T dx_T = \int_{\Omega_T} \mathrm{U}_T \cdot \operatorname{curl}_T \phi_3 dx_T - \int_{\partial\Omega_T} \left(\int_0^1 g dz\right)\phi_3 dx_T. \tag{3.16}$$

Now, since $\operatorname{Curl}_T \mathrm{U}_T \in L^2(\Omega_T)$, then $\mathrm{U}_T \times n_T$ is well defined, we finally deduce that $\int_0^1 g dz = \mathrm{U}_T \times n_T$. Hence, Proposition 3.2 is proved.

Applying Proposition 3.2 to the fields $\mathbf{E}^\varepsilon = (\mathrm{E}_T^\varepsilon, e^\varepsilon)$ and $\mathrm{curl}_\varepsilon \mathbf{E}^\varepsilon = (\theta^\varepsilon, \mathrm{Curl}_T \mathrm{E}_T^\varepsilon)$, we get the following.

Lemma 3.3: There exist subsequences, still denoted, E^ε and θ^ε such that the following weak convergences in $L^2(\Omega)$ hold:

$$E_T^\varepsilon \rightharpoonup E_T, \quad e^\varepsilon \rightharpoonup e, \quad \mathrm{Curl}_T E_T^\varepsilon \rightharpoonup \mathrm{Curl}_T E_T, \quad \theta^\varepsilon \rightharpoonup 0, \tag{3.17}$$

and ET is independent of z. Moreover, the traces satisfy the convergences

$$E_{T|z=k}^\varepsilon \rightharpoonup E_T \text{ weak}, \qquad \theta_{|z=k}^\varepsilon \longrightarrow 0 \text{ strong},$$

$$\int_0^1 E_T^\varepsilon \times n_T dz \rightharpoonup E_T \times n_T \tag{3.18}$$

in $L^2(\Omega_T)$ for k = 1,2 and in $L^2(\partial\Omega_T)$, respectively.

Proof: Lemma 3.1 implies the strong convergence of $\theta^\varepsilon{}_{|z=j}$ to 0. Next, set $U^\varepsilon = \mathrm{curl}_\varepsilon E^\varepsilon$. As $\mathbf{U}^\varepsilon = (\theta^\varepsilon, \mathrm{Curl}_T \mathrm{E}_T^\varepsilon)$ satisfies the conditions of the previous proposition, then $\theta^\varepsilon \rightharpoonup \theta$ weakly in $L^2(\Omega)$ and θ is independent of the variable z. Using again Proposition 3.2, we get $(\theta^\varepsilon \times \mathbf{u}_3)_{|z=k} \rightharpoonup \theta \times \mathbf{u}_3$ weakly in $L^2(\Omega_T)$ for $k = 1,2$. Since $(\theta^\varepsilon \times \mathbf{u}_3)_{|z=k} \to 0$ strongly, then $\theta \equiv 0$ in Ω.

Now, we consider the convergences for P_ε. We have the following.

Lemma 3.4: There exists a subsequence, still denoted by P_ε such that

$$\mathbf{P}^\varepsilon \longrightarrow \mathbf{P} = (P_T, 0) \text{ strong}, \quad \nabla_T \mathbf{P}^\varepsilon \rightharpoonup \nabla_T(P_T, 0), \quad \Pi^\varepsilon \rightharpoonup \Pi \text{ weak},$$

$$\partial_z \mathbf{P}^\varepsilon \longrightarrow 0, \quad E_{\mathrm{eq}}(\mathbf{P}^\varepsilon) \longrightarrow E_{\mathrm{eq}}(P_T) = 2P_T\phi'(|P_T|^2) \text{ strong} \tag{3.19}$$

in $L^2(\Omega)$. Moreover, PT and Π are independent of the variable z and e = 0. Finally, P_T satisfies on $\partial\Omega_T$ the boundary condition $P_T \cdot n_T = 0$.

Proof: Using the bounds in $L^2(\Omega)$ of $\nabla_\varepsilon \mathbf{P}^\varepsilon$ and $\mathbf{P}^\varepsilon$, we deduce that there exists a subsequence such that $\mathbf{P}^\varepsilon \rightharpoonup \mathbf{P} = (P_T, p)$ weakly in $H^1(\Omega)$. Moreover, $\partial_z \mathbf{P}^\varepsilon \to 0$ strongly in $L^2(\Omega)$. It follows that P is independent of z. Furthermore, the pressure Π^ε converges weakly to Π in $L^2(\Omega)$. Next, the trace of P^ε on $\partial\Omega$ converges weakly in $H^{1/2}(\partial\Omega)$ to the trace of P. Since we have $p^\varepsilon(x_T, 1) = p^\varepsilon(x_T, 0) = 0$ and p is independent of z, then p = 0. The trace $\mathbf{P}^\varepsilon \times n \rightharpoonup \mathbf{P} \times n$ weakly in $L^2(\partial\Omega_T \times (0,1))$. We may pass to the limit in the boundary condition to get $\mathbf{P} \cdot n = 0$ on $\partial\Omega$ which gives $P_T \cdot n_T = 0$. We rewrite the Neumann boundary condition in its original form $\mathrm{curl}_\varepsilon P^\varepsilon \times n = 0$. We write $(\theta_1^\varepsilon, \mathrm{Curl}_T \mathrm{P}_T^\varepsilon)$, where θ_1^ε is defined as in Section 3, then the boundary condition becomes

$$\theta_1^\varepsilon(x_T, 1) = \theta_1^\varepsilon(x_T, 0) = 0,$$

$$\theta_1^\varepsilon \times n_T = 0, \quad \mathrm{Curl}_T \mathrm{P}_T^\varepsilon = 0, \quad \text{on } \partial\Omega_T. \tag{1.20}$$

We apply Proposition 3.2 to $\mathbf{U}^\varepsilon = \mathrm{curl}_\varepsilon \mathbf{P}^\varepsilon$. Since we have $\Delta_\varepsilon \mathbf{P}^\varepsilon = \mathrm{curl}_\varepsilon(\mathrm{curl}_\varepsilon \mathbf{P}^\varepsilon)$ because we have $\mathrm{div}_\varepsilon \mathrm{P}^\varepsilon = 0$, then by Lemma 3.1, it follows that $\mathrm{curl}_\varepsilon \mathbf{U}^\varepsilon$ is bounded in $L^2(\Omega)$. Applying Proposition 3.2, we deduce that θ_1, the weak limit of θ_1^ε, is independent of z where θ_1 is the weak limit of θ_1^ε Finally, using the boundary condition satisfied by θ_1^ε at $z = 0$ and $z = 1$, we deduce that $\theta_1 = 0$. Next, we use the bound in $L^2(\Omega)$ of $\nabla_\varepsilon \mathrm{P}_T^\varepsilon$ T to deduce that $\mathrm{Curl}_T \mathrm{P}_T^\varepsilon \rightharpoonup \mathrm{Curl}_T P_T$ and $\mathrm{P}_T^\varepsilon \cdot n_T \rightharpoonup P_T \cdot n_T$ weakly in $H^{1/2}(\partial\Omega)$. To end the proof of the lemma, we will prove that Π is independent of z and $e = 0$. We set $\sigma^\varepsilon = \partial_z((1/\varepsilon)p^\varepsilon)$. The condition $\mathrm{div}_\varepsilon \mathbf{P}^\varepsilon = 0$ is rewritten as $\mathrm{div}_T \mathrm{P}_T^\varepsilon + \sigma^\varepsilon = 0$ and from the equation satisfied by P^ε we deduce that $-\lambda^2\partial_z\sigma^\varepsilon + \partial_z\Pi^\varepsilon = \varepsilon R^\varepsilon$ where the remainder term R^ε is bounded in $L^2(\Omega)$. Since σ^ε is bounded in $L^2(\Omega)$, then passing to the limit we get div $P_T + \sigma = 0$ and $-\lambda^2\partial_z\sigma + \partial_z\Pi = 0$. Since PT is independent of z, then so is σ which implies that $\partial_z\Pi = 0$. Let us consider the equation satisfied by E^ε. We multiply the equation by the test function $\varphi = (0,0,\phi)$ with $\phi \in \mathcal{D}(\Omega)$. Since we have $\mathrm{curl}_\varepsilon \varphi = (\partial_2\phi, -\partial_1\phi, 0)$, then $\mathrm{curl}_\varepsilon \mathbf{E}^\varepsilon \cdot \mathrm{curl}_\varepsilon \varphi = \theta^\varepsilon \cdot \mathrm{curl}_T \phi$, then we get after an integration by parts (recall that the third component of F is 0)

$$\int_\Omega (\zeta_1(\omega)e^\varepsilon - \omega^2 p^\varepsilon)\phi\, dx + \lambda^2 \int_\Omega \theta^\varepsilon \cdot \mathrm{curl}_T \phi\, dx = 0. \tag{3.21}$$

Passing to the limit, we obtain $\zeta_1(\omega)e - \omega^2 p = 0$. Using that $p = 0$, we get $e = 0$.

THE REDUCED PROBLEM

Let us introduce the Hilbert space $H(\mathrm{Curl}_T, \Omega_T) = \{U \in L^2(\Omega_T)^2, \mathrm{Curl}_T U \in L^2(\Omega_T)\}$. We will prove the following main result describing the dimensional reduction of the thin ferroelectric cylinder.

Theorem 4.1: *Let $F \in (L^2(\Omega_T))^2$ be such that $\mathrm{div}_T F = 0$ in Ω_T. Then for $\omega > 0$* fixed, there exists a unique solution $(E,P) \in H(\mathrm{Curl}_T, \Omega_T) \times H(\mathrm{Curl}_T, \Omega_T)$ of the reduced problem

$$\zeta_1(\omega)E + \lambda^2 \mathrm{curl}_T(\mathrm{Curl}_T E) + \iota\omega\mu(\beta_1 + \beta_0)E - \omega^2 P = \iota\omega F,$$

$$\mathrm{Curl}_T E + \iota\omega\mu\beta E \times n_T = 0, \quad \text{on } \partial\Omega_T,$$

$$\zeta_2(\omega)P - \lambda^2\Delta P + \nabla_T\Pi = -b(E_{\mathrm{eq}}(P) - E),$$

$$\mathrm{div}_T P = 0,$$

$$\mathrm{curl}_T P = 0, \quad P \cdot n_T = 0, \quad \text{on } \partial\Omega_T. \tag{4.1}$$

Furthermore, $\mathrm{Curl}_T E, \mathrm{Curl}_T P \in H^1(\Omega_T)$ and the solution is obtained as the limit of the sequence $(E^\varepsilon, P^\varepsilon)$ of the model problem (3.5).

Proof: To prove this theorem, we pass to the limit in the weak formulation of equation (3.5). Since the limit solution (E,P) is independent of z and e = p = 0, we choose test functions of the type $\varphi = (\varphi_1, \varphi_2, 0)$, $\psi = (\psi_1, \psi_2, 0)$) and a scalar function φ which are independent of the variable z. We suppose that $\phi, \varphi, \psi \in \mathcal{D}(\overline{\Omega_T})$ with $\mathrm{div}_T \psi = 0$ and $\psi\, n_T = 0$ on $\partial\Omega$T. Multiplying the first equation by ϕ^*, the second by ψ^*, and the constraint $\mathrm{div}_\varepsilon P^\varepsilon = 0$ by φ^* then integrating by parts we get

$$\int_\Omega (\zeta_1(\omega)\mathrm{E}_T^\varepsilon - \omega^2 \mathrm{P}_T^\varepsilon)\cdot\varphi^* dx + \lambda^2\int_\Omega \mathrm{Curl}_T\,\mathrm{E}_T^\varepsilon\cdot\mathrm{Curl}_T\,\varphi^* dx$$

$$+\imath\mu\omega\beta\int_{\partial\Omega_T\times(0,1)}(\mathrm{E}_T^\varepsilon\times n_T)\cdot(\varphi^*\times n_T)d\sigma + \imath\mu\omega\beta_0\int_{\Omega_T}(\mathrm{E}_T^\varepsilon\times\mathbf{u}_3)\cdot(\varphi^*\times\mathbf{u}_3)(x_T,0)dx_T$$

$$+\imath\mu\omega\beta_1\int_{\Omega_T}(\mathrm{E}_T^\varepsilon\times\mathbf{u}_3)\cdot(\varphi^*\times\mathbf{u}_3)(x_T,1)dx_T = \imath\omega\int_\Omega \mathrm{F}\cdot\varphi^* dx,$$

$$\int_\Omega(\zeta_2(\omega)\mathrm{P}_T^\varepsilon + b(\mathrm{E}_{eq}(\mathrm{P}^\varepsilon)-\mathrm{E}_T^\varepsilon))\cdot\psi^* dx + \lambda^2\int_\Omega \mathrm{Curl}_T\,\mathrm{P}_T^\varepsilon\cdot\mathrm{Curl}_T\,\psi^* dx = 0,$$

$$\int_\Omega \mathrm{P}_T^\varepsilon\cdot\nabla_T\phi^* dx = 0, \tag{4.2}$$

where we set $E_{eq}(\mathbf{P}^\varepsilon) = (\mathrm{E}_{eq}(\mathbf{P}^\varepsilon), e_{eq}(\mathbf{P}^\varepsilon))$. We used $-\Delta_\varepsilon\mathbf{P}^\varepsilon = \mathrm{curl}_\varepsilon^2\,\mathbf{P}^\varepsilon, \mathrm{div}_\varepsilon\,\mathbf{P}^\varepsilon = 0, \mathrm{curl}_\varepsilon\,\mathbf{E}^\varepsilon =$ $(\theta^\varepsilon, \mathrm{Curl}_T\,\mathrm{E}_T^\varepsilon)$, $\mathrm{curl}_\varepsilon\,\varphi = (0,0,\mathrm{Curl}_T\,\varphi)$, and the same properties for P^ε.

Applying our convergence results proved in Section 3 and passing to the limit in the weak formulation, we obtain

$$\int_{\Omega_T}(\zeta_1(\omega)\mathrm{E} - \omega^2\mathrm{P})\cdot\varphi^* dx_T + \lambda^2\int_{\Omega_T}\mathrm{Curl}_T\,\mathrm{E}\cdot\mathrm{Curl}_T\,\varphi^* dx_T$$

$$+\imath\omega\mu(\beta_1+\beta_0)\int_{\Omega_T}(\mathrm{E}\times\mathbf{u}_3)\cdot(\varphi^*\times\mathbf{u}_3)dx_T + \imath\omega\beta\mu\int_{\partial\Omega_T}(\mathrm{E}\times n_T)\cdot(\varphi^*\times n_T)d\sigma$$

$$= \imath\omega\int_{\Omega_T}\mathrm{F}\cdot\varphi^* dx_T,$$

$$\int_{\Omega_T}(\zeta_2(\omega)\mathrm{P} + b(\mathrm{E}_{eq}(\mathrm{P})-\mathrm{E}))\cdot\psi^* dx_T + \lambda^2\int_{\Omega_T}\mathrm{Curl}_T\,\mathrm{P}\cdot\mathrm{Curl}_T\,\psi^* dx_T = 0,$$

$$\int_{\Omega_T}\mathrm{P}\cdot\nabla_T\phi^* dx_T = 0. \tag{4.3}$$

Observe that the condition $\mathrm{div}_T\,\mathrm{P} = 0$ shows that $\mathrm{curl}_T(\mathrm{Curl}_T\,\mathrm{P}) = -\Delta\mathrm{P}$. Our main result is then proved.

CONCLUDING REMARKS

Let us conclude this work by the following remarks. If we impose that the regular part of the polarization P is 0, then $\mathrm{P} = \nabla\phi$ with $\phi = c$ on $\partial\Omega$. The equation satisfied by the polarization field writes in Ω as

$$(\zeta_2(\omega) + 2b\phi'(|\nabla\varphi|^2))\nabla\varphi = bE \tag{5.1}$$

while the electric field E satisfies the Maxwell equation

$$\zeta_1(\omega)E + \lambda^2\,\mathrm{curl}^2\,E - \omega^2\nabla\varphi = \imath\omega F. \tag{5.2}$$

We set for $X\in\mathbb{C}$, $a(|X|^2) = 2b\phi'(|X|^2) - \omega^2 + \imath\omega a_2$. We will show that the map

$$X\in\mathbb{C}\longmapsto H(X) = a(|X|^2)X\in\mathbb{C} \tag{5.3}$$

is onto. The equation $a(|X|^2)X = Y$ gives $|a(r)|^2 r = t$ with $r = |X|^2$ and $t = |Y|^2$ or equivalently as $((2b\phi'(r)-\omega^2)^2 + \omega^2 a_2^2)r = t$. Let

$$\theta(r) = |a(r)|^2 r = (2b\phi'(r)-\omega^2)^2 r + \omega^2 a_2^2 r, \tag{5.4}$$

then we have $\theta'(r) = (2b\phi'(r)-\omega^2)^2 + a_2^2\omega^2 + 4br\phi^{(2)}(r)(2b\phi'(r)-\omega^2)$. It follows that $\theta'(r) = (2b\phi'(r) - \omega^2 + 2br\phi^{(2)}(r))^2 + a_2^2 - 4b^2r^2(\phi^{(2)}(r))^2$. Assuming that $0\le r\phi^{(2)}(r)\le a_2\omega/(2b)$ for all $r\ge 0$, we get $\theta'(r) > 0$ for all $r\ge 0$. Consequently, θ is invertible and for all $t\ge 0$ there exists a unique $r\ge 0$ given by $r = \theta^{-1}(t)$. Hence, for all $Y\in C^3$, the equation $\theta(|X|^2) = |Y|^2$ gives $|X|^2 = \theta^{-1}(|Y|^2)$. Finally, for given $Y\in C^3$, the equation $a(|X|^2)X = Y$ admits a unique solution $X\in C^3$ given by $X = Y/a(\theta^{-1}(|Y^2))$.

Now, let $E\in L^2(\Omega)$, then there exists a unique $U\in L^2(\Omega)$ solution to the equation $a(|U|^2)U = bE$ which is given by

$$U = \frac{bE}{a(\theta^{-1}(b^2|E|^2))}. \tag{5.5}$$

Finally, E should satisfy the nonlinear Maxwell equation

$$\zeta_1(\omega)E+\lambda^2\text{curl}^2 E=\omega^2 b\sigma(|E|^2)E+\imath\omega F,$$
$$\text{curl}(\sigma(|E|^2)E)=0,$$
$$E\times n=0, \tag{5.6}$$

with $\sigma(|E|^2) = 1/a(\theta^{-1}(b^2|E|^2))$. The condition $\text{curl}(\sigma(|E|^2)E) = 0$ allows to show that $\sigma(|E|^2)E$ is a gradient. We will come back to this problem in a forthcoming work. In [5, Section 3], and [6, Section 2.2], the authors introduce the following model equations to describe the dynamic of the time-dependant spontaneous polarization p in a ferroelectric domain Ω (here we use the Dav´ı notations):

$$\rho_m\partial_t^2\text{p}+(\text{D}+\text{G})\partial_t\text{p}-\sigma^2\Delta\text{p}=\frac{\partial W}{\partial \text{p}}(\text{p})+\frac{\partial\varphi}{\partial \text{p}}(\mathbf{F},\text{e},\text{p})-\rho\text{e},\quad \text{in }\Omega\times(0,T),$$
$$\sigma^2\frac{\partial \text{p}}{\partial n}=\text{t},\quad \text{on }\partial\Omega\times(0,T),$$
$$\text{p}(0)=\text{p}_0,\qquad \rho_m\partial_t\text{p}(0)=\text{p}_1. \tag{5.7}$$

The electric field e with p satisfies in $R^3 \times (0,T)$ the electrostatic equations

$$\text{div}(\rho\text{p}+\text{e})=0,\qquad \text{curl}\,\text{e}=0, \tag{5.8}$$

with the natural jump conditions across the boundary $\partial\Omega \times (0,T)$. The parameters appearing in the equation are defined in [5, 6]. It is important to notice that the system is coupled to some elasticity model describing the dynamic of the deformation F (e.g., when we assume that $F = I + \nabla u$ where u is the mechanical displacement, see [6, Section 3]). Next, the nonhomogeneous boundary condition satisfied by p takes into account the density of the electric dipoles. This model is more complete than the one introduced in [1]. If we consider rigid body, then both models are essentially the same. An interesting question is to study the full model satisfied by (e,p,u).

ACKNOWLEDGMENT

The authors appreciate very much the comments and remarks of the referee, his enlightenments about ferroelectric models, and for quoting references [5, 7].

REFERENCES

1. J. M. Greenberg, R. C. MacCamy, and C. V. Coffman, "On the long-time behavior of ferroelectric systems," Physica D: Nonlinear Phenomena, vol. 134, no. 3, pp. 362–383, 1999.
2. H. Ammari and K. Hamdache, "Global existence and regularity of solutions to a system of nonlinear Maxwell equations," Journal of Mathematical Analysis and Applications, vol. 286, no. 1, pp. 51–63, 2003.
3. M. Fabrizio and A. Morro, "Models of electromagnetic materials," Mathematical and Computer Modelling, vol. 34, no. 12-13, pp. 1431–1457, 2001.
4. K. Hamdache and I. Ngningone-Eya, "Existence and low frequency behaviour of time harmonic solutions to a model of ferroelectric materials," Applicable Analysis, vol. 84, no. 7, pp. 721–736, 2005.
5. F. Daví, "On domain switching in deformable ferroelectrics, seen as continua with microstructure," Zeitschrift für angewandte mathematik und Physik, vol. 52, no. 6, pp. 966–989, 2001.
6. F. Daví and P. M. Mariano, "Evolution of domain walls in ferroelectric solids," Journal of the Mechanics and Physics of Solids, vol. 49, no. 8, pp. 1701–1726, 2001.
7. P. Rosakis and Q. Jiang, "On the morphology of ferroelectric domains," International Journal of Engineering Science, vol. 33, no. 1, pp. 1–12, 1995.
8. N. Aïssa, "Regularity and asymptotic behavior of thin ferroelectric materials," submitted.
9. R. Dautray and J.-L. Lions, Analyse mathématique et calcul numérique pour les sciences et les techniques, vol. 2, Masson, Paris, France, 1987, chapter 9.
10. J.-C. Nédélec, Acoustic and Electromagnetic Equations, vol. 144 of Applied Mathematical Sciences, Springer, New York, NY, USA, 2001.
11. L. C. Evans, Partial Differential Equations, vol. 19 of Graduate Studies in Mathematics, American Mathematical Society, Providence, RI, USA, 1998.
12. J.-L. Lions, Quelques méthodes de résolution des problèmes aux limites non linéaires, Dunod, Paris, France, 1969.
13. L. Tartar, "On the characterization of traces of a Sobolev space used Maxwell's equations," in Proceedings of a Meeting, A. Y. Leroux, Ed., Bordeaux, France, November 1997.

CHAPTER 18

STATUS AND PERSPECTIVES OF MULTIFERROIC MAGNETOELECTRIC COMPOSITE MATERIALS AND APPLICATIONS

Haribabu Palneedi, Venkateswarlu Annapureddy, Shashank Priya and Jungho Ryu

Keywords: Multiferroic, magnetoelectric, ferromagnetic, ferroelectric, magnetostrictive, piezoelectric, composites, thin film, thick film, core/shell, Janus-type structure

INTRODUCTION

The multifunctional properties of multiferroics enable the design of novel electronic devices for various sensing, transduction and memory applications. Muliferroic materials are characterized by two or more ferroic orders, such as ferroelectric, ferromagnetic, or ferroelastic, and the interactions between these order parameters, as shown in Figure 1. In multiferroic magnetoelectric (ME) materials, coupling occurs between the magnetic and electric subsystems. This enables the control of dielectric polarization P by a magnetic field H (direct ME (DME) effect: $\Delta P = \alpha_H \Delta H$) and the manipulation of magnetization M by an electric field E (converse ME (CME) effect: $\mu_0 \Delta M = \alpha_E \Delta E$), with μ_0 denoting the vacuum permeability. The ME response is quantified in terms of the ME coupling coefficient (α_H or α_E), which represents the coupling efficiency between the electric and magnetic fields. This is considered as the figure of merit for the strength of ME coupling [1]. This article discusses the present status and recent progress in the development of ME materials and their applications in devices.

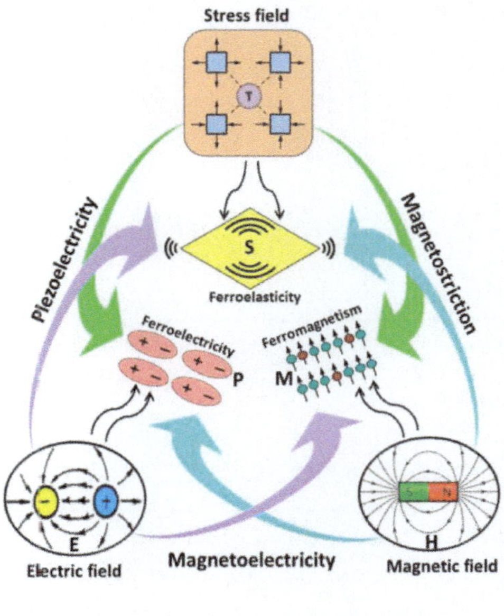

Figure 1. Schematic illustrating magnetic-elastic-electric couplings in multiferroic materials. Here, M is magnetization, S is mechanical strain, and P is dielectric polarization.

Single phase magnetoelectrics are chemically homogenous and isotropic compounds. They exhibit intrinsic ME coupling but require the co-existence of magnetic moments and electric dipoles with long-range ordering. From a fundamental point of view, the coupling between the magnetic and polar sublattices in single phase ME compounds is fascinating. However, due to the mutual exclusion of ferromagnetism and ferroelectricity, only few monolithic ME materials exhibiting non-zero coupling at room temperature have been found so far. Most of the single phase materials possess either low permittivity or low permeability at room temperature and thus exhibit weak ME coupling which hinders their applications. For example, $BiFeO_3$, the archetype of single phase ME compounds, displays

Citation: Haribabu Palneedi, Venkateswarlu Annapureddy, Shashank Priya and Jungho Ryu, Status and Perspectives of Multiferroic Magnetoelectric Composite Materials and Applications, doi:10.3390/act5010009.

good ferroelectricity but weak ferromagnetic properties above room temperature [2].

ME composites, the focus of this review, consist of physically separated magnetic and electric order phases. These composites show coupling with orders of magnitude larger than those found in single phase materials at room temperature [3]. Current understanding suggests that ME coupling in composites occurs extrinsically in three different ways mediated through (i) strain, (ii) charge carrier, and (iii) spin exchange. Among these mechanisms, the strain-mediated ME coupling has been widely studied, while the investigations on the other two mechanisms are still in early stages [4]. In this paper, the ME composites based on the strain-mediated coupling will be discussed in detail, and their applications in potential devices will be highlighted.

The strain-mediated ME effect in composites is a product tensor property and results from the elastic coupling between the piezoelectric and magnetostrictive components as illustrated in Figure 2 [5,6]. In DME coupling, the applied magnetic field generates strain in the magnetic layer via the magnetostriction effect, and this strain is transferred to the piezoelectric layer resulting in an electric displacement or a dielectric polarization through the piezoelectric effect. In CME coupling, an external electric field induces strain in the ferroelectric layer due to the inverse piezoelectric effect, and the strain transferred to the magnetic layer produces a magnetization change or domain reorientation by the piezomagnetic effect.

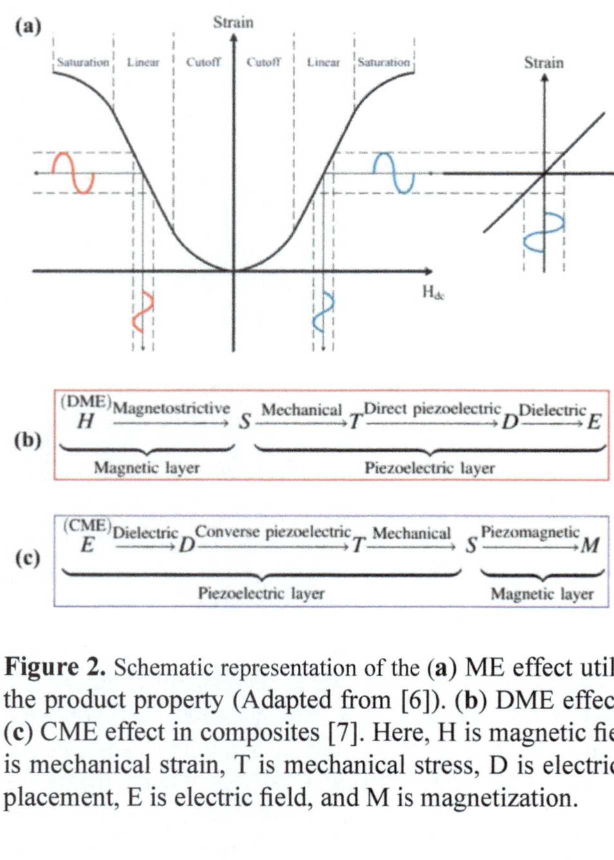

Figure 2. Schematic representation of the (**a**) ME effect utilizing the product property (Adapted from [6]). (**b**) DME effect and (**c**) CME effect in composites [7]. Here, H is magnetic field, S is mechanical strain, T is mechanical stress, D is electric displacement, E is electric field, and M is magnetization.

FACTORS AFFECTING ME COUPLING IN COMPOSITES

In recent years, ME composites have been extensively studied due to the ease of fabrication and design flexibility. ME composites have been developed with a diverse set of materials, properties, and microstructures using a variety of processing techniques. The parameters and variables that should be taken into account for developing high performance composites with strong ME coupling are summarized in this section.

Connectivity and Interface Bonding

The phase connectivity and interfacial bonding of the piezoelectric and magnetostrictive constituents have a strong influence on elastic coupling in the composite and its ME response. Good mechanical bonding at the interface between the composite phases, characterized by the interface coupling factor (k_c), facilitates efficient strain transfer. Though several connectivity schemes have been proposed for designing two-phase composites [8], ME composites have been commonly prepared with 0-3, 1-3, and 2-2 connectivity, as shown in Figure 3a, d, and g, respectively. Here, the numbers represent the connectivity of the magnetic and piezoelectric phases, respectively. In the 0-3 particle-matrix composites, magnetic particles are embedded in the piezoelectric matrix. 1-3 cylinder-matrix composites are formed by embedding magnetic fibers/rods/tubes/wires in the piezoelectric matrix. The particles (in the 0-3 composite) and fibers (in the 1-3 composite) can be either randomly dispersed or periodically aligned [9]. A 2-2 laminate composite consists of alternating magnetic and piezoelectric layers. Such laminates can be prepared in different shapes and geometries, including discs, squares, rectangles, and rings, with different dimensions. They can be arranged as unimorphs and bimorphs as well as bilayered and multilayered structures. Further, for all of the above composites, the volume fraction and dimensions of the constituents can be altered to tailor the properties of the composite. The 2-2 composites preserve the physical characteristics of individual phases, and they are comparatively simpler to fabricate. These composites can be poled to a higher degree since the piezoelectric and the low resistivity magnetic phases are separated. Results of both theoretical and experimental studies have shown that the 2-2 layered composites exhibit higher ME responses compared to the 0-3 and 1-3 composites. There have been other composite structures reported in the literature. Park*et al.* [10] reported the development of a 3-2 structured composite consisting of a $(Ni_{0.6}Cu_{0.2}Zn_{0.2})Fe_2O_4$ [NCZF] phase with 2D connectivity dispersed in a $0.8Pb(Zr_{0.52}Ti_{0.48})O_3$–$0.2Pb(Zn_{1/3}Nb_{2/3})O_3$ [PZNT] matrix. Recently, Gillette *et al.* [11] fabricated a quasi-one-dimensional ME composite by inserting a magnetostrictive wire (FeNi/FeGa/FeCoV) into a PZT tube where the tube-wire interface bonding was made with silver paste. Comparing the results from these prior studies, it can be stated that 2-2 composite structure has inherent advantages in terms of fabrication and performance.

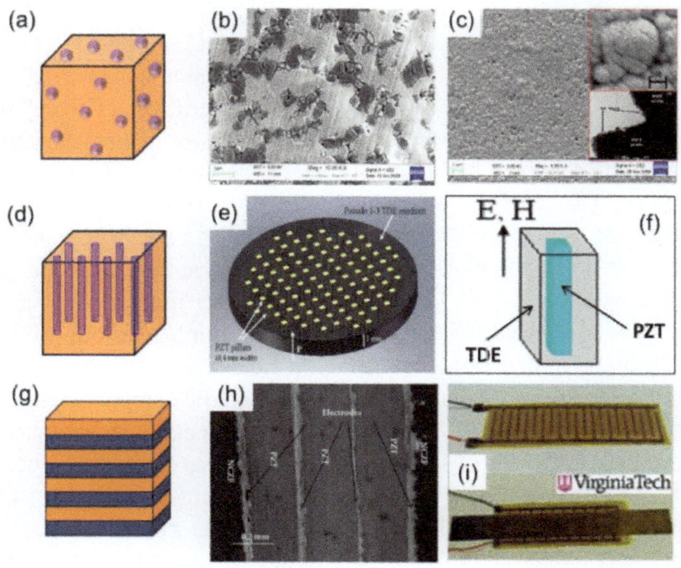

Figure 3: Development of bulk ME composites with different phase connectivity: **(a)–(c)** 0-3 connectivity [13,18]; **(d)–(f)** 1-3 connectivity [20,21]; and **(g)–(i)** 2-2 connectivity [22,24].

Materials and Their Properties

The selection of suitable materials is a primary step in fabricating a composite with good ME response. Various piezoelectric and magnetostrictive materials commonly used for synthesizing ME composites are listed in Table 1. Enhanced piezoelectric and ferroelectric properties can be achieved in ceramics through (a) composition selection (ideally near morphotropic phase boundary (MPB) or polymorphic phase transition (PPT)) and modification (by doping); and (b) microstructure design (via domain engineering and texturing) [12]. Among piezoelectric materials, PZT-based ceramics have been widely employed to fabricate the ME composites due to their low cost, high piezoelectric response, and flexibility in modifying the composition to achieve desired properties for targeted applications. For the magnetic component in ME composites, Terfenol-D with high magnetostriction and Metglas (amorphous Fe-alloy) with high magnetic permeability have been the most used materials. Besides magnetic properties, factors such as the processing temperatures, electrical resistance of the material, magnitude of the bias field, phase connectivity have been considered in choosing the magnetostrictive materials.

Table 1: List of well-known piezoelectric and magnetostrictive materials used as constituents of ME composites

Piezoelectric Phase	Magnetostrictive Phase
Lead-based:	**Metals:**
$Pb(Zr,Ti)O_3$ (PZT)	Fe, Co, Ni
$Pb(Mg_{1/3}Nb_{2/3})O_3$-$PbTiO_3$ (PMN-PT)	**Alloys:**
$Pb(Zn_{1/3}Nb_{2/3})O_3$-$PbTiO_3$ (PZN-PT)	FeNi-based
$Pb(Mg_{1/3}Nb_{2/3})_y(Zr_xTi_{1-x})_{1-y}O_3$ (PMN-PZT)	FeCo-based
$Pb(In_{1/2}Nb_{1/2})O_3$-$Pb(Mg_{1/3}Nb_{2/3})O_3$-$PbTiO_3$ (PIN-PMN-PT)	CoNi-based
Lead-free:	Ni_2MnGa
$BaTiO_3$ (BTO)-based	Permendur (FeCoV)
$(K_{0.5}Na_{0.5})NbO_3$ (KNN)-based	Galfenol (FeGa), FeGaB
$Na_{0.5}Bi_{0.5}TiO_3$ (NBT)-based	Samfenol ($SmFe_2$)
Others:	Terfenol-D ($Tb_{1-x}Dy_xFe_2$)
AlN	Fe-based metallic glasses (FeBSi, FeBSiC, FeCoB,
ZnO	FeCoSi, FeCoSiB, FeCuNbSiB)
(Sr, Ba)Nb_2O_5	**Ceramics:**
$Ba_{1-x}Sr_xTiO_3$ (BSTO)	Fe_3O_4
$Bi_{1-x}Sr_xTiO_3$ (BST)	$Zn_{0.1}Fe_{2.9}O_4$ (ZFO)
$La_3Ga_{5.5}SiO_{14}$ (LGS)	$La_xSr_yMnO_3$ (LSMO)
$La_3Ga_{5.5}Ta_{0.5}O_{14}$ (LGT)	$La_xCa_yMnO_3$ (LCMO)
Polyurethane (PU)	Ferrites or doped Ferrites (e.g., $NiFe_2O_4$ (NFO),
Polyvinylidene difluoride (PVDF)	$CoFe_2O_4$ (CFO), Li ferrite, Cu ferrite, Mn ferrite)

The material properties that are considered important when developing ME composites include electrical properties such as dielectric constant or permittivity (ε_r), dielectric loss (tan δ), Curie temperature (T_C), remnant polarization (P_r), coercive electric field (E_c), piezoelectric strain constant (d_{ij}), piezoelectric voltage constant (g_{ij}), electromechanical coupling factor (k_{ij}), mechanical quality factor (Q_m), and acoustic impedance (Z). The magnetic properties of relevance include magnetic permeability (μ), remnant magnetization (M_r), coercive magnetic field (H_c), magnetostriction (λ_{ij}), piezomagnetic coefficient (q_{ij}), and Curie and Neel temperatures (T_C and T_N). Generally, to obtain a better ME response from the composite, the combination of a piezoelectric material displaying a high piezoelectric voltage constant ($g_{ij} = d_{ij}/\varepsilon$) and low dielectric and piezoelectric losses, and a magnetostrictive material exhibiting a high magnetostriction coefficient (q_{ij}) and good interfacial coupling between these two phase components is desired.

Fabrication of ME Composites

Bulk ME Composites

A widely used method for synthesizing of 0-3 particulate composites consists of sintering a mixture of ceramic oxide powders at high temperature. This ceramic sintering process is simple and the properties of the composite can be conveniently tailored by selecting the constituent phases, their starting particle sizes, and processing parameters. Using this method, mainly magnetic ferrites have been co-processed with piezoelectric ceramics due to their high temperature stability. Figure 3b shows the microstructure of a $Pb(Zr_{0.56}Ti_{0.44})O_3$–$Ni_{0.6}Zn_{0.2}Cu_{0.2}Fe_2O_4$ sample sintered at 1000 °C [13]. At present, the measured ME coupling of 0-3 particulate composites are far lower than the theoretically predicted ones. This can be attributed to several factors: (a) the misfit strain at the interface arising from the thermal expansion mismatch between the piezoelectric and ferrite phases, which reduces the densification; (b) the interdiffusion and/or chemical reactions between the two phases during high-temperature sintering, which deteriorates the piezoelectricity and/or magnetostriction of constituent phases and the strain transfer between two phases; and (c) current leakage due to the low resistivity in the randomly distributed magnetic phase with low percolation threshold, which makes the electric poling of these ME composites quite difficult [2,3]. Some efforts have been made to solve these issues. Hot pressing and spark plasma sintering (SPS) have been utilized to achieve high density while avoiding possible reaction between the constituent phases [14,15]. A modified precipitation method involving a controlled annealing, quenching, and aging cycle [16] has been employed to obtain good dispersion of ferrite ($NiFe_{1.9}Mn_{0.1}O_4$) particles in the matrix (PZT) phase. Core/shell (ferrite/piezoceramic) structured composites have been synthesized to avoid the contact of the conductive ferrite particles during sintering [17]. Unlike the other cases, Islam *et al.* [18] reported a PZT core/ $NiFe_2O_4$ shell ME composite (Figure 3c). To circumvent the problem of high conductivity through the ferrite phase, a 100-mm layer of PZT was used to cover both faces of the sintered pellet.

Very few attempts have been made to fabricate 1-3 type bulk ME composites [19,20,21]. Shi *et al.*[19] prepared a pseudo-1-3 type ME composite consisting of a PZT rod array (with base) and Terfenol-D/epoxy matrix (TDE) using the dice-and-fill technique, where diced PZT sample was filled with an epoxy resin solution containing Terfenol-D particles (Figure 3e). A peak ME response of 6 V/cm·Oe was obtained between 80 and 110 kHz. Lam *et al.* [20] studied the frequency response of PZT-TDE composite, made using the same dice-and-fill method, under magnetic bias field. It was found that the resonance shifted to a lower frequency with increasing magnetic bias field. Ma and colleagues [21] presented a simple 1-3 composite made up of just a single PZT rod embedded in a TDE mixture (Figure 3f), which avoided the need for dicing of the PZT rod arrays. This composite exhibited a maximum ME response of 18.2 V/cm·Oe at 84 kHz. Such a large ME response was attributed to the enhancement of coupling at electromechanical resonance, which assisted the elastic interaction between the TDE medium and PZT rod. Preparation of bulk 2-2 laminate composites has been mainly conducted by epoxy bonding and co-firing [5,22,23,24,25,26]. The most common geometries in laminate composites are either bilayer structure of magnetostrictive and piezoelectric layers or a sandwich structure, whereby the piezoelectric layer is arranged between two magnetostrictive ones. Laminate composites made by bonding magnetostrictive and piezoelectric layers with silver epoxy are usually cured at a lower temperature (80–100 °C). The epoxy acts as a conductive bond and strengthens the interface between the composite layers, although the epoxy layer must be kept as thin as possible to obtain good ME coupling. Various combinations of piezoelectric and magnetostrictive materials can be used to fabricate different types of ME composites with this bonding process as it does not involve high temperatures. However, the mechanically soft and viscoelastic epoxy layer will dampen the generated strain and induce a noise floor, suppressing the interface coupling and thus the ME response. Co-firing is an established process used for sintering multilayer capacitors. Besides strengthening the interface in layered ME composites, the co-firing process also provides the possibility of cost-effective mass production. Figure 3h shows the cross-sectional optical image of a NCZF-stack [0.9 PZT-0.1 PZN]-NCZF trilayer co-fired at 900 °C for 3 h [22]. Despite its advantages, the high temperature co-firing process, similar to the ceramic sintering of 0-3 particulate composites, is challenging owing to the different shrinkage rates, thermal expansion mismatch, and interface inhomogeneities.

Piezoelectric macro-fiber composites (MFCs) offer better flexibility and higher performances than the traditional monolithic piezoelectric ceramics. Compared to a PZT plate, MFCs are easier to elongate and shrink when force is applied. Several ME composites have been developed based on these MFC structures, which employ an interdigitated electrode (IDE) [23,24]. Figure 3i displays picture of an IDE/piezofiber core composite and its implementation in a complete Metglas/ piezofiber ME sensor. This composite structure yielded a giant ME response with an extremely low equivalent magnetic noise. Some polymer-based laminate ME composites prepared using polyvinylidene difluoride (PVDF), a piezopolymer possessing a low ε and a high g, in combination with Metglas or Vitrovac (Fe-based amorphous alloy) and Terfenol-D, have also shown large ME coupling behavior [25,26].

Film-Based ME Composites

As device design trends towards miniaturization and multifunctionality, thin film conformations are highly desired for application in integrated magnetic/electric devices. In film-based ME composites, optimized interface coupling can be achieved through the direct bonding of components. Moreover, large electric fields can be comfortably applied to thin films since they require relatively smaller bias voltages than those needed for bulk samples. This extends the scope of thin films for devices which have limitations on their operating voltages [27]. The greater freedom and flexibility in the fabrication of film-based ME composites render their property tuning, through interfacial coupling and strain engineering, control of crystal structure and orientation, grain size, and layer thickness, as well as chemical modification with a wide range of substituents, *etc.* Composite films can also facilitate the understanding of the physical phenomena involved in ME effects at lower dimensions, and thus enable the design of new types of magnetoelectrics with novel phase structures. A number of ME composite films have been recently prepared using various film deposition methods. Some of these methods can yield excellent thin film epitaxial growth with atomic scale thickness control and coherent interfaces [12]. However, most ME composite films reported in the literature show very low ME properties due to reduced electromechanical parameters as a consequence of substrate clamping.

Since ME coupling occurs through interfacial strain transfer, the ME response of 0-3 type ME nanocomposites can be enhanced by dispersing magnetic nanoparticles (NPs) with large interfacial area in a piezoelectric matrix. Based on this idea, McDannald *et al.* [28] synthesized PZT-CFO nanocomposite films (Figure 4a) via spin coating, and the ME coupling was modulated by controlling the concentration of CFO NPs dispersed in the PZT matrix. A high ME coefficient of 0.549 V/cm·Oe was obtained with a low concentration of CFO NPs, and the subsequent decrease in ME response with higher CFO concentrations was ascribed to the enhanced agglomeration of NPs.

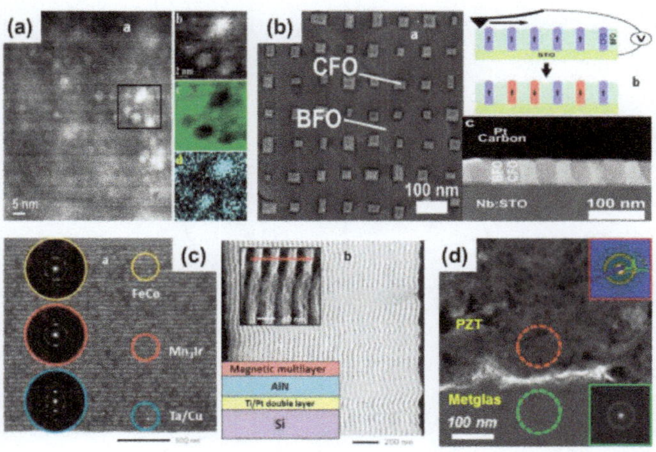

Figure 4. Development of film-based ME composites with (**a**) 0-3 connectivity [28]; (**b**) 1-3 connectivity [30]; (**c**) and (**d**) 2-2 connectivity [33,39].

The 1-3 type ME film composites have mainly been explored at the micro- and nanoscale where they can be easily synthesized through the self-assembly process [29]. Aimon *et al.* [30] developed self-assembled $BiFeO_3$–$CoFe_2O_4$ nanocomposites (Figure 4b) that were templated into ordered structures in which the ferrimagnetic $CoFe_2O_4$ pillars formed square arrays having periods of 60–100 nm in a ferroelectric $BiFeO_3$ matrix. The 1-3 ME composites prepared via self-assembly were reported to exhibit a reduced substrate clamping effect and more efficient strain coupling, as a result of their large interfacial surface area [12]. However, the design and control of such a structure remains a challenge, and their ME coefficients could not be directly measured because of the leakage problem, which resulted from the low resistance of the magnetic pillars penetrating through the films.

Deposition methods that have been used to develop thin film-based 2-2 type ME composites include pulsed laser deposition, chemical vapor deposition, sputtering, molecular beam epitaxy, spin coating, spray pyrolysis, and spin-spray techniques [31,32]. Self-biased ME film composites that show high sensitivity to AC magnetic fields were reported by Lage *et al.* [33] These thin film 2-2 ME composites fabricated by magnetron sputtering on silicon-cantilever substrates consisted of piezoelectric AlN and multilayers having the sequence Ta/Cu/$Mn_{70}Ir_{30}$/$Fe_{50}Co_{50}$ or Ta/Cu/$Mn_{70}Ir_{30}$/$Fe_{70.2}Co_{7.8}Si_{12}B_{10}$ serving as the magnetostrictive component (Figure 4c). Giant ME couplings were reported for ME thin film composites based on $BaTiO_3$-CoFe-$BaTiO_3$, and AlN-$(Fe_{90}Co_{10})_{78}Si_{12}B_{10}$ systems [34,35]. Thick film-based 2-2 layered ME composites have been prepared by employing several deposition techniques such as tape casting [36], aerosol deposition [37], electrophoretic deposition (EPD) [38], granule spray in vacuum deposition (GSV) [39], electrodeposition [40], and electroless deposition [41]. Recently, a unique fabrication approach involving a combination of room temperature deposition (GSV) and localized annealing of PZT thick film by laser radiation has been demonstrated [39] to synthesize PZT/Metglas ME film composites (Figure 4d). Localized heating through selective absorption of laser irradiation in the PZT film not only induced significant improvement in its dielectric and ferroelectric properties but also contributed to a colossal ME output by avoiding thermal damage to the Metglas substrate and preserving its inherent magnetic properties. Alternatively, photonic sintering or flash light irradiation can also be used to induce localized annealing effects in films deposited on thermally sensitive substrates. This method, commonly employed in printed electronics, uses broadband (ultraviolet to infrared) intense light pulses generated from a Xenon flash lamp to control the thermal diffusion in the nanostructured films.

RECENT ADVANCES IN THE DEVELOPMENT OF ME COMPOSITES

Realization of Broadband ME Response with Piezoelectric Anisotropy

In general, depending upon the material properties, the ME coefficient of laminate composites exhibit maxima at a specific DC bias under off-resonance condition. In the resonant condition, ME laminates exhibit a sharp peak around the resonance frequency. Such behavior limits their ability to sense wide ranges of both DC and AC magnetic fields. The broadband/wideband behavior is characterized by flat ME response over a given AC frequency range and DC magnetic bias. Limited efforts have been made to achieve a wide bandwidth around resonance frequency either by combining several PZT/Terfenol-D bilayers through parallel and series electrical connections [42] or by developing composite structures of PZNT/Metglas with a dimensionally gradient architecture [43,44]. These peak broadening approaches have indeed shown some promise, but they only cover a limited range of frequencies with a non-uniform power distribution. Patil *et al.* [45] demonstrated a resonant ME response with a wide bandwidth by adopting tri-layer ME laminates (Ni/PMN-PZT/Ni) having inherent multiple resonance frequencies, attributable to the anisotropic piezoelectric response of the [011]-oriented PMN-PZT single crystal. The required broadening of the resonance frequencies was achieved by serially connecting three ME laminates with different thickness ratios of magnetostrictive and piezoelectric layers. Such a wide bandwidth, along with the giant ME coupling, could be used in multifunctional devices such as broadband energy harvesters or field sensors. Further, Patil *et al.* [46] demonstrated the presence of giant ME voltage coefficients and magnetic-field-direction dependent ME signals at both low frequencies and resonance frequency in a symmetric ME laminate (Metglas/PMN-PT/Metglas). High performance resulted from the anisotropic transverse piezoelectric coefficients of the [011]-oriented PMN-PT single crystal. In another study, Kambale *et al.* [47] designed the ME rectangular unimorph cantilever beam structure consisting of Ni and PMN-PZT single crystal with <001> and <011>-cut crystallographic directions and investigated their mechanical vibration-based energy harvesting behavior. Both the ME voltage coefficient (α_{ME}) and mechanically harvested power output was found to be strongly dependent on the crystallographic cut directions of the PMN-PZT single crystals.

ME Composites with Textured Piezoelectric Ceramics

To achieve high ME coefficients, it is necessary to use a piezoelectric material exhibiting a high magnitude of piezoelectric voltage constant ($g = d/\varepsilon$). However, piezoelectric ceramic compositions having a high d_{33} usually show a low g_{33}, while high g_{33} compositions possess a low d_{33}. Thus, the achievement of both high d and g from a single piezoelectric composition has been challenging. Though domain-engineered, relaxor piezoelectric single crystals such as PMN-PT and PZN-PT exhibit excellent piezoelectric properties, their application has been limited due to the high production cost and small size of useful samples. PZT-based compositions cannot be readily grown in single crystal form because of their incongruent melting behavior [48]. Alternately, textured ceramics based on PMN-PT and PMN-PZT have been prepared using the templated grain growth (TGG) method. Here, the texturing produces an engineered domain state resulting in high d values and the use of oriented and platelet shaped template crystals having low permittivity suppresses the increase in ε of the textured ceramic, thereby achieving low cost high performance piezoelectric ceramics suitable for ME composites. Yan *et al.* [49] reported an enhanced α_{ME} for Metglas/textured PMN-PZT/Metglas laminate compared to Metglas/random PMN-PZT/Metglas, where the maximum α_{ME} was found to be 0.348 V/cm·Oe and 1.49 V/cm·Oe, respectively. Because of the environmental concerns of using lead containing materials, recent investigations have focused on developing ME composites employing lead-free piezoelectric ceramics with compositions based on KNN, NBT, and BT. A large value of maximum α_{ME} (1.32 V/cm·Oe) has been reported from a laminate ME composite of Terfenol-D/Mn-doped NBT–BT single crystal/ Terfenol-D [50].

Self-Biased ME Composites

Most conventional ME composites need a DC bias, to obtain the maximum ME response, and that necessitates the use of permanent magnets or electromagnets resulting in bulky devices and problems of electromagnetic interference. To circumvent these issues, self-biased magnetoelectric (SME) composites that provide sizeable ME coupling under an external AC magnetic field in the absence of a DC magnetic field have been proposed [51]. This would enable device miniaturization and the development of ME composite-based circuit components for integrated electronics and medical applications. The SME composites are categorized into five groups according to their working mechanism: (a) functionally graded FM-based SME; (b) exchange bias-mediated SME; (c) magnetostriction hysteresis-based SME; (d) built-in stress-mediated SME; and (e) non-linear SME composites (Figure 5). Large and tunable SME responses have been reported for both bulk and film-based ME composite systems [33,52]. A recent review by Zhou *et al.* [51] provides a comprehensive coverage of the development and prospects of the SME composites.

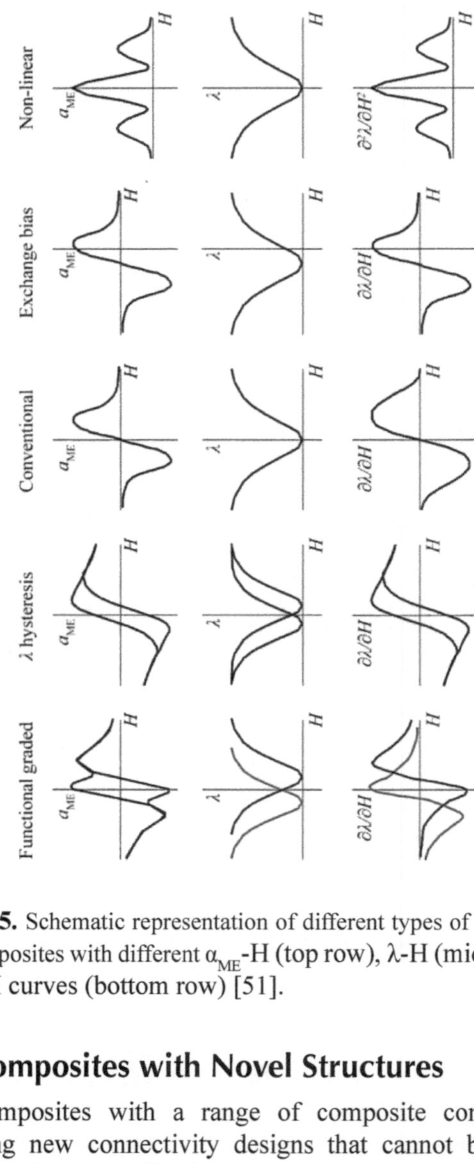

Figure 5. Schematic representation of different types of self-biased ME composites with different α_{ME}-H (top row), λ-H (middle row), and q-H curves (bottom row) [51].

ME Composites with Novel Structures

ME composites with a range of composite connectivity, including new connectivity designs that cannot be readily synthesized by traditional routes are being developed through the fabrication of nanostructured ME materials. In recent years, there have been some studies on the synthesis of ME nanocomposites having core/shell arrangement in the form of nanoparticles, nanowire arrays, and nanotubes. Other 1D composite structures with random and Janus-type arrangements have also been prepared (Figure 6). These nanostructured composites were prepared using different wet chemical synthesis methods. Kalyan *et al.* [53] synthesized $CoFe_2O_4/BaTiO_3$ core/shell nanoparticles (Figure 6a) using a combination of solution processing and high temperature calcination. Rongzheng *et al.* [17] demonstrated the formation of ferrite/perovskite oxide core/shell nanostructures in several multiferroic systems such as $Fe_3O_4/PbTiO_3$, γ-$Fe_2O_3/PbTiO_3$, γ-$Fe_2O_3/Pb(Zr,Ti)O_3$, $CoFe_2O_4/BaTiO_3$, $CoFe_2O_4/PbTiO_3$ and

$CoFe_2O_4$/$Pb(Zr,Ti)O_3$ using a combined hydrothermal and annealing process. For synthesizing ordered arrays of $NiFe_2O_4$/PZT core/shell nanowires (Figure 6b), Ming *et al.* [54] developed a method involving the combination of a modified sol-gel process, electrochemical deposition, and subsequent oxidization in anodized nanoporous alumina membranes. Recently, Andrew *et al.* [55,56] reviewed the efforts in the development of multiferroic nanofibers with core/shell, random, and Janus-type arrangements (Figure 6c–e), fabricated by electrospinning. Nevertheless, some of the challenges in the practical implementation of these ME nanocomposites include forming isolated multiferroic particles that are free of agglomerates, substrate free assembling of nanofibers into ordered structures, and difficulties in accessing their ME properties. Further, in some cases, these nanostructured ME composites will have to be consolidated into a dense form for device applications which may affect the stability of nanostructures.

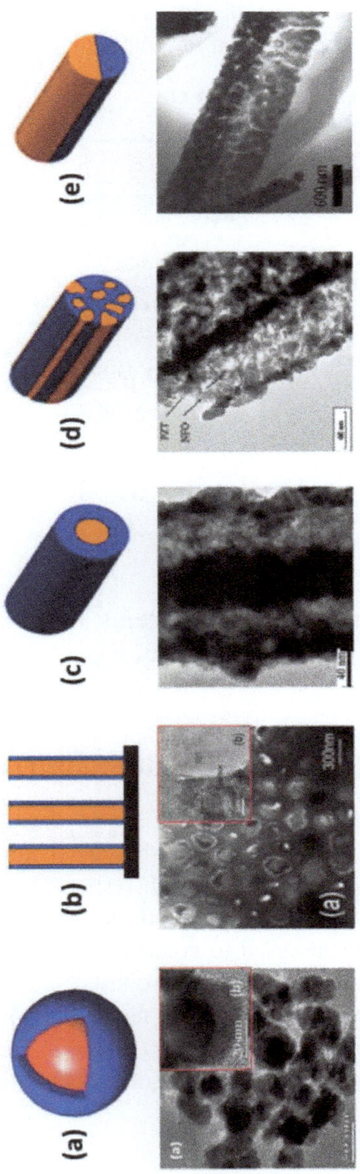

Figure 6. Bi-phasic ME composites with different types of arrangements between the two phases (**a–c**) core/shell arrangement: (**a**) nanoparticles [53]; (**b**) nanowires [54]; and (**c**) nanotubes [57]; (**d**) random arrangement [58]; and (**e**) Janus-type arrangement [56].

CHARACTERIZATION OF ME COUPLING

To understand the effect of multiple parameters and variables (materials, connectivity, operational mode, synthesis method, *etc.*) on the ME coupling, and to predict the performance of ME composites in various applications, investigations on their working behavior are essential. Several characterization tools have been reported in literature on DME and CME coupling, based on the measurement as a function of an applied magnetic field or an electric field over a range of frequencies from DC to 110 GHz. Some of these analysis techniques have been discussed in detail by Srinivasan [59]. The primary tools for studying DME coupling are measurements of the low-frequency ME voltage coefficient, voltage output in response to applied magnetic fields at the bending and electromechanical resonance modes, nonlinear ME effects for large AC magnetic fields, static magnetic field-induced polarization, and magneto-dielectric effects. Studies on the nature of CME coupling include low-frequency ME effects under an applied AC electric field and measurement of the induced magnetic flux in a coil wound around the composite, static electric field E-induced magnetization, E-tuning of inductance, or ferromagnetic resonance. Recently, ME measurements on thin films and nanostructured composites have been made by employing scanning probe microscopy (SPM). Piezoresponse force microscopy (PFM) under an applied magnetic field is used for studies on DME, and magnetic force microscopy (MFM) under an applied voltage is used for analyzing the CME.

From the point of view of functional devices, it is important to control the magnitude of the ME voltage coefficient and its behavior [60]. In a ME composite, the measured DME coupling coefficient is the field conversion ratio between applied H_{ac} and induced E_{ac} under a bias field H_{dc}, $\alpha_{ME} = \delta E_{ac}/\delta H_{ac}$. For such measurements, it is necessary to first pole (DC poling or Corona poling) the ferroelectric phase in the composite. In general, under a constant applied AC magnetic field, the ME coupling coefficient first increases with increasing H_{dc}, reaching a maximum at an optimized DC bias (H_{bias}), and then decreases with further increasing H_{dc}. When the bias field is reversed, the ME voltage shows a 180° phase shift. The H-dependence tracks the slope of λ *versus* H. Saturation of λ at high H leads to $\alpha_{ME} = 0$ [27]. For strain-mediated ME composites, it has been shown that $\alpha_{ME} \propto q_{ij} = (d\lambda_{ij})/dH$. This indicates that the ME coefficient is directly related to the nature of the ferromagnetic phase and the effectiveness of elastic coupling between the two phases. Figure 7a demonstrates the typical behavior of the α_{ME}, λ, q, and their relationship as a function of H_{dc}. The variation of magnetostriction as a function of H and the dependence of α_{ME} on q suggests the need for additional H_{dc}. Usually, the α_{ME} value is higher when H and δH are parallel to each other and to the sample plane, compared to the out-of-plane magnetic fields due to demagnetization. So far, giant ME responses have been reported in a variety of composite systems and the highest α_{ME} was reported for composites with ferromagnetic alloys. Figure 8 summarizes the best values of α_{ME} obtained for different combinations of materials in bulk and film-based ME composites having 0-3, 1-3, and 2-2 connectivity.

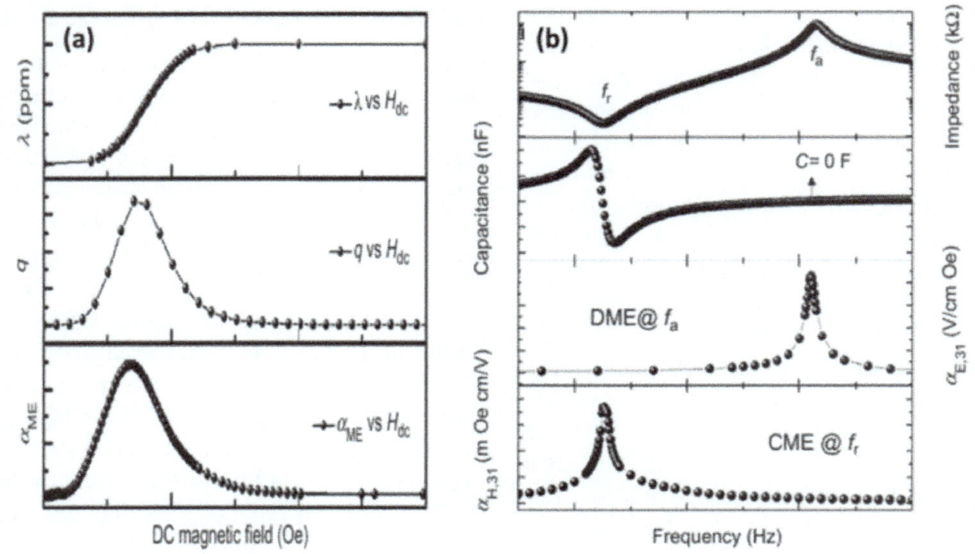

Figure 7. (**a**) Typical behaviors of magnetostriction, piezomagnetic coefficient, and ME coefficient as a function of DC magnetic field [60]; (**b**) Typical behaviors of impedance, capacitance, and DME and CME coefficients as a function of AC frequency [7].

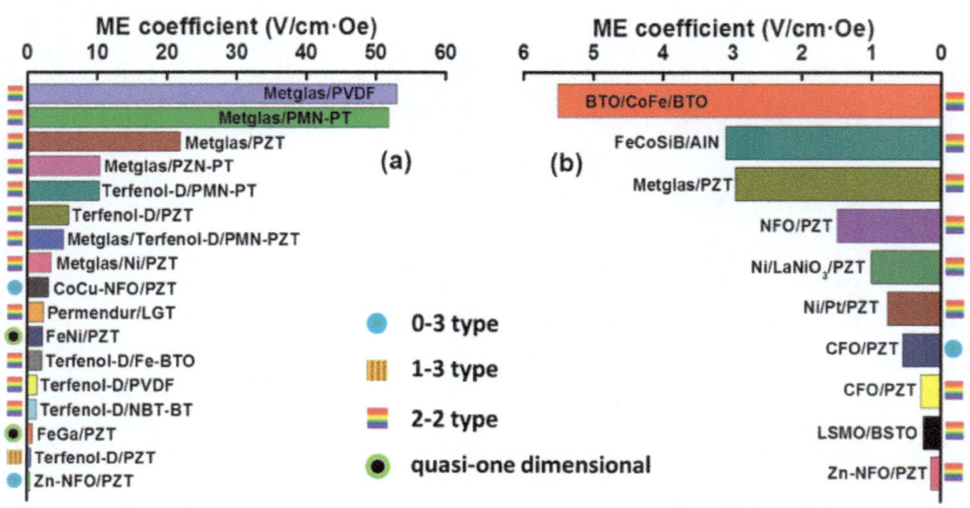

Figure 8. Reported values of off-resonance ME voltage coefficients for various material systems: (**a**) bulk and (**b**) film-based ME composites. Data taken are from Ref. [5,11,21,23,24,25,26,28,34,35,36,37,39,50,52,61,62,63,64,65,66,67,68,69,70, 71].

Another parameter of importance is the change in ME coupling as a function of applied AC magnetic field frequency (ME *vs.* f_{ac}), which depends on the sample dimensions. When AC magnetic field is applied at a frequency corresponding to the electromechanical resonance (EMR) of the piezoelectric phase, or the ferromagnetic resonance (FMR) of the magnetic phase in the ME composite, the ME voltage coefficient shows a peak behavior, with as much as a 100-fold increase in its magnitude compared to the off-resonance condition. Cho *et al.* [7] found that the DME coupling was maximized at the anti-resonance frequency (f_a), while the CME coupling exhibited peak at the EMR frequency (f_r) of the piezoelectric layer, as shown in Figure 7b. This phenomenon was further explained by using piezoelectric constitutive equations and combining them with the frequency-dependent capacitance of the piezoelectric layer. Comparison of both DME and CME coefficients have rarely been performed because it has been challenging to achieve strong DME and CME coupling simultaneously [72].

MAGNETOELECTRIC DEVICES AND APPLICATIONS

Several of the bulk as well as film-based ME composites have been reported to exhibit strong ME responses under both off/on resonance conditions. Based on the type of ME coupling and the mechanisms used to control the order parameters, a variety of applications have been proposed, including magnetic sensors, high-frequency inductors, memory devices, and high-frequency signal processing devices (Table 2). Recent progress on the design and development of some of the ME devices will be discussed in this section.

Table 2. Classification of different ME devices. Adapted from [72]

ME Coupling	Physical Mechanism	ME Devices
Direct ME coupling	H control of electric polarization	Magnetic sensors, current sensors, transformers, gyrators, energy harvesters
	E control of magnetization switching	Spintronics, including random access memories, tunnel junctions
Converse ME coupling	E control of permeability μ	Voltage tunable inductors, tunable band-pass filters, phase shifters
	E control of spin wave	Voltage tunable filters, tunable resonators, phase shifters

Devices Based on the DME Effect

Magnetic Field Sensors

ME material-based sensor devices are considered to be promising alternatives for conventional Hall sensors and giant magnetoresistive (GMR) devices. Because of their passive nature and self-powered operation at room temperature, ME sensors may be able to replace bulky and expensive superconducting quantum interference devices (SQUIDs). ME sensors have the potential to perform biomagnetic measurements analogous to all of those performed by MEG and fMRI. The key requirements for magnetic field sensors in this application are (i) sensitivity of ~pT to fT per √Hz at low frequencies (10^{-2} to 10^3 Hz); and (ii) ambient temperature and wide bandwidth (0.1 to 100 Hz) operation [73]. The direct ME coupling effect, where the ME voltage coefficient is dependent on AC and DC magnetic fields, would allow the ME composites to sense either an AC or DC magnetic field by monitoring the output electrical signals. Since a current passing through a wire generates a magnetic field in the surrounding space, the ME composite can also be used as a current probe for detecting current by monitoring the corresponding magnetic flux.

Various ME laminate-based sensors have been reported in the literature over the last decade to detect minuscule magnetic fields which could have applications ranging from medical imaging to oil prospecting. Efforts have focused on: (a) different categories of sensor construction [24,74,75,76], geometry and other features [77,78,79,80]; (b) packaging of the sensor units; (c) improvements in fabrication techniques [81]; (d) signal processing conditions [82,83]; and (e) gradiometric configurations to reduce environmental noise sources [84]. The multi-push-pull configurations, multilayer configuration, and bimorphs, all of which exhibit an improved ME voltage coefficient, have demonstrated considerable potential for sensing low-frequency magnetic field variations. A push-pull mode ME laminate consists of a symmetric longitudinally poled piezoelectric PMN-PT single crystal and two longitudinally magnetized magnetostrictive Terfenol-D layers. The symmetric nature allows for optimized elastic coupling between consecutive layers. As a result, a large ME coefficient of 30 V/cm·Oe at a resonance of 77.8 kHz was obtained with a magnetic field sensitivity of 136 pT/√Hz [74]. The very low sensitivity of 5.1 pT/√Hz, which is very close to the predicted value of 4.2 pT/√Hz, has been achieved for a Metglas/PMNT laminate structure with a multi-push-pull (M-P-P) mode that is longitudinally magnetized and longitudinally poled, coupled with a low noise charge amplifier [24]. Fang *et al.* [85] very recently proposed a Metglas/Mn-PMNT laminate composite consisting of longitudinal magnetized Metglas layers and different numbers (N) of transversely polarized Mn-PMNT fibers connected in series, as illustrated in Figure 9. An ultralow magnetic field sensitivity of 0.87 pT/√Hz at room temperature was reported for these Metglas/Mn-PMNT fiber laminate composites.

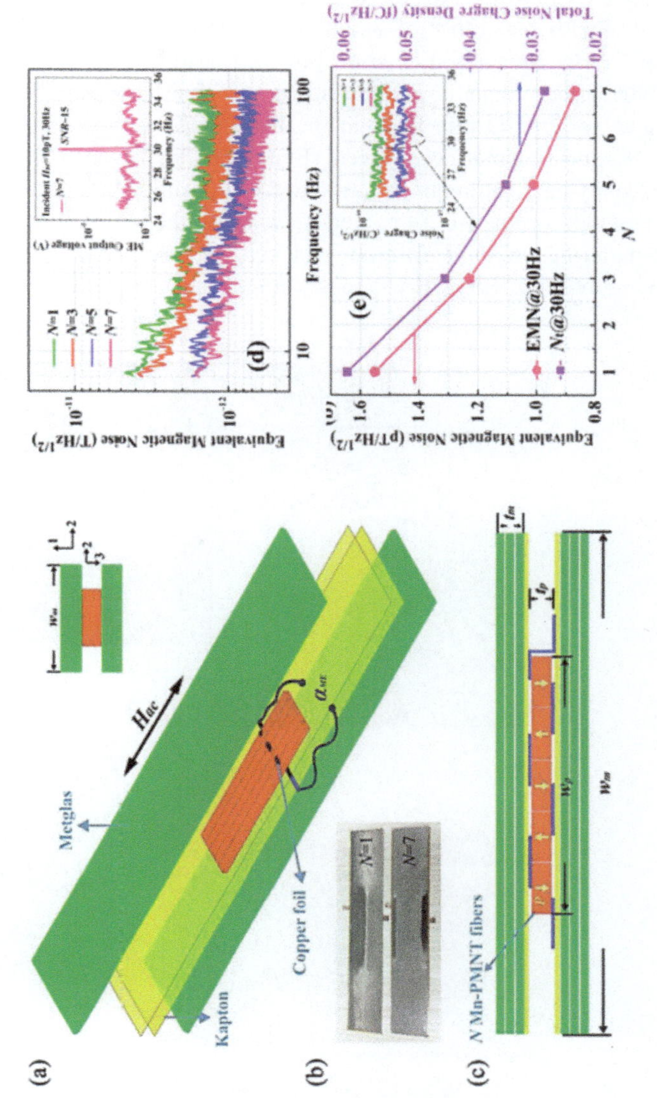

Figure 9. (a)–(c) 3D structure, photograph, cross-sectional schematic diagram, respectively, of the Metglas/Mn-PMNT composite. (d) and (e) Equivalent magnetic noise level and total noise charge density of the ME magnetic sensors [85].

Thin film-based ME sensors enable the fabrication of miniaturized low-cost sensor devices with high sensitivity and high spectral resolution [86]. Thin film-based architecture also provides the capability for designing sensor arrays that can be integrated with other circuit components. Zhao *et al.* [87] reported the fabrication of a thin film ME sensor by sputter depositing a $Fe_{0.7}Ga_{0.3}$ film (1.5 μm thick) over a sol-gel derived PZT film (1.5 μm thick) on a micromachined Si (35 μm thick) cantilever. Substantial improvement in ME coupling was observed due to reduction in substrate clamping by decreasing the Si cantilever thickness (from 180 to 35 μm). The sensor device showed a maximum ME coefficient of 1.81 V/cm·Oe at the electromechanical resonance frequency of 333 Hz and could detect a 2.3×10^{-8} T (2.3×10^{-4} Oe) AC magnetic field with the 50 nV noise floor. Marauska *et al.* [86] developed a ME sensor system with a stack of SiO_2/Ti/Pt/AlN/Cr/FeCoSiB fabricated on a 150 mm thick Si wafer cantilever. The AlN piezoelectric layer (1 μm thick) was grown by reactive pulsed-dc magnetron sputtering, and a $(Fe_{90}Co_{10})_{78}Si_{12}B_{10}$ magnetostricitve layer (2.2

μm thick) was sputter deposited in a magnetic bias field of 10 mT. The ME sensor exhibited a maximum ME coefficient of 1000 (V/m)/(A/m) (= 795.8 V/cm·Oe) under resonance at 2.4 kHz with sensitivity of 780 V/T and noise level above 100 pT/√ Hz. Recently, Lee *et al.* [88] demonstrated all-thin-film ME ac/dc magnetic field sensor arrays made by using PZT and Terfenol-D thin films in the form of 2.98-μm-thick and 300-μm-long microcantilevers (Figure 10a,b). The PZT films, displaying a columnar structure, were deposited by the sol-gel method. Terfenol-D films deposited using the RF magnetron sputtering method formed well crystallized large grains (Figure 10c). The ME voltages were measured at 60 Hz from the cantilevers with the PZT and PZT/Terfenol-D layers with increasing magnetic field between 2×10^{-9} T to 20×10^{-9} T. It was observed that the PZT/Terfenol-D cantilever produced a significant voltage output, while the PZT-only cantilever showed a negligible output (Figure 10d). There was also a gradual increment in the ME voltage of PZT/Terfenol-D in proportion to the DC magnetic field. The minimum detectable DC magnetic field was estimated by applying a DC current of 0.005 μA corresponding to 1×10^{-12} T, and the ME voltage from the sensor measured at 60 Hz AC field in off-resonance mode was found to be 86 μV with a 150-nV noise floor (Figure 10e).

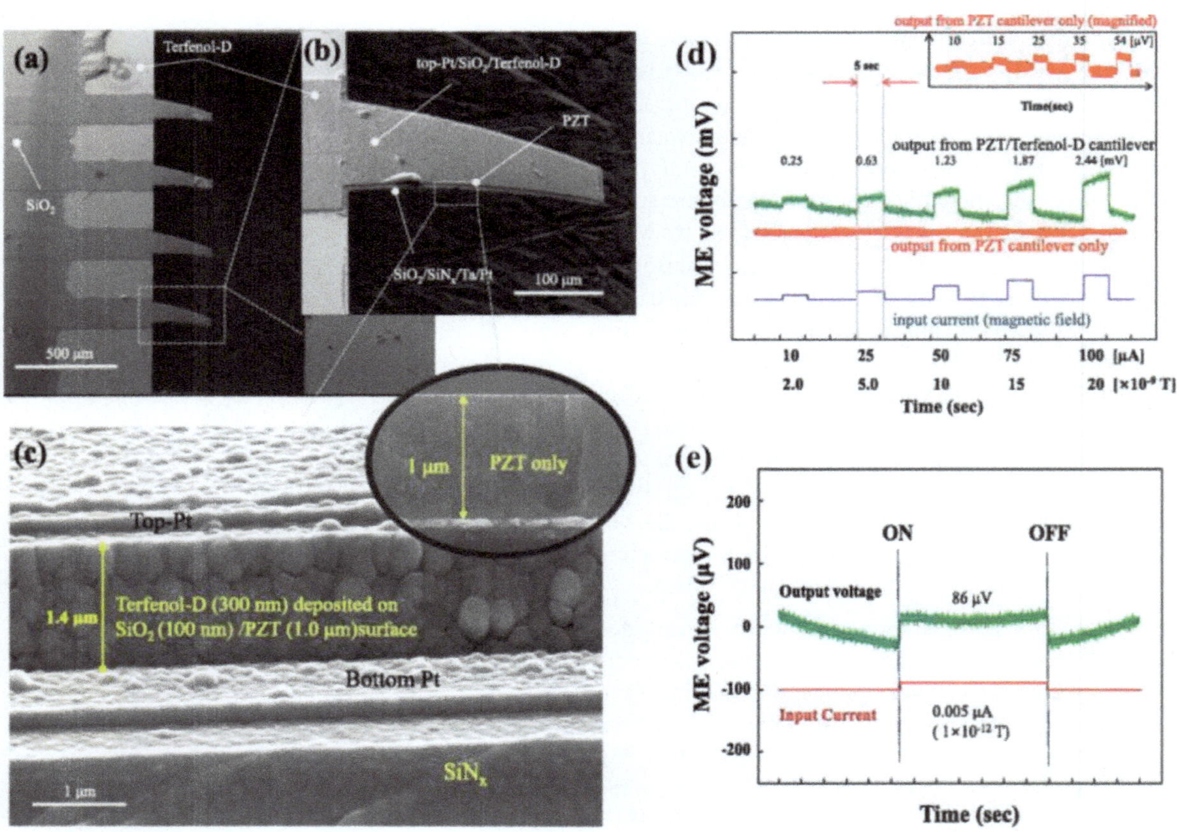

Figure 10. SEM images and ME properties of ME microcantilever array consisting of Terfenol-D and PZT thin film [88].

Electric Current Sensors

Conventional current sensors, which operate by detecting the electric current induced by magnetic fields are best represented by Hall effect and reluctance devices [89]. Hall devices need to be powered by highly stable constant current supplies, and their inherently weak Hall voltages (5 to 40 μV/Oe) impose great demands on signal conditioners. Reluctance devices require being interfaced with highly precise integrators, and real-time measurements are generally inhibited at low frequencies (100 Hz). In contrast, current sensors based on ME composites do not suffer from these problems due to the extrinsic ME effect exhibited by the composites. In principle, a straight wire carrying an AC or a DC current will excite an AC or DC vortex magnetic field around this wire according to Ampère's Law. The strength of the magnetic field depends on the current I in the wire and r, the distance from the wire ($H = I/2\pi r$). Ring-type ME laminates, therefore, have the essential configurations of electric current sensors. For example, Leung *et al.* [90] demonstrated a ring-type electric current sensor (Figure 11) operating in vortex magnetic field detection mode. The sensor design was based on a ring-shaped magnetoelectric laminate consisting of an axially polarized PZT piezoelectric ceramic ring bonded between two circumferentially magnetized epoxy-bonded Terfenol-D short-fiber/NdFeB magnet magnetostrictive composite rings. The sensitivity of the electric current sensor was evaluated, both theoretically and experimentally. It was reported that the output voltage had good linear responses to the electric current. The sensor showed a high off-resonance sensitivity of 12.6 mV/A over a flat frequency range of 1 Hz–30 kHz and a large resonance sensitivity of 92.2 mV/A at the resonance of 67 kHz. Excellent linearity and large current sensitivity of 114.2 mV/A was observed in a Metglas/PZT laminate when measuring low-frequency alternating magnetic fields of 50 Hz [91]. This sensor is ideally suited for power-line current measurement. By combining an ME ring with a piezoelectric transformer structure, Zhang *et al.* [92] achieved a high resonance sensitivity of ~157 mV/A at electromechanical resonance frequency of 62 kHz. Yu *et al.* [93] have recently reported the design and implementation of a cantilever-type device for a current sensor based on two layers of longitudinally magnetized Terfenol-D and one layer of transversely polarized piezoelectric PZT material. The layers were epoxy-bonded, and the wire conducting the current was coiled along the device and the output voltage was obtained across the two surfaces of the PZT. Compared with a ring-type current sensor, this sensor could be installed and maintained without any interruption in the power supply, which was very convenient and practical. The high-sensitivity, power-free, bias-free, and wide-bandwidth nature of the ME current sensor also provided great potential for real-time monitoring of the conditions of engineering systems which have electric current-carrying cables or conductors.

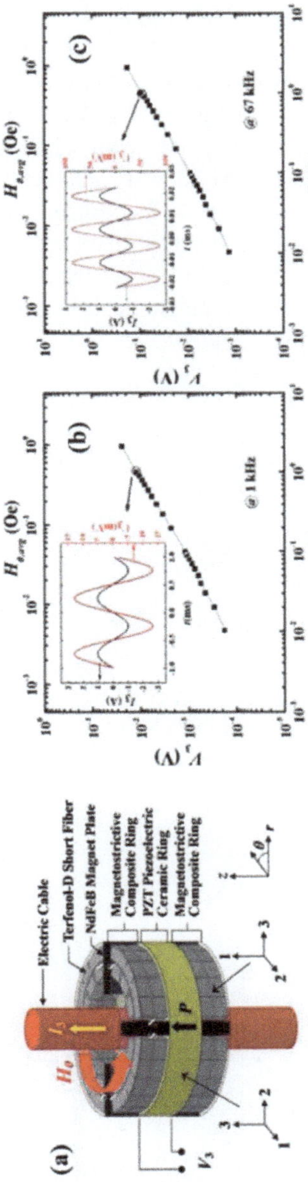

Figure 11. (**a**) Schematic diagram of the proposed ring-type electric current sensor. (**b**) and (**c**) The AC electric voltage (V_3) output from the sensor as a function of both the AC electric current (I_3) and its associated average AC vortex magnetic field ($H_{\theta,avg}$) [90].

Energy Harvesters

Harvesting energy from ambient energy sources such as vibrations, sound, radiofrequency waves, light, temperature gradients, wind, and others is an area of focus for current and next-generation remote monitoring electronic devices and self-powered wireless sensor networks with the goal of improving device lifetime and addressing the limitations of conventional batteries. In addition, the ambient environment is filled with magnetic noise of 50–60 Hz almost everywhere these days. Harvesting this weak and low-frequency magnetic noise (< 1 mT = 10 G) to develop a consistent electricity source remains a difficult challenge. Over the last decade, Ryu group [47,94,95] and other researchers [96,97,98,99 ,100,101,102,103,104,105] have investigated methods to obtain optimum electricity from the tiny magnetic fields in the surroundings, using magneto-mechano-electric (MME) mechanism. The operation mechanism can be described as follows: When the ME composite is placed in an AC magnetic field, the magnetostrictive layer in the composite responds to the mechanical vibration (magneto-mechano coupling), thereby straining the piezoelectric layer, which results in an output voltage across the electrical load through the direct piezoelectric effect (mechano-electric coupling). Due to the existence of the piezoelectric phase in the ME composite, any mechanical oscillation applied to the composite directly creates electrical voltage. Consequently, the MME generator could be used to harvest energy from both the magnetic field and external vibrations at the same time [95]. This sequential operating process is schematically depicted in Figure 12a.

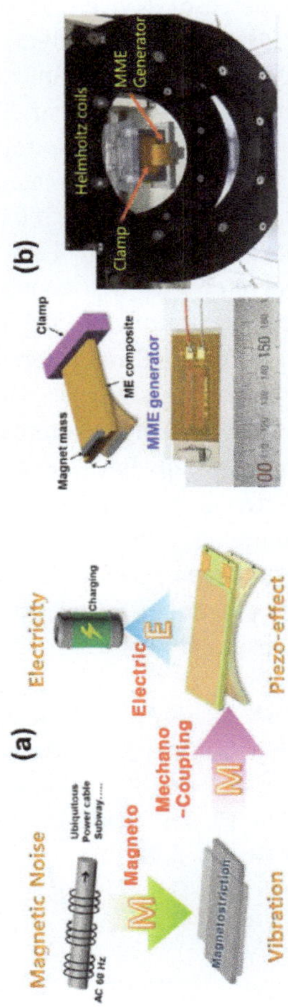

Figure 12. (**a**) Schematic depicting the working principle of Magneto-Mechano-Electric (MME) energy harvester; (**b**)

Schematic and photo of cantilever structured MME energy harvester [95].

By selecting high performance piezoelectric and magnetostrictive materials and by optimizing the composite structure, it is expected that high electric power density can be obtained from a low-frequency magnetic field using the MME generator. Gao *et al.* [103] designed an asymmetrical bi-layered push-pull mode Metglas/Pb(Zr,Ti)O_3 laminate, and obtained a giant αME of more than 400 V/cm·Oe with a tunable resonance frequency in the range of 60 Hz to 220 Hz by tip mass loading. The maximum harvested power output was about 16 μW/Oe with a 6 MΩ resistance load, with a corresponding power density of 200 μW/cm^3 at 60 Hz. Dong *et al.* [104] achieved a power output of 420 μW/Oe across a 50 kΩ load under an AC magnetic field of 1 Oe at about 21 kHz using an ME cantilever based on the push-pull type Metglas/Pb(Zr,Ti)O_3 laminate. The ME laminate configuration, with an optimized beam aspect ratio of 0.86, exhibited an output power gain of 52.5 mW, and a corresponding power density of about 28.5 mW/cm^3 at 30 Hz under 6.9 m/s excitation (0.7 g acceleration) at the resonance frequency, even in the absence of a DC bias magnetic field [105].

Ryu *et al.* [95] demonstrated an MME generator constructed using an anisotropic and flexible piezoelectric Pb($Mg_{1/3}Nb_{2/3}$)O_3–$PbTiO_3$ (PMN–PT) single crystal fiber composite (SFC), a cost-effective magnetostrictive Ni plate and Nd permanent magnetic proof mass (Figure 12b). The flexibility of the SFC ensures the high compliance of the sample, which is ideal for achieving low resonance frequency in a cantilever structure. The flexibility also increases device durability and enables the application of increased strain magnitudes. The Ni plate can be easily self-biased and generates a linear strain response in a low-level magnetic field environment. The performance of the MME generator containing an anisotropic <011> SFC with d_{32} mode under a small noise level magnetic field is shown in Figure 12b. At 60 Hz, and H_{ac}~500 μT, the maximum generated voltage was ~34 Vpp (~12.4 Vrms). The power from the MME generator was high enough to fully charge a 220-μF electrolytic capacitor after rectifying for 3 min. Using the stored power in the charged capacitor, the device was able to turn on 35 commercial high intensity LEDs with a turn on/off frequency of ~1 Hz. Seeking to improve the output power density, many research groups have developed various MME harvesters using different combinations of magnetostrictive and piezoelectric materials; some of these are summarized in Figure 13. These results represent significant advances towards next-generation remote monitoring electronic devices and self-powered wireless sensor networks.

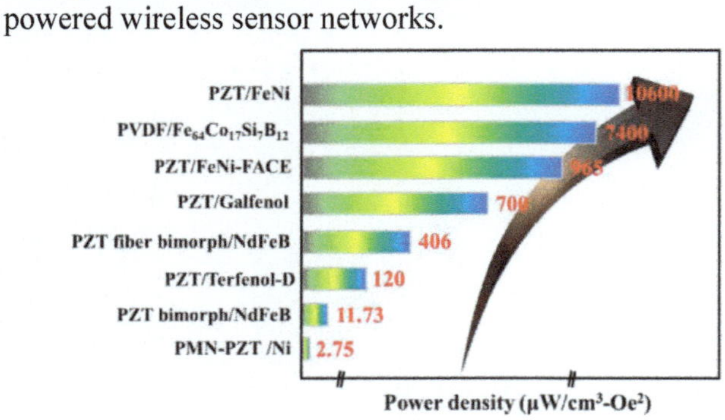

Figure 13. Summary of reported power densities from the MME harvesters made with different composite systems. Data are from references [94,96,97,98,99,100,101,102,103, 104,105].

Efforts have also been made to develop energy harvesters based on ME composite films. Onuta and colleagues [101] developed miniaturized energy harvesters using all-thin-film ME structures consisting of piezoelectric Pb($Zr_{0.52}Ti_{0.48}$)O_3 (PZT) and magnetostrictive $Fe_{0.7}Ga_{0.3}$ layers (Figure 14). The device was fabricated on a micromachined Si cantilever coated with a silicon oxide/nitride/oxide (ONO) stack (3.8 μm thick). The $Fe_{0.7}Ga_{0.3}$ layer (500 nm thick) was sputtered on the Pt-buffered PZT layer (500 nm thick) deposited over the ONO stack. The film-stress in the ONO stack was engineered, and a photo-lithographic process was employed to make an unbent, free-standing cantilever beam structure (950 μm long and 200 μm wide). The chip (6.6 mm × 6.6 mm) containing six cantilever devices was mounted in a vacuum chamber placed between a pair of Helmholtz coils, and it was aligned parallel to the magnetic fields. The results of the energy harvesting measurements from a single ME device are shown in Figure 14b,c. The voltage output from the harvesting device became saturated at higher external loading, while the power output was peaked at a load impedance of 12.5 kΩ (Figure 14b) from which the peak power density was determined to be 0.7 mW/cm^3 (RMS). Further, both the voltage and power outputs have shown saturation plateaus with respect to increasing Hac (Figure 14c), and such behavior was attributed to the saturation of the internal stress of magnetic origin that involves the rotation of the magnetization vector.

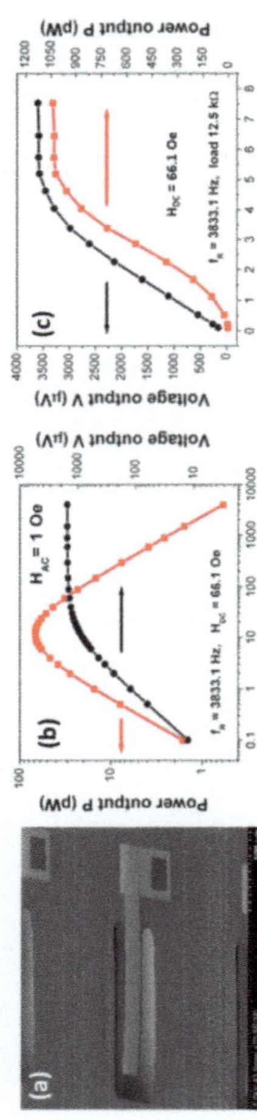

Figure 14. (**a**) SEM micrograph of a thin film ME energy harvester based on PZT/FeGa and (**b,c**) its energy harvesting properties [101].

Magnetic Recording Read Head

Magnetic recording read heads in hard disk drives function based on one of the magneto-resistance (MR) effects, *i.e.*, anisotropic magneto-resistance (AMR), giant magneto-resistance (GMR) or tunneling magneto-resistance (TMR). The MR read head contains a sensing stack along with several additional layers and structures that ensures the proper biasing and shielding of the sensor [106,107]. During operation, the internal resistance of the MR sensor changes due to interaction with the stray fields from the recorded bits of a magnetic recording medium. The change in resistance (ΔR) is translated into a read signal as a voltage amplitude change $\Delta V = I \times \Delta R$. However, it is necessary to pass a constant DC test current through the sensor stack to detect the change in the resistance/amplitude. Higher recording densities can be achieved by reducing the shield-to-shield spacing, which is possible by thinning down the sensor stack. But such a trimming of the sensor size inflicts complications in fabrication and in the operation of the sensor and the read head.

Vopson *et al.* [106] proposed a new design for a magnetic recording read head sensor based on the DME, *i.e.* the magnetically induced ME effect, as shown in Figure 15a,b. The proposed sensor stack incorporates a tri-layer ME composite structure and consists of 7 layers, arranged as seed/AFM/FM/FE/FM/AFM/cap, which is much simplified in comparison with the existing TMR sensor stack, which has around 15 layers. The two AFM layers added adjacent to the FM layers produce the required DC bias magnetic field. This design will facilitate reduction in sensor size without sensitivity loss and will eliminate the need for permanent magnets currently used for horizontal biasing. Since the ME-based sensors operate without using any electric power, there will be no need for the DC test current through the sensor, and the data can be read back directly as an induced voltage. This will reduce power consumption as well as minimize the Joule heating issues that occur in the high resistance TMR sensors.

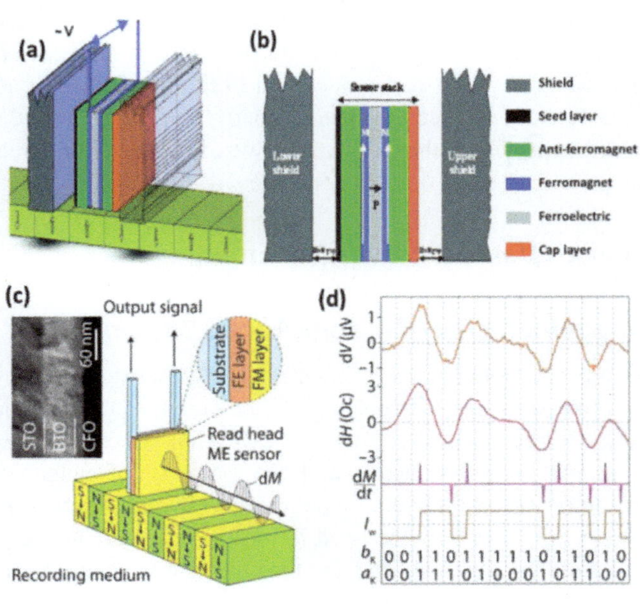

Figure 15. (**a**) and (**b**) 3D diagram and cross-section of the magnetic read head proposed by Vopson *et al.* [106] (**c**) and (**d**) Schematic, TEM image (inset) and voltage output of an ME read-head sensor based on a bilayer CFO/BTO heterostructure grown on a single-crystal STO substrate [107].

Zhang *et al.* [107] developed a prototype ME read head (Figure 15c) fabricated with BTO-NFO and BTO-CFO bilayered heterostructures that was grown using PLD on 0.7% Nb-doped STO single crystal substrates. The ME sensitivity of this structure was measured to be about 0.5 V/Oe at 1 kHz. To illustrate the reading and writing process, a primal 16-bit input signal (a_k) was written on the disk tracks by a writing current (I_w) in non-return-to-zero format after being encoded into a new signal (b_k). These written bits b_k were able to yield a media field δH in the recording medium. A small in-plane δH (about a few Oersted) that was considered to model the media fields (Figure 15d) from the bits in the recording medium was generated using a set of small Helmholtz coils, and a response signal of 1–2 µV was obtained. In the fully developed device case, greatly enhanced outputs (mV) were expected by applying higher amplitudes of media fields (~100 Oe) and by operating the device at the resonance frequency.

Biomedical Applications

Motivated by the advances in multiferroics, ME composites have been suggested for biomedical applications such as wireless endoscopy, minimally invasive surgical tools, and stimulation of functions of living cells [73,108]. The potential use of ME nanoparticles (MENs) as carriers for on-demand drug release and to artificially stimulate the neural activity deep in the brain has also been suggested [109,110].

Wireless capsule endoscopes (WCEs) are often used to examine the gastrointestinal (GI) tract for clinical diagnosis (Figure 16a). Though the WCEs are far less invasive compared to conventional endoscopes, the passive nature of WCEs makes it difficult to control their position and orientation as the capsule moves along the GI tract. Currently, tracking of WCEs is carried out by methods such as radio frequency (RF) triangulation, magnetic tracking, radiation vision, and ultrasound sensing. However, accurate localization and high fidelity tracking of WCEs is still a challenging task. In the magnetic tracking method, a permanent magnet is placed inside the wireless capsule, which is located at an x, y, z location inside the human GI tract with respect to a reference frame. Since the permanent magnet emits a magnetic field, it can be identified using a magnetic sensor, such as a 3-axis magnetic sensor, which can approach nT sensitivity. It has been proposed that multiferroic ME transducers are promising candidates to provide WCE position feedback, by using them in the form of enhanced sensitivity (pT range) magnetic field sensors [73]. By incorporating a MEMS-based ME composite structure into the WCE capsule, power can be generated on demand as required for recharging the battery of the WCE. Further, the use of self-biased ME composites will reduce complexity in the device design. To improve its diagnostic capabilities, focused efforts are needed, particularly to achieve reliable control over the movement and orientation of the capsule within the constraints of the size and shape of the device.

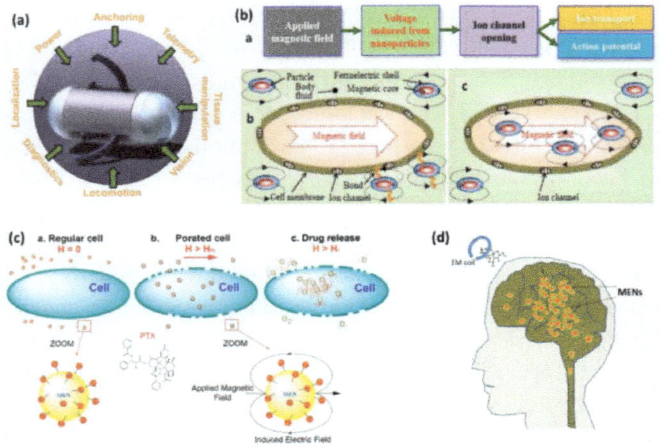

Figure 16. (**a**) Key functionalities that complement the wireless capsule endoscopes [73]; (**b**) Illustration of possible mechanisms of stimulation of ion channels: a chain of actions triggered by the applied magnetic field pulses, external and internal stimulation by uptaken nanoparticles [108]; (**c**) MENs as field-controlled nano-electroporation sites to let the drug through the cancer cell membranes [109]; (**d**) Illustration of the deep brain stimulation approach [110].

Controlling the function of biological macromolecules is of vital importance in health science studies. Approaches used to stimulate cell functions include the use of the heat generated by hysteresis losses in magnetic nanoparticles placed in a high-frequency magnetic field, and the mechanical agitation of magnetic nanoparticles attached to cells using external low-frequency magnetic fields [111,112]. The interaction between electromagnetic fields and biological macromolecules can be understood by studying the ion channels which regulate several cellular processes, such as action potentials in neurons or muscle contraction. Ion channels are membrane proteins that form pores for the controlled exchange of ions across cellular membranes [113]. The common type, voltage-gated ion channels open or close in response to a transmembrane electric field. Ion channel-related disorders have been found to cause several health issues, which are usually treated using pharmacological agents that modify the gating kinetics, or block the channel transiently or permanently. Recently, Kargol *et al.* [108] proposed an innovative approach, based on the use of core/shell structured MENs with a ferromagnetic core and a ferroelectric shell, that would allow the remote control of ion channel gating via externally applied magnetic fields (Figure 16b). In this approach, electric fields in the vicinity of the cells generated by MENs introduced extra- or intracellularly can be locally modified to invoke appropriate conformational changes in the ion channels. The resulting local depolarization or hyperpolarization of the membrane will lead to opening or closing of the ion channels accordingly. Because of the remote way in which the stimulation will be performed, individual cells or selected groups of cells can be targeted, rather than whole tissues.

Targeted drug delivery with adequate high specificity (to tumor cells) remains a formidable task in the treatment of cancer in general, particularly ovarian cancer. Although the survival rates have been improved by intraperitoneal (IP) delivery through a surgically implanted catheter, toxicity and catheter complications have precluded widespread adoption of this invasive means of delivery. Guduru *et al.* [109] addressed this challenge by exploiting the dependence of the membrane's porosity on the electric field, *i.e.* electroporation that can be utilized to trigger drug delivery into the cells. Above a threshold magnetic field (H_{th}), the MENs loaded with the drug and optionally with the biomarker-specific antibodies (for delivery to the tumor cells) can generate localized fields large enough to open up the membrane pores in their proximity and thus allow the delivery of the drug inside the tumor cells. The drug can be released off the MENs by further increasing the field above the second critical value, H_r, necessary for overcoming the drug-MEN binding energy. This hypothesis was testified through *in vitro* studies on human ovarian carcinoma cell (SKOV-3) and healthy cell (HOMEC) lines, where a 30-Oe DC bias was applied to trigger high-specificity uptake of Paclitaxel (PTX) loaded on 30-nm $CoFe_2O_4/BaTiO_3$ core/shell MENs. The drug penetrated through the membrane and completely eradicated the tumor within 24 h without affecting the normal cells.

In the human neural network, chemical and electrical synapses transfer information between adjacent axons and dendrites directly or indirectly through electric field energy. The ability to efficiently control the network at micro- or even nano-scale can enable significant control over important brain functions. Existing non-invasive brain stimulation methods including repetitive transcranial magnetic stimulation (rTMS) and transcranial direct current stimulation (tDCS) are limited in their depth and locality. A potential solution would be to use MENs for non-invasive control of the neural network. In this approach, very low intensity external magnetic field is required to stimulate brain activity at any depth in the brain and the field can be focused to act upon MENs in any particular region of the brain. The external magnetic field generates AC signals in ME nanoparticles that are correlated with the frequency spectrum of the neural activity, which in turn causes neurons in that region to fire at similar frequencies (Figure 16d). Yue *et al.* [110] modeled the effect of MENs to non-invasively stimulate the brain of a patient with Parkinson's disease. Using the optimized values for the concentration of the 20-nm nanoparticles (with α_{ME} of 100 mV/cm·Oe in the aqueous solution) of 3×10^6 particles/cc and excitation frequency of the externally applied 300-Oe magnetic field of 80 Hz, the pulsed sequences of the electric field were brought to the levels comparable to those of healthy people.

Devices Based on the CME Effect

Magnetoelectric Random Access Memory

Ferroelectric random access memory (FeRAM) stores binary information using ferroeletric polarization states, while magnetic random access memory (MRAM) stores data bits using magnetization states and the property of magnetoresistance in multilayers. Applications of these memory devices are limited by the need for destructive read and reset operations in FeRAM, and by inconveniences in the write process in MRAM caused by the requirement for large currents for magnetization reversal. The mutual control of magnetic order and electric polarization in ME materials can be exploited to combine the best features of FeRAM and MRAM, and thus create a novel type of non-volatile RAM, *i.e.* MeRAM with multiple memory states [114]. This approach would offer greater memory density, reduced power consumption, and improved thermal stability. Bibes *et al.* [115] proposed a hybrid MeRAM device that combines the ME coupling in a multiferroic (consisting of FE and AFM phases) layer with the interfacial exchange bias between the multiferroic AFM phase and a FM layer to induce magnetization switching through voltage control. One challenging task in the realization of practical MeRAM devices is the achievement of reversible 180° deterministic switching of magnetization, on which existing magnetic memories, including hard disk drives and MRAMs, rely upon. Heron *et al.* [116] approached this issue through E-field control of exchange bias in multiferroic/magnetic ($CoFe/BiFeO_3$) heterostructures. Xue *et al.* [117] demonstrated that the E-field induced near 180° magnetization switching in the AFM/FM/FE heterostructures through voltage-controlled tuning of exchange bias and through manipulation of coercive fields. The irreversibility of the E-field induced near 180° magnetization switching in AFM/FM/FE heterostructures was resolved by employing a magnetic impulse, which could switch the magnetization back and lead to a continuous magnetization switching [118]. The heterostructure was formed by epoxy gluing of the magnetron sputtering deposited exchange coupled multilayer structure of NiFe/NiCoO/glass to a (011) cut PZN-PT single crystal, with the magnetic easy axis along the [100] or [01−1] PZN-PT crystal directions (Figure 17a). Switching magnetization was very close to the saturation magnetization (from ~M_s to −0.95 M_s along PZN-PT [01−1]; from 0.76 M_s to −0.9 M_s along PZN-PT [100]), which implies that near 180° magnetization reversal was observed in the NiFe/NiCoO/glass/PZN-PT heterostructure. The reported findings are a significant advancement in realizing the voltage writing of magnetic bits in MeRAM.

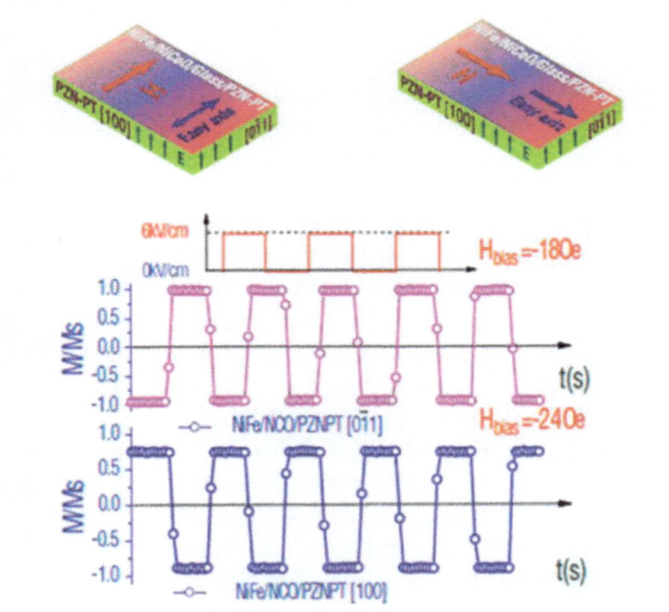

Figure 17. Magnetization switching by E-field in NiFe/NiCoO/glass/PZN-PT heterostructure-based MeRAM [117].

Phase Shifters

Microwave phase shifters are important elements for radar applications, telecommunications, oscillators and phased array antenna systems. A wide variety of phase shifters based on semiconductors, ferrites, and ferroelectrics have been developed [119,120,121]. The ferrite-based phase shifters are based on the Faraday rotation of electromagnetic radiation in magnetized ferrite rods in waveguides. Large magnetic bias fields are required for phase tuning, which involves huge power dissipation; consequently they cannot be miniaturized in size or made compatible with integrated circuit technologies. A second class of microwave phase shifters is based on ferroelectric materials. Their distinguishing features are fast electric tunability and low power consumption. However, such phase shifters are very lossy at frequencies above 1–5 GHz. A ferrite-ferroelectric layered structure, such as a ME composite, opens up the possibility of dual-tunable microwave devices, which offers higher efficiency, lower noise, compact size and lightweight compared to conventional microwave deceives. Following theoretical investigations, Ustinov *et al.* [122] reported the development of such ME devices. Their ME dual phase shifter based on FMR is a bilayer ME composite consisting of a ferrite YIG layer with a thickness of 5.7 μm and a ferroelectric BST layer with a thickness of 500 μm. The schematic structure of the cross-section of a ME dual phase shifter is displayed in Figure 18a. The electric field control of the phase shift arises through a strong ME coupling in between the ferrite and ferroelectric layers. For an applied electric field E = 20 kV/cm, a maximum differential phase shift, $\Delta\varphi$ = 650° was achieved. The estimated figure of merit of the phase shifter was about 25 °/dB. An electric field tunable YIG/PZT phase shifter was also designed and characterized by Tatarenko *et al.* [123]. For E = 5–8 kV/cm and $\Delta\varphi$ = 90–1800, an insertion loss of 1.5–4 dB was obtained. The observed insertion loss was somewhat closer to the desirable 0.5 dB needed for practical applications. Recently, bilayers of hexaferrites and piezoelectric PZT were used for novel design of phase shifters. For a strip-line Zn_2Y/PMN-PT phase shifter, data on the differential phase shift *versus* E indicated a differential phase shift of 500 for E = 12 kV/cm, a linear variation in the phase shift with E, and an insertion loss of 4–8 dB [124].

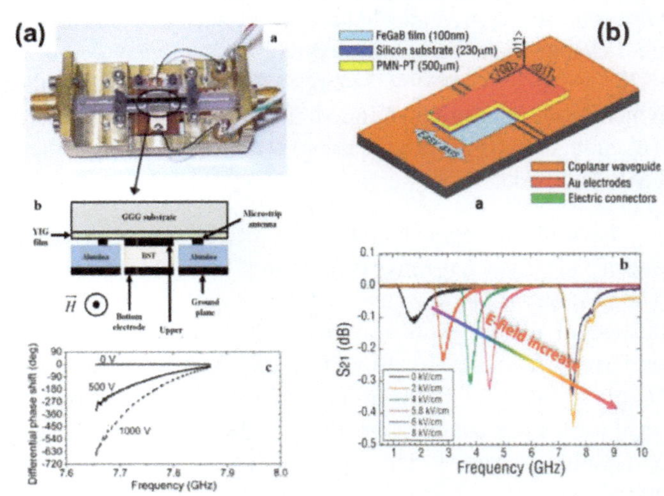

Figure 18. (**a**) Experimental prototype of the YIG/BST phase shifter, schematic of cross-section of the device and the YIG/BST layered structure, and electrically induced differential phase shift [122]; (**b**) Schematic of sample configuration and microwave measurement setup, and electric field dependence of the transmission coefficient (S_{21}) spectra of the FeGaB/PZN-PT resonator [132,133].

Resonators

Resonators are front-end elements in phased array radars, and are also used in filters and phase shifters. At the ferromagnetic resonance (FMR), a ferromagnetic material is subjected to high-frequency excitations [125]. FMR devices based on soft magnetic materials are utilized in microwave signal processing devices, resonators, band-pass/band-stop filters and phase shifters. Such devices require a magnetic field to operate and to achieve frequency tuning, which creates disadvantages like slow operation, high noise levels, large power consumption, and limited miniaturization. By replacing ferrite materials with ME composites, electrical tuning of FMR is possible. Fetisov *et al.* [126] experimentally demonstrated an electric field tunable 1–10 GHz microwave planar resonator using a YIG/PZT composite. The device consisted of a single micro strip (50 mm thick and 3 mm long) deposited onto an Al_2O_3 substrate. For a signal input of 2–10 GHz, 0.1 mW of power was applied to the microstrip transducer. Low input power was chosen to prevent heating of the sample due to power absorption at the FMR. For zero applied electric field (E = 0), the spectra contained a well-defined FMR absorption peak at about 3.5 GHz, with a maximum insertion loss of 45 dB and a 3-dB line-width of $\Delta f = 3.4$ MHz. The off-resonance loss could be partially due to ferroelectric losses in the PZT. When the electric field E = 10 kV/cm was applied, the FMR peak shifted by 40 MHz to a lower frequency. When the electric field E = −10 kV/cm was applied by reversing the voltage applied to the PZT, the peak up-shifted by 38 MHz. The converse ME coupling constant $A = \Delta f/E \approx 4$ MHz cm/kV, and the voltage tuning for E = ±10 kV/cm was only about 2% of the central frequency. Li *et al.* [127] reported X-band resonator studies using epitaxial nickel ferrite ($NiFe_2O_4$) films (thickness of about 2.0 μm) deposited on PZN-PT or PMN-PT substrates using direct liquid injection chemical vapor deposition. A strong ME coupling resulting in large shifts in the FMR profile was observed, and were due to electrostatic field-induced anisotropic magnetic field changes. The data indicated an ME coefficient $A = \delta H/E$ = 20 Oe·cm/kV (or 60 MHz cm/kV) at 9 GHz and 11 GHz, which is an order of magnitude higher than that for YIG/PZT. The main drawback with the use of ferrite material is the large FMR line-width.

Recently, hexagonal ferrites have been used for resonators over a wide frequency range from 15 to 110 GHz [128,129,130]. The Y-($Ba_2Zn_2Fe_{12}O_{22}$:ZnY) and Z-type ($Ba_3Co_2Fe_{24}O_{41}$) hexagonal ferrites with easy-plane magnetic anisotropy are preferred for use at 10–40 GHz [128]. M-type hexaferrites, $MFe_{12}O_{19}$ (M = Ba, Sr) with uniaxial magnetic anisotropy are ideal for the frequency range 40–75 GHz [129]. The Al doped M-type hexaferrites further increase the resonance frequency, and are suitable for 50–110 GHz devices [130]. Tatarenko *et al.* [131] reported the converse ME effect over the 8–25 GHz range in bilayers of single crystal ZnY and polycrystalline PZT or single crystal PMN-PT. The resonator was tuned by 120 MHz with E = 12 kV/cm, and corresponding ME coupling strength A was about 10 MHz cm/kOe. Lou *et al.* [132,133] reported a layered ferroic FeGaB and single crystal piezoelectric PZN-PT-based resonator. They observed that a large electric field induced an effective magnetic anisotropy field of 750 Oe with a narrow FMR line-width of 50 Oe at X-band. This resonator achieved a large electric field tunable FMR frequency range between 1.75 and 7.57 GHz even at zero magnetic bias field, as shown in Figure 18b, which corresponds to a mean tunable frequency per unit electric field of about 970 MHz cm/kOe. The giant tunable magnetic field and FMR frequency of the FeGaB/PZN-PT resonator makes it a promising candidate for wide-band electric field tunable microwave devices. Compared to conventional tunable microwave magnetic devices, which are tuned by magnetic fields, these electrostatically tunable microwave multiferroic devices are much less noisy, and more energy efficient, compact, and lightweight.

Inductors

One of the three fundamental components for electronic circuits, tunable inductors, find widespread use in various applications such as communication systems and power electronics. Most tunable inductors are magnetically tuned by using electromagnets. Since electromagnets are typically bulky, noisy and energy consuming, they are inconvenient for use in such applications. Efforts have been made to develop electronically tunable inductors that have large tunability, high quality factors, and low energy consumption. For example, microelectromechanical systems (MEMS)-based tunable inductors can have very high quality factors, although they have a limited tunable range of 20% and are difficult to fabricate. The use of ME composites as a core material could be advantageous for tunable inductor applications since tuning magnetic properties by electric field is possible even at low fields. A variety of ME composite-based inductors have been made using different structures. Tunability is calculated from inductance (L) using the formula $[\Delta L/L_0 = (L - L_0)/L_0] \times 100$. Fang *et al.* [134] have reported a PZT/MnZn/PZT composite inductor that showed a small tunable inductance range $\Delta L/L_{min}$ of up to 20%. Lou *et al.* [135] reported a composite consisting of a combination of PZT slab and two layers of amorphous Metglas ribbons with a large tunable inductance range $\Delta L/L_{min}$ of up to 450%, together with improved quality factors (Figure 19a). This tunability of inductance and increased quality factor were due to strong ME coupling in the composite core, which led to electric field-induced permeability change. However, due to the relatively large thickness of the Metgals ribbons (~25 μm), excessive eddy current loss limits the operational frequency range of the conventional Metglas-based inductors to be <100–200 kHz, and leads to low quality factors of these inductors at microwave frequencies. Recently, Lin *et al.* [136] investigated the effect of ferromagnetic layer thickness on the performance of ME inductors. The inductance tunable range of $\Delta L/L_{min}$ of 370% was achieved, together with a significant 3-fold enhancement in quality factor, with an increase in the operational frequency range. To miniaturize the inductor, a ring inductor was designed by Mandal *et al.* [137] using insulating single phase $Bi_{0.7}Dy_{0.3}FeO_3$ multiferroic material, but the tunability of the inductor was only about 18%.

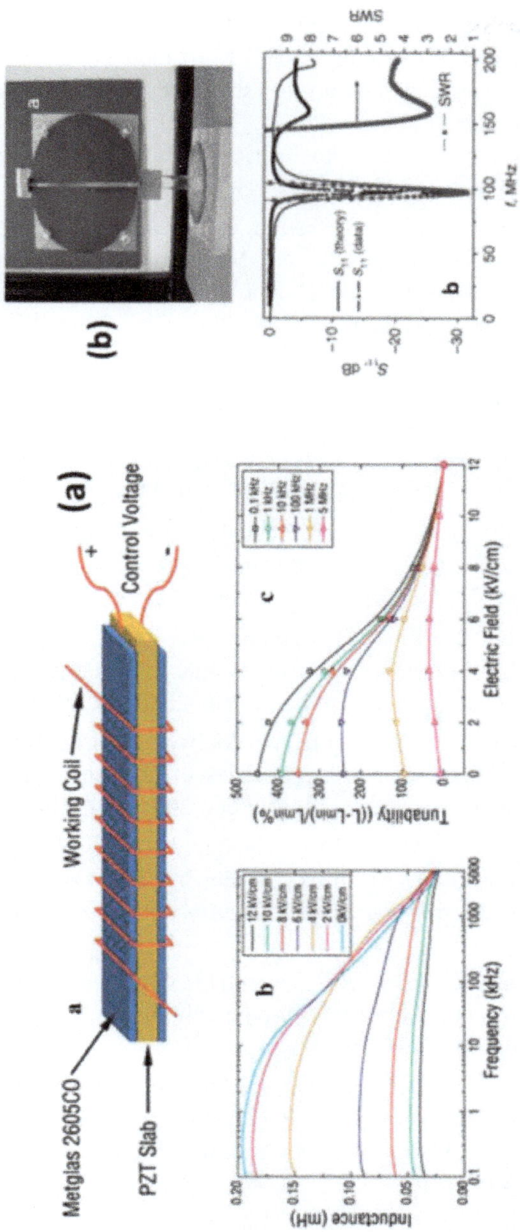

Figure 19. (**a**) Schematic of the magnetoelectric inductor, inductance spectra, and inductance tunability at different frequencies and electric field [135]; (**b**) Miniature microstrip antenna on NZFO/BST composite substrate, measured and estimated return loss and measured SWR for the antenna [140].

ME Antenna

In recent years, reduction in antenna size, which is suitable for lower frequency bands where wavelength is larger, has received significant attention because of the demands of numerous mobile communication systems. Several approaches have been devised to achieve antenna size reductions [138,139]. The first is to use capacitive or inductive loadings and/or meandered lines to obtain slow-wave resonance. The main drawbacks to these approaches are increased ohmic losses and narrow bandwidths. The second is to reduce the wavelength of the structure using dielectric materials, but this is prone to surface-wave excitations and corresponding losses of efficiency. A new approach was recently found that uses a novel engineered ME composite material. A half-wavelength resonant microstrip antenna on an ME composite of nickel zinc ferrites ($Ni_{1-x}Zn_xFe_2O_4$:NZFO) and ferroelectric BST was made by Petrov *et al.* [140]. They concentrated on the miniaturization of the 100-MHz antenna, and its effects on impedance matching, bandwidth and efficiency (Figure 19b). The ME composite disc with a diameter of 22 cm and thickness of 0.85 cm was metallized on one side, and the microstrip had a length of 22 cm and width of 0.65 cm. A vector network analyzer was used for measurements of return and transmission losses. The measured SWR at resonance was close to 1.3 and indicated good impedance matching with free space. The absence of reflection at the boundary between the substrate of the antenna and the surrounding medium reduced the energy absorbed in the substrate. The helical antennas were fabricated using Co_2Z type hexaferrite and ferrite substrates [141,142]. The resonant frequency and bandwidth at 10 dB for the antennas were measured to be 195 MHz and 27 MHz, and 209 MHz and 41 MHz, respectively. A helical antenna was also fabricated on (Co/Ti) doped BaM substrate (45 × 11 × 3.8 mm^3), and the antenna was then mounted on an FR4 board (10 × 5 mm^2) with an 8 × 5 mm^2 copper ground [143]. A 50-Ω coaxial cable was used to feed the antenna. The antenna performance was characterized with a network analyzer and a Wheeler cap. The antenna resonant frequency shifted from 231 to 201 MHz with increasing permeability from 4.51 to 8.58. This data indicates that the factor of miniaturization of the antenna is about 7% with a permeability of 8.58. It is worth noting here that the low-loss (Co/Ti) doped BaM is an excellent soft magnetic material for miniature antenna applications in a very high-frequency (30–300 MHz) range.

SUMMARY

In recent years, much attention has been paid towards exploiting the promising technological potential of magnetoelectric coupling in materials. Although there has been tremendous progress in the development of both bulk and film-based ME composites, efforts are likely to continue in the pursuit of desired levels of performance. To realize strong ME coupling and high performance ME composites it is essential to select the appropriate combination of piezoelectric and magnetostrictive materials with better properties; adopt a suitable fabrication approach and configuration of the composite constituents; optimize interfacial coupling; reduce the noise contributions to the output signal; tune and control other dynamic parameters; and achieve a comprehensive understanding of the contributing factors and physical phenomena by theoretical modeling. To effectively employ the coupling across interfaces in ME composites and to achieve giant ME effects, the nature of the interfaces should be unraveled. ME-based devices are compact, light-weight, fast in response, less noisy, and energy-efficient and thus are viable alternatives for some existing conventional electrical and magnetic devices. Although some prototype devices based on bulk ME composites have been demonstrated, challenges still exist in terms of obtaining control over materials properties and fabrication, and optimization of the device performance. Further, the reliability of the ME composite systems needs to be evaluated in practical device designs. While much work remains to be done to accomplish on-chip integration of the ME devices, there is significant scope, and many avenues can be explored.

ACKNOWLEDGMENTS

This work was supported by the Global Frontier R&D Program (Grant Nos. 2013M3A6B1078872) on Center for Hybrid Interface Materials (HIM) of the National Research Foundation (NRF) of Korea funded by the Ministry of Science, ICT & Future Planning, Korea and internal R&D program of Korea Institute of Materials Science (KIMS). S.P. would like to acknowledge the financial support from Office of Basic Energy Science, Department of Energy, USA through Grant No. DE-FG02-06ER46290.

REFERENCES

1. Wang, Y.; Hu, J.; Lin, Y.; Nan, C.-W. Multiferroic magnetoelectric composite nanostructures.*NPG Asia Mater.* 2010, *2*, 61–68.
2. Priya, S.; Islam, R.; Dong, S.; Viehland, D. Recent advancements in magnetoelectric particulate and laminate composites. *J. Electroceram.* 2007, *19*, 149–166.
3. Nan, C.-W.; Bichurin, M.I.; Dong, S.; Viehland, D.; Srinivasan, G. Multiferroic magnetoelectric composites: Historical perspective, status, and future directions. *J. Appl. Phys.* 2008, *103*, 031101.
4. Vaz, C.A.F.; Hoffman, J.; Ahn, C.H.; Ramesh, R. Magnetoelectric coupling effects in multiferroic complex oxide composite structures. *Adv. Mater.* 2010, *22*, 2900–2918.
5. Ryu, J.; Priya, S.; Uchino, K.; Kim, H.-E. Magnetoelectric Effect in Composites of Magnetostrictive and Piezoelectric Materials. *J. Electroceram.* 2002, *8*, 107–119.
6. Fang, Z.; Zhang, Q.; Datta, S. Ultra sensitive magnetic sensors integrating the giant magnetoelectric effect with advanced microelectronics. Ph.D. Thesis, Pennsylvania State University, University Park, PA, USA, December 2011.
7. Cho, K.-H.; Priya, S. Direct and converse effect in magnetoelectric laminate composites. *Appl. Phys. Lett.* 2011, *98*, 232904.
8. Newnham, R.E.; Skinner, D.P.; Cross, L.E. Connectivity and piezoelectric-pyroelectric composites. *Mater. Res. Bull.* 1978, *13*, 525–536.
9. Nan, C.-W.; Jia, Q. Obtaining ultimate functionalities in nanocomposites: Design, control, and fabrication. *MRS Bull.* 2015, *40*, 719–724.
10. Park, C.S.; Ryu, J.; Choi, J.J.; Park, D.S.; Ahn, C.W.; Priya, S. Giant Magnetoelectric Coefficient in 3–2 Nanocomposite Thick Films. *Jpn. J. Appl. Phys.* 2009, *48*, 080204.
11. Gillette, S.M.; Fitchorov, T.; Obi, O.; Jiang, L.; Hao, H.; Wu, S.; Chen, Y.; Harris, V.G. Effects of intrinsic magnetostriction on tube-topology magnetoelectric sensors with high magnetic field sensitivity. *J. Appl. Phys.* 2014, *115*, 17C734.
12. Yan, Y.; Priya, S. Multiferroic Magnetoelectric Composites/Hybrids. In *Hybrid and Hierarchical Composite Materials*; Kim, C.-S., Randow, C., Sano, T., Eds.; Springer International Publishing: Cham, Switzerland, 2015; pp. 95–160.
13. Islam, R.; Rong, C.-B.; Liu, J.P.; Priya, S. Effect of gradient composite structure in cofired bilayer composites of $Pb(Zr_{0.56}Ti_{0.44})O_3$–$Ni_{0.6}Zn_{0.2}Cu_{0.2}Fe_2O_4$ system on magnetoelectric coefficient. *J. Mater. Sci.* 2008, *43*, 6337–6343.
14. Srinivasan, G.; DeVreugd, C.P.; Flattery, C.S.; Laletsin, V.M.; Paddubnaya, N. Magnetoelectric interactions in hot-pressed nickel zinc ferrite and lead zirconante titanate composites. *Appl. Phys. Lett.* 2004, *85*, 2550–2552.
15. Jiang, Q.H.; Shen, Z.J.; Zhou, J.P.; Shi, Z.; Nan, C.-W. Magnetoelectric composites of nickel ferrite and lead zirconnate titanate prepared by spark plasma sintering. *J. Eur. Ceram. Soc.*2007, *27*, 279–284.
16. Islam, R.A.; Priya, S. Synthesis of high magnetoelectric coefficient composites using annealing and aging route. *Int. J. Appl. Ceram. Tech.* 2006, *3*, 353–363.
17. Liu, R.; Zhao, Y.; Huang, R.; Zhao, Y.; Zhou, H. Multiferroic ferrite/perovskite oxide core/shell nanostructures. *J. Mater. Chem.* 2010, *20*, 10665–10670.
18. Islam, R.A.; Bedekar, V.; Poudyal, N.; Liu, J.P.; Priya, S. Magnetoelectric properties of core-shell particulate nanocomposites. *J. Appl. Phys.* 2008, *104*, 104111.
19. Shi, Z.; Nan, C.W.; Zhang, J.; Cai, N.; Li, J.-F. Magnetoelectric effect of $Pb(Zr,Ti)O_3$ rod arrays in a $(Tb,Dy)Fe_2$/epoxy medium. *Appl. Phys. Lett.* 2005, *87*, 012503.
20. Lam, K.H.; Lo, C.Y.; Chan, H.L.W. Frequency response of magnetoelectric 1–3-type composites. *J. Appl. Phys.* 2010, *107*, 093901.
21. Ma, J.; Shi, Z.; Nan, C.W. Magnetoelectric properties of composites of single $Pb(Zr,Ti)O_3$ rods and Terfenol-D/Epoxy with a single-period of 1–3-type structure. *Adv. Mater.* 2007, *19*, 2571–2573.
22. Islam, R.A.; Ni, Y.; Khachaturyan, A.G.; Priya, S. Giant magnetoelectric effect in sintered multilayered composite structures. *J. Appl. Phys.* 2008, *104*, 044103.
23. Dong, S.; Zhai, J.; Li, J.; Viehland, D. Near-ideal magnetoelectricity in high-permeability magnetostrictive/piezofiber laminates with a (2–1) connectivity. *Appl. Phys. Lett.* 2006, *89*, 252904.
24. Wang, Y.; Gray, D.; Berry, D.; Gao, J.; Li, M.; Li, J.; Viehland, D. An extremely low equivalent magnetic noise magnetoelectric sensor. *Adv. Mater.* 2011, *23*, 4111–4114.
25. Mori, K.; Wuttig, M. Magnetoelectric coupling in Terfenol-D/polyvinylidenedifluoride composites. *Appl. Phys. Lett.* 2002, *81*, 100–101.
26. Silva, M.; Reis, S.; Lehmann, C.S.; Martins, P.; Lanceros-Mendez, S.; Lasheras, A.; Gutiérrez, J.; Barandiarán, J.M. Optimization of the Magnetoelectric Response of Poly(vinylidene fluoride)/Epoxy/Vitrovac Laminates. *ACS Appl. Mater. Inter.* 2013, *5*, 10912–10919.
27. Lawes, G.; Srinivasan, G. Introduction to magnetoelectric coupling and multiferroic films. *J. Phys. D Appl. Phy.* 2011, *44*, 243001.
28. McDannald, A.; Staruch, M.; Sreenivasulu, G.; Cantoni,

C.; Srinivasan, G.; Jain, M. Magnetoelectric coupling in solution derived 3–0 type $PbZr_{0.52}Ti_{0.48}O_3$:x$CoFe_2O_4$nanocomposite films. *Appl. Phys. Lett.* 2013, *102*, 122905.

29. Zheng, H.; Wang, J.; Lofland, S.E.; Ma, Z.; Mohaddes-Ardabili, L.; Zhao, T.; Salamanca-Riba, L.; Shinde, S.R.; Ogale, S.B.; Bai, F.; *et al.* Multiferroic $BaTiO_3$-$CoFe_2O_4$ nanostructures.*Science* 2004, *303*, 661–663.

30. Aimon, N.M.; Kim, D.H.; Sun, X.; Ross, C.A. Multiferroic behavior of templated $BiFeO_3$-$CoFe_2O_4$ self-assembled nanocomposites. *ACS Appl. Mater. Interf.* 2015, *7*, 2263–2268.

31. Srinivasan, G. Layered multiferroic composites. In *Composite Magnetoelectrics*; Srinivasan, G., Priya, S., Sun, N.X., Eds.; Woodhead Publishing: Cambridge, UK, 2015; pp. 55–70.

32. Zhou, Z.; Sun, N.X. Multiferroic nanostructures. In *Composite Magnetoelectrics*; Srinivasan, G., Priya, S., Sun, N.X., Eds.; Woodhead Publishing: Cambridge, UK, 2015; pp. 71–86.

33. Lage, E.; Kirchhof, C.; Hrkac, V.; Kienle, L.; Jahns, R.; Knöchel, R.; Quandt, E.; Meyners, D. Exchange biasing of magnetoelectric composites. *Nat. Mater.* 2012, *11*, 523–529.

34. Greve, H.; Woltermann, E.; Quenzer, H.-J.; Wagner, B.; Quandt, E. Giant magnetoelectric coefficients in $(Fe_{90}Co_{10})_{78}Si_{12}B_{10}$-AlN thin film composites. *Appl. Phys. Lett.* 2010, *96*, 182501.

35. Nair, S.S.; Pookat, G.; Saravanan, V.; Anantharaman, M.R. Lead free heterogeneous multilayers with giant magneto electric coupling for microelectronics/ microelectromechanical systems applications. *J. Appl. Phys.* 2013, *114*, 064309.

36. Srinivasan, G.; Rasmussen, E.T.; Gallegos, J.; Srinivasan, R.; Bokhan, Y.I.; Laletin, V.M. Magnetoelectric bilayer and multilayer structures of magnetostrictive and piezoelectric oxides. *Phys. Rev. B.* 2001, *64*, 214408.

37. Kambale, R.C.; Ryu, J.; Patil, D.R.; Chai, Y.S.; Kim, K.H.; Yoon, W.-H.; Jeong, D.-Y.; Park, D.-S.; Kim, J.-W.; Choi, J.-J.; *et al.* Colossal magnetoelectric response of PZT thick films on Ni substrates with a conductive $LaNiO_3$ electrode. *J. Phys. D Appl. Phy.* 2013, *46*, 092002.

38. Zhou, D.; Jian, G.; Zheng, Y.; Gong, S.; Shi, F. Electrophoretic deposition of $BaTiO_3$/ $CoFe_2O_4$multiferroic composite films. *Appl. Surf. Sci.* 2011, *257*, 7621–7626.

39. Palneedi, H.; Maurya, D.; Kim, G.-Y.; Priya, S.; Kang, S.-J.L.; Kim, K.-H.; Choi, S.-Y.; Ryu, J. Enhanced off-resonance magnetoelectric response in laser annealed PZT thick film grown on magnetostrictive amorphous metal substrate. *Appl. Phys. Lett.* 2015, *107*, 012904.

40. Pan, D.A.; Bai, Y.; Chu, W.Y.; Qiao, L.J. Ni–PZT–Ni trilayered magnetoelectric composites synthesized by electro-deposition. *J. Phys. Condens. Matter.* 2008, *20*, 025203.

41. Bi, K.; Wang, Y.G.; Wu, W.; Pan, D.A. The large magnetoelectric effect in Ni–lead zirconium titanate–Ni trilayers derived by electroless deposition. *J. Phys. D Appl. Phy.* 2010, *43*, 132002.

42. Yu, H.; Zeng, M.; Wang, Y.; Wan, J.G.; Liu, J.-M. Magnetoelectric resonance-bandwidth broadening of Terfenol-D/epoxy-Pb(Zr,Ti)O_3 bilayers in parallel and series connections.*Appl. Phys. Lett.* 2005, *86*, 032508.

43. Park, C.-S.; Ahn, C.-W.; Ryu, J.; Yoon, W.-H.; Park, D.-S.; Kim, H.-E.; Priya, S. Design and characterization of broadband magnetoelectric sensor. *J. Appl. Phys.* 2009, *105*, 094111.

44. Park, C.-S.; Priya, S. Broadband/wideband magnetoelectric response. *Adv. Condens. Matter Phys.* 2012, *2012*, 323165.

45. Patil, D.R.; Kambale, R.C.; Chai, Y.; Yoon, W.-H.; Jeong, D.-Y.; Park, D.-S.; Kim, J.-W.; Choi, J.-J.; Ahn, C.-W.; Hahn, B.-D.; *et al.* Multiple broadband magnetoelectric response in thickness-controlled Ni/ 011 $Pb(Mg_{1/3}Nb_{2/3})O_3$-Pb(Zr,Ti)O_3 single crystal/Ni laminates. *Appl. Phys. Lett.* 2013, *103*, 052907.

46. Patil, D.R.; Chai, Y.; Kambale, R.C.; Jeon, B.-G.; Yoo, K.; Ryu, J.; Yoon, W.-H.; Park, D.-S.; Jeong, D.-Y.; Lee, S.-G.; *et al.* Enhancement of resonant and non-resonant magnetoelectric coupling in multiferroic laminates with anisotropic piezoelectric properties. *Appl. Phys. Lett.*2013, *102*, 062909.

47. Kambale, R.C.; Yoon, W.-H.; Park, D.-S.; Choi, J.-J.; Ahn, C.-W.; Kim, J.-W.; Hahn, B.-D.; Jeong, D.-Y.; Lee, B.-C.; Chung, G.-S.; *et al.* Magnetoelectric properties and magnetomechanical energy harvesting from stray vibration and electromagnetic wave by $Pb(Mg_{1/3}Nb_{2/3})O_3$-Pb(Zr,Ti)O_3 single crystal/Ni cantilever. *J. Appl. Phys.* 2013, *113*, 204108.

48. Priya, S.; Yang, S.C.; Maurya, D.; Yan, Y. Recent advances in piezoelectric and magnetoelectric materials phenomena. In *Composite Magnetoelectrics*; Srinivasan, G., Priya, S., Sun, N.X., Eds.; Woodhead Publishing: Cambridge, UK, 2015; pp. 103–157.

49. Yan, Y.; Cho, K.-H.; Maurya, D.; Kumar, A.; Kalinin, S.; Khachaturyan, A.; Priya, S. Giant energy density in 001 -textured $Pb(Mg_{1/3}Nb_{2/3})O_3$-$PbZrO_3$-$PbTiO_3$ piezoelectric ceramics.*Appl. Phys. Lett.* 2013, *102*, 042903.

50. Wang, Y.; Zhao, X.; Jiao, J.; Zhang, Q.; Di, W.; Luo, H.; Leung, C.M.; Or, S.W. Lead-free magnetoelectric laminated composite of Mn-doped $Na_{0.5}Bi_{0.5}TiO_3$–$BaTiO_3$ single crystal and $Tb_{0.3}Dy_{0.7}Fe_{1.92}$ alloy. *J. Alloy Compd.* 2010, *496*, L4–L6.

51. Zhou, Y.; Maurya, D.; Yan, Y.; Srinivasan, G.; Quandt, E.; Priya, S. Self-biased magnetoelectric composites: An overview and future perspectives. *Energy Harvest. Syst.* 2016, *3*, 1–42.

52. Mandal, S.K.; Sreenivasulu, G.; Petrov, V.M.; Srinivasan, G. Magnetization-graded multiferroic composite and magnetoelectric effects at zero bias. *Phys. Rev. B.* 2011, *84*, 014432.

53. Raidongia, K.; Nag, A.; Sundaresan, A.; Rao, C.N.R. Multiferroic and magnetoelectric properties of core-shell $CoFe_2O_4$@$BaTiO_3$ nanocomposites. *Appl. Phys. Lett.* 2010, *97*, 062904.

54. Liu, M.; Li, X.; Imrane, H.; Chen, Y.; Goodrich, T.; Cai,

Z.; Ziemer, K.S.; Huang, J.Y.; Sun, N.X. Synthesis of ordered arrays of multiferroic $NiFe_2O_4$-$Pb(Zr_{0.52}Ti_{0.48})O_3$ core-shell nanowires.*Appl. Phys. Lett.* 2007, *90*, 152501.

55. Starr, J.D.; Andrew, J.S. Janus-type bi-phasic functional nanofibers. *Chem. Comm.* 2013, *49*, 4151–4153.

56. Andrew, J.S.; Starr, J.D.; Budi, M.A.K. Prospects for nanostructured multiferroic composite materials. *Scripta Mater.* 2014, *74*, 38–43.

57. Xie, S.; Ma, F.; Liu, Y.; Li, J. Multiferroic $CoFe_2O_4$-$Pb(Zr_{0.52}Ti_{0.48})O_3$ core-shell nanofibers and their magnetoelectric coupling. *Nanoscale* 2011, *3*, 3152–3158.

58. Xie, S.H.; Li, J.Y.; Liu, Y.Y.; Lan, L.N.; Jin, G.; Zhou, Y.C. Electrospinning and multiferroic properties of $NiFe_2O_4$–$Pb(Zr_{0.52}Ti_{0.48})O_3$ composite nanofibers. *J. Appl. Phys.* 2008, *104*, 024115.

59. Srinivasan, G. Magnetoelectric characterization techniques. In *Composite Magnetoelectrics*; Srinivasan, G., Priya, S., Sun, N.X., Eds.; Woodhead Publishing: Cambridge, UK, 2015; pp. 41–54.

60. Zhou, Y.; Apo, D.J.; Sanghadasa, M.; Bichurin, M.; Petrov, V.M.; Priya, S. Magnetoelectric energy harvester. In *Composite Magnetoelectrics*; Srinivasan, G., Priya, S., Sun, N.X., Eds.; Woodhead Publishing: Cambridge, UK, 2015; pp. 161–207.

61. Dong, S.; Zhai, J.; Xing, Z.; Li, J.; Viehland, D. Giant magnetoelectric effect (under a dc magnetic bias of 2 Oe) in laminate composites of FeBSiC alloy ribbons and $Pb(Zn_{1/3}Nb_{2/3})O_3$–7%$PbTiO_3$ fibers. *Appl. Phys. Lett.* 2007, *91*, 022915.

62. Park, C.-S.; Cho, K.-H.; Arat, M.A.; Evey, J.; Priya, S. High magnetic field sensitivity in $Pb(Zr,Ti)O_3$–$Pb(Mg_{1/3}Nb_{2/3})O_3$ single crystal/Terfenol-D/Metglas magnetoelectric laminate composites. *J. Appl. Phys.* 2010, *107*, 094109.

63. Ramana, M.V.; Reddy, N.R.; Sreenivasulu, G.; Kumar, K.V.S.; Murty, B.S.; Murthy, V.R.K. Enhanced mangnetoelectric voltage in multiferroic particulate $Ni_{0.83}Co_{0.15}Cu_{0.02}Fe_{1.9}O_{4-\delta}$/$PbZr_{0.52}Ti_{0.48}O_3$ composites—dielectric, piezoelectric and magnetic properties. *Curr. Appl. Phys.* 2009, *9*, 1134–1139.

64. Sreenivasulu, G.; Fetisov, L.Y.; Fetisov, Y.K.; Srinivasan, G. Piezoelectric single crystal langatate and ferromagnetic composites: Studies on low-frequency and resonance magnetoelectric effects. *Appl. Phys. Lett.* 2012, *100*, 052901.

65. Chen, Y.; Gillette, S.M.; Fitchorov, T.; Jiang, L.; Hao, H.; Li, J.; Gao, X.; Geiler, A.; Vittoria, C.; Harris, V.G. Quasi-one-dimensional miniature multiferroic magnetic field sensor with high sensitivity at zero bias field. *Appl. Phys. Lett.* 2011, *99*, 042505.

66. Zhang, N.; Fan, J.; Rong, X.; Cao, H.; Wei, J. Magnetoelectric effect in laminate composites of $Tb_{1-x}Dy_xFe_{2-y}$ and Fe-doped $BaTiO_3$. *J. Appl. Phys.* 2007, *101*, 063907.

67. Islam, R.A.; Kim, H.; Priya, S.; Stephanou, H. Piezoelectric transformer based ultrahigh sensitivity magnetic field sensor. *Appl. Phys. Lett.* 2006, *89*, 152908.

68. Feng, M.; Wang, J.-J.; Hu, J.-M.; Wang, J.; Ma, J.; Li, H.-B.; Shen, Y.; Lin, Y.-H.; Chen, L.-Q.; Nan, C.-W. Optimizing direct magnetoelectric coupling in $Pb(Zr,Ti)O_3$/Ni multiferroic film heterostructures. *Appl. Phys. Lett.* 2015, *106*, 072901.

69. Tahmasebi, K.; Barzegar, A.; Ding, J.; Herng, T.S.; Huang, A.; Shannigrahi, S. Magnetoelectric effect in $Pb(Zr_{0.95}Ti_{0.05})O_3$ and $CoFe_2O_4$ heteroepitaxial thin film composite. *Mater. Des.*2011, *32*, 2370–2373.

70. Martínez, R.; Kumar, A.; Palai, R.; Srinivasan, G.; Katiyar, R.S. Observation of strong magnetoelectric effects in $Ba_{0.7}Sr_{0.3}TiO_3$/$La_{0.7}Sr_{0.3}MnO_3$ thin film heterostructures. *J. Appl. Phys.* 2012, *111*, 104104.

71. Park, C.-S.; Khachaturyan, A.; Priya, S. Giant magnetoelectric coupling in laminate thin film structure grown on magnetostrictive substrate. *Appl. Phys. Lett.* 2012, *100*, 192904.

72. Sun, N.X.; Srinivasan, G. Voltage control of magnetism in multiferroic heterostructures and devices. *SPIN* 2012, *2*, 1240004.

73. Paluszek, M.; Avirovik, D.; Zhou, Y.; Kundu, S.; Chopra, A.; Montague, R.; Priya, S. Magnetoelectric composites for medical application. In *Composite Magnetoelectrics*; Srinivasan, G., Priya, S., Sun, N.X., Eds.; Woodhead Publishing: Cambridge, UK, 2015; pp. 297–327.

74. Dong, S.; Zhai, J.; Bai, F.; Li, J.-F.; Viehland, D. Push-pull mode magnetostrictive /piezoelectric laminate composite with an enhanced magnetoelectric voltage coefficient.*Appl. Phys. Lett.* 2005, *87*, 062502.

75. Gao, J.; Das, J.; Xing, Z.; Li, J.; Viehland, D. Comparison of noise floor and sensitivity for different magnetoelectric laminates. *J. Appl. Phys.* 2010, *108*, 084509.

76. Li, M.; Wang, Z.; Wang, Y.; Li, J.; Viehland, D. Giant magnetoelectric effect in self-biased laminates under zero magnetic field. *Appl. Phys. Lett.* 2013, *102*, 082404.

77. Wang, Y.; Hasanyan, D.; Li, M.; Gao, J.; Li, J.; Viehland, D. Equivalent magnetic noise in multi-push-pull configuration magnetoelectric composites: Model and experiment. *IEEE Trans. Ultrason. Ferroelectr. Freq. Control.* 2013, *60*, 1227–1233.

78. Das, J.; Gao, J.; Xing, Z.; Li, J.F.; Viehland, D. Enhancement in the field sensitivity of magnetoelectric laminate heterostructures. *Appl. Phys. Lett.* 2009, *95*, 092501.

79. Gao, J.; Gray, D.; Shen, Y.; Li, J.; Viehland, D. Enhanced dc magnetic field sensitivity by improved flux concentration in magnetoelectric laminates. *Appl. Phys. Lett.* 2011, *99*, 153502.

80. Wang, Y.; Li, M.; Hasanyan, D.; Gao, J.; Li, J.; Viehland, D. Geometry-induced magnetoelectric effect enhancement and noise floor reduction in Metglas/piezofiber sensors.*Appl. Phys. Lett.* 2012, *101*, 092905.

81. Li, M.; Berry, D.; Das, J.; Gray, D.; Li, J.; Viehland, D. Enhanced sensitivity and reduced noise floor in magnetoelectric laminate sensors by an improved lamination process. *J. Amer. Ceram. Soc.* 2011, *94*, 3738–3741.

82. Wang, Y.J.; Gao, J.Q.; Li, M.H.; Shen, Y.; Hasanyan, D.;

Li, J.F.; Viehland, D. A review on equivalent magnetic noise of magnetoelectric laminate sensors. *Philos. Trans. R. Soc. London Ser. A* 2014, *372*.

83. Gao, J.; Wang, Y.; Li, M.; Shen, Y.; Li, J.; Viehland, D. Quasi-static (f<10−2 Hz) frequency response of magnetoelectric composites based magnetic sensor. *Mater. Lett.* 2012, *85*, 84–87.

84. Shen, Y.; Gao, J.; Shen, L.; Gray, D.; Li, J.; Finkel, P.; Viehland, D.; Zhuang, X.; Saez, S.; Dolabdjian, C. Analysis of the environmental magnetic noise rejection by using two simple magnetoelectric sensors. *Sens. Actuators A Phys.* 2011, *171*, 63–68.

85. Fang, C.; Jiao, J.; Ma, J.; Lin, D.; Xu, H.; Zhao, X.; Luo, H. Significant reduction of equivalent magnetic noise by in-plane series connection in magnetoelectric Metglas/Mn-doped $Pb(Mg_{1/3}Nb_{2/3})O_3$-$PbTiO_3$ laminate composites. *J. Phys. D Appl. Phy.* 2015, *48*, 465002.

86. Stephan, M.; Robert, J.; Henry, G.; Eckhard, Q.; Reinhard, K.; Bernhard, W. MEMS magnetic field sensor based on magnetoelectric composites. *J. Micromech. Microeng.* 2012, *22*, 065024.

87. Zhao, P.; Zhao, Z.; Hunter, D.; Suchoski, R.; Gao, C.; Mathews, S.; Wuttig, M.; Takeuchi, I. Fabrication and characterization of all-thin-film magnetoelectric sensors. *Appl. Phys. Lett.*2009, *94*, 243507.

88. Lee, D.G.; Kim, S.M.; Yoo, Y.K.G.; Han, J.H.; Chun, D.W.; Kim, Y.-C.; Kim, J.; Hwang, K.S.; Kim, T.S.; Jo, W.W.; *et al.* Ultra-sensitive magnetoelectric microcantilever at a low frequency.*Appl. Phys. Lett.* 2012, *101*, 182902.

89. Ramsden, E. *Hall-Effect Sensors: Theory and Applications*; Newnes Publications: Burlington, NJ, USA, 2006; pp. 1–9.

90. Leung, C.M.; Or, S.W.; Zhang, S.; Ho, S.L. Ring-type electric current sensor based on ring-shaped magnetoelectric laminate of epoxy-bonded $Tb_{0.3}Dy_{0.7}Fe_{1.92}$ short-fiber/NdFeB magnet magnetostrictive composite and Pb(Zr, Ti)O_3 piezoelectric ceramic. *J. Appl. Phys.*2010, *107*, 09D918.

91. Lu, C.; Li, P.; Wen, Y.; Yang, A.; Yang, C.; Wang, D.; He, W.; Zhang, J. Magnetoelectric composite Metglas/PZT-based current sensor. *IEEE Trans. Magn.* 2014, *50*, 1–4.

92. Zhang, S.; Leung, C.M.; Kuang, W.; Or, S.W.; Ho, S.L. Concurrent operational modes and enhanced current sensitivity in heterostructure of magnetoelectric ring and piezoelectric transformer. *J. Appl. Phys.* 2013, *113*, 17C733.

93. Yu, X.; Lou, G.; Chen, H.; Wen, C.; Lu, S. A slice-type magnetoelectric laminated current sensor. *IEEE Sens. J.* 2015, *15*, 5839–5850.

94. Patil, D.R.; Zhou, Y.; Kang, J.-E.; Sharpes, N.; Jeong, D.-Y.; Kim, Y.-D.; Kim, K.H.; Priya, S.; Ryu, J. Anisotropic self-biased dual-phase low frequency magneto-mechano-electric energy harvesters with giant power densities. *APL Mater.* 2014, *2*, 046102.

95. Ryu, J.; Kang, J.-E.; Zhou, Y.; Choi, S.-Y.; Yoon, W.-H.; Park, D.-S.; Choi, J.-J.; Hahn, B.-D.; Ahn, C.-W.; Kim, J.-W.; *et al.* Ubiquitous magneto-mechano-electric generator. *Energy Environ. Sci.* 2015, *8*, 2402–2408.

96. Zhou, Y.; Apo, D.J.; Priya, S. Dual-phase self-biased magnetoelectric energy harvester. *Appl. Phys. Lett.* 2013, *103*, 192909.

97. Liu, G.; Ci, P.; Dong, S. Energy harvesting from ambient low-frequency magnetic field using magneto-mechano-electric composite cantilever. *Appl. Phys. Lett.* 2014, *104*, 032908.

98. Lasheras, A.; Gutiérrez, J.; Reis, S.; Sousa, D.; Silva, M.; Martins, P.; Lanceros-Mendez, S.; Barandiarán, J.M.; Shishkin, D.A.; Potapov, A.P. Energy harvesting device based on a metallic glass/PVDF magnetoelectric laminated composite. *Smart Mater. Struct.* 2015, *24*, 065024.

99. Li, P.; Wen, Y.; Liu, P.; Li, X.; Jia, C. A magnetoelectric energy harvester and management circuit for wireless sensor network. *Sens. Actuators A. Phys.* 2010, *157*, 100–106.

100. Han, J.; Hu, J.; Wang, S.X.; He, J. Magnetic energy harvesting properties of piezofiber bimorph/NdFeB composites. *Appl. Phys. Lett.* 2014, *104*, 093901.

101. Onuta, T.-D.; Wang, Y.; Long, C.J.; Takeuchi, I. Energy harvesting properties of all-thin-film multiferroic cantilevers. *Appl. Phys. Lett.* 2011, *99*, 203506.

102. Bian, L.; Wen, Y.; Li, P.; Gao, Q.; Zheng, M. Magnetoelectric transducer with high quality factor for wireless power receiving. *Sens. Actuators A Phys.* 2009, *150*, 207–211.

103. Gao, J.; Hasanyan, D.; Shen, Y.; Wang, Y.; Li, J.; Viehland, D. Giant resonant magnetoelectric effect in bi-layered Metglas/Pb(Zr,Ti)O_3 composites. *J. Appl. Phys.* 2012, *112*, 104101.

104. Dong, S.; Zhai, J.; Li, J.F.; Viehland, D.; Priya, S. Multimodal system for harvesting magnetic and mechanical energy. *Appl. Phys. Lett.* 2008, *93*, 103511.

105. Cho, K.-H.; Park, H.-Y.; Heo, J.S.; Priya, S. Structure-performance relationships for cantilever-type piezoelectric energy harvesters. *J. Appl. Phys.* 2014, *115*, 204108.

106. Vopsaroiu, M.; Blackburn, J.; Muniz-Piniella, A.; Cain, M.G. Multiferroic magnetic recording read head technology for 1Tbit/in.^{2} and beyond. *J. Appl. Phys.* 2008, *103*, 07F506.

107. Zhang, Y.; Li, Z.; Deng, C.; Ma, J.; Lin, Y.; Nan, C.-W. Demonstration of magnetoelectric read head of multiferroic heterostructures. *Appl. Phys. Lett.* 2008, *92*, 152510.

108. Kargol, A.; Malkinski, L.; Caruntu, G. Biomedical applications of multiferroic nanoparticles. In *Advanced Magnetic Materials*; Malkinski, L., Ed.; InTech: Rijeka, Croatia, 2012; pp. 89–118.

109. Guduru, R.; Liang, P.; Runowicz, C.; Nair, M.; Atluri, V.; Khizroev, S. Magneto-electric nanoparticles to enable field-controlled high-specificity drug delivery to eradicate ovarian cancer cells. *Sci. Rep.* 2013, *3*, 2953.

110. Yue, K.; Guduru, R.; Hong, J.; Liang, P.; Nair, M.; Khizroev, S. Magneto-electric nano-particles for non-invasive brain stimulation. *PLoS One* 2012, *7*, e44040.

111. Dobson, J. Remote control of cellular behaviour with magnetic nanoparticles. *Nat. Nano* 2008,*3*, 139–143.

112. Huang, H.; Delikanli, S.; Zeng, H.; Ferkey, D.M.; Pralle, A. Remote control of ion channels and neurons through magnetic-field heating of nanoparticles. *Nat. Nano.* 2010, *5*, 602–606.

113. Ashcroft, F.M. *Ion Channels and Disease*; Academic Press: San Diego, CA, USA, 2000; pp. 1–2.

114. Vopson, M.M. Fundamentals of multiferroic materials and their possible applications. *Crit. Rev. Solid State Mater. Sci.* 2015, *40*, 223–250.

115. Bibes, M.; Barthelemy, A. Multiferroics: Towards a magnetoelectric memory. *Nat. Mater.*2008, *7*, 425–426.

116. Heron, J.T.; Trassin, M.; Ashraf, K.; Gajek, M.; He, Q.; Yang, S.Y.; Nikonov, D.E.; Chu, Y.H.; Salahuddin, S.; Ramesh, R. Electric-field-induced magnetization reversal in a ferromagnet-multiferroic heterostructure. *Phys. Rev. Lett.* 2011, *107*, 217202.

117. Xue, X.; Zhou, Z.; Peng, B.; Zhu, M.; Zhang, Y.; Ren, W.; Ren, T.; Yang, X.; Nan, T.; Sun, N.X.;*et al.* Electric field induced reversible 180° magnetization switching through tuning of interfacial exchange bias along magnetic easy-axis in multiferroic laminates. *Sci. Rep.* 2015, *5*, 16480.

118. Liu, M.; Lou, J.; Li, S.; Sun, N.X. E-field control of exchange bias and deterministic magnetization switching in AFM/FM/FE multiferroic heterostructures. *Adv. Funct. Mater.*2011, *21*, 2593–2598.

119. White, J.F. Review of semiconductor microwave phase shifters. *Proc. IEEE* 1968, *56*, 1924–1931.

120. Boyd, C.R. A 60 GHz dual-mode ferrite phase shifter. In Proceedings of the 1982 IEEE MTT-S International Microwave Symposium Digest, Dallas, TX, USA, 15–17 June 1982; pp. 257–259.

121. Zhao, Z.; Wang, X.; Choi, K.; Lugo, C.; Hunt, A.T. Ferroelectric Phase Shifters at 20 and 30 GHz. *IEEE Trans. Microw. Theory Tech.* 2007, *55*, 430–437.

122. Ustinov, A.B.; Srinivasan, G.; Kalinikos, B.A. Ferrite-ferroelectric hybrid wave phase shifters.*Appl. Phys. Lett.* 2007, *90*, 031913.

123. Tatarenko, A.S.; Srinivasan, G.; Bichurin, M.I. Magnetoelectric microwave phase shifter. *Appl. Phys. Lett.* 2006, *88*, 183507.

124. Tatarenko, A.S.; Srinivasan, G. A strain engineered voltage tunable millimeter-wave ferrite phase shifter. *Microwave Opt. Technol. Lett.* 2011, *53*, 261–264.

125. Kittel, C. On the theory of ferromagnetic resonance absorption. *Phys. Rev.* 1948, *73*, 155–161.

126. Fetisov, Y.K.; Srinivasan, G. Electric field tuning characteristics of a ferrite-piezoelectric microwave resonator. *Appl. Phys. Lett.* 2006, *88*, 143503.

127. Li, N.; Liu, M.; Zhou, Z.; Sun, N.X.; Murthy, D.V.B.; Srinivasan, G.; Klein, T.M.; Petrov, V.M.; Gupta, A. Electrostatic tuning of ferromagnetic resonance and magnetoelectric interactions in ferrite-piezoelectric heterostructures grown by chemical vapor deposition. *Appl. Phys. Lett.*2011, *99*, 192502.

128. Srinivasan, G. Microwave and millimeter-wave multiferroic devices. In *Composite Magnetoelectrics*; Srinivasan, G., Priya, S., Sun, N.X., Eds.; Woodhead Publishing: Cambridge, UK, 2015; pp. 241–264.

129. Wang, S.G.; Yoon, S.D.; Vittoria, C. Microwave and magnetic properties of double-sided hexaferrite films on (111) magnesium oxide substrates. *J. Appl. Phys.* 2002, *92*, 6728–6732.

130. Ustinov, A.B.; Tatarenko, A.S.; Srinivasan, G.; Balbashov, A.M. Al substituted Ba-hexaferrite single-crystal films for millimeter-wave devices. *J. Appl. Phys.* 2009, *105*, 023908.

131. Tatarenko, A.S.; Murthy, D.V.B.; Srinivasan, G. Hexagonal ferrite-piezoelectric composites for dual magnetic and electric field tunable 8–25 GHz microstripline resonators and phase shifters. *Microw. Opt. Technol. Lett.* 2012, *54*, 1215–1218.

132. Lou, J.; Reed, D.; Pettiford, C.; Liu, M.; Han, P.; Dong, S.; Sun, N.X. Giant microwave tunability in FeGaB/lead magnesium niobate-lead titanate multiferroic composites. *Appl. Phys. Lett.* 2008, *92*, 262502.

133. Lou, J.; Liu, M.; Reed, D.; Ren, Y.; Sun, N.X. Giant electric field tuning of magnetism in novel multiferroic FeGaB/Lead Zinc Niobate–Lead Titanate (PZN-PT) Heterostructures. *Adv. Mater.* 2009, *21*, 4711–4715.

134. Fang, X.; Zhang, N.; Wang, Z.L. Converse magnetoelectric effects on heterotype electrostrain-piezopermeability composites. *Appl. Phys. Lett.* 2008, *93*, 102503.

135. Lou, J.; Reed, D.; Liu, M.; Sun, N.X. Electrostatically tunable magnetoelectric inductors with large inductance tenability. *Appl. Phys. Lett.* 2009, *94*, 112508.

136. Hwaider, L.; Jing, L.; Yuan, G.; Hasegawa, R.; Ming, L.; Howe, B.; Jones, J.; Brown, G.; Sun, N.X. Voltage tunable magnetoelectric inductors with improved operational frequency and quality factor for power electronics. *IEEE Trans. Magn.* 2015, *51*, 1–5.

137. Mandal, M.; Duttagupta, S.P.; Palkar, V.R. Study of multiferroic $Bi_{0.7}Dy_{0.3}FeO_3$ based tunable ring inductor. *J. Phys. D Appl. Phy.* 2013, *46*, 325001.

138. Garg, R. *Microstrip Antenna Design Handbook*; Artech House: Boston, MA, USA, 2001.

139. Kim, D. Planar magneto-dielectric metasubstrate for miniaturization of a microstrip patch antenna. *Microw. Opt. Technol. Lett.* 2012, *54*, 2871–2874.

140. Petrov, R.V.; Tatarenko, A.S.; Pandey, S.; Srinivasan, G.; Mantese, J.V.; Azadegan, R. Miniature antenna based on magnetoelectric composites. *Electron. Lett.* 2008, *44*, 506–508.

141. Bae, S.; Hong, Y.K.; Lee, J.J.; Jalli, J.; Abo, G.S.; Lyle, A.; Seong, W.M.; Kum, J.S. Low loss Z-type barium ferrite (Co_2Z) for terrestrial digital multimedia broadcasting antenna application. *J. Appl. Phys.* 2009, *105*, 07A515.

142. Bae, S.; Hong, Y.K.; Lee, J.J.; Abo, G.S.; Jalli, J.; Park, J.H.; Seong, W.M.; Kum, J.S.; Ahn, W.K.; Park, S.H. Miniaturized broadband ferrite T-DMB antenna for mobile-phone applications.*IEEE Trans. Magn.* 2010, *46*, 2361–2364.

143. Lee, J.; Hong, Y.-K.; Lee, W.; Abo, G.S.; Park, J.; Neveu, N.; Seong, W.-M.; Park, S.-H.; Ahn, W.-K. Soft M-type hexaferrite for very high frequency miniature antenna applications. *J. Appl. Phys.* 2012, *111*, 07A520.

CHAPTER 19

FERROELECTRICS UNDER THE SYNCHROTRON LIGHT

Luis E. Fuentes-Cobas, María E. Montero-Cabrera, Lorena Pardo and Luis Fuentes-Montero

Keywords: Synchrotron X-rays diffraction, scattering and absorption, field-induced transformations, local versus global order, strain and texture analysis

INTRODUCTION

Materials with ferro- and piezoelectric properties are globally required, respectively, in information technology and in electromechanical transduction. Ferroelectric crystals and ceramics are also piezoelectric. Applied materials science faces today the challenge of finding lead-free ferroelectric materials with high ferro-piezoelectric properties [1]. One of the most common structures of ferroelectric oxides is the well-known perovksite-type arrangement [2], of general formula ABO_3. It consists of a three-dimensional arrangement of vertex-sharing oxygen octahedra, in between which large A cations are located and whose center is occupied by small B cations. The deformation from the cubic prototype structure, the off-center displacement of the A and B cations and the oxygen octahedra tilting cause the displacement of the centers of positive and negative charges in the unit cell and build-up the spontaneous polarization, P_s, of this type of ferroelectrics.Figure 1 represents the Glazer notation [3] for octahedra tilting in three of the crystal symmetries considered in the investigation of ferro- piezoelectric perovskites. Another important symmetry in ferroelectric crystals is *P4mm*, with Galzer notation $a^0a^0c^0$ [4]. This latter case looks like *Pm-3m*when viewed along the *z* axis.

Space Group	Rhombohedral (R3c)	Tetragonal (P4bm)	Cubic (Pm$\bar{3}$m)
Structure	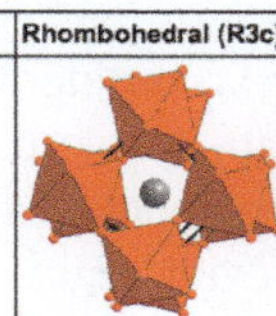	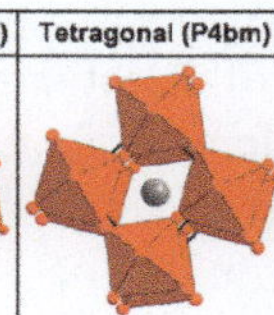	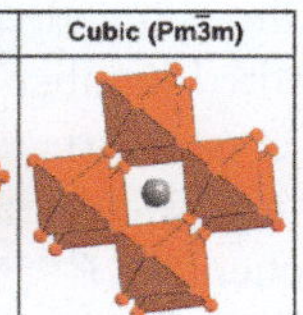
Glazer System	$a^-a^-a^-$	$a^0a^0c^+$	$a^0a^0a^0$
a,c,α,β,γ	5.489,13.51,90,90,120	5.518,3.907,90,90,90	3.914,3.914,90,90,90

Figure 1. Structural properties of the rhombohedral *R3c*, tetragonal *P4bm*, and cubic *Pm-3m* phases reported for $Bi_{0.5}Na_{0.5}TiO_3$ (BNT). The lattice parameters for the rhombohedral case are given in the hexagonal system. Reproduced with permission from [5].

The history of the understanding of the structural origin of ferroelectricity in $BaTiO_3$ (BT) illustrates the importance of the development of new and more powerful crystallographic tools. Up to the 1940s, $BaTiO_3$ ceramics were known as intriguing ceramics. Atomic-level explanations for their polarization phenomena were lacking. They were polycrystalline and, in virgin state, isotropic/non-ferroelectric. However, they were capable of being poled and afterward they became ferro and piezoelectric. At the time, the domain structure of $BaTiO_3$ was proposed, phase transformations among various crystal systems were reported [6] and the first models for the poling process were suggested [7]. The following important question remained unsolved. Does the $BaTiO_3$ tetragonal ferroelectric structure show the necessary inversion symmetry break-down or not? Available photographic XRD methods were not capable of revealing with the required detail the atomic positions in BT [8]. Around that time, the first Geiger-counter measurements of diffracted intensities were essayed at the Philips X-ray laboratories in Irvington (NY). H.T. Evans,

Citation: Luis E. Fuentes-Cobas, María E. Montero-Cabrera, Lorena Pardo and Luis Fuentes-Montero, Ferroelectrics under the Synchrotron Light, doi:10.3390/ma9010014.

from MIT, went to Irvington and examined a BT sample in the recently invented "diffractometer". His result was the first experimental detection of inversion symmetry break-down in $BaTiO_3$ [9]. Further refinement of the BT crystal structure was possible in the following years, thanks to another invention that was just emerging and growing exponentially in impact: the computer [10].

Recent structural descriptions of ferroelectric perovskites often refer to their pseudo-symmetries, due to the existence of local deviations from the global symmetry, arising from the many possible local displacement of the ionic positions and chemical order [11,12].

Regarding the interaction of ferroelectric perovskites with electric fields, it is well known that poling causes increasing orientation of ferroelectric domains as the electric field and time increases until reaching saturation of the remnant polarization, P_r. In this way, poling a polycrystal ceramic creates a macroscopic, non-centrosymmetric ∞mm symmetry, which leads to piezoelectric activity. Only the reorientation of the 180° domains, in which the polarization has the same direction and opposite sense at both sides of the domain wall, can undergo a reorientation process without a crystal strain. A non-180° domain reorientation is accompanied by a crystal strain. The remnant polarization of poled ferroelectric polycrystals, P_r, and consequently, their piezoelectric performance, is determined by the existence of domains in which the spontaneous polarization of the crystal is oriented in a limited number of directions according to the symmetry of the crystal structure (tetragonal: $P_r^{max} = 0.83\ P_s$, with six <001> equivalent directions of the spontaneous polarization; rhombohedral: $P_r^{max} = 0.87\ P_s$, with eight allowed <111> equivalent directions and orthorhombic: $P_r^{max} = 0.91\ P_s$, with 12 allowed <110> equivalent directions [13,14]). Those solid solution systems with the so called Morphotropic Phase Boundary (MPB), for which composition there is symmetries coexistence, are desired. This increases the number of equivalent directions ofP_s and, consequently, increases the P_r value. It is also possible to increase P_r over the values of the classical randomly oriented ceramics by the processing of textured ceramics oriented in the poling direction. In ferroelectric ceramics, electric field-induced structural transitions have been observed [15,16,17,18,19]. Of particular interest for strain derived applications are those in $Bi_{0.5}Na_{0.5}TiO_3$(BNT)-based compositions that take place from a weakly polar cubic pseudo- symmetry to ferro-piezoelectrically active phases of lower symmetry.

In comparison with the Evans research on BT, the crystallography arsenal has grown impressively. The resources available to characterize the phenomena of interest include electron microscopy, X-ray and neutron diffraction, Raman spectroscopy and properties measurements, among others. Particularly, presently available synchrotron X-rays beams are millions of times more intense than in those of the 1950s, the types of radiation–matter interaction used today with research purposes is not limited to diffraction, the detection systems show pixel resolution and energy discrimination. Thanks to the mentioned advanced, today, departures from model symmetries (at local and global levels) in real-world materials can be detected with several orders of magnitude higher sensitivity. Finally, the computer capabilities are simply not comparable with the ones that helped decipher the $BaTiO_3$ structure. The aim of the present article is to deliver a compact review of the applicability of synchrotron radiation techniques to the study of ferroelectric ceramics. Emphasis will be put on diffraction (well-defined peaks), diffuse scattering and absorption spectroscopy. Lead-free compounds and topics related with crystal symmetry, texture, strain and the effect of electric field on crystal- and microstructures will be highlighted.

X-RAY DIFFRACTION AND SCATTERING

One-Dimensional Measurements

X-ray diffraction (XRD) is a useful, yet complex, tool that has been extensively devoted to studying the evolution of the crystal structure of ferroelectrics under the action of an electric-field using conventional diffraction (CuKα radiation tubes) [20,21,22,23,24] and synchrotron radiation sources [25,26]. It must be noticed that poling and depoling of ferroelectric ceramics can be also achieved by the application of mechanical loads to the sample and there is also extended parallel literature that uses the same structure, strain and texture determination tools and models [27] as those used for the electrical poling, a review of which is outside the scope of this work.

To understand the relevance and uniqueness of the experiments on ferroelectrics using synchrotron radiation, a number of fundamental concepts of powder X-ray diffraction [28] must be bear in mind. First, a polycrystalline sample should contain thousands of crystallites. Such it is the case of a ferroelectric ceramic thin disk or plate with a surface to explore of tens of mm^2 that is typically constituted by ceramic grains of a few microns size. It should be mentioned that X-ray diffraction provides information from so-called coherent diffraction domains and not from ceramic grains. A coherent diffraction domain is the largest region in three-dimensional space that satisfies the periodic translation of the crystal unit-cell [29]. Grains in ferroelectric ceramics are formed by several crystallites and each crystallite may be formed by different ferroelectric domains. Ferroelectric domains show diverse orientations of the spontaneous polarizations. In a ferroelectric, coherent diffraction domains coincide with ferroelectric domains.

Proper use of XRD data from a polycrystalline sample requires the consideration of all the experimentally observable diffraction peaks.

The most commonly used geometry in laboratory powder diffraction is the parafocussing Bragg-Brentano configuration. In this arrangement, the X-ray tube is fixed and the incident- and diffracted- beam slits are located in points of a circle that is centered on a point at the flat surface of the sample. Divergent X-rays from the source hit the sample at different points on its surface. During the diffraction process, the X-rays are refocused at the diffracted-beam slit. This arrangement provides the best combination of intensity, peak shape, and angular resolution.

Often ferroelectric ceramic thin disks are poled perpendicularly to their faces ("thickness poling" [30]). When analyzing thickness poled disks in this geometry, the electric field effect on the material that is actually probed is limited to the action on a small fraction of crystallites, those contributing to the intersection of the detector scan with the Debye cone, whose d_{hkl} is parallel to the electric field. This is a relevant condition since it provides information on the stronger changes that the structure of the ferroelectric material may undergo under the electric field action.

Synchrotron radiation is the brightest light on earth. It is the single most powerful tool available to X-ray crystallographers. X-ray beams are generated by electrons flying at nearly the speed of light in a (roughly) circular loop, guided by strong magnetic fields [31]. Synchrotron radiation is also characterized by its tunability in energy (from E ~ eV to MeV), with a high degree of monochromatization and collimation. In synchrotron radiation diffraction, the wavelengths used are frequently smaller than those of X-ray tubes, which allows scanning to higher Q ($=4\pi\cdot\sin\theta/\lambda$) with high counting rates. All these characteristics allow the volume representativeness, geometric resolution and statistics of synchrotron experiments to be significantly better than when using X-ray tubes.

Synchrotron intensity is not constant in time. A correction that is often applied consists of a normalization based on the readings of the incident beam monitor.

To take advantage of the high intensity, collimation and monochromaticity of their X-ray beams, synchrotron X-ray diffractometers preferably use the so-called parallel-beam configuration. Synchrotron X-rays do not diverge, unlike those generated from an X-ray tube. This geometry allows spot sizes on the micron scale (producing micro-diffraction) and eliminates the sample displacement systematic error of diffraction experiments. On the other hand, in the Bragg-Brentano arrangement the area bathed by X-rays is larger. In the case of inhomogeneous surfaces, Bragg-Brentano may offer better averaging.

An example of some diffraction peaks measured using a Cu-cathode X-ray tube and synchrotron radiation is shown in Figure 2. The improved resolution makes clear the advantages of the synchrotron X-ray diffraction.

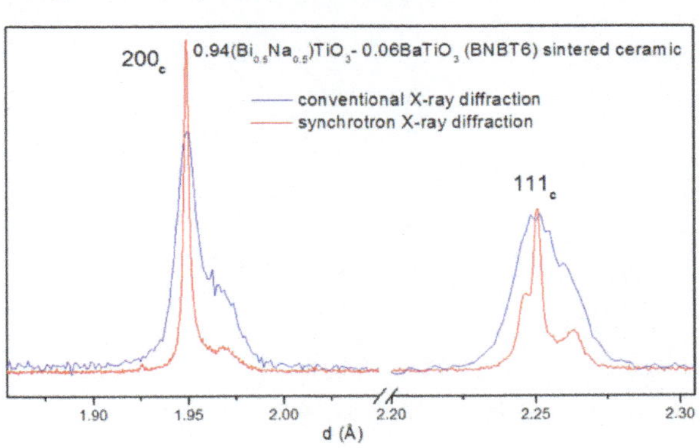

Figure 2. Diffraction peaks from sintered BNBT6 bulk ceramic from a powder flat specimen measured using a conventional Cu-cathode X-ray tube (step 0.05 2θ° and 5 s counting time) and synchrotron radiation at MCX beamline Elettra Sincrotrone, Trieste (Italy) (step 0.005 2θ° and 2 s counting time).

Recent Work on PZT and $BaTiO_3$ [12,32]

Structures and phase diagrams of ferroelectric ceramics have been studied since the origin of their discovery. Nevertheless, recent studies with synchrotron radiation and other techniques keep on revealing finer details of classical compositions.

$PbZr_{1-x}Ti_xO_3$ (PZT) is one of the most important and widely used piezoelectric materials. The study of its local and average structures is of fundamental importance in understanding the origin of its high-performance piezoelectricity. In 1999, using high-resolution synchrotron powder diffraction, Noheda, *et al.* discovered the existence of a monoclinic phase in the Zr-rich region of the PZT phase diagram [33]. In recent years, new significant details of the PZT monoclinic ordering have been observed by the combined use of synchrotron light and neutrons [12,34]. The nature of the monoclinic phase across the Zr-rich and morphotropic phase boundary region of the PZT phase diagram is described in detail in a fresh study by Zhang, *et al.* [12]. Long-range average rhombohedral and both long- and short-range monoclinic regions coexist at all compositions. In addition, a boundary between a monoclinic (M_A) structure and another monoclinic (M_B) structure has been established. Both monoclinic structures, at room temperature, belong to the space group*Cm* and in both the Pb cations are located in a {110} symmetry plane. In a rhombohedral arrangement, the displacements of the Pb cations with respect to the cubic structure would be in a <111> direction. In the M_A phase, this shift departs from <111> on the way to a <001> direction (symmetry tends to tetragonal). In M_B, the Pb displacement deviates towards <101> (symmetry tends to orthorhombic). A refined PZT phase diagram (Figure 3) is suggested.

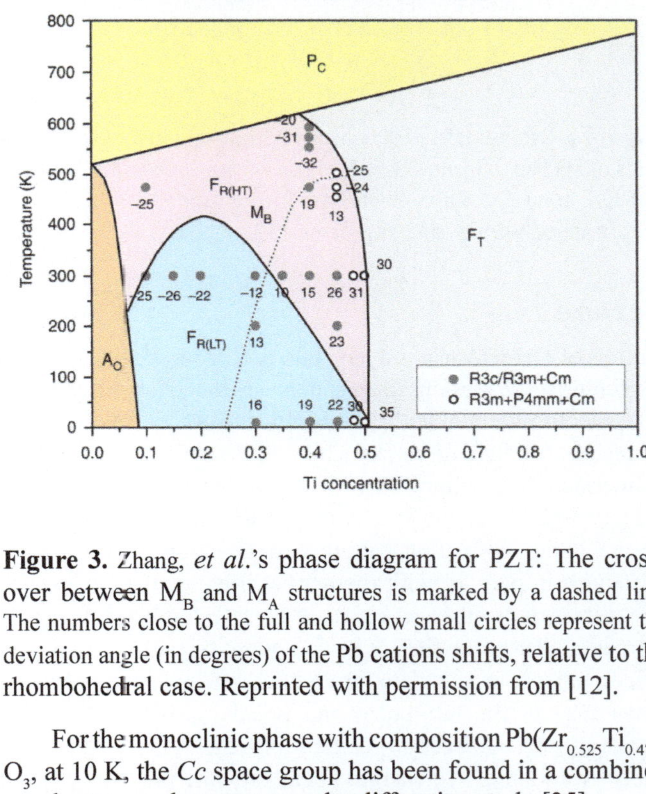

Figure 3. Zhang, *et al.*'s phase diagram for PZT: The cross-over between M_B and M_A structures is marked by a dashed line. The numbers close to the full and hollow small circles represent the deviation angle (in degrees) of the Pb cations shifts, relative to the rhombohedral case. Reprinted with permission from [12].

For the monoclinic phase with composition $Pb(Zr_{0.525}Ti_{0.475})O_3$, at 10 K, the *Cc* space group has been found in a combined synchrotron and neutron powder diffraction study [35].

Also recently, Kalyani, *et al.* [32] published their finding, based on high resolution synchrotron XRD (Figure 4) and other techniques, that in $BaTiO_3$ at room temperature a subtle monoclinic *Pm*phase coexists with the tetragonal *P4mm* phase These results propose that $BaTiO_3$ at room temperature is within an instability regime, and is structurally more akin to lead-based morphotropic phase boundary systems such as PZT, PMN-PT, PZN-PT instead of $PbTiO_3$. The results by Kalyani and collaborators offer a new perspective to understand the anomalous piezoelectric and dielectric responses in poled single crystals and polycrystalline and chemically substituted

$BaTiO_3$ systems which are increasingly becoming important as lead-free piezoelectric materials.

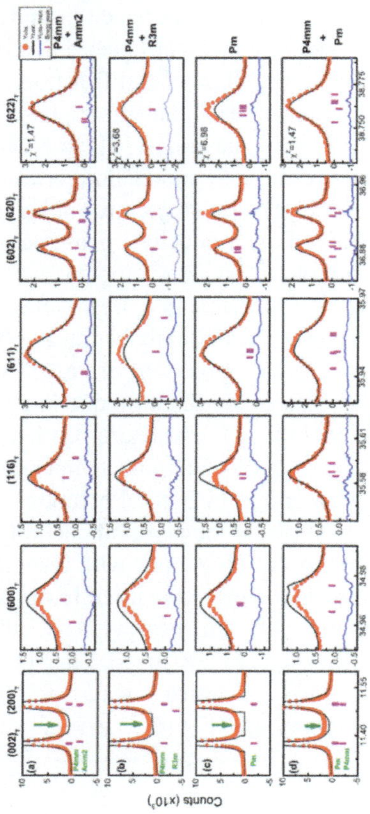

Figure 4. Rietveld fitted synchrotron high-resolution XRD profiles of $BaTiO_3$ with models (**a**) *P4mm* + *Amm*2, (**b**) *P4mm* +*R3m*, (**c**) *Pm*, and (**d**) *P4mm* + *Pm*. Best fitting model is *P4mm* +*Pm*. Reproduced with permission from [32].

Texture

Poling of ferroelectric polycrystals is a process that creates a preferential orientation, since it increases the volume fraction of ferroelectric domains with their polarizations as close as possible to the electric field direction. This is done at the expense of the reduction of the volume of the less favorably aligned domains. As a consequence, there is a growth in the intensity of the peaks at the X-ray diffraction pattern corresponding to interplanar distances in parallel with the applied field at the expense of the decrease of others corresponding to interplanar distances perpendicular to the field. The volume of ferroelectric domain alignment can be calculated from the changes in the integrated intensities in the non-poled and poled samples of selected peaks of the diffraction pattern [13,20,21,23,24,25,26,27]. Deconvolution of degenerated peaks for the purpose of peak intensity calculation can be performed by a variety of software [36]. For calculations, correct profile shape functions must be considered [37]. More information related with texture analysis by means of synchrotron XRD will be given in the section about two-dimensional measurements, below.

Diffuse Scattering: PDF Analysis

Diffraction analysis leads to long-range space-and time-averaged high-resolution structure determination via information contained in the diffraction peaks. Material information contained in the background is frequently neglected. However, every point in a diffraction pattern, including the background, contains structural information. Local (static or dynamic) atomic positions deviations from the reference points decrease the peak intensities, in the amount characterized by the Debye-Waller factor. The intensities not appearing in the Bragg peaks contribute to the so-called diffuse scattering. Careful measurement and data processing of (diffraction + diffuse scattering) data leads to the so-called pair distribution function [38] PDF = *G(r)*, containing important information regarding the local environment of atoms in the investigated materials. *G(r)*represents the probability (relative to the average electron density) of finding pairs of atoms separated by the distance *r*. In the particular case of ferroelectrics, differences between local and long-range ordering may be significant. Figure 5 and Figure 6 show the results of a PDF determination by means of an X-ray total elastic scattering experiment [39]. Temperature-dependent powder diffraction data from the $(1-x)Na_{0.5}Bi_{0.5}TiO_{3-}xBaTiO_3$ (BNBT) family ($0 \leq x \leq 0.08$) were collected on heating between 120 and 800 K at the BW5 beamline of the DORIS-III facility in Deutsches Elektronen-Synchrotron (DESY) using an incident beam energy of 100 keV (λ = 0.12398 Å), which provided a maximum reciprocal-space *Q* value of 25 Å^{-1}. Data processing was performed with program PDFgetX3 [40].Figure 5 shows the low-*r* range of the BNBTs PDF and Figure 6 shows a wider interval using a color-based representation of PDF intensities.

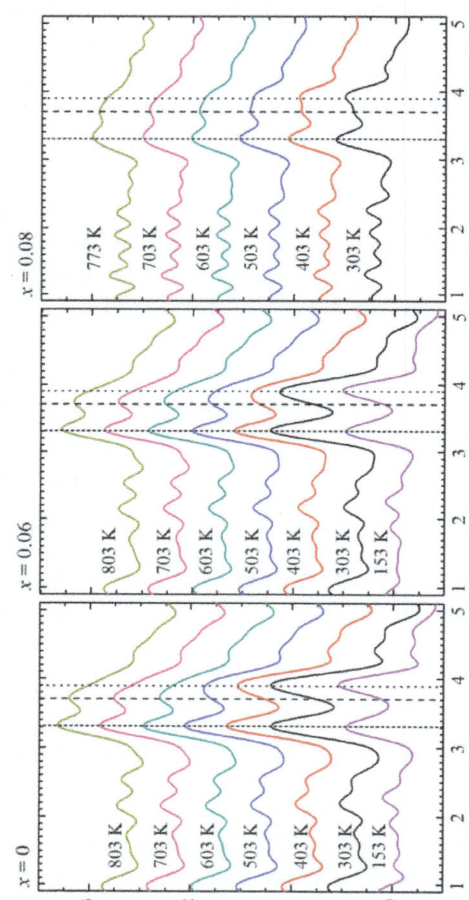

Figure 5. Temperature-dependent X-ray PDF patterns of BNBT for *x* = 0, 0.06, and 0.08. The short-dashed lines (3.3 Å) mark the feature corresponding to A-B distances; the dashed (3.7 Å) and dotted lines (3.9 Å) mark the two components in the feature related to A-A and B-B distances. Reproduced with permission from [39].

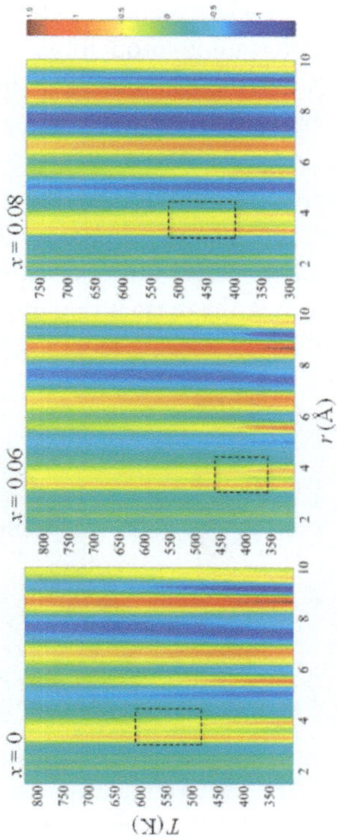

Figure 6. Temperature evolution of X-ray PDF intensities for BNBT for $x = 0$, 0.06, and 0.08; the dashed rectangles mark the temperature ranges of the most pronounced changes in PDFs. Reproduced with permission from [39].

Some qualitative observations on the cation-cation correlations can be performed from the determined PDFs. The peak at 3.3 Å represents the coupling between the A and B subsystems, basically Bi-Ti distances. These distances remain nearly unchanged both as a function of composition and temperature. The incorporation of Ba affects the A-A and B-B contributions to the PDF around 3.8 Å. The observed variations in mentioned interactions correlate with the enhancement of ferro-piezoelectric properties in the MPB of the BNBT system.

PDF results are consistent with those obtained by Raman spectroscopy.

Two-Dimensional Measurements

The diffraction/scattering research field is changing significantly with the profusion of 2D detectors. Two-dimensional measurements represent a meaningful gain in information from diffraction/scattering experiments. Here, we first describe a few representative 2D detectors and the advantages of using one or another type. Afterwards, we proceed to present a sampler of current research trends using 2D measurements.

Two-Dimensional Image Plate

Two-dimensional Image Plate detectors were initially developed for X-ray Radiography, and then they found applications in X-ray diffraction, in conventional equipment and synchrotron radiation beamlines. Their operating principle is similar to that of a photographic plate emulsion. X-rays impinge on a film containing a phosphor coating which is excited and stores a latent image. Then, the image is "revealed" by scanning the plate with a He-Ne laser beam across it, which produces photoluminescence. The luminescence intensity is read by photodetectors, and the phosphor is erased by exposing it to visible light. The spatial and temporal resolution of the detector depends on the characteristics of the phosphor, laser and associated electronics.

MAR345 image plate detectors are used in synchrotrons for X-ray dispersion and diffraction studies in transmission and grazing incidence geometries. The phosphor consists of a 0.4 mm coating of $BaFBr:Eu_2$. An example of its use in grazing incidence XRD to study ferroelectric materials is the work of Torres, *et al* [41], performed at the Stanford Synchrotron Radiation Lightsource (SSRL) Beamline 11-3 (Figure 7a). The use of this kind of detectors for the measurement of crystallographic texture is described in [42].

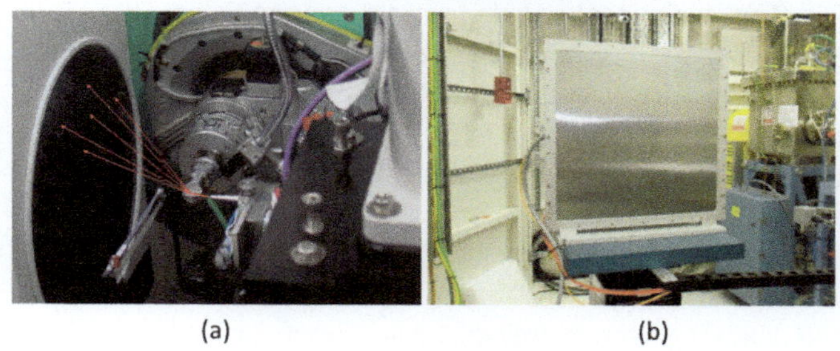

Figure 7. (**a**) MAR345 detector at SSRL beamline 11-3. X-ray trajectory is represented by red arrows; (**b**) Pixium RF4343 detector, covered with Al front plate, at Diamond Light Source beamline I12.

Charge-Coupled Device (CCD)

The charge-coupled device (CCD) detector consists of an array of metal-oxide-semiconductor capacitors (CMOS). These capacitors detect visible light through a transparent layer of polycrystalline silicon electrodes. Most CCD X-rays detectors record incident photons indirectly through scintillators that detect X-rays and produce visible light. The visible light is then detected by the CCD [43]. The Diamond Light Source (DLS) beamline I12 is equipped with an X-ray imaging camera for radiography and tomography, based on an array of scintillators and CMOS cameras [44].

The Pixium type detector, employed in synchrotron radiation XRD and related techniques, is a CMOS camera with a CsI scintillator. For instance, the Pixium RF 4343 detector conversion of X-ray to visible light is based on a columnar crystalline CsI scintillator array, which offers a high X-ray absorption, due to its high average mass atom number. The visible light is detected by an amorphous silicon photodiode array. In Figure 7b, the detector from the beamline I12 at Diamond is presented. This type of detector is suitable for two-dimensional powder diffraction, Laue-diffraction, total scattering and SAXS experiments [44,45].

Single-Photon—Counting Mode

New detectors have been developed on the concept of single-photon-counting mode [46]. This operation is based on the total absorption of X-ray photon by a scintillator or a semiconductor, which is connected to a cell of collection electrodes. Each cell represents one pixel. The current registered by the pixel is proportional to the energy deposited from the incident X-ray photon. For photons below 100 keV, most events deposit all the energy in a single pixel. In other words, every X-ray photon is directly converted into an electrical signal and counted by the detector system. This is the case of Pilatus detectors [47].

There is a next generation of PILATUS detector, also composed of semiconductor sensor, CMOS readout chip, and readout electronics. X-rays deposit their energy in a pixelated silicon sensor that is typically 320 μm thick [48].

Grazing Incidence 2D-XRD: Texture Analysis

Texture plays an important role in polycrystal physical properties. A detailed discussion about the influence of texture on piezoelectricity can be found in [49].

The main tools for measuring textures are based on neutrons', X-rays' and electrons' diffraction. The neutron sources require access to nuclear (relatively hard to reach) facilities, but provide the most representative results in terms of observed volume [50]. Electron microscopes, in image and diffraction modes, offer grain-by-grain characterizations of texture, but represent local techniques, of low statistical significance [51]. XRD analysis of textures is accessible and leads to average results from significant volumes [52]. With synchrotron radiation the analyzed volumes are greater than with laboratory diffractometers. Particularly, with 2D detection, the exposure times can be shortened dramatically.

Though being a non-ferroelectric, one of the main applications of fibers or nanorods and polycrystalline films of the wurzite type structure ZnO semiconductors lays on their piezoelectric properties.

Figure 8 and Figure 9 exemplify the correlation of electron microscopy and synchrotron 2D-XRD analyses of texture in piezoelectric thin layers [53]. ZnO nanorods were synthesized by aerosol assisted chemical vapor deposition onto TiO_2 covered borosilicate glass substrates. Figure 8 shows the rod morphology of the ZnO obtained nanocrystals. The SEM evaluation of the nanorods orientation dispersion was performed by analysis of Figure 8.

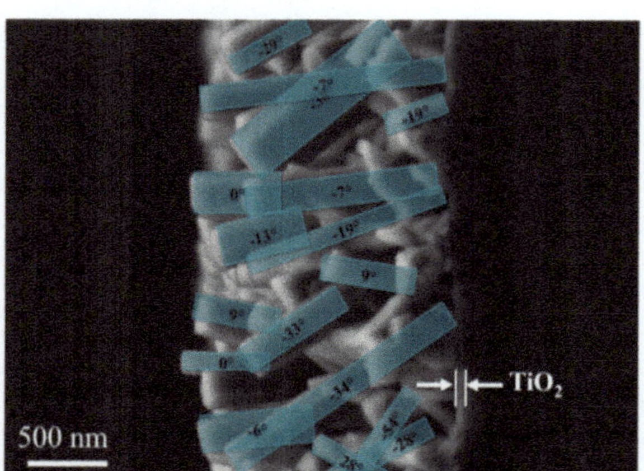

Figure 8. SEM micrograph of textured ZnO thin layer on TiO_2/borosilicate glass. Cross section view. Image exhibits several ZnO nanorods and their axis inclination from substrate normal. Reproduced with permission from [53].

The sample observed by SEM was also examined at beamline 11-3 of SSRL. The grazing incidence experiment was as shown in Figure 7a. The experimental 2D-XRD pattern is shown in Figure 9a. Diffraction data interpretation was by means of computer-aided modelling. Software package ANAELU [54] was used to simulate 2D diffraction patterns produced by the proposed axially symmetric orientation distributions. In program ANAELU, texture is represented by a Gaussian-shaped inverse pole figure $R(\varphi) \propto exp{-}(\varphi/\Omega)^2$. φ is the angle between reciprocal direction h and preferred orientation h_0. Ω characterizes the orientation distribution width. Figure 9b shows the best-fitting modelled output by ANAELU.

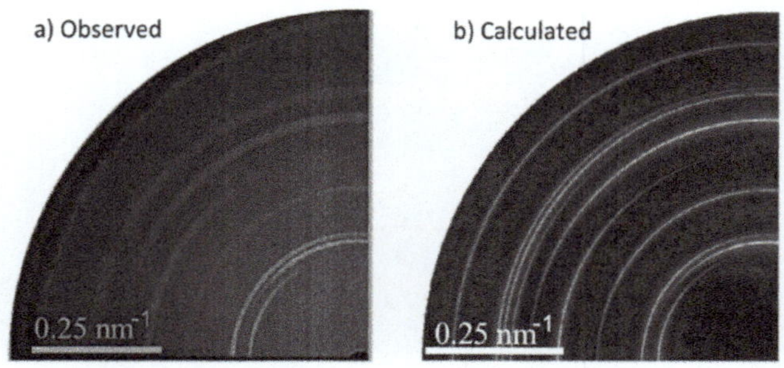

Figure 9. 2D-GIXRD patterns of ZnO nanorods: (**a**) as observed at SSRL beamline 11.3 MAR345 detector; and (**b**) showing the texture simulation with program ANAELU [54]. Reproduced with permission from [53].

SEM and 2D-XRD analyses converged to the conclusion of a preferred growth in [001] direction with a distribution width $\Omega = (20 \pm 2)°$.

2D-Diffuse Scattering: Reciprocal Space Mapping

Real crystals normally show deviations from ideal periodicity. These imperfections reduce the intensity of the diffraction maxima and generate intensity distributions around the reciprocal space nodes. Depending on the nature, dimensionality and importance of these imperfections, the diffuse scattering distributions can form streaks, planes or rods connecting the reciprocal nodes. The diffuse radiation is produced by static disorder (chemical heterogeneities, microdeformations) and/or kinematic disorder (vibration, phonons). The study of these phenomena by means of so-called reciprocal space maps has been significantly favored by the growing capabilities for measuring 2D images of diffuse scattering [31].

Going back to lead-free ferroelectrics, the work by Chen, *et al.* [55] on $0.93(Bi_{1/2}Na_{1/2})TiO_3$-$0.07BaTiO_3$ (BNBT7) illustrates the usefulness of 2D mapping of scattered synchrotron X-rays. In the mentioned contribution, textured samples of BNBT7 were obtained from powder at 1300 °C through spontaneous nucleation at the critical temperature by a cooling rate at 1 °C/h. Figure 10a shows the room-temperature XRD obtained in conventional laboratory equipment. The sample shows an intense dual-component [(001) + (100)] texture, as can be seen from the absence of the 110 and 111 reflections. Figure 10b shows in detail the analysis of the reflection cluster at $2\theta \approx 46°$. The appearance of the reflection denoted aT demonstrates the presence of a certain population of (100) planes parallel to the sample surface. The cT (002) peak, on the other hand, exhibiting such an important intensity (despite its smaller multiplicity factor) proves that the main texture component represents a [001] polarization out of the plane of the sample. Proper modelling of the observed spectrum requires the consideration of a (scarcely visible) rhombohedral (R) phase and also of a surface artifact to explain a small hump on the low-angle side of the peak cluster (approximately at 2θ ≈ 44°). This hump will be discussed in more detail later.

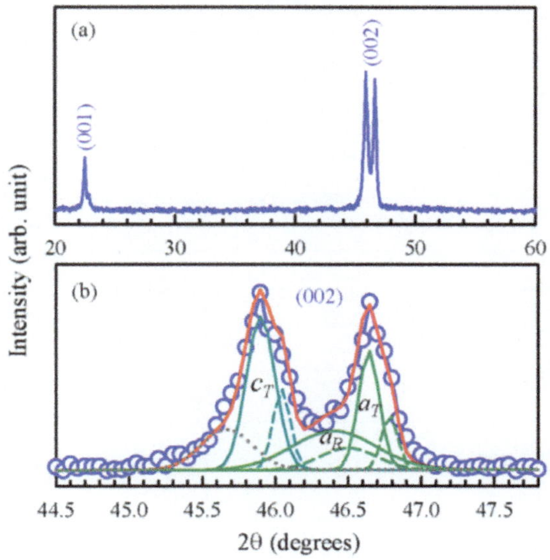

Figure 10. (**a**) XRD spectrum; and (**b**) fits of the (002)c peak for BNB7T crystal. The solid and dashed lines correlate to $K\alpha_1$ and $K\alpha_2$ radiations. The red line is the sum of fitting curves. The dotted line is a contribution from the surface layer. Reproduced with permission from [55].

Figure 11 shows the synchrotron-measured reciprocal space mapping of the considered peak cluster. The two signals from the T phase and the one from the R component are now clearly displayed by the three contour centers in the two-dimensional reciprocal domain. According to Chen and collaborators, the broad reflections in R and T mapping contours imply nanostructures, which are often expected in relaxor ferroelectric and ferroelastic crystals to accommodate the lattice distortions of different structures. Since nanostructures are usually smaller than the coherent length of X-ray radiation, diffracted waves from individual nanostructures can coherently superimpose and thus broaden the diffraction peaks.

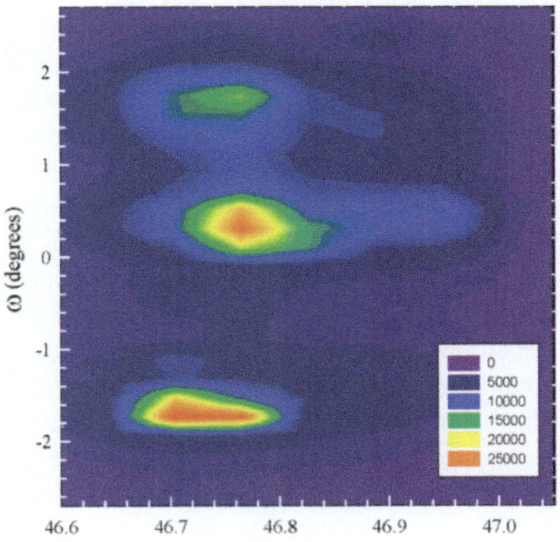

Figure 11. (002)c XRD reciprocal space mapping. The mapping intensity is on a log scale. Reproduced with permission from [55].

XRD With In-Situ Applied Electric Field

The development of diffraction experiments under *in-situ* application of electric fields to ceramic specimens at synchrotron radiation facilities has expanded considerably in the last decade [56,57,58,59,60,61,62,63,64,65,66,67,68] . In the literature, one can find a diversity of configurations regarding the relative orientation of the electric field (E) and the diffraction vectors k_0 (incident), k(diffracted) and Q (scattering) = k–k_0. Figure 12 describes some representative experimental arrangements. Table 1 summarizes relations among angles in the presented experimental setups.

Table 1. Angles between electric field **E** and scattering vector **Q**

Setup (Figure 12)	Reference	Angle φ between Q and E
Symmetric reflection (12a)	Present report	0
	[62]	90°
Asymmetric transmission (12b)	[68] E // k_0	$\varphi = \omega - \theta$
	[69] $E \perp k_0$	
2D-XRD (12c)	[57,58]	$\cos\varphi = \cos\theta \cdot \cos\alpha$ [54]

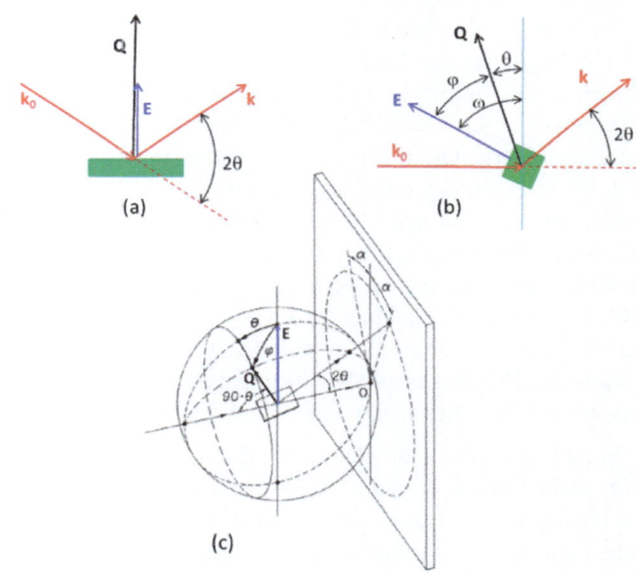

Figure 12. Experimental arrangements for diffraction measurements under electric fields. (**a**) One-dimensional diffraction, symmetric reflection (θ–2θ) with applied electric field perpendicular to sample surface; (**b**) Transmission setup with control of the angle between scattering vector Q and field E [69]. The condition ω = 90° represents the case of electric field parallel to the incident beam [68]; (**c**) Generic two-dimensional (2D) detection of diffracted/scattered X-rays. Characteristic directions and angles in the 2D experiment are represented. $Q$is the scattering vector, θ is the Bragg angle, α is the azimuth on the 2D-XRD pattern and φ is the angle between Q and E. For α = 0, φ equals the Bragg angle. If θ→0, then φ→ α. Figure 12c adapted from [54].

The transmission geometry (Figure 12b,c) is used with neutrons and high energy X-rays, possibly with area detectors. In Table 2, we summarize characteristic data of some facilities that are active in the considered field.

Table 2. Some synchrotrons active in research on ferroelectricity

Synchrotron	Beamline	Energy, Wavelength	Diffraction Geometry	Detector	References
HASYLAB, DESY. Hamburg, Germany	B2	25 keV, 0.5 Å	Transmission with parallel electric field	position-sensitive image-plate	[25,56,68]
European Synchrotron	ID15B	87.7 keV, 0.14 Å	Transmission with transverse electric field	2D detector (Pixium 4700)	[15,37,57,59].
Radiation Facility (ESRF). Grenoble, France	ID11	80.5 keV, 0.155 Å	Transmission with transverse electric field or mechanical loads	FReLoN4M area detector	[26,60,64,65]
	BM28-XMaS	9.8 keV, 1,26 Å	Electric field applied parallel to the explored surface	MAR CCD camera	[62]
Elettra Sincrotrone, Trieste, Italy	MCX	13 keV, 0.95 Å	4-circle Huber goniometer	2D detector, MAR345	[70]
Swiss Light Source (SLS). Villigen, Switzerland	MS - X04SA: Materials Science	28 keV 0.443Å	static field in the beam direction	solid-state silicon microstrip, MYTHEN detector	[61]
Advanced Photon Source at Argonne Nat. Lab. Lemont, IL, USA	5-BM-D	65 keV 0.1907 Å	Transmission with transverse electric field	2D detector, MAR345	[63]
Synchrotron Radiation Research Institute (SPring-8). Japan	BL02B1	35 keV 0.35 Å	Transmission, three-axis goniometer for single-crystal orientation	large cylindrical two-dimensional imaging plate (IP) camera	[66,67]

As a remarkable result of the experimental effort carried out with high energy, high brightness and high resolution diffraction measurements, it is now well documented that the response to the electric field, the modification of crystal symmetry, is orientation-dependent. That is to say that the departure from the equilibrium symmetry of each crystallite depends on its initial orientation with respect to the electric field. The origin of this is that ferroelectric domain orientation under the action of the electric field generates anisotropic strains in ferroelectric ceramics [26,64,65] acting locally on the original structure. The orientation-dependent response of the crystal to the field has been also documented by conventional X-ray and neutron diffraction experiments [5,23,71].

In particular, the giant electric-field-induced macroscopic strain characteristic of many BNT-based ceramics [1] has been shown to arise from a combination of the electric-field-induced phase transformation, the induced texture due to ferroelectric domain reorientation, and electric-field-induced crystal strain [58]. The strain associated with the induced phase transformation is volumetric or dilatational, resulting in an equivalent strain in all sample dimensions. Strains induced by both the domain reorientation, or domain wall movement, and field-induced crystal strains, however, are considered to cause shear strains resulting from changes in the spontaneous polarization direction and are therefore mostly volume-conserving, *i.e.*, they mainly lead to local cation displacements [58].

Rietveld Analysis of Structures under an Applied Electric Field

The Rietveld method is based on the assumption that the researcher is capable of representing mathematically the structural features of the analyzed system and the effect of such features on the interaction between the material and the X-ray beam. Rietveld programs are constantly evolving with the objective to refine their representation of diffraction/scattering phenomena and thus allow increasingly exact characterizations of crystallographic systems.

For example, homogeneous strains produce peak shifts and inhomogeneous strains cause peak broadening [28]. Full-pattern Rietveld analysis requires consideration of so-called anisotropic peak broadening and peak displacements. This possibility is implemented in currently distributed programs [72]. Examples of Rietveld analysis for refinement of diffraction patterns using a single crystallographic model with anisotropic lattice strains can be found in references [16,73].

The crystallite-orientation dependent action of the electric-field requires careful selection of the starting model for the texture. To characterize real cases, single-component March-Dollase model may result an over-simplification [31]. The superposition of a number of texture components is also considered in available programs. Fitting a texture model to multiple diffraction patterns taken at different sample orientations [74] can avoid some of the limitations of the single-peak fitting and single-orientation texture methods. Also, strain and texture can be co-refined simultaneously [75]. This means that electric field induced piezoelectric strain and the full orientation-dependent crystal structure of ferroelectric ceramics could also be refined under *in situ* electric fields if sufficient three-dimensional diffraction data could be measured using 2D detectors. Indeed, some facilities are available and there is nowadays a general trend of development of synchrotron sample environments to study the next-generation of field-driven device physics [69,76,77]. A useful approach for the structure refinement from single X-ray diffraction patterns obtained for ceramic bodies [78] is adding to a starting model—representing the prevailing symmetry—a number of secondary phases to simultaneously account with coexisting equilibrium symmetries. Such is the case of some BNT-based ceramics even in the unpoled state [11,79,80]. For BNT and BNT-based and other lead-free ferroelectric ceramics, this strategy has been widely used [70,81,82,83,84,85,86,87].

A single diffraction pattern from a θ–2θ scan using a flat specimen can be related with 2D experiments. For the case of high-energy synchrotron radiation (small wavelength, small Bragg angle), the symmetric 1D experiment provides approximately the same information as the data recorded in a transmission 2D experiment for $\alpha = 0$ (Figure 12c). The 1D experiment is more sensitive to surface effects at low Bragg angles, which explains why one might do one or the other depending on if surface effects are the focus of the experiment. In literature, one can find informative studies that carry on Rietveld structural refinement of single conventional or synchrotron X-ray diffraction patterns with flat specimens and coupled θ–2θ scans, of ferroelectrics after the application of electric field [70,73,78,81,82,83,84,85,86,87,88,89].

The validity of such a structure refinement resides in that this represents the simplest but a sufficient interpretation of the obtained diffraction data. When applied consistently to a given diffraction geometry [78,89], it does offer a method for comparing structures in a series of samples from single diffraction patterns, which allows understanding electric field effects and their relationship with the functional properties of the ferroelectric ceramics.

X-RAY ABSORPTION FINE STRUCTURE SPECTROSCOPY

General

X-ray absorption fine structure (XAFS) spectroscopy is another useful synchrotron technique for understanding ferroelectric systems and many other materials. The considered phenomenon is the amplitude modulation of the X-ray absorption coefficient in energies near and above an absorption edge of a given chemical element. It appears only when atoms are in crystals or molecules. Experimentally, it involves the measurement of the X-ray absorption coefficient as a function of photon energy above the K or L absorption edges of a study atom. These experiments require tuned synchrotron radiation. The absorption edge energies used for XAFS are in the range of tenths to tens keV.

The observed XAFS effects are conventionally divided into two energy zones, nearby the absorption edge (so called X-ray Absorption Near Edge Structure—XANES) and the extended one (Extended X-ray Absorption Fine Structure—EXAFS), about 50–80 eV above the absorption edge. An example of an experimental XAFS spectrum of an oxide material is presented in Figure 13. These regions contain different information on the chemical states of the absorbing element and neighboring atoms. Both phenomena offer information on the local atomic coordination to distances normally of 0.5 nm and sometimes until 0.8 nm. XANES offers information on oxidation state of the absorbing element. EXAFS allows the determination of interatomic distances, coordination numbers and degree of thermal and/or structural disorder of the local structure (first few atomic coordination shells) surrounding the absorbing atom. The information is averaged over all absorbing atoms of the same type in the studied sample. The XAFS phenomena are consequence of the X-ray photoelectric effect in the studied material, and the further behavior of the scattered photoelectron. These techniques can be applied to almost any element of the periodic table regardless crystallinity state (samples can be amorphous or crystalline) or concentration, *i.e.*, XANES and EXAFS limits of detection are 10 ppm and 100 ppm, respectively.

The XAFS spectrum (E) is defined phenomenologically [90] as the normalized, oscillatory part of the x-ray absorption coefficient above a given absorption edge, *i.e.*,

$$\chi(E) = [\mu(E) - \mu_0(E)]/\Delta\mu_0, \quad (1)$$

where $\mu_0(E)$ is the smoothly varying background absorption function, that would occur if the absorbing atoms were far apart (so-called "bare atoms") and $\Delta\mu_0$ is the height of the absorption edge as normalization factor. Further, the XAFS spectrum is transformed for obtaining the information it provides.

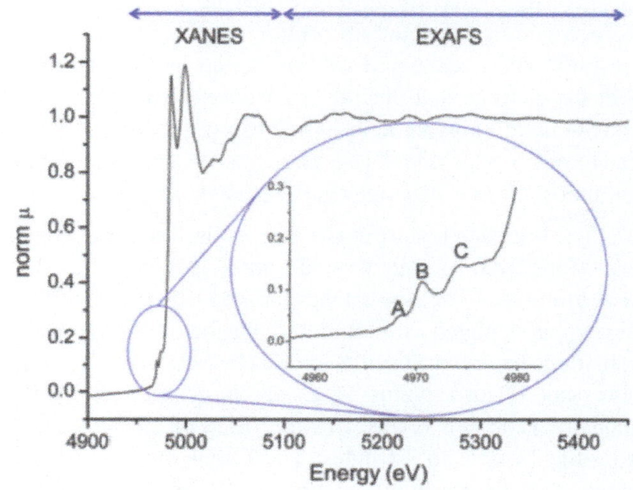

Figure 13. Normalized XAFS spectrum of the Ti K edge of the centrosymmetric perovskite $LaFeNiTiO_3$, recorded at Stanford Synchrotron Radiation Lightsource. XANES and EXAFS regions are presented. The inset corresponds to the pre-edge of XANES zone. The meaning of peaks A, B, and C is given in the text.

Descriptions, fundamentals, experimental methods, data reduction and interpretations of XAFS is well documented in books [91,92,93] and in many articles. The application of XAFS in ferroelectricity studies is also well supported. The reader is referred to the text recently published "X-ray Absorption Fine Structure Applied to Ferroelectrics" [94], where the foundation and merits of the method are described. The content includes the significant XANES effects for the study of ferroelectric materials, the experimental methods and XAFS spectra processing. It offers a review representing XAFS applied to ferroelectrics, from the explanation of the displacive or order-disorder nature of the materials $PbTiO_3$ and $BaTiO_3$, to those dedicated to relaxors and Aurivillius oxides.

XAFS spectroscopy helps in studying ferroelectric materials by analyzing experimental results in both XANES and EXAFS regions. In the EXAFS region the information consists of interatomic distances, coordination numbers and a parameter related to the disorder of the local structure around the absorbing atom. The EXAFS averaged spectrum of each absorption edge of a study element is processed to retrieve the dependence of absorption coefficient on the photoelectron wave number and then to generate a function related to the radial distribution function around atoms of the given element. Particularly interesting is the interpretation of experiments in which absorption edges of two or more different elements are measured. In these cases, the whole set of modeled coordination numbers, interatomic distances and Debye-Waller factors must fulfill compatibility conditions that lead to auto-consistent global solutions [95].

Here, we will briefly outline the observed features on the XANES spectra and the bases of the involved phenomena.

Many ferroelectric materials include in their structure transition metal elements. The effect of the transition metal oxidation state and the cation site symmetry in the XANES spectra is illustrated inFigure 14. When the oxidation state rises, there is a shifting of the absorption edge towards increasing energies. This increase in oxidation state is often associated with cation sites that deviate from the centrosymmetric, and transitions at energies below the absorption edge ("pre-edge transitions") intensify. These phenomena have been reported for almost all transition metals [96,97,98].

The pre-edge feature exists as a result of the photoelectron transitions from 1*s* state to unoccupied excited states. In the inset of Figure 13 are shown the characteristic peaks of the pre-edge zone, depicted as A, B and C. The nature of the pre-edge transitions has been carefully studied by Vedrinskii, *et al.* [99]. The peak A corresponds to a quadrupole transition of low intensity. Peak B is related to a transition of the photoelectron from the 1*s* state to 3*d* states. These transitions in the metal state (oxidation 0) do not take place because the metal atoms are in a very symmetrical environment and the change in the *l* quantum number is 2 (quadrupole transition). However, for tetrahedral or distorted octahedral geometries of increasing oxidation states, 3*d* state mixes with the 2*p* ligand oxygen and the transition *s-p* (dipole transition) is not hindered, resulting in the increasingly intense peak observed in the XANES spectrum. The intensity of the peak C is attributed to transitions of the photoelectron from the 1*s* state to 3*d* levels of neighboring octahedra.

Vedrinskii, *et al.* [99] have obtained the Expression (2) to evaluate the displacement of the titanium cation from the centrosymmetric position. The area under the peak B $\langle I_B \rangle$ is directly proportional to the mean square displacement $\langle \delta^2 \rangle$ of Ti and inversely proportional to the power of the lattice parameter a of the perovskite cell:

$$\langle IB \rangle = K \langle \delta 2 \rangle a5.5 \qquad (2)$$

This expression and the overall explanation given by [99] of pre-edge transitions play a significant role in the application of XAFS to the study of the ferroelectric perovskite-like structure.

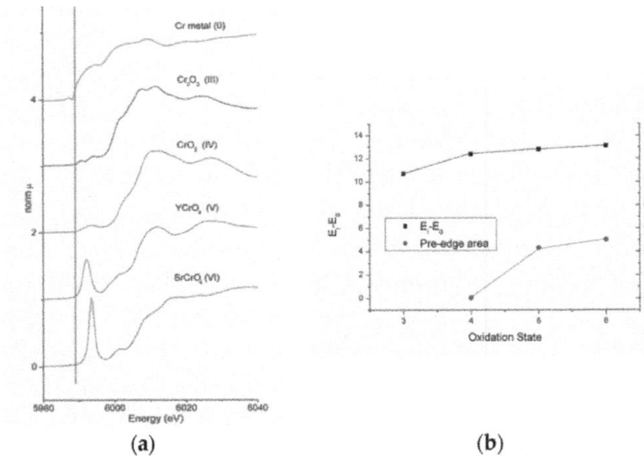

Figure 14. (**a**) XANES spectra of Cr compounds of increasing Cr oxidation states. The vertical line is the position of the first inflection (E_0) in the edge on Cr metal spectrum. Spectra are vertically shifted for clarity; (**b**) Oxidation state dependence of the main absorption edge position (first inflection—E_i) and pre-edge peak area from the spectra presented at (a) [100].

The nature of the pre-edge transitions has been carefully studied also by [101,102,103]. In particular, Yamamoto [103] observed that the pre-edge peaks in K-edge spectra for transition metals with 4*d* electrons, and LI-edge of 5*d* elements should be analogous to those for 3*d* metals.

Technological interest in ferro-piezoelectric ceramics requests characterizing their structures and interactions at the atomic coordination scale. The studies referred to below, aimed at understanding the local-structure basis of polarization, show sustained interest in ferroelectric prototypes ($PbTiO_3$, $BaTiO_3$, PZT) as well as in novel highly demanded materials (lead-free ferroelectrics).

$PbTiO_3$-Based Ceramics Studied by XAFS

A case is seen in the work of Mesquita, Michalowicz and Mastelaro [104], which presents the role of the replacement of Pb by La in the local and electronic structure of PZT and its influence on the ferroelectric behavior. For this purpose, XANES spectra of Ti and O of ceramics $Pb_{1-x}La_xZr_{0.4}Ti_{0.6}O_3$ (PLZT100x) with compositions $x = 0$, 0.05, 0.11, 0.12, 0.13, 0.14, 0.15, 0.21 were studied. Figure 1 in their work shows the spectra resulting from the Ti K-pre-edge. It shows the feature peaks A, B and C, similar to those in Figure 13. The feature B corresponds to the dipole transition occurring in transition metals surrounded by oxygens in distorted symmetry. By determining the peak areas designated as B, and applying a simplified version of Equation (2), reported in [101,102], the authors [104] assessed the Ti cation displacement from the centrosymmetric structure, descending from 0.42 Å for $x = 0$–0.28 Å for $x = 0.21$. In that paper, they conclude that the replacement of Pb by La leads to the ferroelectric phase transition from normal to relaxor. Other works devoted to study $PbTiO_3$ and PZT by XAFS are [105,106].

$BaTiO_3$-Based Ceramics Studied by XAFS

The paper of Levin, *et al.* [107] remarkably presents the use of feature B in the pre-edge region of Ti K-edge to study the local structure of the solid solutions type $Ba(Ti,Zr)O_3$ in different concentrations. Ti and Zr cations behavior is characterized by the measurement of Ti K-edge XANES and Zr K-edge EXAFS. It should be noted that is not possible to acquire the Ti K-edge EXAFS spectrum when Ba is present in the sample, because the Ba LIII-edge energy is greater but close to that of Ti K-edge and the Ba absorption edge signal is much larger than that of the Ti in that region. Figure 15 shows the XANES spectra presented in [107]. The article uses the Expression (2) [99] to evaluate the displacement of the titanium cation from centrosymmetric position, as shown in Figure 15a. They compute the structures by density-functional theory (DFT) and use XRD data for clarifying how the Ti cations are locally ordered in dilute systems, and compare these results to those obtained from the peak B area of the Ti pre-edge XANES. The comparison is presented in Figure 16. The DFT and XRD results presented in [107] suggests that when Ti or Zr cations occupy the B site in a relatively isolated environment, the perovskite host lattice somewhat expands but the area I_B of the pre-edge peak B (red circles in Figure 16a) decreases, that is, the B cation remains nonpolar. On the contrary, when there are neighboring Ti ions, these atoms are displaced off center, especially in the ⟨111⟩ direction. The pseudocubic cell with displacements closer to the ⟨111⟩ directions does exhibit stronger Ti 3*d*–O 2*p* hybridization and, consequently, larger areas of the pre-edge feature B. Authors [107] concluded that relatively large concentrations of the same type B atoms are needed to create enough Ti–Ti or Zr–Zr pairs for ensuring a perceptible B type polarization.

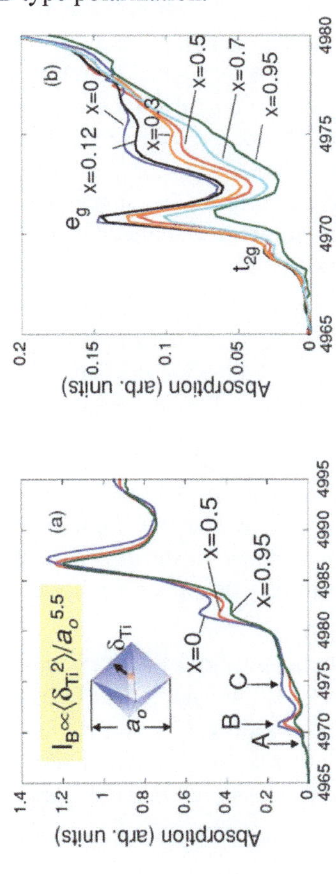

Figure 15. Ti K-absorption edge XANES spectra from Ba(Ti,Zr)O_3 samples studied in [107]. (**a**) Spectra for selected compositions, indicating the pre-edge features A, B and C. Expression (2) from [99] and structural parameters are also shown; (**b**) Ti pre-edge features for all measured compositions, where systematic variation in areas of features B and C is especially noteworthy. Reprinted from [107]. Copyright by the American Physical Society, 2011.

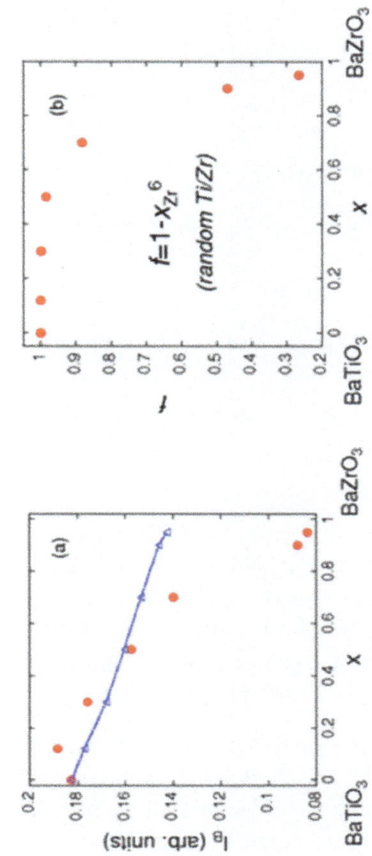

Figure 16. Compositional dependence of the area of pre-edge peak B in Ti K-edge XANES spectra (red circles). (**a**) The blue dashed line/triangles indicate the peak B area reduction according to the $a^{5.5}$ term in Equation (2) if a values are equal to the XRD lattice parameter; (**b**) Calculated compositional dependence of the fraction f of Ti cations having one or more Ti as a next-nearest neighbor for a random distribution of Ti and Zr cations occupying the sites B in perovskites. Reprinted figure with permission from [107]. Copyright by the American Physical Society, 2011.

The ($Bi_{0.5}Na_{0.5}$)TiO_3 *Family Studied by XAFS*

XAFS spectroscopy has also been applied to the study of BNBT's family, with the aim of studying the effects of substitution of Ti by Zr [108], and the replacement of sodium by potassium [109]. The paper by Blanchard, *et al.* [108] devotes marked interest, again, in the behavior of the Ti K-edge and Zr LIII-edge XANES spectra, focusing on five important photoelectron transitions. Figure 17, extracted from the article, presents the Ti K-edge XANES spectra of the ($Bi_{0.5}Na_{0.5}$)$Ti_{1-x}Zr_xO_3$ for x= 0.0, 0.1, 0.2, 0.3, 0.4 and 0.5. For comparison, Figure 17 shows also XANES spectra for $BaTi_{1-x}Zr_xO_3$ in the same composition interval, similar to those studied in [107].

The first results noticed from XANES spectra in Figure 17 is that the doping of both ceramics types with Zr does not induce a shift in the energy of the main jump of the Ti K-edge. Nonetheless, Zr doping does have influence in the increase or decrease of dipole-nature peak in the pre-edge, denoted by B2 in Figure 17.

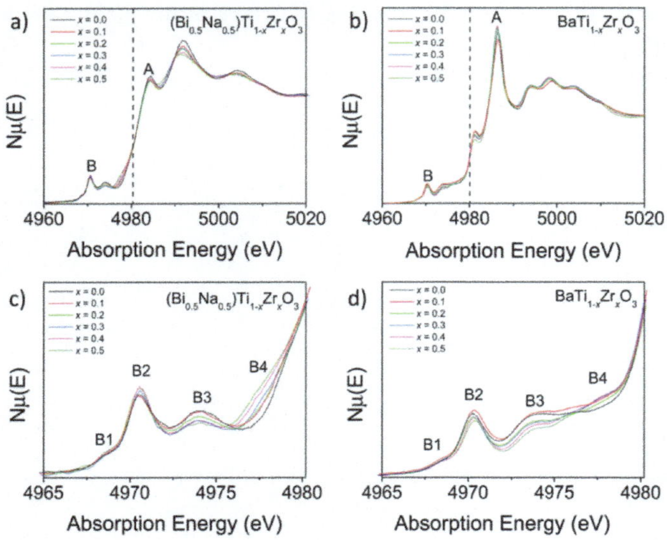

Figure 17. The Ti K-edge XANES spectra of representative members of the (**a**) $(Bi_{0.5}Na_{0.5})Ti_{1-x}Zr_xO_3$; and (**b**) $BaTi_{1-x}Zr_xO_3$ solid solutions. Major features of the Ti K-edge pre-edge region of the $(Bi_{0.5}Na_{0.5})Ti_{1-x}Zr_xO_3$ and $BaTi_{1-x}Zr_xO_3$solid solutions are highlighted in (**c**) and (**d**). Reproduced under a Creative Commons Attribution 3.0 Unreported License.

The work [108] associates the decrease in B2 intensity in $BaTi_{1-x}Zr_xO_3$ with the trend of Ti atoms to locate in a more centrosymmetric position inside the octahedron, while the $(Bi_{0.5}Na_{0.5})Ti_{1-x}Zr_xO_3$ ceramic displays the opposite trend. The changes in the intensities of the peaks B3 and B4 when Zr concentration increases are attributed to transitions of the photoelectron from the 1*s* state to 3*d* levels of neighboring octahedra, and then their behavior suggests that $Ti^{4+}$and Zr^{4+} cations are randomly distributed and there is no clustering of TiO_6 and ZrO_6 octahedra.

The XAFS analysis of the BNBT group of compounds, derived from BNT, is analyzed in detail below.

CASE STUDY: THE BNBT SYSTEM

Introduction

We devote our last section to the intensively studied system $(1-x)(Na_{0.5}Bi_{0.5})TiO_3-xBaTiO_3$, commonly denoted $(1-x)$ BNT–xBT or BNBT100x. Despite the extended literature on the structure–property correlation in this family of materials, there exists nowadays some discussion about the equilibrium structures at room temperature of the compositions near and at the MPB, especially in the $0.04 < x < 0.07$ interval, as well as about their changes with the electric filed. We summarize representative publications and divulge our results and considerations.

As a starting reference, we reproduce in Figure 18 the BNBT room temperature phase diagram, with consideration of electric field application, given by Ma, *et al* [110] and obtained by Transmission Electron Microscopy (TEM) and Selected Area Electron Diffraction (SAED) techniques.

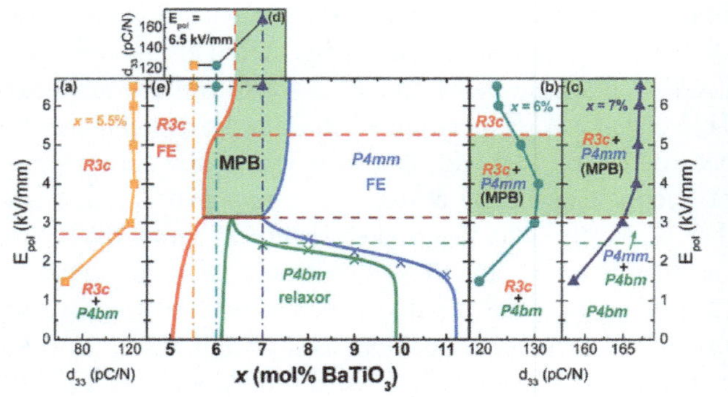

Figure 18. Electric field modulated BNBT phase diagram. Ferroelectric *R3c*, relaxor *P4bm* and ferroelectric *P4mm* regions in the composition-electric field space are identified. Plots of the piezoelectric coefficient d_{33} as function of the electric field intensity, for selected compositions, are included. Reproduced with permission from [110].

In the diagram, the following elements of the MPB are worth being noticed. For $x = 0.04$, the rhombohedral *R3c* symmetry (ferroelectric) remains independent of the applied electric field. For $x= 0.06$, in absence of electric field, a pseudocubic global symmetry, weakly polar at the nanoscale and consisting of polar nanoregions of tetragonal *P4bm* symmetry (ferrielectric) embedded in a cubic matrix and a global rhombohedral *R3c* phases coexist, with predominance of the pseudocubic component. Under an intermediate electric field, a different tetragonal phase, ferroelectric *P4mm*, showing domains with lamellar morphology substitutes the *P4bm* fraction. If an intense electric field is applied, fully poled sample shows only rhombohedral *R3c* symmetry. For $x = 0.07$ and zero electric field, the sample shows only the same pseudocubic global symmetry as for $x = 0.06$. If an electric field is applied, *P4mm* appears in the range of moderate applied field and *R3c* appears for intense applied field, where a mixture of *P4mm* and *R3c* exists.

Diverse variants to the Ma, *et al* representation have been published [15,19,68,70,83,111]. In the following, we revisit the $0.04 < x < 0.07$ interval. Following the general purpose of the article, diffraction and absorption techniques are considered.

Diffraction

We comment first the $x = 0.04$ and $x \geq 0.07$ cases, where our proposed models are similar to those just revised, and then the $x = 0.06$ composition, where the diversity of models in the literature is larger.

For BNBT4, the possibility of a monoclinic *Cc* structure has been proposed [83]. At the time of this writing, this model had not been confirmed by other publications. The work [70] agrees with Ma, *et al.* and delivers a quantitative characterization of the considered *R3c* structure.

A summary of last-mentioned data is included in Table 4 and Table 5, below.

The particular case of a poled BNBT7 ceramic has been studied in detail by [15]. A synchrotron high energy XRD study showed the splitting of the cubic perovskite 200 peak and its evolution with the α angle (Figure 12). In this study, the considered structures were referred to as pseudo-cubic, for zero E-field, and tetragonal *P4mm* for intense electric field.

Anisotropic lattice strains were detected.

Last mentioned experiment is compatible with the one by Ma, *et al.* It is worth explaining here the relationship between the *Pm-3m* cubic, the *P4bm* and the *P4mm* tetragonal models.

The cell geometry of the considered *P4bm* model of the polar (ferrielectric) nanoregions of unpoled BNBT6 and BNBT7 is also the high temperature symmetry for BNT. This is significantly different from the *P4mm* model (ferroelectric) associated with poled BNBT7 (and BNBTs with $x > 0.10$). In *P4mm*, TiO_4 tetrahedra are stretched and cell parameter c is measurably larger than a.Consequently, 00*l* diffraction peaks split from *h*00 ones. In *P4bm*, tetrahedra are tilted ($a^0a^0c^+$, in Glazer notation) and cell parameters are such that the positions of diffraction peaks coincide with the ones generated by the cubic *Pm-3m* perovskite structure. No peak splitting occurs. The differentiating feature of the unpoled BNBT6 and BNBT7 *P4bm* diffraction pattern is the appearance of superlattice peaks. These peaks are extremely weak. They are practically non-observable by X-ray (including synchrotron light) diffraction. The electron microscopy and neutron diffraction experiments by [110,111] revealed the mentioned superlattice peaks.

Figure 19 shows the X-ray and neutron calculated diffraction patterns of BNBT7, modeled with the Kitanaka [111] *P4bm* model. The patterns have been shifted to facilitate observation. Notice the mentioned details.

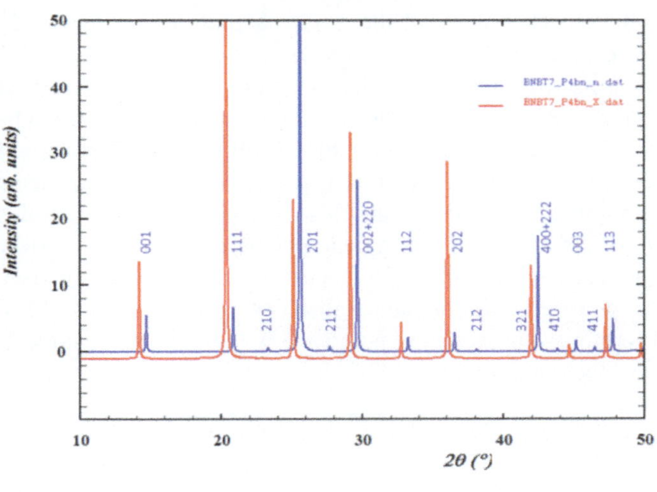

Figure 19: X-ray (red) and neutron (blue) diffraction patterns for tetragonal *P4bm* BNBT. Wavelength: 1 Å. Tetragonal indexing of peaks. Indexes in the lower level correspond to superlattice reflections.

By means of XRD experiments, it is difficult to distinguish tetragonal *P4bm* from cubic *Pm-3m*symmetry [112]. On the one hand, the conjunction of the neutron diffraction [5] (Figure 20), TEM and SAED information [110,113,114,115] unambiguously document the existence of polar phases at the nanoscale. On the other hand, the dielectric relaxor characteristics [116], of BNBT6 and 7 ceramics must be taken into account to select the coexistence of nanoregions of tetragonal *P4bm*symmetry (ferrielectric) embedded in a cubic *Pm-3m* (non-polar) matrix as the equilibrium room temperature symmetry for these two ceramics.

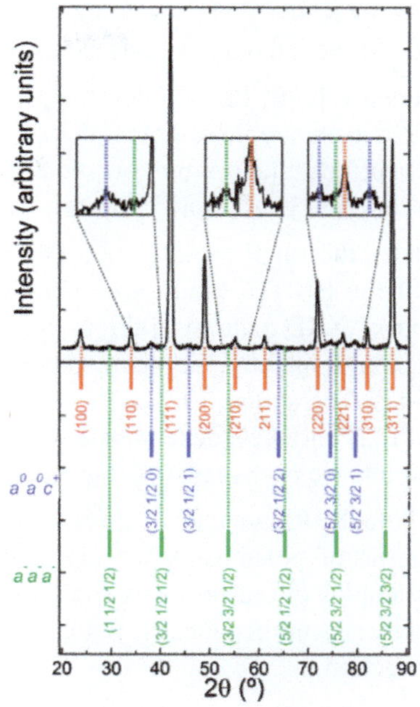

Figure 20. High resolution neutron powder diffraction pattern collected at a wavelength of 1.622 Å. Peak indexing shows contributions from the $a^0a^0c^+$ (*P*4*bm*) and $a^-a^-a^-$ (*R*3*c*) tilt-systems, respectively. Reproduced with permission from [5].

Some early reports using conventional XRD, see, for example, [19], on poled BNBT6 and BNBT7 ceramics associated the relatively complex profile of *h00* peaks (cubic indexing) as the characteristic 00*l*/*h*00 *P4mm* tetragonal peak splitting for BNBTs with $x > 0.11$ (Figure 18).

Recent neutron, electron and XRD synchrotron experiments [68] characterized the structural differences between BNBT6, and BNBT7. By extension of the currently accepted peak splitting association with *P4mm* tetragonal symmetry, this work reports that BNBT6, in the unpoled condition, shows a coexistence of rhombohedral *R3c* and tetragonal *P4bm* global ferroelectric symmetries. Upon application of an electric field, the *P4bm* component evolves to *P4mm*.

Alternatively, a Rietveld analysis of synchrotron X-ray diffraction patterns [87] of unpoled BNBT6 ceramics from nanopowders [117] led to the conclusion that the relatively complex structure of *h00* peaks of this material involved the coexistence of three components. According to these authors, the analyzed BNBT6 ceramic shows a major cubic *Pm-3m* phase, polar at local level as revealed by XANES, that produces the most intense, sharp perovskite peaks. This model is in agreement with both early XRD studies [112] and recent TEM and SAED studies [110,113,114]. Besides this primary phase, two other components were identified. The presence of a rhombohedral *R3c* phase was demonstrated by the appearance of the characteristic 113 (hexagonal notation) weak peak. The third phase was ascribed to significantly wide peaks (See Figure 21), humps, appearing at the low-angle side of the sharp and intense cubic *Pm-3m* global structure peaks. It was modeled as nano-sized crystallites (~20nm).

Due to the dielectric relaxor behavior of this BNBT6 ceramic [116], related with the existence of polar nanoregions (PNRs), humps [87] have been associated with PNRs, presumably of local *P4bm*symmetry [87]. PNRs were observed by TEM in BNT6 [110,114] and have been also detected in

PLZT, PMN-PT, PZN and PMN by force microscopy [118] and neutron radial function distributions (RFDs) [119].

Recent literature [120] has pointed that, as the assumed size of the PNR is comparable or below the coherence length for X-rays and neutrons, the information about PNRs obtained by diffraction/scattering is of limited accuracy.

To elucidate the applicability of the three phases' interpretation of the BNBT6 (and close compositions at the MPB) synchrotron XRD patterns [87] as an alternative to the *P4mm* tetragonal symmetry, the nature of the low-angle side humps in BNT-based ceramics [70,78,82,83,84,85,87,121] and single-crystals [122] and their dependence with respect to the main perovskite intense peaks must be clarified.

We report here the Rietveld analysis of synchrotron X-ray diffraction patterns of poled BNBT6 ceramics, identical to those unpoled samples of our previous study [87]. Poling has been performed by a saturation electric field, applied *ex-situ*, in the thickness direction.

Poled bulk ceramics after electrodes soft removal and crushed powder from these samples were studied. Materials handling is described in [123].

Since the electric-field action on the structure and properties of constituent crystallites has been found irreversible at room temperature for this composition, both specimens are expected to have the same crystal structure. The experiment aimed to shine light on the nature of the component ascribed to the repeatedly observed humps in BNT-based ceramics [70,78,82,83,84,85,87,121] and single-crystals [122].

High Q and high counting statistics synchrotron diffraction experiments were carried out using a coupled θ–2θ scan (from 10° to 100° 2θ) with flat specimen at the four circle diffractometer of the MCX line at Elettra Sincrotrone [124] using an X-ray beam of 13.048 keV (λ = 0.9500 Å). Longer data collection times in the 2θ final segment allowed measuring the weak signals of this region with statistical errors $\sigma_{rel} \approx 2\%$. Prior to any calculation, the different segments are normalized considering the measuring conditions and readings of the incident beam intensity monitor.

Figure 21 shows the pattern obtained for the poled ceramic and the corresponding Rietveld analysis (program Fullprof [125]) plot with reliability factors. The pattern was refined for the coexistence of a global rhombohedral *R3c* phase and nano-sized crystallites, modeled as cubic (*Pm-3m*) symmetry.

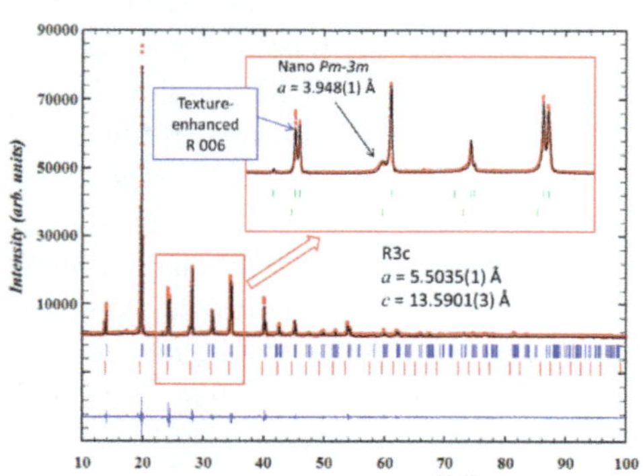

Figure 21. Synchrotron X-ray diffraction pattern and Rietveld analysis of a hot-pressed and recrystallized BNBT6 ceramic flat specimen after poling with electric field perpendicular to the explored surface. Reliability factors: *R*p = 5.63, *R*wp = 8.06. Inset shows some relevant peaks of the ceramic perovskite structure consisting on a rhombohedral *R3c* symmetry and a nanosized cubic symmetry, associated with the humps that appear at the low angle side of the main perovskite peaks.

Comparison of Figure 21 with the corresponding XRD patterns in [87] shows that, under polarization, the global rhombohedral *R3c* phase increases at the expenses of the weakly distorted pseudocubic/tetragonal (*Pm-3m/P4bm*) symmetry, a major component in the non-poled sample. This scheme follows the non-ergodic relaxor to ferroelectric field-induced phase transition in BNBT6 [116,126,127].

It must be noticed that, in our studies, humps take place both in hot-pressed and recrystallized as well as in sintered ceramics [128] and both in unpoled [87] and poled samples (Figure 21). Consequently, we can discard the origin of this feature of the pattern as due to texture or other processing parameters of a given ceramic specimen.

When comparing the results of the poled ceramic with our previous results in non-poled ceramic samples, it is noticeable that the integrated intensities of the humps remain virtually unchanged after poling. If the humps were associated with the diffraction peaks of the major perovskite component, for example, as the tetragonal 002/200 doublet, such humps should increase with polarization, as a sign of non-180° domain reorientation. This effect has not been observed.

It must also be noted that the relative areas of the humps with respect to those of the rhombohedral peaks (Figure 22) decrease as the 2θ angle increases in the poled sample. This suggests that humps are the signature of a surface-related component. We are exploring a volume of a depth of few tenths of ceramic grains. Humps are the manifestation of a structural feature at the surface of that explored volume. Experiments in modified lead titanate and lead zirconate-titanate ceramics showed that there exists a near-electrodes surface layer (few microns) unaffected by the electric poling field due to space charge effects and domain-wall blocking mechanisms [129]. Therefore, humps in non-poled and poled samples could well have the same origin.

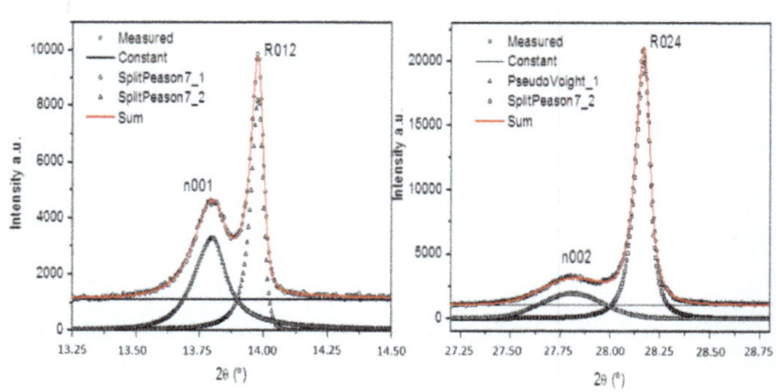

Figure 22. Magnification of two peaks showing the 2θ angular dependence of the intensity of the humps. Peak fitting was carried out using the software described in [36].

Figure 23 shows the experimental pattern and the Rietveld plots obtained for the powder after crushing the poled ceramic. As expected, the preferential orientation corresponding to a poled ceramic is lost. The figure inset reveals that the pattern no longer shows humps at the low angle side of the main perovskite peaks. Therefore, this result, though initially it may be a surprise, is not completely unexpected. It is also noticeable that the lattice parameters of the rhombohedral structure changes very little from the ceramic to the crushed sample.

It seems clear from this result that the experimental condition of random crystallite orientations virtually eliminates the structural contribution to the diffraction pattern that causes the humps. This suggests that such structural contribution is a highly directional one related with the experimental conditions of the measurement. In our poled ceramic experiment, only interplanar distances that are parallel to the electric field are detected. This means that humps are associated to crystallites whose morphology is nanosized, specifically in the direction of the electric field.

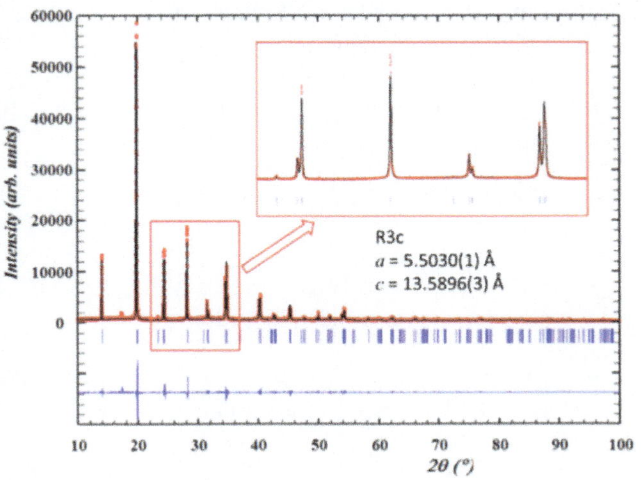

Figure 23. Synchrotron X-ray diffraction pattern and Rietveld analysis of powder flat specimen obtained by crushing the poled BNBT6 ceramic. Reliability factors: *R*p = 6.22, *R*wp = 8.61. Inset shows some relevant peaks of the rhombohedral *R*3*c*symmetry of the powder structure.

Humps could be the signature of planar defects such as those observed in BNBT5 [130] or be assigned to a near surface layer considered for relaxor ferroelectrics to accommodate the lattice distortion near the surface [131,132,133].

In our three-component model for the studied non-poled (ergodic relaxor) [87] and poled (ferroelectric) BNBT6 ceramics (present report), our plausible explanation of the experimental results is that humps reflect the lamellar crystallites of tetragonal (*P*4*mm*) symmetry observed by TEM in poled BNBT6 specimens [110]. The transition from the *P*4*bm* to the *P*4*mm* tetragonal symmetries was observed by TEM at the first poling stages of BNBT6 [110]. These lamellar *P*4*mm*crystallites could appear in the unpoled ceramic as a result of the intergranular stress arising during sintering of the synthetic powder (mechanical poling). They are un-affected by the field at the explored near-electrode surface of the poled ceramic in our measuring geometry. These lamellar crystallites would have a few tenths of nanometer size in the direction of the electric field, and their large faces would be oriented parallel to the sample surface. A negligible volume fraction of near-electrode surface, highly oriented, lamellar tetragonal phase that most probably cause the humps of the diffraction pattern was observed in our experiments on BNBT6. At the same time, we can unambiguously discard through this study the coexistence of rhombohedral and tetragonal ferroelectric global phases in BNBT6.

As a result of this study, we found a virtually single-phase, ferroelectric global rhombohedral (*R*3*c*) phase in BNBT6. The results of our Rietveld refinement are included in Table 4 and Table 5, below.

The structural characterization given to the considered ceramic is very much related with the piezoelectric properties of this material. The absence of coexisting global ferroelectric phases for BNBT6 composition prevents an enhanced remnant polarization in the material and the enhanced electromechanical properties, by similarity with what happens with the lead titanate-zirconate one [2], expected but not found for this composition of the MPB of the BNT-BT system.

XAFS

It has been well established that the MPB in the BNBT phase diagram represents the most worthy domain for the search of competitive ferro-piezoelectric properties. To our knowledge, the first Ti K-edge XANES study of this kind of materials was presented in the work [87]. The general conclusion in that work was that the phases present in the material are a globally pseudocubic *Pm*-3*m* perovskite and an averaged ferroelectric rhombohedral *R*3*c* phase. In [87] the comparison of the pre-edge peak type B of BNBT6 suggests that even in the unpoled state, BNBT6 samples, both powder and ceramic, have spontaneous polarization arising from the off-displacement of the Ti^{4+}ion from the center of the oxygen octahedra.

The general behavior of absorption coefficient from the lead-free material BNBT4 is presented inFigure 24. The XANES spectrum corresponds to the powder sample, prepared before pressing. In the same figure appears the modeling of the A, B and C features for the BNBT4 powder. Fitting was performed with Lorentzian shape for the B peak and the area was adjusted until satisfactory fitting was obtained.

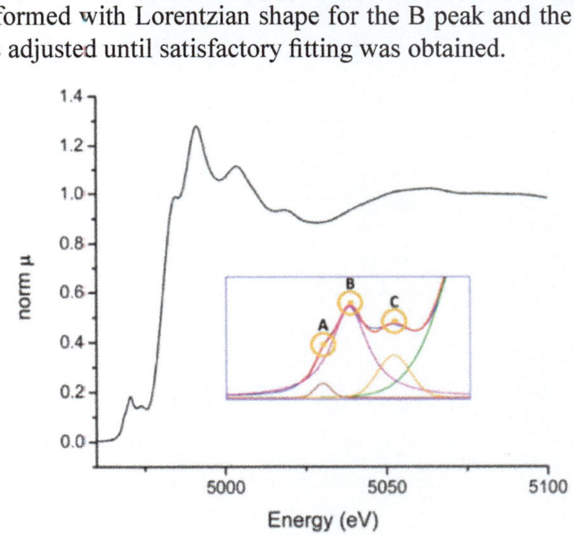

Figure 24. Normalized XANES spectrum of the Ti K edge of a BNBT4 powder sample, recorded at Stanford Synchrotron Radiation Lightsource. The inset corresponds to A, B and C modeled Lorentzian peaks of the pre-edge of XANES zone.

Table 3 shows the values of the pre-edge peak B areas and uncertainties, expressed as standard deviations, from the reference compounds, a $PbTiO_3$ thin film, BNBT4-powder and ceramic and the BNBT6-powder and ceramic presented in [87]. The same table presents an assessment of the proportionality constant K of Expression (2) for the calculation of the root mean square displacement $\langle \delta^2 \rangle^{1/2}$ of Ti atom from the centrosymmetric position. This evaluation was done by applying Expression (2) to the modeled area of pre-edge B features $\langle I_B \rangle$ of the $PbTiO_3$ thin film and the average Ti displacement reported in [134] and to the $\langle I_B \rangle$ corresponding to the Ti displacement resulting of the structure presented in Table 4 below. This value K may be useful for evaluating the displacement of Ti from centrosymmetry in other experiments.

Table 3. XANES results for BNBT100*x*

Sample	$\langle I_B \rangle$ (eV)	$\langle \delta^2 \rangle$ (A^2)	K (eV $A^{3.5}$)
$PbTiO_3$	1.0 (2)	0.308 (4) [a]	$6.5 (1) \times 10^4$
BNBT4-powder	0.95 (9)	–	–
BNBT4-ceramic	0.86 (8)	–	–
BNBT6-powder	1.02 (9)	0.069 (6) [b]	$1.1 (2) \times 10^6$
BNBT6-ceramic	0.8 (1)	–	–

[a] reference [134]; [b] diffraction component of present work. Uncertainties expressed as standard deviations.

Therefore, the table shows that BNBT4 and BNBT6 samples have, within uncertainties, practically the same areas and they are equal to that of the $PbTiO_3$ thin film. These results show that all BNBT samples have practically equal Ti^{4+} cation displacement from symmetry center and confirm that even in the unpoled state, the BNBT samples—powder and ceramic—have spontaneous polarization arising from the off-displacement of the Ti^{4+} ion from the center of the oxygen octahedra. It should be taken into account, on one hand, that XANES depends of interatomic distances, and on the other hand, that our measurement has been performed on polycrystalline samples. Then, the pre-edge feature does not show easily if the displacement is toward the vertex or the face of the octahedron, especially in the case of a small translation. The different order of magnitude of the K value presented in Table 2 has its origin, probably, in experimental differences. The XAFS signal of the thin film should be less intense than that of the powder, and nevertheless, it produces an area $\langle I_B \rangle$ comparable to that of BNBT samples.

BNBT MPB Summary

Integrating bibliographic review, diffraction and XAFS experiments, the characterization we propose for the BNBT morphotropic phase boundary is as follows.

Ceramic XRD patterns occasionally show humps caused by nanometric thickness lamellae (NTL), surface artifacts of tetragonal symmetry, lamellar morphology and nanometric thickness, oriented parallel to the sample surface.

- BNBT4 is essentially *R*3*c* + NTL;
- Unpoled BNBT6 contains *R*3*c* and pseudocubic global phase + NTL;
- Unpoled BNBT7 contains pseudocubic global phase + NTL;
- So-called pseudocubic phase consists of *P*4*bm* 3D nanodomains embedded in a cubic matrix;
- Under poling, BNBT6 shows *R*3*c* symmetry and BNBT7 becomes predominantly *P*4*mm*;
- Crystallographic data of considered representative phases are summarized in Table 4 andTable 5.

Table 4. BNBT*x* (*x* = 4, 6, 7) space groups and lattice parameters

Material			Structure/Symmetry	*a* (Å)	*c* (Å)	Source
BNBT4 unpoled			*R*3*c* + NTL	5.4966(2)	13.5290(6)	[70]
BNBT6	Unpoled		*R*3*c* *P*4*bm* relaxor + NTL	5.505(1) ≅5.5172	13.598(1) ≅3.9010	[87]
	Poled	Ceramic	*R*3*c* + NTL	5.5035(1)	13.5901(3)	Present report
		Milled	*R*3*c*	5.5030(1)	13.5896(3)	
BNBT7 unpoled			*P*4*bm* relaxor + NTL	5.5229(1)	3.9063(1)	[111]

Table 5. BNBT*x* (*x* = 4, 6, 7) atomic coordinates. To facilitate comparison with other reports, all the atomic coordinates have been referred to the most commonly used coordinate system

Material	Atom	Wyck	*x*	*y*	*z*	SOF	ITF (B)	Source
BNBT4	Bi, Na, Ba	6a	0	0	0.250	0.48	Anisotropic factors given in [70]	[70]
*R*3*c*	Ti	6a	0	0	−0.0043(1)	0.04		
unpoled	O	18b	0.1307(1)	0.3404(1)	0.0843(1)	1		
BNBT6	Bi, Na, Ba	6a	0	0	0.250	0.47	2.88(3)	Present report
*R*3*c*	Ti	6a	0	0	−0.0050(4)	0.06	1.92(6)	
poled, milled	O	18b	0.146(1)	0.355(1)	0.0593(4)	1	2.7(2)	
BNBT7	Bi, Na, Ba	2b	0	0.5	0.553(2)	0.465	Anisotropic factors given in [111]	[111]
*P*4*bm*	Ti	2a	0	0	0	0.07		
unpoled	O	2a	0	0	0.512(1)	1		
	O	4c	0.263(1)	0.237(1)	0.030(1)	1		

GENERAL SUMMARY

Summary of considered synchrotron analysis techniques.

Techniques			Revealed structural features
Diffraction/scattering	1D $I = I(2\theta)$	High-resolution powder 1D-XRD; Rietveld analysis; Total scattering; Powder 1D with *in situ*variations of temperature and electric field	Qualitative and quantitative phase analysis; Global crystal structure (lattice parameters, space group, atomic positions); Microstructure (crystal and domain sizes, strain condition, texture); Electric-field induced phase transformations; Radial distribution function
	2 D $I = I(2\theta, \alpha)$	Debye ring analysis; Diffuse scattering; Single and polycrystal 2D reciprocal space investigations with *in situ* variations of temperature and electric field	Texture; Chemical and structural local order-disorder phenomena; Crystallite and ferroelectric domains size and shape; Orientation dependence of the effect of electric field on structure, microstructure, polarization and strain conditions
XAFS	XANES	Pre-edge peak intensities; Main edge transitions intensities	Randomly oriented local polarization; Phase transitions; Lattice strain; Oxidation state; Density of states, Magnetization
	EXAFS	Nearest neighbor distances (elemental specific); Coordination numbers	Radial distribution function around absorbing atom

ACKNOWLEDGMENTS

Authors are grateful to Diamond, Stanford and Elettra synchrotron lightsources for multifaceted support. C. Galassi is acknowledged for the BNBT ceramic samples from nanopowders. Funding from Project CONACYT 257912 "Representación y pronóstico de las propiedades físicas de los materiales mono- y policristalinos" is recognized.

AUTHOR CONTRIBUTIONS

Luis E. Fuentes-Cobas coordinated the article project and revised the physics of diffraction and diffuse scattering, María E. Montero-Cabrera characterized the application of XAFS to ferroelectrics, Lorena Pardo reviewed the diffraction under electric field and analyzed the structure-ferroelectricity relationship and Luis Fuentes-Montero described 2D experiments and data processing.

REFERENCES

1. Rödel, J.; Webber, K.G.; Dittmer, R.; Jo, W.; Kimura, M.; Damjanovic, D. Transferring lead-free piezoelectric ceramics into application. *J. Eur. Ceram. Soc.* 2015, *35*, 1659–1681.
2. Haertling, G.H. Ferroelectric ceramics: History and technology. *J. Am. Ceram. Soc.* 1999, *82*, 797–818.
3. Glazer, A. The classification of tilted octahedra in perovskites. *Acta Crystallogr. Sect. B Struct. Crystallogr. Cryst Chem.* 1972, *28*, 3384–3392.
4. Stokes, H.T.; Kisi, E.H.; Hatch, D.M.; Howard, C.J. Group-theoretical analysis of octahedral tilting in ferroelectric perovskites. *Acta Crystallogr. Sect. B Struct. Sci.* 2002, *58*, 934–938.
5. Simons, H.; Daniels, J.; Jo, W.; Dittmer, R.; Studer, A.; Avdeev, M.; Rodel, J.; Hoffman, M. Electric-field-induced strain mechanisms in lead-free 94%$(Bi_{1/2}Na_{1/2})TiO_3$–6%$BaTiO_3$. *Appl. Phys. Lett.* 2011, *98*, 082901.
6. Kay, H.F.; Vousden, P. XCV. Symmetry changes in barium titanate at low temperatures and their relation to its ferroelectric properties. *London Edinburgh Dublin Philos. Mag. J. Sci.* 1949,*40*, 1019–1040.
7. Von Hippel, A. Ferroelectricity, domain structure, and phase transitions of barium titanate.*Rev. Mod. Phys.* 1950, *22*.
8. Kay, H.; Wellard, H.; Vousden, P. Atomic positions and optical properties of barium titanate.*Nature* 1949, *163*, 636–637.
9. Evans, H. The crystal structure of tetragonal barium

titanate. *Acta Crystallogr.* 1951, *4*, 377.

10. Evans, H.T. An X-ray diffraction study of tetragonal barium titanate. *Acta Crystallogr.* 1961,*14*, 1019–1026.

11. Levin, I.; Reaney, I.M.; Anton, E.M.; Jo, W.; Rödel, J.; Pokorny, J.; Schmitt, L.A.; Kleebe, H.J.; Hinterstein, M.; Jones, J.L. Local structure, pseudosymmetry, and phase transitions in $Na_{1/2}Bi_{1/2}TiO_3$–$K_{1/2}Bi_{1/2}TiO_3$ ceramics. *Phys. Rev. B* 2013, *87*.

12. Zhang, N.; Yokota, H.; Glazer, A.M.; Ren, Z.; Keen, D.; Keeble, D.S.; Thomas, P.A.; Ye, Z.-G. The missing boundary in the phase diagram of $PbZr1-xTixO_3$. *Nature Commun.* 2014, *5*.

13. Subbarao, E.C.; McQuarrie, M.C.; Buessem, W.R. Domain effects in polycrystalline barium titanate. *J. Appl. Phys.* 1957, *28*, 1194–1200.

14. Baerwald, H.G. Thermodynamic theory of ferroelectric ceramics. *Phys. Rev.* 1957, *105*, 480–486.

15. Daniels, J.E.; Jo, W.; Rodel, J.; Jones, J.L. Electric-field-induced phase transformation at a lead-free morphotropic phase boundary: Case study in a 93%($Bi_{0.5}Na_{0.5}$) TiO_3–7%$BaTiO_3$piezoelectric ceramic. *Appl. Phys. Lett.* 2009, *95*.

16. Hinterstein, M.; Knapp, M.; Holzel, M.; Jo, W.; Cervellino, A.; Ehrenberg, H.; Fuess, H. Field-induced phase transition in $Bi_{1/2}Na_{1/2}TiO_3$-based lead-free piezoelectric ceramics. *J. Appl. Crystallogr.* 2010, *43*, 1314–1321.

17. Royles, A.; Bell, A.; Jephcoat, A.; Kleppe, A.; Milne, S.; Comyn, T. Electric-field-induced phase switching in the lead free piezoelectric potassium sodium bismuth titanate. *Appl. Phys. Lett.* 2010, *97*.

18. Rubio-Marcos, F.; Reinosa, J.J.; Vendrell, X.; Romero, J.J.; Mestres, L.; Leret, P.; Fernández, J.F.; Marchet, P. Structure, microstructure and electrical properties of Cu^{2+} doped (K,Na,Li)(Nb,Ta,Sb)O_3 piezoelectric ceramics. *Ceram. Int.* 2013, *39*, 4139–4149.

19. Picht, G.; Töpfer, J.; Hennig, E. Structural properties of $(Bi_{0.5}Na_{0.5})_{1-x}Ba_xTiO_3$ lead-free piezoelectric ceramics. *J. Eur. Ceram. Soc.* 2010, *30*, 3445–3453.

20. Mendiola, J.; Pardo, L. A XRD study of 90° domains in tetragonal PLZT under poling.*Ferroelectrics* 1984, *54*, 199–202.

21. Mendiola, J.; Pardo, L.; Carmona, F.; Gonzalez, A. Study of poling on calcium modified $PbTiO_3$ piezoceramics. *Ferroelectrics* 1990, *109*, 125–130.

22. Bai, F.; Wang, N.; Li, J.; Viehland, D.; Gehring, P.M.; Xu, G.; Shirane, G. X-ray and neutron diffraction investigations of the structural phase transformation sequence under electric field in 0.7Pb($Mg_{1/3}Nb_{2/3}$)–0.3$PbTiO_3$ crystal. *J. Appl. Phys.* 2004, *96*, 1620–1627.

23. Jones, J.L.; Slamovich, E.B.; Bowman, K.J. Domain texture distributions in tetragonal lead zirconate titanate by X-ray and neutron diffraction. *J. Appl. Phys.* 2005, *97*.

24. Zhang, X.; Lei, C.; Chen, K. Ferroelectric 90° domain evaluation in tetragonal Pb($Mg_{1/3}Nb_{2/3}$)O_3–$PbTiO_3$ ceramics. *J. Am. Ceram. Soc.* 2005, *88*, 335–338.

25. Endriss, A.; Hammer, M.; Hoffmann, M.J.; Kolleck, A.; Schneider, G.A. Microscopic and macroscopic ferroelectric–ferroelastic and piezoelectric behavior of PZT ceramics. *J. Eur. Ceram. Soc.* 1999, *19*, 1229–1231.

26. Hall, D.; Steuwer, A.; Cherdhirunkorn, B.; Mori, T.; Withers, P. A high energy synchrotron X-ray study of crystallographic texture and lattice strain in soft lead zirconate titanate ceramics. *J. Appl. Phys.* 2004, *96*, 4245–4252.

27. Rogan, R.C.; Üstündag, E.; Clausen, B.; Daymond, M.R. Texture and strain analysis of the ferroelastic behavior of Pb(Zr,Ti)O_3 by *in situ* neutron diffraction. *J. Appl. Phys.* 2003, *93*, 4104–4111.

28. Cullity, B.; Stock, S. *Elements of X-ray Diffraction*, 3rd ed.; Prentice-Hall: Upper Saddle River, NJ, USA, 2001.

29. Guinebretière, R. *X-Ray Diffraction by Polycrystalline Materials*; ISTE Ltd.: London, UK, 2013.

30. Newnham, R.E. *Properties of Materials: Anisotropy, Symmetry, Structure: Anisotropy, Symmetry, Structure*; Oxford University Press: Oxford, UK, 2004.

31. Fuentes-Cobas, L. Synchrotron radiation diffraction and scattering in ferroelectrics. In*Multifunctional Polycrystalline Ferroelectric Materials*; Pardo, L., Ricote, J., Eds.; Springer: Dordrecht, The Netherlands, 2011; Volume 140, pp. 217–280.

32. Kalyani, A.K.; Khatua, D.K.; Loukya, B.; Datta, R.; Fitch, A.N.; Senyshyn, A.; Ranjan, R. Metastable monoclinic and orthorhombic phases and electric field induced irreversible phase transformation at room temperature in the lead-free classical ferroelectric $BaTiO_3$. *Phys. Rev. B* 2015, *91*.

33. Noheda, B.; Cox, D.; Shirane, G.; Gonzalo, J.; Cross, L.; Park, S. A monoclinic ferroelectric phase in the Pb $(Zr_{1-x}Ti_x)O_3$ solid solution. *Appl. Phys. Lett.* 1999, *74*, 2059–2061.

34. Gorfman, S.; Keeble, D.S.; Glazer, A.M.; Long, X.; Xie, Y.; Ye, Z.-G.; Collins, S.; Thomas, P.A. High-resolution X-ray diffraction study of single crystals of lead zirconate titanate. *Phys. Rev. B* 2011, *84*.

35. Solanki, R.S.; Mishra, S.K.; Senyshyn, A.; Yoon, S.; Baik, S.; Shin, N.; Pandey, D. Confirmation of the monoclinic *Cc* space group for the ground state phase of $Pb(Zr_{0.525}Ti_{0.475})O_3$(PZT525): A combined synchrotron X-ray and neutron powder diffraction study. *Appl. Phys. Lett.* 2013,*102*.

36. Wojdyr, M. Fityk: A general-purpose peak fitting program. *J. Appl. Crystallogr.* 2010, *43*, 1126–1128.

37. Daniels, J.; Jones, J.; Finlayson, T. Characterization of domain structures from diffraction profiles in tetragonal ferroelastic ceramics. *J. Phys. D Appl. Phys.* 2006, *39*.

38. Laulhé, C.; Hippert, F.; Bellissent, R.; Simon, A.; Cuello, G.J. Local structure in $BaTi_{1-x}Zr_xO_3$relaxors from neutron pair distribution function analysis. *Phys. Rev. B* 2009, *79*.

39. Datta, K.; Richter, A.; Göbbels, M.; Neder, R.; Mihailova, B. Atomistic origin of huge response functions at the morphotropic phase boundary of (1−*x*)$Na_{0.5}Bi_{0.5}TiO_3$–*x*$BaTiO_3$. *Phys. Rev. B*2014, *90*.

40. Juhás, P.; Davis, T.; Farrow, C.L.; Billinge, S.J.

PDFgetX3: a rapid and highly automatable program for processing powder diffraction data into total scattering pair distribution functions. *J. Appl. Crystallogr.* 2013, *46*, 560–566.

41. Torres, M.; Pardo, L.; Ricote, J.; Fuentes-Cobas, L.E.; Rodriguez, B.J.; Calzada, M.L. Fabricating ordered functional nanostructures onto polycrystalline substrates from the bottom-up. *J. Nanopart. Res.* 2012, *14*, 1–10.
42. Wcislak, L.; Schneider, J.R.; Tschentscher, T.; Klein, H.; Bunge, H.J. Hard X-ray texture measurements with an on-line image plate detector. *Nucl. Instrum. Methods Phys. Res. Sect. A*2001, *467–468*, 1257–1260.
43. He, B.B. *Two-Dimensional X-Ray Diffraction*; John Wiley & Sons: Hoboken, NJ, USA, 2009.
44. Drakopoulos, M.; Connolley, T.; Reinhard, C.; Atwood, R.; Magdysyuk, O.; Vo, N.; Hart, M.; Connor, L.; Humphreys, B.; Howell, G.; *et al.* I12: The joint engineering, environment and processing (JEEP) beamline at diamond light source. *J. Synchrotron Radiat.* 2015, *22*, 828–838.
45. Daniels, J.E.; Drakopoulos, M. High-energy X-ray diffraction using the Pixium 4700 flat-panel detector. *J. Synchrotron Radiat.* 2009, *16*, 463–468.
46. Veerappan, C.; Richardson, J.; Walker, R.; Li, D.-U.; Fishburn, M.W.; Maruyama, Y.; Stoppa, D.; Borghetti, F.; Gersbach, M.; Henderson, R.K. A 160 × 128 single-photon image sensor with on-pixel 55ps 10b time-to-digital converter. In Proceedings of the Solid-State Circuits Conference Digest of Technical Papers (ISSCC), 2011 IEEE International, San Francisco, CA, USA, 20–24 February 2011; pp. 312–314.
47. Henrich, B.; Bergamaschi, A.; Broennimann, C.; Dinapoli, R.; Eikenberry, E.F.; Johnson, I.; Kobas, M.; Kraft, P.; Mozzanica, A.; Schmitt, B. PILATUS: A single photon counting pixel detector for X-ray applications. *Nucl. Instrum. Methods Phys. Res. Sect. A* 2009, *607*, 247–249.
48. Johnson, I.; Bergamaschi, A.; Buitenhuis, J.; Dinapoli, R.; Greiffenberg, D.; Henrich, B.; Ikonen, T.; Meier, G.; Menzel, A.; Mozzanica, A.; *et al.* Capturing dynamics with Eiger, a fast-framing X-ray detector. *J. Synchrotron Radiat.* 2012, *19*, 1001–1005.
49. Fuentes-Cobas, L.E.; Muñoz-Romero, A.; Montero-Cabrera, M.E.; Fuentes-Montero, L.; Fuentes-Montero, M.E. Predicting the coupling properties of axially-textured materials.*Materials* 2013, *6*, 4967–4984.
50. Wenk, H.-R.; Lutterotti, L.; Vogel, S. Texture analysis with the new HIPPO TOF diffractometer. *Nucl. Instrum. Methods Phys. Res. Sect. A* 2003, *515*, 575–588.
51. Cruz-Gandarilla, F.; Salcedo-Garrido, A.; Avalos, M.; Bolmaro, R.; Baudin, T.; Cabañas-Moreno, J.; Dorantes-Rosales, H. EBSD characterization of an IF steel processed by Accumulative Roll Bonding. In Proceedings of the 17th International Conference on Textures of Materials (ICOTOM), Dresden, Germany, 24–29 August 2014.
52. Reyes-Ruiz, C.; Figueroa, I.A.; Braham, C.; Cabrera, J.M.; Alfonso, I.; Gonzalez, G. Texture and lattice distortion study of an Al-6061-T6 alloy produced by ECAP. *Mater. Trans.* 2015, *56*, 1781–1786.
53. Sáenz-Trevizo, A.; Amézaga-Madrid, P.; Fuentes-Cobas, L.; Pizá-Ruiz, P.; Antúnez-Flores, W.; Ornelas-Gutiérrez, C.; Pérez-García, S.; Miki-Yoshida, M. Microstructural, chemical and textural characterization of ZnO nanorods synthesized by aerosol assisted chemical vapor deposition. *Mater. Charact.* 2014, *98*, 215–221.
54. Fuentes-Montero, L.; Montero-Cabrera, M.E.; Fuentes-Cobas, L. The software package ANAELU for X-ray diffraction analysis using two-dimensional patterns. *J. Appl. Crystallogr.*2010, *44*, 241–246.
55. Chen, C.-S.; Chen, P.-Y.; Tu, C.-S. Polar nanoregions and dielectric properties in high-strain lead-free $0.93(Bi_{1/2}Na_{1/2})TiO_3$–$0.07BaTiO_3$ piezoelectric single crystals. *J. Appl. Phys.* 2014,*115*.
56. Schönau, K.A.; Knapp, M.; Kungl, H.; Hoffmann, M.J.; Fuess, H. *In situ* synchrotron diffraction investigation of morphotropic $PbZr_{1-x}Ti_xO_3$ under an applied electric field.*Phys. Rev. B* 2007, *76*.
57. Daniels, J.E.; Pramanick, A.; Jones, J.L. Time-Resolved Characterization of Ferroelectrics Using High-Energy X-Ray Diffraction. In Proceedings of the 17th IEEE International Symposium on Applications of Ferroelectrics, Santa Fe, NM, USA, 24–27 February 2009.
58. Daniels, J.E.; Jo, W.; Rödel, J.; Honkimäki, V.; Jones, J.L. Electric-field-induced phase-change behavior in $(Bi_{0.5}Na_{0.5})TiO_3$–$BaTiO_3$–$(K_{0.5}Na_{0.5})NbO_3$: A combinatorial investigation. *Acta Mater.* 2010, *58*, 2103–2111.
59. Pramanick, A.; Damjanovic, D.; Daniels, J.E.; Nino, J.C.; Jones, J.L. Origins of electro-mechanical coupling in polycrystalline ferroelectrics during subcoercive electrical loading. *J. Am. Ceram. Soc.* 2011, *94*, 293–309.
60. Hall, D.A.; Azough, F.; Middleton-Stewart, N.; Cernik, R.J.; Freer, R.; Mori, T.; Kungl, H.; Curfs, C. *In-Situ* X-ray diffraction study of ferroelectric domain switching in orthorhombic NKN ceramics. *Funct. Mater. Lett.* 2010, *3*, 31–34.
61. Hinterstein, M.; Rouquette, J.; Haines, J.; Papet, P.; Knapp, M.; Glaum, J.; Fuess, H. Structural description of the macroscopic piezo- and ferroelectric properties of lead zirconate titanate.*Phys. Rev. Lett.* 2011, *107*.
62. Wooldridge, J.; Ryding, S.; Brown, S.; Burnett, T.L.; Cain, M.G.; Cernik, R.; Hino, R.; Stewart, M.; Thompson, P. Simultaneous measurement of X-ray diffraction and ferroelectric polarization data as a function of applied electric field and frequency. *J. Synchrotron Radiat.*2012, *19*, 710–716.
63. Fancher, C.M.; Iamsasri, T.; Blendell, J.E.; Bowman, K.J. Effect of crystallographic texture on the field-induced-phase transformation behavior of $Bi_{0.5}Na_{0.5}TiO_3$–$7BaTiO_3$–$2K_{0.5}Na_{0.5}NbO_3$. *Mater. Res. Lett.* 2013, *1*, 156–160.
64. Daniel, L.; Hall, D.; Webber, K.; King, A.; Withers, P. Identification of crystalline elastic anisotropy in PZT ceramics from *in-situ* blocking stress measurements. *J. Appl. Phys.* 2014,*115*.

65. Daniel, L.; Hall, D.; Koruza, J.; Webber, K.; King, A.; Withers, P. Revisiting the blocking force test on ferroelectric ceramics using high energy X-ray diffraction. *J. Appl. Phys.* 2015, *117*.

66. Moriyoshi, C.; Hiramoto, S.; Ohkubo, H.; Kuroiwa, Y.; Osawa, H.; Sugimoto, K.; Kimura, S.; Takata, M.; Kitanaka, Y.; Noguchi, Y. Synchrotron radiation study on time-resolved tetragonal lattice strain of $BaTiO_3$ under electric field. *Jpn. J. Appl. Phys.* 2011, *50*.

67. Ogino, M.; Noguchi, Y.; Kitanaka, Y.; Miyayama, M.; Moriyoshi, C.; Kuroiwa, Y. Polarization rotation and monoclinic distortion in ferroelectric $(Bi_{0.5}Na_{0.5})TiO_3$–$BaTiO_3$ single crystals under electric fields. *Crystals* 2014, *4*, 273–295.

68. Hinterstein, M.; Schmitt, L.A.; Hoelzel, M.; Jo, W.; Rödel, J.; Kleebe, H.-J.; Hoffman, M. Cyclic electric field response of morphotropic $Bi_{1/2}Na_{1/2}TiO_3$-$BaTiO_3$ piezoceramics. *Appl. Phys. Lett.* 2015, *106*.

69. Hinterstein, M.; Hoelzel, M.; Rouquette, J.; Haines, J.; Glaum, J.; Kungl, H.; Hoffman, M. Interplay of strain mechanisms in morphotropic piezoceramics. *Acta Mater.* 2015, *94*, 319–327.

70. Fuentes-Cobas, L.E.; Pardo, L.; Montero-Cabrera, M.E.; Plaisier, J.R.; García, A.; Brebøl, K.; Mercadelli, E.; Galassi, C. The $0.96(Bi_{0.5}Na_{0.5})TiO_3$–$0.04BaTiO_3$ crystal structure: A high-Q, high-counting statistics synchrotron diffraction analysis. *Cryst. Res. Technol.* 2014, *49*, 190–194.

71. Simons, H.; Daniels, J.E.; Studer, A.J.; Jones, J.L.; Hoffman, M. Measurement and analysis of field-induced crystallographic texture using curved position-sensitive diffraction detectors. *J. Electroceram.* 2014, *32*, 283–291.

72. Rodriguez-Carvajal, J. Recent developments of the prrogram FULLPROF. *Comm. Powder Diffr. IUCr. Newsl.* 2001, *26*, 12–19.

73. Tutuncu, G.; Motahari, M.; Daymond, M.R.; Ustundag, E. A modified Rietveld method to model highly anisotropic ceramics. *Acta Mater.* 2012, *60*, 1494–1502.

74. Matthies, S.; Lutteroti, L.; Wenk, H.R. Advances in texture analysis from diffraction spectra. *J. Appl. Crystallogr.* 1997, *30*, 31–42.

75. Lutterotti, L.; Chateigner, D.; Ferrari, S.; Ricote, J. Texture, residual stress and structural analysis of thin films using a combined X-ray analysis. *Thin Solid Films* 2004, *450*, 34–41.

76. Fertey, P.; Alle, P.; Wenger, E.; Dinkespiler, B.; Cambon, O.; Haines, J.; Hustache, S.; Medjoubi, K.; Picca, F.; Dawiec, A. Diffraction studies under *in situ* electric field using a large-area hybrid pixel XPAD detector. *J. Appl. Crystallogr.* 2013, *46*, 1151–1161.

77. Oddershede, J.; Majkut, M.; Cao, Q.; Schmidt, S.; Wright, J.P.; Kenesei, P.; Daniels, J. Quantitative grain-scale ferroic domain volume fractions and domain switching strains from three-dimensional X-ray diffraction data. *J. Appl. Crystallogr.* 2015, *48*, 882–889.

78. Fuentes-Cobas, L.E.; Montero-Cabrera, M.E.; Garcia, A.; Brebol, K.; Mercadeli, E.; Galassi, C.; Pardo, L. Efecto del campo eléctrico en la estructura cristalina de cerámicas submicroestructuradas de $(1–x)(Bi_{0.5}Na_{0.5})TiO_3$–$xBaTiO_3$ con composición próxima a la frontera morfotrópica. In Proceedings of the X Reunión Nacional de Electrocerámica, Madrid, Spain, 13–15 June 2011; p. 30.

79. Rao, B.N.; Ranjan, R. Electric-field-driven monoclinic-to-rhombohedral transformation in $Na_{1/2}Bi_{1/2}TiO_3$. *Phys. Rev. B Condens. Matter Mater. Phys.* 2012, *86*.

80. Gröting, M.; Kornev, I.; Dkhil, B.; Albe, K. Pressure-induced phase transitions and structure of chemically ordered nanoregions in the lead-free relaxor ferroelectric $Na_{1/2}Bi_{1/2}TiO_3$. *Phys. Rev. B* 2012, *86*.

81. Garg, R.; Rao, B.N.; Senyshyn, A.; Krishna, P.; Ranjan, R. Lead-free piezoelectric system $(Na_{0.5}Bi_{0.5})TiO_3$–$BaTiO_3$: Equilibrium structures and irreversible structural transformations driven by electric field and mechanical impact. *Phys. Rev. B* 2013, *88*.

82. Aksel, E.; Forrester, J.S.; Kowalski, B.; Deluca, M.; Damjanovic, D.; Jones, J.L. Structure and properties of Fe-modified $Na_{0.5}Bi_{0.5}TiO_3$ at ambient and elevated temperature. *Phys. Rev. B*2012, *85*.

83. Usher, T.M.; Forrester, J.S.; dela Cruz, C.R.; Jones, J.L. Crystal structure of $0.96(Na_{0.5}Bi_{0.5}TiO_3)$–$0.04(BaTiO_3)$ from combined refinement of x-ray and neutron diffraction patterns. *Appl. Phys. Lett.* 2012, *101*.

84. Cheng, S.-Y.; Shieh, J.; Lu, H.-Y.; Shen, C.-Y.; Tang, Y.-C.; Ho, N.-J. Structure analysis of bismuth sodium titanate-based A-site relaxor ferroelectrics by electron diffraction. *J. Eur. Ceram. Soc.* 2013, *33*, 2141–2153.

85. Carter, J.; Aksel, E.; Iamsasri, T.; Forrester, J.S.; Chen, J.; Jones, J.L. Structure and ferroelectricity of nonstoichiometric $(Na_{0.5}Bi_{0.5})TiO_3$. *Appl. Phys. Lett.* 2014, *104*.

86. Iamsasri, T.; Tutuncu, G.; Uthaisar, C.; Wongsaenmai, S.; Pojprapai, S.; Jones, J.L. Electric field-induced phase transitions in Li-modified $Na_{0.5}K_{0.5}NbO_3$ at the polymorphic phase boundary. *J. Appl. Phys.* 2015, *117*.

87. Montero-Cabrera, M.E.; Pardo, L.; García, A.; Fuentes-Montero, M.E.; Ballinas-Casarrubias, M.L.; Fuentes-Cobas, L.E. The global and local symmetries of nanostructured ferroelectric relaxor $0.94(Bi_{0.5}Na_{0.5})TiO_3$–$0.06BaTiO_3$. *Ferroelectrics* 2014, *469*, 50–60.

88. Rao, B.N.; Datta, R.; Chandrashekaran, S.S.; Mishra, D.K.; Sathe, V.; Senyshyn, A.; Ranjan, R. Local structural disorder and its influence on the average global structure and polar properties in $Na_{0.5}Bi_{0.5}TiO_3$. *Phys. Rev. B* 2013, *88*.

89. Ehmke, M.C.; Khansur, N.H.; Daniels, J.E.; Blendell, J.E.; Bowman, K.J. Resolving structural contributions to the electric-field-induced strain in lead-free $(1–x)Ba(Zr_{0.2}Ti_{0.8})O_3$–$x(Ba_{0.7}Ca_{0.3})TiO_3$ piezoceramics. *Acta Mater.* 2014, *66*, 340–348.

90. Rehr, J.J.; Albers, R.C. Theoretical approaches to X-ray absorption fine structure. *Rev. Mod. Phys.* 2000, *72*, 621–654.

91. Teo, B.K. *EXAFS: Basics Principles and Data Analysis*, 1st ed.; Springer-Verlag: Berlin Heidelberg, Germany,

New York, Tokyo; Bell Telephone Laboratories, Incorprorated: Murray Hill, NJ, USA, 1986.

92. Koningsberger, D.C.; Prins, R.E. *X-Ray Absorption: Principles, Applications, Techniques of EXAFS, SEXAFS, and XANES*; John Wiley and Sons: New York, NY, USA, 1988; Volume 92, p. 684.

93. Calvin, S. *XAFS for Everyone*; CRC Press: Boca Raton, FL, USA, 2013; p. 457.

94. Lee, J.H.; Choi, H.J.; Lee, D.; Kim, M.G.; Bark, C.W.; Ryu, S.; Oak, M.A.; Jang, H.M. Variations of ferroelectric off-centering distortion and 3*d*-4*p* orbital mixing in La-doped $BiFeO_3$multiferroics. *Phys. Rev. B* 2010, *82*.

95. García-Guaderrama, M.; Montero-Cabrera, M.E.; Morán, E.; Alario-Franco, M.A.; Fuentes-Cobas, L.E.; Macías-Ríos, E.; Esparza-Ponce, H.E.; Fuentes-Montero, M.E. Long-and short-range structure of ferrimagnetic iron—Chromium maghemites. *Inorg. Chem.* 2015, *54*, 11200–11208.

96. Wong, J.; Lytle, F.W.; Messmer, R.P.; Maylotte, D.H. K-edge absorption spectra of selected vanadium compounds. *Phys. Rev. B* 1984, *30*, 5596–5610.

97. Eustace, D.A.; McComb, D.W.; Craven, A.J. Probing magnetic order in EELS of chromite spinels using both multiple scattering (FEFF8.2) and DFT (WIEN2k). *Micron* 2010, *41*, 547–553.

98. Farges, F. *Ab initio* and experimental pre-edge investigations of the Mn K-edge XANES in oxide-type materials. *Phys. Rev. B* 2005, *71*.

99. Vedrinskii, R.V.; Kraizman, V.L.; Novakovich, A.A.; Demekhin, P.V.; Urazhdin, S.V. Pre-edge fine structure of the 3*d* atom K X-ray absorption spectra and quantitative atomic structure determinations for ferroelectric perovskite structure crystals. *J. Phys. Condes. Matter* 1998, *10*, 9561–9580.

100. Montero-Cabrera, M.E.; Sáez-Puche, R.; Climent-Pascual, E.; Morán, E.; Fuentes, M.E.; Fuentes-Cobas, L.E. XAFS characterization of zircon and scheelite $RCrO_4$ (R = Er and Y) polymorphs. In Proceedings of the 14th International Conference on X-Ray Absorption Fine Structure (XAFS XIV), Camerino, Italy, 26–31 July 2009; p. PS1 110.

101. Shanthakumar, P.; Balasubramanian, M.; Pease, D.M.; Frenkel, A.I.; Potrepka, D.M.; Kraizman, V.; Budnick, J.I.; Hines, W.A. X-ray study of the ferroelectric $Ba_{0.6}Sr_{0.4}(YTa)_{0.03}Ti_{0.94}O_3$. *Phys. Rev. B* 2006, *74*.

102. Frenkel, A.I.; Pease, D.M.; Giniewicz, J.; Stern, E.A.; Brewe, D.L.; Daniel, M.; Budnick, J. Concentration-dependent short-range order in the relaxor ferroelectric $(1–x)Pb(Sc,Ta)O_3–xPbTiO_3$. *Phys. Rev. B* 2004, *70*.

103. Yamamoto, T. Assignment of pre-edge peaks in K-edge X-ray absorption spectra of 3*d*transition metal compounds: electric dipole or quadrupole? *X-Ray Spectrom.* 2008, *37*, 572–584.

104. Mesquita, A.; Michalowicz, A.; Mastelaro, V.R. Local order and electronic structure of $Pb_{1-x}La_xZr_{0.40}Ti_{0.60}O_3$ materials and its relation with ferroelectric properties. *J. Appl. Phys.*2012, *111*, 104110–104116.

105. Mesquita, A.; Michalowicz, A.; Mastelaro, V. XANES measurements probing the local order and electronic structure of $Pb_{1-x}Ba_xZr_{0.40}Ti_{0.60}O_3$ ferroelectric materials. *J. Alloys Compd.*2015, *640*, 355–361.

106. Kempet, W.; Marungsri, B.; Yimniran, R.; Klysubun, W.; Pojprapai, S. Investigation of the unit cell distortion in PZT ceramic via *in-situ* XAS technique. *Ferroelectrics* 2013, *453*, 106–112.

107. Levin, I.; Cockayne, E.; Krayzman, V.; Woicik, J.C.; Lee, S.; Randall, C.A. Local structure of Ba(Ti,Zr) O_3 perovskite-like solid solutions and its relation to the band-gap behavior. *Phys. Rev. B* 2011, *83*.

108. Blanchard, P.E.; Liu, S.; Kennedy, B.J.; Ling, C.D.; Zhang, Z.; Avdeev, M.; Jang, L.-Y.; Lee, J.-F.; Pao, C.-W.; Chen, J.-L. Studying the effects of Zr-doping in $(Bi_{0.5}Na_{0.5})TiO_3$ via diffraction and spectroscopy. *Dalton Trans.* 2014, *43*.

109. Liu, S.; Blanchard, P.E.R.; Zhang, Z.; Kennedy, B.J.; Ling, C.D. Long- and short-range structure studies of KBT-KBZ solid-solutions using synchrotron radiation. *Dalton Trans.*2015, *44*, 10681–10688.

110. Ma, C.; Guo, H.; Beckman, S.P.; Tan, X. Creation and destruction of morphotropic phase boundaries through electrical poling: A case study of lead-free $(Bi_{1/2}Na_{1/2})TiO_3–BaTiO_3$Piezoelectrics. *Phys. Rev. Lett.* 2012, *109*.

111. Kitanaka, Y.; Ogino, M.; Hirano, K.; Noguchi, Y.; Miyayama, M.; Kagawa, Y.; Moriyoshi, C.; Kuroiwa, Y.; Torii, S.; Kamiyama, T. Crystal structural analyses of ferrielectric tetragonal $(Bi_{1/2}Na_{1/2})TiO_3–7\%BaTiO_3$ powders and single crystals. *Jpn. J. Appl. Phys.* 2013, *52*.

112. Ranjan, R.; Dviwedi, A. Structure and dielectric properties of $(Na_{0.50}Bi_{0.50})_{1-x}Ba_xTiO_3$: $0 \leq x \leq 0.10$. *Solid State Commun.* 2005, *135*, 394–399.

113. Dorcet, V.; Trolliard, G. A transmission electron microscopy study of the A-site disordered perovskite $Na_{0.5}Bi_{0.5}TiO_3$. *Acta Mater.* 2008, *56*, 1753–1761.

114. Ma, C.; Tan, X. Phase diagram of unpoled lead-free $(1–x)$ $(Bi_{1/2}Na_{1/2})TiO_3–xBaTiO_3$ ceramics.*Solid State Commun.* 2010, *150*, 1497–1500.

115. Schmitt, L.A.; Kleebe, H.-J. Single grains hosting two space groups—A transmisision electron microscopy study of a lead-free ferroelectric. *Funct. Mater. Lett.* 2010, *3*, 55–58.

116. Pardo, L.; Mercadelli, E.; Garcia, A.; Brebol, K.; Galassi, C. Field-induced phase transition and relaxor character in submicrometer-structured lead-free $(Bi_{0.5}Na_{0.5})_{0.94}Ba_{0.06}TiO_3$piezoceramics at the morphotropic phase boundary. *IEEE Trans. Ultrason. Ferroelectr. Freq. Control* 2011, *58*, 1893–1904.

117. Pardo, L.; García, A.; Brebol, K.; Mercadelli, E.; Galassi, C. Piezoelectric properties of lead-free submicron-structured $(Bi_{0.5}Na_{0.5})_{0.94}Ba_{0.06}TiO_3$ ceramics from nanopowders. *Smart Mater. Struct.* 2010, *19*.

118. Kholkin, A.L.; Kiselev, D.A.; Bdikin, I.K.; Sternberg, A.; Dkhil, B.; Jesse, S.; Ovchinnikov, O.; Kalinin, S.V. Mapping disorder in polycrystalline relaxors: A piezoresponse force microscopy

approach. *Materials* 2010, *3*, 4860–4870.

119. Jeong, I.-K.; Darling, T.; Lee, J.-K.; Proffen, T.; Heffner, R.; Park, J.; Hong, K.; Dmowski, W.; Egami, T. Direct observation of the formation of polar nanoregions in Pb $(Mg_{1/3}Nb_{2/3})O_3$using neutron pair distribution function analysis. *Phys. Rev. Lett.* 2005, *94*.

120. Dittmer, R.; Jo, W.; Webber, K.G.; Jones, J.L.; Rödel, J. Local structure change evidenced by temperature-dependent elastic measurements: Case study on $Bi_{1/2}Na_{1/2}TiO_3$-based lead-free relaxor piezoceramics. *J. Appl. Phys.* 2014, *115*.

121. Jo, W.; Rodel, J.; Lee, J.-S.; Baik, Y.-H.; Park, C. Effect of substitution of K for Na on the ferroelectric stability and properties of $(Bi_{1/2}Na_{1/2})TiO_3$-$BaTiO_3$-$(K_{0.5}Na_{0.5})NbO_3$. *Funct. Mater. Lett.* 2010, *3*, 41–44.

122. Zhang, Q.; Zhao, X.; Sun, R.; Luo, H. Crystal growth and electric properties of lead-free NBT-BT at compositions near the morphotropic phase boundary. *Phys. Status Solidi A* 2011, *208*, 1012–1020.

123. Pardo, L.; García, A.; Brebøl, K.; Mercadelli, E.; Galassi, C. Characterization of Nanostructured Phases and Peculiar Phase Transitions in BNBT Lead-Free Piezoceramics.*Adv. Sci. Tech.* 2014, *90*, 12–18.

124. Rebuffi, L.; Plaisier, J.R.; Abdellatief, M.; Lausi, A.; Scardi, P. MCX: A synchrotron radiation beamline for X-ray diffraction line profile analysis. *Z. Anorg. Allg. Chem.* 2014, *640*, 3100–3106.

125. Rodríguez-Carvajal, J. Recent advances in magnetic structure determination by neutron powder diffraction. *Phys. B* 1993, *192*, 55–69.

126. Craciun, F.; Galassi, C.; Birjega, R. Electric-field-induced and spontaneous relaxor-ferroelectric phase transitions in $(Na_{1/2}Bi_{1/2})(1-x)Ba_xTiO_3$. *J. Appl. Phys.* 2012, *112*.

127. Groszewicz, P.B.; Breitzke, H.; Dittmer, R.; Sapper, E.; Jo, W.; Buntkowsky, G.; Rödel, J. Nanoscale phase quantification in lead-free $(Bi_{1/2}Na_{1/2})TiO_3$–$BaTiO_3$ relaxor ferroelectrics by means of NMR. *Phys. Rev. B* 2014, *90*.

128. Gonzalez, A.M.; Pardo, L.; Montero-Cabrera, M.E.; Fuentes-Cobas, L.E. Analysis of the rhombohedral and tetragonal symmetries coexistence in lead-free $0.94(Bi_{0.5}Na_{0.5})TiO_3$–$0.06BaTiO_3$ ceramics from nanopowders. In Proceedings of Piezo 2015, Maribor, Slovenija, 25–28 January 2015. *Adv. Appl. Ceram.* accepted.

129. De Frutos, J.; Jimenez, B. Evolution of the spatial distribution of polarization in lead-calcium ferroelectric ceramics. *Ferroelectrics* 1992, *126*, 341–345.

130. Mi, S.-B.; Jin, X.-W.; Cheng, S.; Liu, M.; Lu, L.; Jia, C.-L. Atomic structures of planar defects in $0.95(Na_{0.5}Bi_{0.5})TiO_3$–$0.05BaTiO_3$ lead-free piezoelectric thin films. *J. Alloys Compd.* 2015, *636*, 183–186.

131. Anthoniappen, J.; Tu, C.S.; Chen, P.Y.; Chen, C.S.; Chiu, S.J.; Lee, H.Y.; Ting, Y.; Wang, S.F.; Chai, C.K. Structural phase stability and electric field induced relaxor–ferroelectric phase transition in $(1-x)(Bi_{0.5}Na_{0.5})TiO_3$–$xBaTiO_3$ ceramics. *J. Alloys Compd.* 2015, *618*, 120–126.

132. Xu, G.; Hiraka, H.; Ohwada, K.; Shirane, G. Dual structures in PZN-*x*PT ferroelectric relaxors. *Appl. Phys. Lett.* 2004.

133. Chang, W.S.; Lim, L.C.; Yang, P.; Miao, H.; Tu, C.-S.; Chen, Q.; Soh, A.K. Tetragonal micro/nanotwins in $0.91Pb(Zn_{1/3}Nb_{2/3})O_3$–$0.09PbTiO_3$ revealed by reciprocal space mapping. *Appl. Phys. Lett.* 2009, *94*.

134. Nelmes, R.J.; Piltz, R.O.; Kuhs, W.F.; Tun, Z.; Restori, R. Order-disorder behaviour in the transition of $PbTiO_3$. *Ferroelectrics* 1990, *108*, 165–170.

CHAPTER 20

ANTI-FERROELECTRIC CERAMICS FOR HIGH ENERGY DENSITY CAPACITORS

Aditya Chauhan, Satyanarayan Patel, Rahul Vaish, and Chris R. Bowen

Keywords: Anti-ferroelectric, energy storage, capacitor, bulk ceramics

INTRODUCTION

As the world actively tries to reduce their dependency on fossil based energy reserves, the successful production, distribution, and storage of electricity will form a cornerstone for the development and growth of society and technology in the coming century. The generation of electricity from renewable and non-conventional resources is being investigated on a large scale and much attention is being paid to the topic of electrochemical energy conversion. However, equal importance is being directed to the development of suitable technologies for efficient storage and distribution of generated electricity. At the forefront of this technology are two main contenders, namely batteries and capacitors. Despite being a relatively well-known concept, the importance of electrical energy storage will continue to be of interest as electricity is the primary means for powering much of our technology. Thus, researchers are working to develop suitable materials with increased energy and power density that are capable of meeting the requirements of modern electric machines and devices [1].

Capacitors as a means for storage of electric energy form an indispensible part of all modern electronic and electrical devices; ranging from electric powered automobiles, drive trains and motors, to mobile communication devices and microwave generation [2,3]. Batteries have a high energy storage density but suffer from a lack of desired power output, while capacitors possess relatively lower energy density but are capable of exhibiting a high power output [2,3]. The main reason behind these differences is the mode of energy storage. Batteries convert electrical energy into chemical energy which is converted back into electrical energy through an electrochemical cell [4,5]. Thus, the rate of up-conversion and re-release of the energy is dictated by the rate of chemical reaction. Capacitors, on the other hand, utilize electrical energy to polarize a dielectric material and thus store energy in the form of localized electric fields/dipole moments existing within the material [2,3]. This energy can be built up over a course of several charging cycles and can be released rapidly to create electrical pulses of required power. This makes for some interesting applications and hence capacitors are employed for a number of devices requiring a high power electric supply [2,3,4,5,6,7]. Regardless of their function, an ideal capacitor is expected to possess a number of basic requirements; a capacitor should possess high power density, high energy density, good discharge efficiency, and low dielectric losses associated with its practical applications [6,7]. Other important requirements include thermal stability and endurance against electrical and mechanical fatigue [6,7]. Therefore, the development of a high energy density electrical capacitor is much sought after as it has the potential to revolutionize the field of power electronics, allowing for rapid miniaturization of modern electronic devices and improving the performance of electric vehicles.

In this regard, much progress has been made in the form of electrochemical supercapacitors [7,8,9,10,11,12]. These are "wet" capacitors which rely on the formation of electrical double layers in a dielectric liquid that can be utilized, when desired, to facilitate electron movement through an external circuit [7,11,12]. This field of research has been the subject of tremendous growth over the last decade and variants have been successfully commercialized for specialized applications. However, the issue of chemical corrosion, fabrication

Citation: Aditya Chauhan, Satyanarayan Patel, Rahul Vaish and Chris R. Bowen, Anti-Ferroelectric Ceramics for High Energy Density Capacitors, doi:10.3390/ma8125439.

complexity, lack of suitable electrodes, and the danger of spills remains on issue in the current-state-the-art in electrochemical supercapacitors [7,11,12]. Further, their application is limited to devices which allow sufficient space for successful installation and maintenance of such capacitors. As a consequence, this technology is difficult to downscale to suit the continued need for miniaturization in modern electronics. Thus, the need for a suitable material to overcome these shortcomings is still being sought [7,11,12].

A possible route to tackling the issue of providing a power supply to smaller electronic devices lies with the use of solid-state dielectric materials. The maximum energy storage density per unit volume of any dielectric material used for a capacitor application can be estimated using the following relation [13]:

$$U = \frac{1}{2} \cdot \epsilon_0 \cdot \epsilon_r \cdot E_b{}^2 \quad (1)$$

Here, U represents the stored energy density per unit volume of the material while ϵ_0, ϵ_r and E_b represent the dielectric permittivity of free space, relative permittivity of the material, and breakdown electric field, respectively. The nature of Equation (1) indicates that materials possessing high dielectric constant (polarizability) and high dielectric breakdown strength are materials of interest for energy storage applications [13]. Furthermore, the energy density of a material increases with the square of the electric field as opposed to linear increase with respect to dielectric constant [13]. Hence, it would appear to be highly beneficial to have a material which can sustain large electric field intensities so as to improve its energy storage characteristics. In this regard, the use of glassy materials as a dielectric medium for capacitor applications is of interest since they possess high dielectric breakdown strengths (~10^9 V/m) [14,15,16,17]. They are also relatively easy to process in a required shape, along with low fabrication costs and good thermal stability [17]. All these factors indicate that glasses are promising materials for electrical energy storage. However, this particular approach to the storage of electrical energy at higher electric fields has some disadvantages; when high electric fields are utilized for energy storage applications the device requires special insulating and shielding materials to protect nearby components from electrical discharge [18,19]. This inevitably raises the overall equipment cost. Secondly, special wiring and installation is required to allow safe charging and discharging of the electrical capacitor when working at high operating voltages. Thirdly, at higher electric fields there is an increased risk of dielectric breakdown and this raises several safety concerns [20]. "Theoretically" for the same amount of voltage, a material with high dielectric constant will possess more energy density and conversely, for the same energy density lower dielectric constant would require a higher electric potential. Anti-ferroelectrics (AFE) materials possess significantly higher dielectric constants than glassy materials which are linear dielectrics. Therefore, the same amount of energy can be stored at a lower value of electric field making it a safer option. A higher voltage runs the risk of corona leakage, electrical aging/fatigue, catastrophic dielectric breakdown and insulation failure. Hence, an anti-ferroelectric (AFE) material with similar energy density is safer for energy storage than linear dielectrics. Furthermore, since glass possesses a poor level of polarizability, the application of a high electric field (in the order of ~10–12 MV/cm) is required to store utilizable energy [21]. An exception to this rule is observed with nano-structured TiO_2 which have been reported for exceptional energy storage characteristics possessing a dielectric constant of 110 with a dielectric breakdown strength of 1400 kV/cm. These figures' combined yield an effective energy storage density of 14 J/cm^3 [22]. However, in order to realize this high energy density, an electric field of substantially high intensity is required which again translates back to the original plight of high electric potential. It is for similar reasons that the use of (ferroelectric/dielectric) polymers as capacitor materials have not gained much confidence [23,24,25]. In addition, polymers have relatively poor thermal stability compared to glasses even though ferroelectric co-polymers have been reported to possess larger dielectric constants in comparison to glasses [25,26,27,28]. The third possible class of material in this application is dielectric and ferroelectric thin films [29,30]. Due to their low defect concentration, thin films are capable of withstanding high electric fields [31] and hence possess exceptionally large volume specific energy storage density [32]. However, the absolute energy stored is low for practical purposes and hence such devices are used in configurations having multiple storage layers, which is essentially a device problem. Therefore, thin films have been excluded from discussion in this article. Current research is focused on development of materials which are capable of achieving high energy storage density at relatively low electric fields. In this regard, several other categories of dielectric materials are being extensively explored.

The majority of materials used for fabrication of electrical capacitors consist of linear dielectrics [33], ferroelectrics [34,35], relaxor ferroelectrics [32,36], and anti-ferroelectrics. Figure 1 provides a graphical representation of each of the material classes and their energy storage characteristics as indicated on a polarization versus electric field (P-E) hysteresis loop [37]. The hatched area represents dischargeable energy while the area within the hysteresis loop represents dielectric losses. Further, a ratio of these energies can be used to determine the energy storage efficiency of the material. Discussion for all these factors will be provided in detail in this article. Glass is a good example of linear dielectric which has already been discussed earlier (Figure 1a). Linear dielectrics are often characterized by their low dielectric constant and high dielectric breakdown strength. In contrast, the second most popular choice for capacitor applications has been ferroelectric materials which possess a large dielectric constant but suffer from lower dielectric breakdown strength. Additionally, ferroelectric materials suffer from dielectric losses during charging and discharging cycles as a result of the ferroelectric hysteresis (Figure 1b). Nevertheless, with modern advances in the field of materials chemistry and fabrication techniques it is now possible to tune the dielectric response of such materials [38,39]. In addition, the energy densities compared to high voltage linear dielectrics have been reported [38,39]. Relaxor ferroelectrics are considered to be dipolar glasses displaying a lack of long-range order with respect to ferroelectric domains. As a consequence they possess almost negligible remnant polarization and coercive electric field while also displaying a lack of conceivable ferroelectric hysteresis (Figure 1c). Thus, they are gaining popularity over conventional ferroelectric materials as a suitable candidate for capacitor applications [40,41]. However, material systems which display relaxor characteristics are relatively rare [41] and are mostly lead-based, for example $xPb(Mg_{1/3}Nb_{2/3})O_3$-(1 – x)$PbTiO_3$. Hence, due to growing environmental concern towards reducing the use of lead in ferroelectric materials, a more suitable material is being sought.

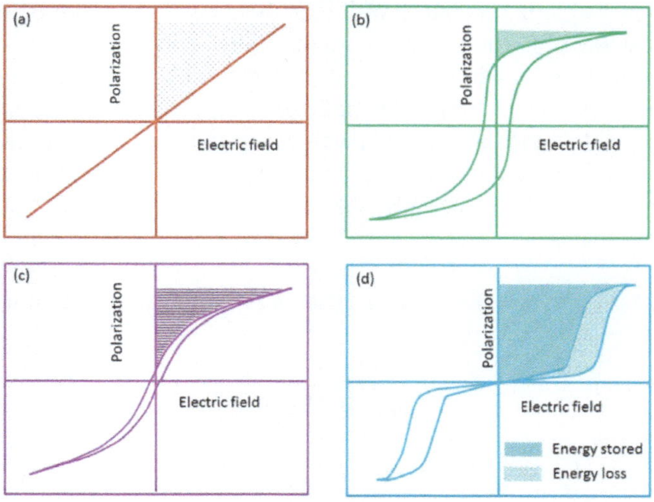

Figure 1. Figure represents the typical polarization versuselectric field (P-E) hysteresis loops and energy storage characteristics of the four classes of solid dielectric materials namely **(a)** linear; **(b)** ferroelectric; **(c)** relaxor ferroelectric; **(d)** anti-ferroelectric (demonstration only; not to scale).

Anti-ferroelectric materials possess relatively larger energy storage density, have lower values of remnant polarization and coercive electric field and faster discharge rates for dissipating stored electrical energy, due to ferroelectric to anti-ferroelectric phase transition [42,43]; see Figure 1d. Due to the lack of ferroelectric domains at low electric field, AFE materials do not possess any considerable hysteresis and have low remnant polarization, coercive electric field, and dielectric loss. However, at higher electric fields, the anti-parallel dipoles are aligned to form a ferroelectric phase. Thus, these materials are capable of storing considerable energy at higher material efficiencies and exhibit faster discharge rates. Appropriate details for the same have been provided in subsequent sections. Thus, upon considering the associated facts with each material family, a consensus can be achieved that anti-ferroelectric materials are a natural choice as potential candidates for electrical energy storage capacitors.

Anti-ferroelectric ceramics in their innate state are characterized by the presence of anti-parallel dipolar alignment [37,44,45]. This arrangement of polar moment in the material creates a complete absence of any observable ferroelectric, piezoelectric, and pyroelectric properties. However, under the application of a sufficiently high intensity electric field, the dipoles rotate to align in the direction of the applied external field [38,39]. This is followed by a first order reversible anti-ferroelectric to ferroelectric phase transformation. The magnitude of the electric field required for the phase change depends on material composition and external parameters such as temperature and stress. Thus, at high fields, the anti-ferroelectric material behaves as a ferroelectric material. Since the transformation is reversible, the ferroelectric phase reverts back to its original anti-ferroelectric form when the applied electric field is reduced below a certain critical value [43,46]. The low hysteresis losses coupled with the reversible phase transformation have been associated with high power and energy density in anti-ferroelectric materials [37,47,48]. Furthermore, the loss of ferroelectric characteristics at low electric field is also responsible for a fast discharge rate and high discharge efficiency in these materials. Thus, anti-ferroelectric compositions have the potential to become high performing contenders in the field of commercial capacitor applications.

Excellent reviews are present in the literature dealing with development and applications of a wide range of anti-ferroelectric materials [49,50]. However, little or no attention has been paid to the most promising attribute of this material class: high energy storage capacitors. Further, the amount of research conducted in this direction is also limited due to a lack of variety in materials with anti-ferroelectric properties [51,52]. However, with advancements in the field of fabrication and characterization techniques and the discovery of novel lead-free ferroelectric materials, research in this field can be reignited [53]. We have attempted to undertake this review article and provide the necessary spark by presenting a timely assessment on the subject. This article aims to review and highlight the relevant scientific advancements that have been made with respect to utilization and development of anti-ferroelectric materials for electric energy storage applications. The structure of this paper is as follows: the article begins with a general introduction discussing the need for high energy density capacitors, the present solutions being used to address this problem and a brief discussion of various advantages of anti-ferroelectric materials for high energy storage applications. This is followed by a general description of anti-ferroelectricity and important anti-ferroelectric materials. The remainder of the paper is divided into two subsections, the first of which presents various physical routes for enhancing energy storage density while the latter section describes how chemical routes can be used to accomplish the same. This is followed by conclusions and future prospects and challenges which need to be addressed in this particular field.

ANTI-FERROELECTRICITY AND ASSOCIATED MATERIALS

AFE materials are similar to ferroelectric materials and largely belong to the perovskite family represented by the nominal chemical formula of ABO_3; here A represents an alkali or alkaline earth metal while B belongs to the transition group elements [54,55]. This structure gives rise to the BO_6octahedra where the transition group element is surrounded by six oxygen atoms [54,55,56]. When the atom is eccentrically placed, it causes the separation of the charge centers, giving rise to a permanent dipole vector within the crystal. Large numbers of such dipoles are aligned together in a ferroelectric material giving rise to ferroelectric domains [54,55]. In a bulk material, many such domains are present, arranged in random directions leading to an electrically neutral material [54,55,56]. However, the application of an external electric field can align dipoles in a predetermined manner, as shown in Figure 2 [56]. In AFE materials, adjacent layers of octahedra have dipoles aligned in opposite directions (anti-parallel), in the absence of electric field [50]. Thus, even at microscopic level they appear to lack any domain structure and display a complete lack of electrical functionality. However, when an external electric field beyond a certain threshold is applied to AFE materials, the dipoles are rotated and give rise to a domain structure, as well as ferroelectric behavior [43]. This is evident from the double hysteresis loop (Figure 1d) which is often used to identify the AFE materials. Thus, an AFE phase can be converted into a ferroelectric phase upon the application of electric field of suitable strength. Consequently, these materials lack a major part of low-field hysteresis which is associated with conventional FE materials and thus possess lower dielectric losses which greatly increases the energy storage efficiency. Furthermore, the AFE to FE transition is accompanied by a volumetric expansion in both longitudinal and transverse directions and has been reported to be sensitive to both applied temperature and stress [43,46]. Table 1 provides a list of the known bulk AFE materials and their associated properties as reported in the literature. While it is not comprehensive, care has been taken to acknowledge any and all relevant results reported in the literature with regards to employment of AFE materials for energy storage applications. Another important point to be mentioned here is that the list provides energy densities which were obtained for a very specific material geometry and processing parameters. Hence, it is entirely possible to obtain values different than those reported here, based on the material's shape (area/thickness), size, and processing.

Table 1. Important bulk anti-ferroelectric materials, with various physical/chemical modifications for enhancing energy storage density, as reported in literature

Material Name	Physical/Chemical Modifications	Energy Density (J/cm^3)	Electric Field (kV/cm)	Reference
$(Pb_{0.97}La_{0.02})(Zr_{0.97}Ti_{0.03})O_3$	3 wt % glass (PbO-B_2O_3-SiO_2-ZnO)	3.1	581	[57]
$(Pb_{0.97}La_{0.02})(Zr_{0.97}Ti_{0.03})O_3$	without glass	1.4	581	[57]
$Pb_{0.99}Nb_{0.02}[(Zr_{0.57}Sn_{0.43})_{1-y}Ti_y]_{0.98}O_3$	with half electrode	1.30	70	[47]
$Pb_{0.99}Nb_{0.02}[(Zr_{0.57}Sn_{0.43})_{1-y}Ti_y]_{0.98}O_3$	with full electrode	1.19	70	[47]
$(Pb_{0.94-x}La_{0.04}Ba_x)[(Zr_{0.60}Sn_{0.40})_{0.841}Ti_{0.16}]O_3$	$x = 0$ (0 MPa)	0.35	30	[58]
$(Pb_{0.94-x}La_{0.04}Ba_x)[(Zr_{0.60}Sn_{0.40})_{0.841}Ti_{0.16}]O_3$	$x = 0.02$ (20 MPa)	0.38	30	[58]
$Pb_{0.99}Nb_{0.02}[(Zr_{0.60}Sn_{0.40})_{0.95}Ti_{0.05}]O_3$	-	0.62	56	[59]
$Pb_{0.97}La_{0.02}(Zr_{0.56}Sn_{0.35}Ti_{0.09})O_3$	4 wt % glass (CdO-Bi_2O_3-PbO-ZnO-Al_2O_3-B_2O_3-SiO_2)	3.3	150	[60]
$Pb_{0.97}La_{0.02}(Zr_{0.56}Sn_{0.35}Ti_{0.09})O_3$	without glass	1.9	110	[60]
$(Pb0_{.97}La_{0:02})(Zr_{0.92}Sn_{0.05}Ti_{0:03})O_3$	3% glass ($0.8PbO$-$0.2B_2O_3$)	7.4	475	[61]
$(Pb0_{.97}La_{0:02})(Zr_{0.92}Sn_{0.05}Ti_{0:03})O_3$	without glass	4.5	320	[61]
$Pb_{0.97}La_{0.02}(Zr_{0.95}Ti_{0.05})O_3$	-	12.4	1120	[62]
$(Pb_{0.85}Ba_{0.08}Sr_{0.03}La_{0.03})(Zr_{0.74}Sn_{0.22}Ti_{0.04})$	-	1.2	100	[63]
$[(Bi_{1/2}Na_{1/2})_{0.94}Ba_{0.06}]La_{0.8}Zr_{0.2}TiO_3$	-	1.58	85	[64]
$(Ba_{0.1}La_{0.02})(Zr_{0.675}Sn_{0.275}Ti_{0.05})O_3$	-	2.05	70	[65]
$0.75(_{0.80}Bi_{1/2}Na_{1/2}TiO_3$-$0.20Bi_{1/2}K_{1/2}TiO_3)$-$0.25SrTiO_3$	-	0.84	100	[66]
$Pb_{0.97}La_{0.02}(Zr_{0.50}Sn_{0.45}Ti_{0.05})O_3$	-	5.6	400	[67]
$(Pb_{0.858}Ba_{0.1}La_{0.02}Y_{0.008})(Zr_{0.65}Sn_{0.3}Ti_{0.05})O_3$-$(Pb_{0.97}La_{0.02})(Zr_{0.9}Sn_{0.05}Ti_{0.05})O_3$	-	4.65	200	[68]
$0.89Bi_{0.5}Na_{0.5}TiO_3$-$0.06BaTiO_3$-$0.05K_{0.5}Na_{0.5}NbO_3$	double stage sintering	0.90	100	[69]
$(Pb_{0.858}Ba_{0.1}La_{0.02}Y_{0.008})(Zr_{0.65}Sn_{0.3}Ti_{0.05})O_3$-$(Pb_{0.97}La_{0.02})(Zr_{0.9}Sn_{0.05}Ti_{0.05})O_3$	spark plasma sintering	6.40	275	[70]
$(Pb_{0.858}Ba_{0.1}La_{0.02}Y_{0.008})(Zr_{0.65}Sn_{0.3}Ti_{0.05})O_3$-$(Pb_{0.97}La_{0.02})(Zr_{0.9}Sn_{0.05}Ti_{0.05})O_3$	conventional sintering	4.65	200	[70]
$(Na_{1-x}Ca_x)(Nb_{1-x}Zr_x)O_3$ $x = 0.04$	conventional sintering	0.91	130	[71]
$(Pb_{0.92}La_{0.04}Ba_{0.02})[(Zr_{0.60}Sn_{0.40})_{0.84}Ti_{0.16}]O_3$	at 90 MPa	0.91	60	[43]
$0.91(Bi_{0.5}Na_{0.5})TiO_3$-$0.07BaTiO_3$-$0.02(K_{0.5}Na_{0.5})NbO_3$	at 100 MPa	0.387	60	[37]
$(Pb_{0.96}La_{0.04})(Zr_{0.90}Ti_{0.10})O_3$	at 100 MPa	0.698	60	[43]

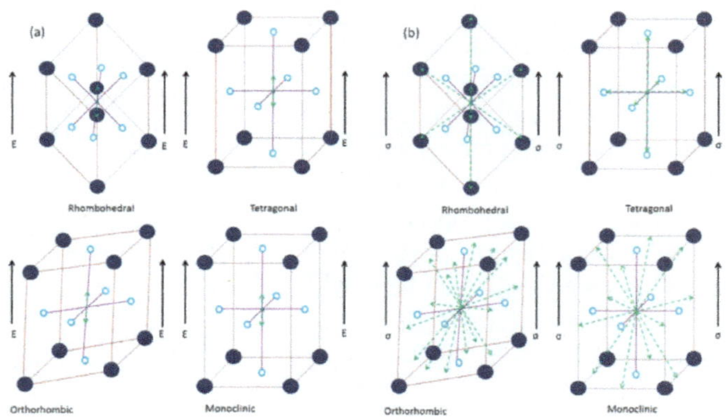

Figure 2. Graphical representation of the mechanism and various switching directions associated with different crystallographic structures under (a) ferroelectric; (b) ferroelastic domain rotation, respectively.

Due to their high field transformation, low hysteresis, low remnant polarization and high energy storage density, AFE materials have been proposed as leading contenders for solid-state capacitor applications [50,51,52,66,72]. However, since the materials typically belong to the perovskite ceramic family they suffer from low dielectric breakdown strength [73,74]. This serves to limit the maximum amount of energy that can be stored in these materials which ultimately falls behind that reported for linear dielectrics. Hence, efforts have been made to increase the energy storage density in AFE. This approach can be subdivided into two main techniques namely physical and chemical routes which attempt to reduce hysteresis, increase the dielectric constant, or enhance the dielectric breakdown strength of the AFE. The following sections will discuss in detail how each of these methods is accomplished and the relevant studies which have reportedly employed these techniques to enhance the energy storage density in AFE materials.

ENHANCING ENERGY STORAGE CAPACITY IN ANTI-FERROELECTRIC MATERIALS: PHYSICAL ROUTES

As explained previously, Equation (1) can be used to evaluate the energy storage density of a dielectric material. However, another popular method for determining the energy storage density in ferroelectric, relaxor ferroelectric, and anti-ferroelectric materials is to numerically integrate the area under the polarization versus electric field (P-E) hysteresis curves (as in Figure 1). The numerical representation for this is given as [43,75]:

$$U=\oint E\cdot dP \quad (2)$$

Here E is applied electric field and P is polarisation. The expression in Equation (2) can be used to determine the energy density of most ferroelectric materials at their saturation electric field values. Furthermore, since ferroelectric materials are prone to hysteresis losses, another important factor known as material efficiency needs to be determined according to the following relation [37]:

$$\eta=U/U+U_1 \quad (3)$$

Here the symbols η and U_1 represent the material efficiency and dielectric losses respectively. It can be clearly understood from Equation (3) that if dielectric losses can be reduced, it leads to improved energy storage characteristics. It has already been established in the literature that an AFE to FE transition is sensitive to both ambient temperature and applied pre-stress to the material [76,77,78]. Hence, the physical methods are concentrated on how to utilize these effects in order to delay the transition, so as to increase the energy density or reduce hysteresis losses and thereby increase material efficiency. These have been discussed in detail as follows.

Compressive Pre-Stresses

When an electric field is applied to an AFE material it continues to energize the anti-parallel dipole alignment until the threshold intensity is reached, at which the FE characteristic is initiated. This is accompanied by a large volumetric expansion and corresponding strain [79,80]. The process is reversed when the applied electric field is lowered below a certain critical point. This difference in forward and backward switching electric fields is known as ΔE [79,80]. This phenomenon is similar to the ferroelectric domain rotation as observed in conventional FE materials. Therefore, assuming AFE to be initial phase and FE to be final phase for forward process, the minimum energy required to initiate domain formation can be dictated using the Helmholtz free energy density (A) for ferroelectric materials as follows [81]:

$$dA=\sigma d\epsilon+EdD-SdT \quad (4)$$

Here, σ, ϵ, E, D, T and S denote stress, strain, electric field intensity, electric displacement, temperature, and entropy, respectively. From Equation (4) it can be deduced that the equilibrium can be shifted in the favor of a preferred state by a suitable application of stress, electric field or temperature. Therefore, Equation (4) can be rewritten to represent a phase change as follows [82]:

$$dA^{\alpha\to\beta}=\sigma d\epsilon^{\alpha\to\beta}+EdD^{\alpha\to\beta}+TdS^{\alpha\to\beta} \quad (5)$$

Here, α and β indicate the initial (AFE) and final (FE) phases of material. This effect has been reported to drive a phase change in a large number of studies [46,59,83,84,85,86,87]. It can also be inferred from Equations (4) and (5) that if $d\epsilon \neq 0$, an externally applied mechanical stress will influence the phase transformation. The nature of influence of compressive stresses on any structural transition depends on $d\epsilon$. As is the case of AFE to FE phase change, it is accompanied by volumetric expansion in both lateral and axial directions. Thus, the application of a compressive stress or mechanical confinement to the material will shift the value of electric field required to drive the transition to a higher magnitude [58,78]. This phenomenon can be explained on the basis of the competing nature of mechanical and electrical domain switching modes. In its innate state, a bulk ferroelectric ceramic consists of randomly aligned domains. Thus, the material as a whole does not exhibit any spontaneous polarization or ferroelectric activity. However, when the same material is subjected to an electric field of increasing magnitude, domain realignment is initiated. Initially, minor domain wall motion is observed which favors the growth of preferentially aligned domains [43,46,86,87]. However, upon being subjected to higher magnitude of electric field, the domains themselves are rotated in the direction of the applied electric field. This phenomenon is known as ferroelectric domain switching and is generally associated with 180° domain rotation [37,43,46,58,62,78]. Conversely, when a poled or activated, ferroelectric material is subjected to compressive stress, the domains realign to face away from the direction of stress applied; termed ferroelastic domain switching [43,86,87]. The movement of the domains is such that it minimizes structural energy to accommodate the additional strain being generated in the material. Ferroelastic domain rotations are inherently non-180° in nature and the switching direction has been observed to be dependent on the crystallographic structure of the material [37,43,46,58,62,78]. A graphical description of the ferroelectric and ferroelastic domain switching (directions) associated with different crystallographic structures is shown in Figure 2. It can be clearly seen from Figure 2 that each crystal structure has several possible switching sites available to it under electric field or stress. The application of an electric field generally forces the movement of a charge center in such a manner that it minimizes the electrical strain of the material, through the formation of electrical dipoles [37,43,46,58,62,78]. This is generally accomplished by displacing the charge centers along the direction of applied electric field (180° switching). Conversely, when such a poled material is subjected to the application of a compressive stress the charge centers are displaced in such a manner that it minimizes the strain energy of the system. This is accomplished by shifting the "mobile" charge center to one of the available side sites (non-180° switching) [37,43,46,58,62,78].

Thus, when a material is subjected to a combined electro-mechanical loading, the domain alignment is decided by the equilibrium value of the competing fields. It is possible to utilize this phenomenon to also create stress-induced domain pinning in AFE materials [58,78]. Since hysteresis is the equivalent energy utilized by the electric field in rotating the dipoles, domain pinning counters this effect directly. With increasing magnitude of applied compressive stress, a growing number of dipoles are "pinned" to their position. Consequently, the switchable part of polarization is reduced and this is manifested in the form of reduced hysteresis. The phenomenon has been demonstrated using Figure 3 where Figure 3a shows the two states of ferroelectric hysteresis observed in AFE materials in the absence and presence of a compressive stress. It can be observed that the application of compressive stress leads to a loss of saturation polarization with a simultaneous reduction of hysteresis. The inset shows the corresponding nature of field induced ferroelectric domains. Figure 3b shows P-E hysteresis curves for $(Pb_{0.96}La_{0.04})(Zr_{0.90}Ti_{0.10})O_3$ AFE material under various compressive stresses [88]. However, it is to be noted that ferroelectric and ferroelastic switching act to counter each other.

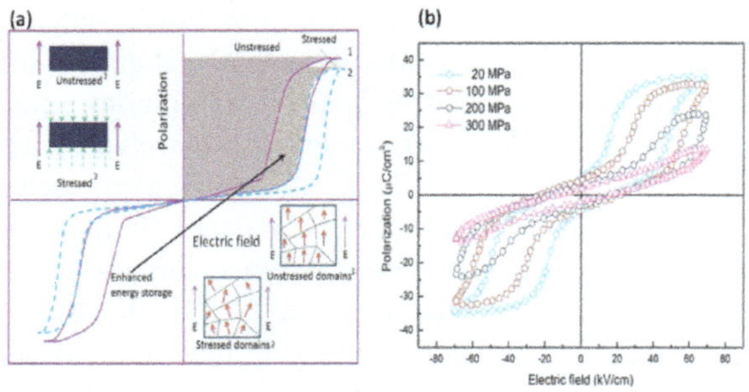

Figure 3. Graphical representation for the technique of physical confinement to increase the energy storage density in anti-ferroelectric materials. (**a**) Displays how the hysteresis loop changes to yield enhanced recoverable energy under uniaxial compressive stress while (**b**) displays P-E hysteresis loops for $(Pb_{0.96}La_{0.04})(Zr_{0.90}Ti_{0.10})O_3$ material under compressive loading [88].

Thus, if the magnitude of one of the fields is increased beyond a certain critical value it will completely override the effect of other. In such a situation, no phase change can be initiated and thus material performance begins to degrade.

In a study reported by Xu et al. it was observed that in bulk AFE ceramics having a nominal composition of $(Pb_{0.94-x}La_{0.04}Ba_x)[(Zr_{0.60}Sn_{0.40})_{0.84}Ti_{0.16}]O_3$, the energy storage density could be increased through mechanical confinement in (x =) 0.02 and 0.04 compositions [58]. Barium modification of the material was seen to reduce the domain size and make the pre-stress less effective in preventing domain polarization alignment. However, for enhancing the energy density, an optimum value of compressive stress is required, in accordance with Equation (5). This was also observed for composition of x = 0.02 where the maximum energy storage density was observed at 70 MPa (~14.8% increase) and any further increase in pre-stress led to a decrease in energy storage density.

Our group has attempted to enhance the energy storage density in multiple AFE ceramics through the application of compressive pre-stress. We undertook systematic investigations to examine the effect of directional confinement on the energy storage density of $0.91(Bi_{0.5}Na_{0.5})TiO_3$-$0.07BaTiO_3$-$0.02(K_{0.5}Na_{0.5})NbO_3$ bulk AFE ceramic [37]. Hysteresis behavior was analyzed as a function of varying levels of compressive stress and operational temperature. It was observed that a peak energy density of 0.387 J/cm³ was obtained at 100 MPa applied stress (25 °C), while a maximum energy density of 0.568 J/cm³ was obtained for the same stress at 80 °C. These values are indicative of a significant, 25% and 84%, improvement in the value of stored energy compared to an unloaded material. A further study [37] clearly revealed that under the application of a compressive stress, the material performance increased rather than decreased, as the operating temperature was increased. The study was also extended to analyze the discharge efficiency under various loading conditions. The material's discharge efficiency was also improved considerably owing to thermal and mechanical fields. The unloaded sample exhibited an efficiency of 37%, which was later improved to a value of 49% for 250 MPa at room temperature and 72% for 100 MPa at 80 °C. A peak efficiency of 90.42% was observed for measurement conditions of 140 °C and 250 MPa applied stress [37].

The second analysis [43] was based on bulk $(Pb_{1-x}La_x)(Zr_{0.90}Ti_{0.10})_{1-x/4}O_3$ AFE ceramics. Mechanical confinement was observed to provide an increase in energy storage density and efficiency by approximately 38% and 25%, respectively, for the x = 4 composition [43]. The highest recoverable energy density of 0.698 J/cm³ was achieved under a compressive stress of 100 MPa and electric field of 60 kV/cm. In this composition, the storage density could be improved from 0.536 J/cm³ to 0.704 J/cm³ for a temperature increase of 25 to 120 °C, respectively. Similarly, the storage efficiency also increased from 41% to 67% with an increase in temperature from 25 to 120 °C. This is due to fact that when the ferroelectric/anti-ferroelectric material is heated, it relaxes and disorients the domains. This reduces the reversible part of ferroelectric switching, which ultimately decreases the area under the hysteresis loop. Therefore, the energy loss of the material decreases while simultaneously increasing the storage density and efficiency. In contrast, a sufficiently high temperature also reduces saturation polarization and, in addition, temperature also affects the dielectric constant of the material. In summary, the application of elevated temperature (120 °C) can increase approximately 31% and 25% storage density and efficiency, respectively, as compared to normal storage density and efficiency (25 °C) in x = 4 composition.

Self-Clamping Using Electrodes

It is often difficult to create a pre-stress in ceramic capacitors, especially when they form an integral part of an integrated circuit. From the literature it is evident that a compressive stress can be utilized to greatly improve the performance of AFE based capacitors [37,43,58,76,78,89]. Thus, in order to overcome the limitation of applying pre-stresses an alternative method has been proposed, known as self-confinement, using patterned electrodes. The mechanism has been explained using Figure 4. Figure 4a,b describe the symmetrical arrangements of full and half electrodes in a circular parallel plate ceramic capacitor.

When an electric field is applied to this arrangement, the fully electroded area undergoes volumetric expansion as a consequence of the AFE to FE phase transformation. However, un-electroded section remains unaltered, leading to a strain mismatch between different volumes of the ceramic (see Figure 4b); which is similar to inducing a compressive pre-stress within the material. One of the notable attempts in this regard is that reported by Young et al. [47] who used a concentric patterned electrode to induce an effective compressive stress of 30 MPa in the electroded portion of the disk; this helped to increase the energy storage density. The study was supported by a phase-field model and was carried out using bulk samples of $Pb_{0.99}Nb_{0.02}[(Zr_{0.57}Sn_{0.43})_{1-y}Ti_y]_{0.98}O_3$ AFE ceramics, the P-E loops are shown in Figure 4c [47]. When compared to symmetrically coated samples, an increase of ~10% in energy density was observed from 1.19 to 1.30 J/cm^3 for y = 0.06 samples.

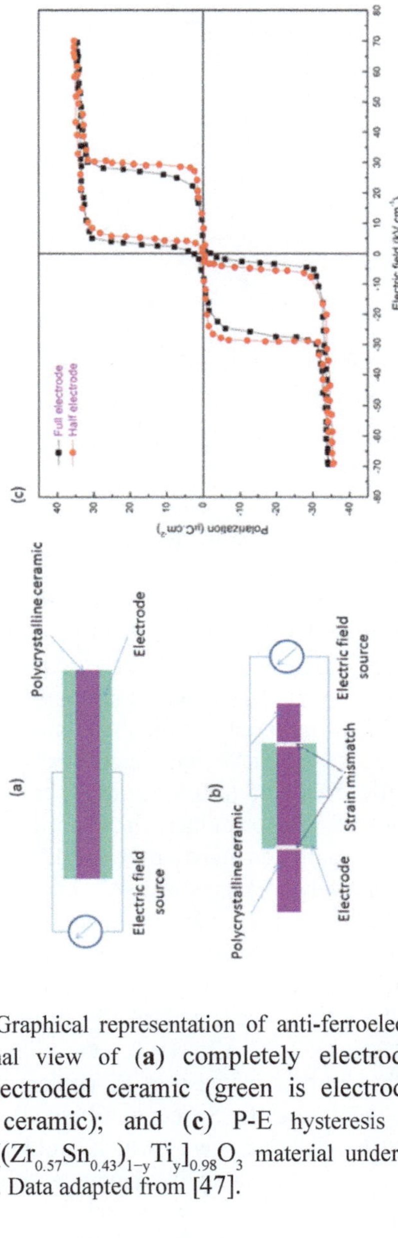

Figure 4. Graphical representation of anti-ferroelectric ceramics' cross-sectional view of (**a**) completely electroded ceramic; (**b**) half electroded ceramic (green is electrode (hatched), purple is ceramic); and (**c**) P-E hysteresis loops for of $Pb_{0.99}Nb_{0.02}[(Zr_{0.57}Sn_{0.43})_{1-y}Ti_y]_{0.98}O_3$ material under induced self-confinement. Data adapted from [47].

ENHANCING ENERGY STORAGE CAPACITY IN ANTI-FERROELECTRIC MATERIALS: CHEMICAL ROUTES

Glass Incorporation and Internal Clamping

Another effective method to increase the energy storage density within AFE ceramics is to incorporate chemically inert glass into the ceramic matrix. Glasses have been added in ferroelectric ceramics in order to reduce the sintering temperature using a viscous sintering approach [90,91]. Glasses have a high electric breakdown strength [21,92,93,94] and careful selection and addition of glasses in ferroelectric ceramics can enhance their energy storage properties. This technique is especially effective if the glass phase resides at the grain boundaries. Grain boundaries in a bulk ceramic are physical discontinuities and act as points of stress concentration owing to the high defect concentration existing on the surface of the grains. Most dielectric failures in ferroelectrics are an indirect consequence of defect concentration at grain boundaries [95] and if the glass can fill the grain boundary discontinuity, it can enhance the electric breakdown strength. However, glass additions can also drastically affect shape, size and growth rate of grains in the ceramics which are responsible for controlling various physical properties [96]. The addition of a glass phase can also produce the required mechanical confinement. Figure 5 shows the polarization behavior of a ferroelectric ceramic (Figure 5a,c) and ferroelectric with added glass (Figure 5b,d) under an electric field. In this situation, domains are also oriented in the direction of applied field. However, they possess some degree of independent orientation which depends on glass content and crystal structure, such as rhombohedral/tetragonal, of the ceramic composite. Internal clamping of domains can be induced due to the presence of glass. This is due to the fact that under an applied electric field, AFE ceramics are strained due to a FE phase transformation. However, glass is non-ferroelectric and remains uneffected and an internal stress is generated between the glass and ceramics phases. Hence, it can be concluded that the addition of a glass leads to ferroelastic domain switching away from the applied electric field. The orientation is in the energetically preferred direction(s) and ultimately reduces the number of switching sites available which helps reduce dielectric losses. Figure 5e shows a typical P-E loop for CdO-Bi_2O_3-PbO-ZnO-$A_{12}O_3$-B_2O_3-SiO_2glass addition and energy storage performance of $Pb_{0.97}La_{0.02}(Zr_{0.56}Sn_{0.35}Ti_{0.09})O_3$ [60]. The glass was added to serve the dual purpose of lowering the sintering temperature and improving the dielectric breakdown strength of the fabricated samples. However, it was noted that with increasing glass content the electric field required for AFE to FE phase transformation also increased, while ΔE decreased simultaneously; this may be a direct consequence of the internal clamping phenomenon. Glass additions were used to lower the sintering temperature from ~1300 to 1050 °C where high densities (>93%) were achieved due to viscous sintering and a reduction of grain size. Furthermore, the incorporation of glass was reported to yield several direct benefits such as decreasing dielectric loss and improving discharge efficiency. The P-E loops of glass containing ceramics also showed a progressive lack of squareness and slanted loops were obtained. It was observed that the sample containing 4 wt % glass and sintered at 1130 °C displayed a large energy storage density of 3.3 J/cm^3, with a high discharge efficiency of 80%.

In other work, glass additions have also been investigated by Chen et al. who fabricated 50 µm-thick $(Pb_{0.97}La_{0.02})(Zr_{0.92}Sn_{0.05}Ti_{0.03})O_3$ antiferroelectric (AFE) thick films with different amounts of $0.8PbO$-$0.2B_2O_3$ glass additions [61]. The films were fabricated by a screen printing method on an alumina substrate with a platinum electrode. It was reported that the incorporation of a glass phase helped to enhance the dielectric breakdown strength of the thick films which in turn yielded higher energy storage densities. Upon incorporation of 3 wt % glass, the films were able to withstand an electric field intensity of 475 kV/m with a reported energy density of 7.4 J/cm^3 and a discharge efficiency of 55%.

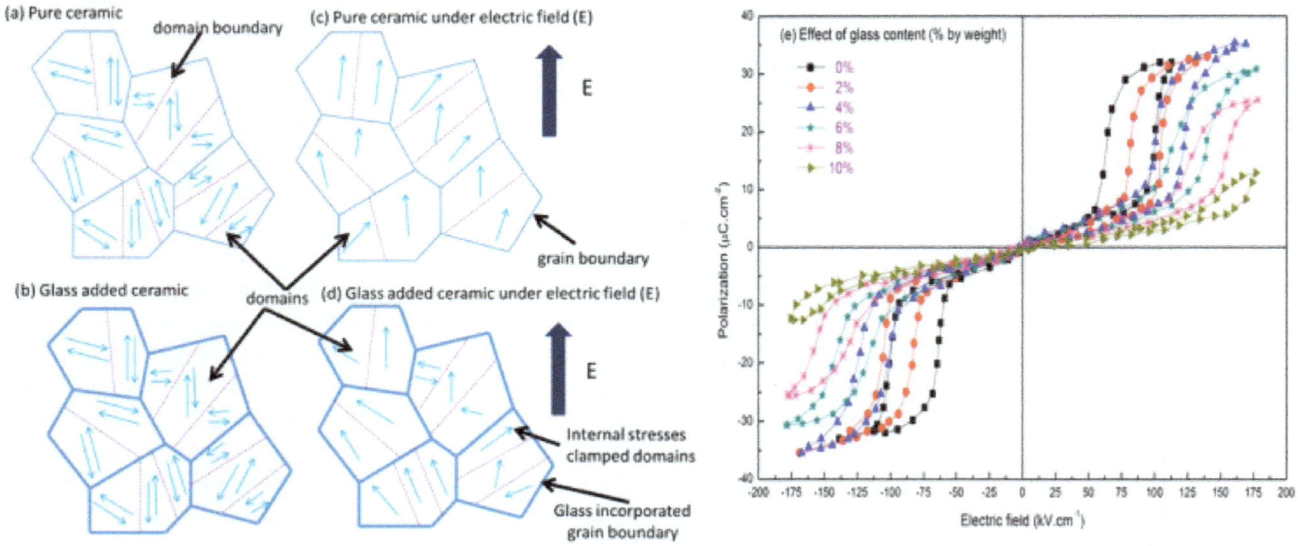

Figure 5. The effect of internal clamping at a microscopic level due to glass incorporation, in anti-ferroelectric material for (**a**) pure ceramic; (**b**) ceramic with glass incorporation; (**c**) pure ceramic under electric field; and (**d**) glass-ceramic composite under electric field and their domain behavior; (**e**) P-Ehysteresis loops for $Pb_{0.97}La_{0.02}(Zr_{0.56}Sn_{0.35}Ti_{0.09})O_3$ glass ceramic composite [60].

Furthermore, the saturation polarization was improved with up to 3 wt % glass addition with a pristine film having a maximum polarization of 26.2 μC/cm^2 while a 3 wt % glass film having a higher polarization of 34.8 μC/cm^2. Hao et al. have reported the effect of glass addition on microstructure and energy storage characteristics on $(Pb_{0.97}La_{0.02})(Zr_{0.97}Ti_{0.03})O_3$ AFE thick films [57]. They used $PbO\text{-}B_2O_3\text{-}SiO_2\text{-}ZnO$ glass ranging from 0 to 5 wt % and fabricated the composites using a screen printing method. The addition of glass was reported to favor the formation of a denser microstructure. Samples containing 3 wt % glass was reported to possess the highest average breakdown strength of 581 kV/cm, which is 140% larger than pristine films. The dielectric constant of the composite was also improved compared to pure perovskite phase and an energy density of 3.1 J/cm^3 was reported, which is more than twice that of a pure ceramic sample. However, the discharge efficiency was observed to decrease with increasing electric field intensity and dropped from a maximum value of 72% at 258 kV/cm to 35% at 581 kV/cm. Thus, the application of a high intensity electric field is not necessarily advantageous for obtaining improved energy density. Glass addition was also reported to impart a good thermal stability to the ceramics within the temperature range of 20 to 120 °C which is also an additional benefit of glass incorporation within the ceramic.

Doping and Chemical Modifications

Tailoring of the material chemistry in the form of solid-solutions, heterovalent and isovalent doping has been extensively reported in the literature to tune the response of any functional material [97,98,99,100,101,102]. The same has also been proposed for enhancing the energy storage characteristics in various AFE materials, where doping has been proven highly effective. In a study by Jiang et al. they investigated the effect of Zr:Sn ratio on the energy storage characteristics of lead lanthanum zirconate stannate titanate AFE ceramics having the nominal composition of $(Pb_{0.87}Ba_{0.1}La_{0.02})(Zr_xSn_{0.95-x}Ti_{0.05})O_3$ (PBLZST, x = 0.6–0.8) [65]. The ceramics were fabricated through the conventional solid-state reaction technique. It was observed that the composition corresponding to x = 0.675 achieved a high value of saturation polarization of 43.5 μC/cm^2 with a correspondingly large energy storage density of 2.05 J/cm^3 and energy efficiency of 68.5% (1.4 J/cm^{3}of recoverable energy). This value was approximately seven times larger than other compositions reported within the same study. It was reported that when the Zr:Sn ratio increased from 0.625 to 0.675, especially from 0.65 to 0.675, the B sites are occupied by the Zr^{4+} ions of a larger radius with a noticeable increase of the unit cell volume and an enlargement of the B site ions' activity space and relative displacement. This in turn leads to the increase of saturation polarization. In contrast to the above result, when the Zr content increased from 0.675 to 0.775, the unit cell volume increases, the increased unit cell volume is so small that the activity space and relative displacement of B site ions are decreased [65]. As a result, saturation polarization decreases with a further increase of Zr content. It was also observed that increasing the Sn content helped to lower the polarization back-switching energy which could help increase the energy efficiency. This study by Jiang et al. is an excellent example of how modification in the relative concentration of solid solutions can increase the energy density. A similar investigation was performed by Chen et al.who described the composition dependent energy storage performance in $Pb_{0.97}La_{0.02}(Zr_xSn_{0.95-x}Ti_{0.05})O_3$ (PLZST) ($0.5 < x < 0.9$) AFE thick films [67]. The films displayed tetragonal and orthogonal AFE phases, depending upon the chemical composition. The effect of Sn content and temperature on the dielectric losses was investigated and as the Sn content was increased, the electric field required for AFE to FE transformation decreased and the energy storage density increased. The maximum energy storage density of 5.6 J/cm^3 (30 °C) and 4.7 J/cm^3 (120 °C) with corresponding energy efficiency of 67% and 73% were obtained in $Pb_{0.97}La_{0.02}(Zr_{0.5}Sn_{0.45}Ti_{0.05})O_3$ ceramic.

With respect to the same family of materials, Xu et al. reported the effect of Ba content on the stress sensitivity of AFE to FE phase transition in $(Pb_{0.94-x}La_{0.04}Ba_x)[(Zr_{0.60}Sn_{0.40})_{0.84}Ti_{0.16}]O_3$ ceramics [58]. Investigations were carried out by monitoring the electric field induced polarization and longitudinal strains under various compressive pre-stresses. The incorporation of Ba could significantly supress the stress sensitivity of the phase transition. It was also shown that the incorporation of Ba at the "A" site of the perovskite structure could reduce the transition field hysteresis; i.e., the energy losses generated from repeated transitioning between AFE and FE states. This could further increase the energy and discharge efficiency. The addition of Ba disrupts large ferroelastic AFE domains, reduces the lattice distortion, and promotes the relaxor behavior; Ba therefore suppresses the ferroelastic contribution and leads to the phase transitions becoming insensitive to compressive stresses.

In a study by Ye et al. the modification of $0.80Bi_{0.5}Na_{0.5}TiO_3$-$0.20Bi_{0.5}K_{0.5}TiO_3$ lead-free AFE ceramics has been examined by doping with $SrTiO_3$ (ST) [66]. Although the study is reported as doping, the extremely large values (over 25 mol %) of incorporation could be aptly described as a solid-solution. The ceramics were prepared by a two-step sintering procedure and the effect of first sintering temperature and ST incorporation was investigated. The addition of 25 mol % ST could induce a double hysteresis loop (as in Figure 1d), which also significantly improved the energy storage characteristics with a maximum value of 0.84 J/cm³. The energy density could be further increased by optimizing the sintering temperature. It was observed that when the first sintering temperature is maintained at 1190 °C, the energy density could be further improved to a maximum value of 0.97 J/cm³. The incorporation of ST also led to an increase in the density of the sintered samples, which also improved the dielectric breakdown strength of the ceramics.

Wang et al. discussed the effect of Zr incorporation into the lead-free $[(Bi_{0.5}Na_{0.5})_{0.94}Ba_{0.06}]La_{1-x}Zr_xTiO_3$ AFE system [64]. Zr incorporation improved several ferroelectric attributes including polarization and energy density. A peak saturation polarization of 37.5 μC/cm² was attained, while a peak energy density of 1.58 J/cm³ was observed for samples containing x = 0.02 Zr. The use of Zr doping helped to decrease both the remnant and saturation polarization which led to a reduced squareness of the hysteresis loops. It was noted that the slanted hysteresis loops thus obtained helped to increase the functional endurance of the ceramics against repeated application and removal of the actuating electric field.

The effect of Ba and Sr co-doping on the energy storage density of PLZST bulk ceramics having composition $(Pb_{0.85}Ba_{0.08}Sr_{0.03}La_{0.03})(Zr_{0.74}Sn_{0.22}Ti_{0.04})$ was studied by Wang et al. [63]. They reported that co-doping with Ba and Sr in a specific ratio could simultaneously lower the Curie temperature to 40 °C and transitioning hysteresis (ΔE) while also increasing the dielectric breakdown strength to 100 kV/cm; this resulted in a high energy density of 1.2 J/cm³ which was several times higher than the pristine composition. Co-doping was attempted to reduce squareness of the hysteresis loops and obtain slanted loops at a higher electric field. The effect created was similar to that demonstrated by glass incorporation (Figure 5e) to obtain slanted loops. However, no direct evidence of higher dielectric breakdown strength was provided and conclusions were only based on the slimness of the hysteresis loops, which could supposedly endure higher electric field.

Chemical modification in the form of doping and solid solutions has also been reported to influence the lattice parameters of AFE and field induced FE phase. This also has a direct bearing on the energy density of the concerned material. For example, in a study by Gao et al.investigations have been made with respect to the c/a (lattice parameters) ratio on the energy storage density on lead-free (0.9 − x) $Bi_{0.5}Na_{0.5}TiO_3$-$xBaTiO_3$-$0.1K_{0.5}Na_{0.5}NbO_3$ (BNT-BT-KNN) bulk ceramics [103]. The samples were fabricated using conventional solid-state reaction technique having x = 0.060, 0.063, 0.066 and 0.069 compositions. X-ray diffraction was used to determine the lattice parameters of the sintered samples which was then correlated with the energy density obtained from P-E hysteresis measurements taken at an electric field of 5 kV/mm. It was observed that compositions having x = 0.063 had the largest c/a ratio of 0.70887 and had a correspondingly large polarization switching of 20.6 μC/cm², with a maximum energy density of 0.424 J/cm³. It was reasoned that by adjusting the BNT/BT ratio in the given compositions, the cell parameters of the dominating tetragonal phase could be altered. This in turn would result in a greater tilting of the octahedra phase, thereby yielding an enhanced polarization switching.

Chen et al. attempted to fabricate slim-loop ferroelectric ceramics that were focussed for high-power pulse capacitor applications [104]. The researchers systematically investigated the effect of partial replacement of A and B site ions in $Pb(Zr,Sn,Ti)O_3$ with Ba, La and Nb respectively. This was undertaken in order to reduce the hysteresis losses, remnant, and coercive parameters, respectively so that a high energy density could be obtained. The bulk ceramics were prepared by a conventional solid-state reaction technique. It was also observed that doping imparted excellent fatigue resistance as the stored charge was observed to degrade very slightly even after the application of 1000 cycles (from 230.49 to 192.96 μC). The PZST42/40/18 ceramics reportedly possessed a maximum energy density of 0.410J/cm³ at 6kV/mm while having a discharge efficiency of 76% after a thousand cycles.

Zhang et al. attempted to increase the energy density by fabricating $(Pb_{0.858}Ba_{0.1}La_{0.02}Y_{0.008})(Zr_{0.65}Sn_{0.3}Ti_{0.05})O_3$-$(Pb_{0.97}La_{0.02})(Zr_{0.9}Sn_{0.05}Ti_{0.05})O_3$ (PBLYZST-PLZST) AFE ceramics with varying phase ratios [68]. The variations were investigated with respect to structure, dielectric and energy storage characteristics of the ceramics. It was reported that with increasing PLZST content, the AFE temperature stability of the composites was improved. Composites with 50% PLZST displayed a considerable FE to AFE transition at an electric field of 130 kV/cm and a high polarization of 49.8 μC/cm²; this resulted in an overall energy storage density of 4.65 J/cm³.

In earlier work, Campbell et al. have attempted to improve the overall energy storage characteristics of fabricated device using a stacked multi-layer capacitor design for modified $PbZrO_3$ AFE ceramics [105]. Both theoretical and experimental investigations were carried out with respect to DC measurements. An effective relative permittivity (dielectric constant) of ~4300 could be achieved and the device displayed a maximum energy storage density of 0.5 J/cm³. This was a rather non-conventional and hybrid approach where two different methods of multi-layer capacitor and chemical modification was used in conjunction to obtain higher energy density. Another non-conventional approach was recently reported by Guan et al. where they reported the fabrication and characterization of an AFE-like polymer for high energy density [106]. A poly(vinylidenefluoride-co-trifluoroethylene-co-chlorotrifluoroethylene)-graftpolystyrene (P(VDF-TrFE-CTFE)-g-PS) graft copolymer with 14 wt % PS side chains was synthesized using a solution growth method. The co-polymer was reported to possess "propeller-like" double hysteresis loops corresponding to AFE materials. However, in conventional PVDF and its various co-polymers, double loops have been reported to only exist at lower electric fields. In their work, Guan and co-workers claimed to confine PVDF lamellar crystals with a low polarizability PS interfacial layer. This could allow the AFE phase to be manifested at electric fields as high as 400 MV/m. The proposed method was also thought to be responsible for significantly reducing the hysteresis losses, thereby increasing the effective energy density. A maximum stored energy density of 20 J/cm^3 was observed at an applied electric field of 500 MV/m. A similar approach was reported by Liet al. where PVDF co-polymers were grafted by poly(ethyl methacrylate) (PEMA) in a fashion similar to as described above [107]. The work was aimed at stabilizing the AFE phase at high poling fields (>500 MV/m). Grafting was reported to enhance the discharge energy density while lowering energy loss. The explanation provided by the authors stated that PEMA side chains were responsible for crystalline impediment and a polarization confinement effect. A highest energy density of 14 J/cm^3 was reported at a discharge efficiency of 70% (550 MV/m) for sample containing 22 wt % PMA.

INFLUENCE OF OTHER FACTORS

In addition to the discussions provided above, several other factors are also responsible for increasing the energy density in various morphologies of AFE ceramics. These include, but are not limited to, the effect of orientation, temperature and electric field. Hao et al. reported the energy storage performance of (100)-oriented $Pb_{0.97}La_{0.02}(Zr_{0.95}Ti_{0.05})O_3$ AFE films [62]. The films were prepared by sol-gel process and deposited on a Pt-buffered silicon substrate, having a thickness of approximately 1.7 μm. In this study it was highlighted that the effect of orientation could increase the saturation polarization. Additionally, it was reported that the stored energy density could be scaled with electric field to yield a maximum value of 12.4 J/cm^3 at an electric field of 1120 kV/cm. Although it was not highlighted in the study, our observations also revealed that a significant reduction in hysteresis losses was observed when the temperature was raised from 20 to 100 °C, this is also expected to yield a larger energy storage density.

Ding et al. reported the enhancement of energy storage density in lead-free $0.89Bi_{0.5}Na_{0.5}TiO_3$-$0.06BaTiO_3$-$0.05K_{0.5}Na_{0.5}NbO_3$ ceramics [69]. This was achieved through a two-step sintering technique. The proposed method yielded a fine grain size and more uniform microstructure as opposed to those processed by single-step sintering. The maximum energy storage density obtained was 0.9 J/cm^3, which was 300% larger than conventionally sintered ceramics. Zhang et al.have reported the beneficial effects of spark plasma sintering (SPS) technique for improving energy storage density in $(Pb_{0.858}Ba_{0.1}La_{0.02}Y_{0.008})(Zr_{0.65}Sn_{0.3}Ti_{0.05})O_3$-$(Pb_{0.97}La_{0.02})(Zr_{0.9}Sn_{0.05}Ti_{0.05})O_3$(PBLYZST–PLZST) AFE ceramics [70]. It was reported that SPS is helpful to supress the diffusion behavior between the tetragonal and orthorhombic phases. Consequently, a high energy storage density of 6.4 J/cm^3 was observed for a 50% PLZST sample with a material efficiency of 62.4%.

A unique study by Chen et al. attempted to elucidate the scaling behavior of energy density in $Pb_{0.99}Nb_{0.02}[(Zr_{0.60}Sn_{0.40})_{0.95}Ti_{0.05}]O_3$ AFE bulk ceramics [59]. Investigations were carried out with respect to electric field intensity and AC frequency of operation. The study was especially beneficial for providing scaling laws for relationships between applied electric field and stored energy density. It was observed that in low fields or high fields regions, the energy densities grow gradually with E increasing or f (frequency) decreasing. However, as E approaches a critical value the energy densities rise steeply with both E and f.

Zhang et al. attempted to illustrate the charge-discharge properties of La-modified $Pb(Zr, Sn, Ti)O_3$ ceramic capacitors as a function of varying electric fields [89]. Direct measurements were made for pulse discharge current-time curves under different electric fields. Upon increasing the electric field from 3 to 3.5 kV/mm, large increments in polarization and discharge current were obtained. Under an electric field of 4 kV/mm, the first current peak was found to surpass 1.1 kA, and more than 80% electric charge stored can be released in <65 ns. In addition, the fabricated ceramics were reported to withstand 2000 charge-discharge cycling with no significant degradation of properties.

DISADVANTAGES ASSOCIATED WITH ANTI-FERROELECTRIC MATERIALS

Despite their numerous advantages and superior performance there are a few factors which serve to limit their practical application as energy storage materials. The foremost among them is the large piezoelectric noise. Piezoelectric noise is the unwanted field-driven actuation of a piezoelectric material. When subjected to large electric field intensity AFE materials respond with an AFE to FE phase transformation which is accompanied by a large strain response. For AFE materials this actuation generally results in a strong unwanted vibration which creates the requirement for damping or isolation lest it begin to harm the structure altogether. Hence, one of the biggest concerns in this field has been to curb or reduce the piezoelectric response, especially for applications requiring alternating current voltages.

Another major problem associated with use of AFE materials for energy storage application is hysteresis and performance degradation. Like all ferroelectric materials, AFEs also suffer from polarization fatigue wherein the magnitude of saturation polarization decreases with increasing number of electric cycles. Also known as functional fatigue, it could also simultaneously result in lower dielectric constant and piezoelectric response with a gradual increase in remnant polarization and dielectric loss. Unlike FE materials the fatigue behavior in AFE materials is rather less explored and proper theories exist which can accurately model the onset or advancement of fatigue in AFE ceramics. But despite all their disadvantages, it has been often reported that the fatigue performance of AFE materials is, in general, better than most FE materials. Nevertheless, if these two main disadvantages could be overcome, it would seriously boost the prospects for the development of a practical AFE based capacitor.

CONCLUDING REMARKS

As modern electronic devices aim for simultaneous miniaturization of size and increase in performance, the requirements for high energy density capacitors will continue to grow. It has been argued that among the potential solid dielectric materials, anti-ferroelectric materials possess the best combination of properties for the development of solid-state capacitors for future electronic applications among other areas. The interest in AFE materials stems from the lack of hysteresis (coercive field, remnant polarization, dielectric losses), while also possessing high energy storage density and discharge efficiency. However, despite evident advantages, little has been done to fully harness the potential of AFE materials. In this paper, we have presented a timely review of the important technological advancements that have been made in this field to date. This has been primarily aimed at inspiring interest towards research in the field of energy storage through the use of AFE materials. Discussion has been provided for the origin of anti-ferroelectricity in bulk materials and how this is best employed for energy storage applications. This was followed by discussion of various physical and chemical parameters that can be used for enhancing and tuning the energy storage density in bulk and thick film AFE materials. The foremost physical methods include the use of mechanical confinement (compressive pre-stress) and self-confinement while the primary chemical techniques employ glass additions and doping. After a thorough review of the literature present in this area along with the contemporary advancements recently achieved, the authors are of the opinion that AFE materials possess significant potential for the development efficient solid state capacitors. We have only uncovered a small fraction of the same and rigorous research is needed to expedite successful commercial ventures.

CHALLENGES AND FUTURE PROSPECTS

Despite their evident advantages, much effort is still required to successfully bridge the gap between laboratory demonstrations and large scale commercial production of AFE ceramic capacitors. There remains a lack of suitable lead-free AFE material which can perform to a similar level of the best lead-based AFE ceramics. In the remarkable progress towards the development of lead-free AFE materials, recent work by Shimizu et al. deserve special attention [71]. As is evident from this review, and as other excellent reviews reported in the literature [3,50,53,80,94], there is a need to develop suitable AFE materials. The majority of materials which display useful AFE characteristics are lead-based while the few lead-free compositions perform relatively poorly. Shimizu and co-workers have proposed a novel family of AFE systems which could be potentially employed for various applications including pulse power application and power electronics. These ceramics belong to the nominal composition of $xCaZrO_3$-$(1-x)$ $NaNbO_3$ where $(0 \leq x \leq 0.10)$ and the AFE characteristics can be stabilized in $NaNbO_3$ by lowering the tolerance factor. This was achieved by simultaneous substitution of Zr and Ca ions in place of Na and Nb sites respectively. This helps lower the polarizability and tolerance factor in the solid solution while maintaining charge neutrality. Transmission electron microscopy analysis revealed that in samples containing x= 4 and 5, only AFE domains were observed as opposed to pure (undoped) ceramic which displayed a mixture of AFE and FE phases, at room temperature. Samples having $0.02 \leq x \leq 0.05$ exhibited double hysteresis loops, indicating AFE behavior. The study was also supported by first principle analysis and upon subsequent investigation by our group, we observed that the ceramics possessed an impressive energy storage density of 0.91 J/cm³. These are very promising results and they have pointed towards new fields of investigations which will help us to expand the AFE materials currently available for storage applications.

In recent work, Dong et al. reported the pressure, temperature, and electric field dependence of phase transformations in Nb modified 95/5 $PbZrTiO_3$ bulk ceramics [108]. Here, the bulk ceramics were subjected to electric field hysteresis under the application of varying temperature and applied hydrostatic pressure. An interesting observation was that under a sufficiently high pressure (>300 MPa at 25 °C) a FE to AFE phase transformation can be induced. The double hysteresis loops became more pronounced on increasing the pressure up to 500 MPa while any significant increase in temperature or electric field was observed to hinder the formation of this pressure induced AFE phase. Figure 6 shows the P-E hysteresis loops of the bulk ceramics, obtained as a function of hydrostatic pressure at the temperatures of 25 and 125 °C. In addition, Figure 7 shows the energy storage density of these ceramics as a function of stress and temperature. A maximum energy density of 0.55 J/cm³ was obtained at a temperature of 125 °C for an applied pressure of 450 MPa. The study holds promising results and must be extended to other compositions to see if there is potential to improve the energy storage density in such ceramics.

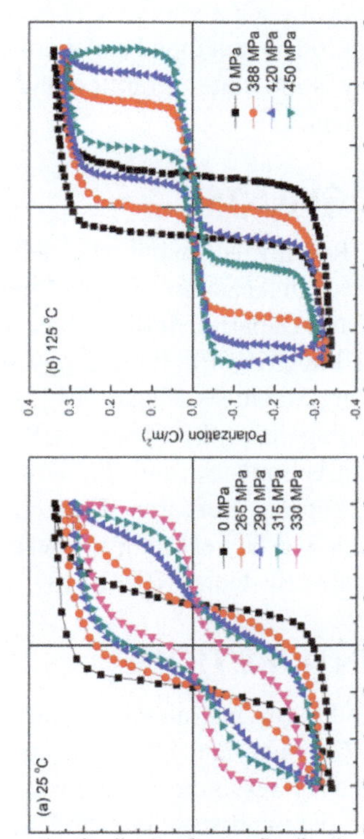

Figure 6. P-E hysteresis loops of the Nb modified 95/5 $PbZrTiO_3$ bulk ceramics, obtained as a function of hydrostatic pressure at the temperatures of (**a**) 25 °C; (**b**) 125 °C. Data adapted from [108].

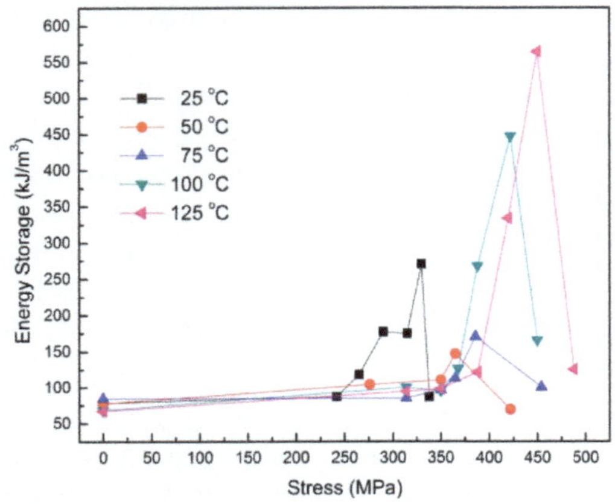

Figure 7. Energy storage density of Nb modified 95/5 PbZ-rTiO_3bulk ceramics as a function of operating temperature and applied hydrostatic pressure. Data adapted from [108].

With the global initiative to phase out lead from ferroelectric and anti-ferroelectric ceramics for technological applications, efforts are required to develop and fabricate new lead-free materials with the required dielectric and AFE properties. Furthermore, extensive research is also needed in order to develop new and economic methods for processing and fabrication of AFE materials which will help to improve their energy storage characteristics. Finally, the AFE to FE transformation is accompanied by large volumetric strain response which is often unwanted behavior and termed as "piezoelectric noise" [109,110]. This can lead to mechanical failure, degradation, and reduces the functional and mechanical endurance of AFE ceramics. Efforts are also required to reduce or remove this unwanted field driven strain response so that the practical appeal of such ceramics can be enhanced. AFE materials have significant untapped potential as high energy and power density solid-state ceramics and present ample scope of improvement.

ACKNOWLEDGMENTS

One of the authors (Rahul Vaish) acknowledges support from the Indian National Science Academy (INSA), New Delhi, India, through a grant by the Department of Science and Technology (DST), New Delhi, under INSPIRE faculty award-2011 (ENG-01) and INSA Young Scientists Medal-2013. Chris R. Bowen acknowledges funding from the European Research Council under the European Union's Seventh Framework Programme (FP/2007-2013)/European Research Council (ERC) Grant Agreement no. 320963 on Novel Energy Materials, Engineering Science and Integrated Systems (NEMESIS).

AUTHOR CONTRIBUTIONS

Aditya Chauhan and Satyanarayan Patel contributed equally to this work. Aditya Chauhan and Stayanarayan Patel collected the data, created the figures and drafted the manuscript. Rahul Vaish analyzed the data and structured the manuscript. Chris R. Bowen has done technical editing and provided valuable conceptual advice.

REFERENCES

1. Rashid, M.H. Power Electronics: Circuits, Devices, and Applications, 3rd ed.; Pearson Education: New Delhi, India, 2003.
2. Karden, E.; Ploumen, S.; Fricke, B.; Miller, T.; Snyder, K. Energy storage devices for future hybrid electric vehicles. J. Power Sources **2007**, 168, 2–11.
3. Hadjipaschalis, I.; Poullikkas, A.; Efthimiou, V. Overview of current and future energy storage technologies for electric power applications. Renew. Sustain. Energ. Rev. **2009**, 13, 1513–1522.
4. Dunn, B.; Kamath, H.; Tarascon, J.M. Electrical energy storage for the grid: A battery of choices. Science **2011**, 334, 928–935.
5. Divya, K.; Østergaard, J. Battery energy storage technology for power systems—An overview. Electr. Power Syst. Res. **2009**, 79, 511–520.
6. Lukic, S.M.; Cao, J.; Bansal, R.C.; Rodriguez, F.; Emadi, A. Energy storage systems for automotive applications. IEEE Trans. Ind. Electron. **2008**, 55, 2258–2267.
7. Zhang, Y.; Feng, H.; Wu, X.; Wang, L.; Zhang, A.; Xia, T.; Dong, H.; Li, X.; Zhang, L. Progress of electrochemical capacitor electrode materials: A review. Int. J. Hydrog. Energy **2009**, 34, 4889–4899.
8. Shukla, A.; Sampath, S.; Vijayamohanan, K. Electrochemical supercapacitors: Energy storage beyond batteries. Curr. Sci. **2000**, 79, 1656–1661.
9. Sharma, R.; Rastogi, A.; Desu, S. Pulse polymerized polypyrrole electrodes for high energy density electrochemical supercapacitor. Electrochem. Commun. **2008**, 10, 268–272.
10. Arbizzani, C.; Mastragostino, M.; Soavi, F. New trends in electrochemical supercapacitors. J. Power Sources **2001**, 100, 164–170.
11. Wang, G.; Zhang, L.; Zhang, J. A review of electrode materials for electrochemical supercapacitors. Chem. Soc. Rev. **2012**, 41, 797–828.
12. Sharma, P.; Bhatti, T. A review on electrochemical double-layer capacitors. Energy Convers. Manag. **2010**, 51, 2901–2912.
13. Cianchetti, M.; Mattoli, V.; Mazzolai, B.; Laschi, C.; Dario, P. A new design methodology of electrostrictive actuators for bio-inspired robotics. Sens. Actuators B Chem. **2009**, 142, 288–297.
14. Gorzkowski, E.; Pan, M.J.; Bender, B.; Wu, C. Glass-ceramics of barium strontium titanate for high energy density capacitors. J. Electroceram **2007**, 18, 269–276.
15. Luo, J.; Du, J.; Tang, Q.; Mao, C. Lead sodium niobate glass-ceramic dielectrics and internal electrode structure for high energy storage density capacitors. IEEE Trans. Electron Dev. **2008**,55, 3549–3554.
16. Rangarajan, B. Nanophase Glass Ceramics for Capacitive Energy Storage. Ph.D. Thesis, The Pennsylvania State University, University Park, PA, USA, December 2009.
17. Dong, G.; Ma, S.; Du, J.; Cui, J. Dielectric properties and energy storage density in ZnO-doped $Ba_{0.3}Sr_{0.7}TiO_3$ ceramics. Ceram. Int. **2009**, 35, 2069–2075.
18. Hara, M.; Gerhold, J. Electrical insulation specification

and design method for superconducting power equipment. Cryogenics **1998**, 38, 1053–1061.

19. Hassenzahl, W.V.; Hazelton, D.W.; Johnson, B.K.; Komarek, P.; Noe, M.; Reis, C.T. Electric power applications of superconductivity. Proc. IEEE **2004**, 92, 1655–1674.
20. Rawlings, R.; Wu, J.; Boccaccini, A. Glass-ceramics: Their production from wastes—A review.J. Mater. Sci **2006**, 41, 733–761.
21. Smith, N.J.; Rangarajan, B.; Lanagan, M.T.; Pantano, C.G. Alkali-free glass as a high energy density dielectric material. Mater. Lett. **2009**, 63, 1245–1248.
22. Chao, S.; Dogan, F. Processing and dielectric properties of TiO_2 thick films for high-energy density capacitor applications. Int. J. Appl. Ceram. Technol. **2011**, 8, 1363–1373.
23. Ahmad, S. Device applications of band-structure-engineered nanomaterials current status and future trend-review. Int. J. Nanoelectron. Mater. **2015**, 8, 129–202.
24. Randall, C.A.; Ogihara, H.; Kim, J.R.; Yang, G.Y.; Stringer, C.S.; Trolier-McKinstry, S.; Lanagan, M. High temperature and high energy density dielectric materials. In Proceedings of the Pulsed Power Conference, Washington, DC, USA, 28 June–5 July 2009; pp. 346–351.
25. Zou, Y.; Han, L.; Yuan, G.; Liu, B.; Zhao, X.; Tian, B.; Wang, J.; Sun, S.; Sun, J.; Meng, X. Enhanced ferroelectric and dielectric properties of the P(VDF-TrFE)/Ag nanoparticles composite thin films. J. Mater. Sci. Mater. **2014**, 25, 3461–3465.
26. Barber, P.; Balasubramanian, S.; Anguchamy, Y.; Gong, S.; Wibowo, A.; Gao, H.; Ploehn, H.J.; Zur Loye, H.C. Polymer composite and nanocomposite dielectric materials for pulse power energy storage. Materials **2009**, 2, 1697–1733.
27. Nyholm, L.; Nyström, G.; Mihranyan, A.; Strømme, M. Toward flexible polymer and paper-based energy storage devices. Adv. Mater. **2011**, 23, 3751–3769.
28. Li, J.; Claude, J.; Norena-Franco, L.E.; Seok, S.I.; Wang, Q. Electrical energy storage in ferroelectric polymer nanocomposites containing surface-functionalized $BaTiO_3$nanoparticles. Chem. Mater. **2008**, 20, 6304–6306.
29. Hao, X.; Zhai, J.; Yao, X. Improved energy storage performance and fatigue endurance of Sr-doped $PbZrO_3$ antiferroelectric thin films. J. Am. Ceram. Soc. **2009**, 92, 1133–1135.
30. Ma, B.; Kwon, D.K.; Narayanan, M.; Balachandran, U. Dielectric properties and energy storage capability of antiferroelectric $Pb_{0.92}La_{0.08}Zr_{0.95}Ti_{0.05}O_3$ film-on-foil capacitors. J. Mater. Res. **2009**, 24, 2993–2996.
31. Choi, K.J.; Biegalski, M.; Li, Y.; Sharan, A.; Schubert, J.; Uecker, R.; Reiche, P.; Chen, Y.; Pan, X.; Gopalan, V. Enhancement of ferroelectricity in strained $BaTiO_3$ thin films. Science **2004**,306, 1005–1009.
32. Wang, Y.; Zhou, X.; Chen, Q.; Chu, B.; Zhang, Q. Recent development of high energy density polymers for dielectric capacitors. IEEE Trans. Dielecte. Electr. Insul. **2010**, 17, 1036–1042.
33. Rangarajan, B.; Jones, B.; Shrout, T.; Lanagan, M. Barium/lead-rich high permittivity glass-ceramics for capacitor applications. J. Am. Ceram. Soc. **2007**, 90, 784–788.
34. Ogihara, H.; Randall, C.A.; Trolier-McKinstry, S. High-energy density capacitors utilizing $0.7BaTiO_3$-$0.3BiScO_3$ Ceramics. J. Am. Ceram. Soc. **2009**, 92, 1719–1724.
35. Haertling, G.H. Ferroelectric ceramics: History and technology. J. Am. Ceram. Soc. **1999**, 82, 797–818.
36. Chu, B.; Zhou, X.; Ren, K.; Neese, B.; Lin, M.; Wang, Q.; Bauer, F.; Zhang, Q. A dielectric polymer with high electric energy density and fast discharge speed. Science **2006**, 313, 334–336.
37. Chauhan, A.; Patel, S.; Vaish, R. Mechanical confinement for improved energy storage density in BNT-BT-KNN lead-free ceramic capacitors. AIP Adv. **2014**, 4, 087106.
38. Setter, N.; Damjanovic, D.; Eng, L.; Fox, G.; Gevorgian, S.; Hong, S.; Kingon, A.; Kohlstedt, H.; Park, N.; Stephenson, G. Ferroelectric thin films: Review of materials, properties, and applications. J. Appl. Phys **2006**, 100, 051606.
39. Ramesh, R.; Spaldin, N.A. Multiferroics: Progress and prospects in thin films. Nat. Mater.**2007**, 6, 21–29.
40. Chu, B.; Zhou, X.; Neese, B.; Zhang, Q.; Bauer, F. Relaxor ferroelectric poly(vinylidene fluoride-trifluoroethylene-chlorofluoroethylene) terpolymer for high energy density storage capacitors. IEEE Trans. Dielecte. Electr. Insul. **2006**, 13, 1162–1169.
41. Ortega, N.; Kumar, A.; Scott, J.; Chrisey, D.B.; Tomazawa, M.; Kumari, S.; Diestra, D.; Katiyar, R. Relaxor-ferroelectric superlattices: High energy density capacitors. J. Phys. Condens. Matter. **2012**, 24, 445901.
42. Wawrzała, P.; Korzekwa, J. Charge-discharge properties of PLZT x/90/10 ceramics.Ferroelectrics **2013**, 446, 91–101.
43. Patel, S.; Chauhan, A.; Vaish, R. Enhancing electrical energy storage density in anti-ferroelectric ceramics using ferroelastic domain switching. Mater. Res. Express **2014**, 1, 045502.
44. Kittel, C. Theory of antiferroelectric crystals. Phys. Rev. **1951**, 82, 729–732.
45. Lieb, E.H. Exact solution of the F model of an antiferroelectric. Phys. Rev. Lett. **1967**, 18, 1046–1048.
46. Patel, S.; Chauhan, A.; Vaish, R. A technique for giant mechanical energy harvesting using ferroelectric/antiferroelectric materials. J. Appl. Phys. **2014**, 115, 084908.
47. Young, S.; Zhang, J.; Hong, W.; Tan, X. Mechanical self-confinement to enhance energy storage density of antiferroelectric capacitors. J. Appl. Phys. **2013**, 113.
48. Gao, F.; Dong, X.; Mao, C.; Liu, W.; Zhang, H.; Yang, L.; Cao, F.; Wang, G. Energy-storage properties of $0.89Bi_{0.5}Na_{0.5}TiO_3$-$0.06BaTiO_3$-$0.05K_{0.5}Na_{0.5}NbO_3$ lead-free anti-ferroelectric ceramics. J. Am. Ceram. Soc. **2011**, 94, 4382–4386.

49. Haertling, G.H. PLZT electrooptic materials and applications—A review. Ferroelectrics **1987**,75, 25–55.

50. Hao, X.; Zhai, J.; Kong, L.B.; Xu, Z. A comprehensive review on the progress of lead zirconate-based antiferroelectric materials. Prog. Mater. Sci. **2014**, 63, 1–57.

51. Hao, X. A review on the dielectric materials for high energy-storage application. J. Adv. Dielectr. **2013**, 3, 1330001.

52. Liu, H.; Dkhil, B. A brief review on the model antiferroelectric $PbZrO_3$ perovskite-like material. Z. Kristallogr. Cryst. Mater. **2011**, 226, 163–170.

53. Panda, P. Review: Environmental friendly lead-free piezoelectric materials. J. Mater. Sci.**2009**, 44, 5049–5062.

54. Castelli, I.E.; Jacobsen, K.W. Designing rules and probabilistic weighting for fast materials discovery in the Perovskite structure. Model. Simul. Mater. Sci. Eng. **2014**, 22, 055007.

55. Kresin, V.Z.; Knight, W. Pair Correlations in Many-Fermion Systems, 1st ed.; Springer, PlenumPress: NewYork, NY, USA, 1998.

56. Chauhan, A.; Patel, S.; Vaish, R. Effect of directional mechanical confinement on the electrical energy storage densityin $68Pb(Mn_{1/3}Nb_{2/3})O_{3-3}2PbTiO_3$ single crystals. Ferroelectrics **2015**,478, 40–53.

57. Hao, X.; Wang, P.; Zhang, X.; Xu, J. Microstructure and energy-storage performance of PbO-B_2O_3-SiO_2-ZnO glass added $(Pb_{0.97}La_{0.02})(Zr_{0.97}Ti_{0.03})O_3$ antiferroelectric thick films. Mater. Res. Bull. **2013**, 48, 84–88.

58. Xu, Y.; Guo, H.; Liu, X.; Feng, Y.; Tan, X. Effect of Ba content on the stress sensitivity of the antiferroelectric to ferroelectric phase transition in (Pb,La,Ba,)(Zr,Sn,Ti) O_3 ceramics. J. Am. Ceram. Soc. **2014**, 97, 206–212.

59. Chen, X.; Cao, F.; Zhang, H.; Yu, G.; Wang, G.; Dong, X.; Gu, Y.; He, H.; Liu, Y. Dynamic hysteresis and scaling behavior of energy density in $Pb_{0.99}Nb_{0.02}(Zr_{0.60}Sn_{0.40})_{0.95}Ti_{0.05}O_3$antiferroelectric bulk ceramics. J. Am. Ceram. Soc. **2012**, 95, 1163–1166.

60. Chen, S.; Yang, T.; Wang, J.; Yao, X. Effects of glass additions on the dielectric properties and energy storage performance of $Pb_{0.97}La_{0.02}(Zr_{0.56}Sn_{0.35}Ti_{0.09})O_3$ antiferroelectric ceramics. J. Mater. Sci. Mater. **2013**, 24, 4764–4768.

61. Chen, S.; Yang, T.; Wang, J.; Yao, X. Effects of glass additions on energy storage performance of $(Pb_{0.97}La_{0.02})(Zr_{0.92}Sn_{0.05}Ti_{0.03})O_3$ antiferroelectric thick films. J. Adv. Dielectr. **2013**, 3, 1350012.

62. Hao, X.; Yue, Z.; Xu, J.; An, S.; Nan, C.W. Energy-storage performance and electrocaloric effect in (100)-oriented $Pb_{0.97}La_{0.02}(Zr_{0.95}Ti_{0.05})O_3$ antiferroelectric thick films. J. Appl. Phys.**2011**, 110, 064109.

63. Wang, J.; Yang, T.; Chen, S.; Li, G. High energy storage density performance of Ba, Sr-modified lead lanthanum zirconate titanate stannate antiferroelectric ceramics. Mater. Res. Bull. **2013**, 48, 3847–3849.

64. Wang, Y.; Lv, Z.; Xie, H.; Cao, J. High energy-storage properties of $(Bi_{1/2}Na_{1/2})_{0.94}Ba_{0.06}La_{(1-x)}Zr_xTiO_3$ lead-free anti-ferroelectric ceramics. Ceram. Int. **2014**, 40, 4323–4326.

65. Jiang, S.; Zhang, L.; Zhang, G.; Liu, S.; Yi, J.; Xiong, X.; Yu, Y.; He, J.; Zeng, Y. Effect of Zr: Sn ratio in the lead lanthanum zirconate stannate titanate anti-ferroelectric ceramics on energy storage properties. Ceram. Int. **2013**, 39, 5571–5575.

66. Ye, J.; Liu, Y.; Lu, Y.; Ding, J.; Ma, C.; Qian, H.; Yu, Z. Enhanced energy-storage properties of $SrTiO_3$ doped $(Bi_{1/2}Na_{1/2})TiO_3$-$(Bi_{1/2}K_{1/2})TiO_3$ lead-free antiferroelectric ceramics. J. Mater. Sci. Mater. **2014**, 25, 4632–4637.

67. Chen, S.; Wang, X.; Yang, T.; Wang, J. Composition-dependent dielectric properties and energy storage performance of $(Pb,La)(Zr,Sn,Ti)O_3$ antiferroelectric ceramics. J. Electroceram.**2014**, 32, 307–310.

68. Zhang, L.; Jiang, S.; Fan, B.; Zhang, G. High energy storage performance in $(Pb_{0.858}Ba_{0.1}La_{0.02}Y_{0.008})(Zr_{0.65}Sn_{0.3}Ti_{0.05})O_3$-$(Pb_{0.97}La_{0.02})(Zr_{0.9}Sn_{0.05}Ti_{0.05})O_3$ anti-ferroelectric composite ceramics. Ceram. Int. **2015**, 41, 1139–1144.

69. Ding, J.; Liu, Y.; Lu, Y.; Qian, H.; Gao, H.; Chen, H.; Ma, C. Enhanced energy-storage properties of $0.89Bi_{0.5}Na_{0.5}TiO_3$-$0.06BaTiO_3$-$0.05K_{0.5}Na_{0.5}NbO_3$ lead-free anti-ferroelectric ceramics by two-step sintering method. Mater. Lett. **2014**, 114, 107–110.

70. Zhang, L.; Jiang, S.; Fan, B.; Zhang, G. Enhanced energy storage performance in $(Pb_{0.858}Ba_{0.1}La_{0.02}Y_{0.008})(Zr_{0.65}Sn_{0.3}Ti_{0.05})O_3$-$(Pb_{0.97}La_{0.02})(Zr_{0.9}Sn_{0.05}Ti_{0.05})O_3$ anti-ferroelectric composite ceramics by Spark Plasma Sintering. J. Alloy Compd. **2015**, 622, 162–165.

71. Shimizu, H.; Guo, H.; Reyes-Lillo, S.E.; Mizuno, Y.; Rabe, K.M.; Randall, C.A. Lead-free antiferroelectric: $xCaZrO_3$-$(1 - x)NaNbO_3$ system ($0 \leq x \leq 0.10$). Dalton Trans. **2015**, 44, 10763–10772.

72. Borkar, H.; Singh, V.N.; Singh, B.P.; Tomar, M.; Gupta, V.; Kumar, A. Room temperature lead-free relaxor–antiferroelectric electroceramics for energy storage applications. RSC Adv.**2014**, 4, 22840–22847.

73. Scott, J.; Melnick, B.; McMillan, L.; de Araujo, C.P. Dielectric breakdown in high-ε films for ULSI DRAMs. Integr. Ferrorelectr. **1993**, 3, 225–243.

74. Lee, F.Y.; Jo, H.R.; Lynch, C.S.; Pilon, L. Pyroelectric energy conversion using PLZT ceramics and the ferroelectric–ergodic relaxor phase transition. Smart Mater. Struct. **2013**, 22, 025038.

75. Fletcher, N.; Hilton, A.; Ricketts, B. Optimization of energy storage density in ceramic capacitors. J. Phys. D Appl. Phys. **1996**, 29, 253.

76. Xu, Z.; Feng, Y.; Zheng, S.; Jin, A.; Wang, F.; Yao, X. Phase transition and dielectric properties of La-doped $Pb(Zr,Sn,Ti)O_3$ antiferroelectric ceramics under hydrostatic pressure and temperature. J. Appl. Phys. **2002**, 92, 2663–2667.

77. Zhang, N.; Liao, L.; Feng, Y.; Xu, Z. Temperature dependence of electrical properties of lead lanthanum zirconate stannate titanate ceramics. Ferroelectrics **2010**, 409, 27–32.

78. Xu, Y.; Yan, Y.; Young, S.E.; Feng, Y. Influence of

perpendicular compressive stress on the phase transition behavior in (Pb,La,Ba,)(Zr,Sn,Ti)O_3 antiferroelectric ceramics. Ceram. Int.**2015**, 42, 721–726.

79. Ranjan, R.; Pandey, D. Antiferroelectric phase transition in $(Sr_{1-x}Ca_x)TiO_3$: II. X-ray diffraction studies. J. Phys. Condens. Matter. **2001**, 13.

80. Tan, X.; Frederick, J.; Ma, C.; Aulbach, E.; Marsilius, M.; Hong, W.; Granzow, T.; Jo, W.; Rödel, J. Electric-field-induced antiferroelectric to ferroelectric phase transition in mechanically confined $Pb_{0.99}Nb_{0.02}(Zr_{0.57}Sn_{0.43})_{0.94}Ti_{0.06}$ $_{0.98}O_3$. Phys. Rev. B **2010**, 81.

81. Dong, W.D.; Finkel, P.; Amin, A.; Lynch, C.S. Giant electro-mechanical energy conversion in 011 cut ferroelectric single crystals. Appl. Phys. Lett. **2012**, 100.

82. Valadez, J.; Sahul, R.; Alberta, E.; Hackenberger, W.; Lynch, C. The effect of a hydrostatic pressure induced phase transformation on the unipolar electrical response of Nb modified 95/5 lead zirconate titanate. J. Appl. Phys **2012**, 111.

83. Marsilius, M.; Frederick, J.; Hu, W.; Tan, X.; Granzow, T.; Han, P. Mechanical confinement: An effective way of tuning properties of piezoelectric crystals. Adv. Funct. Mater. **2012**, 22, 797–802.

84. Chen, W.; Lynch, C.S. A micro-electro-mechanical model for polarization switching of ferroelectric materials. Acta Metall. **1998**, 46, 5303–5311.

85. Hwang, S.C.; Lynch, C.S.; McMeeking, R.M. Ferroelectric/ferroelastic interactions and a polarization switching model. Acta Metall. Mater. **1995**, 43, 2073–2084.

86. Patel, S.; Chauhan, A.; Vaish, R. Enhanced energy harvesting in commercial ferroelectric materials. Mater. Res. Express **2014**, 1.

87. Patel, S.; Chauhan, A.; Vaish, R. Analysis of high-field energy harvesting using ferroelectric materials. Energ. Technol. **2014**, 2, 480–485.

88. Peláiz-Barranco, A.; Hall, D. Influence of composition and pressure on the electric field-induced antiferroelectric to ferroelectric phase transformation in lanthanum modified lead zirconate titanate ceramics. IEEE Trans. Ultrason. Ferroelectr. **2009**, 56, 1785–1791.

89. Zhang, H.; Chen, X.; Cao, F.; Wang, G.; Dong, X.; Hu, Z.; Du, T. Charge-discharge properties of an antiferroelectric ceramics capacitor under different electric fields. J. Am. Ceram. Soc.**2010**, 93, 4015–4017.

90. Holand, W.; Beall, G.H. Glass Ceramic Technology, 2nd ed.; John Wiley & Sons: New York, NY, USA, 2012.

91. Shri Prakash, B.; Varma, K.B.R. Effect of the addition of B_2O_3 and BaO-B2O_3-SiO_2 glasses on the microstructure and dielectric properties of giant dielectric constant material $CaCu_3Ti_4O_{12}$. J. Solid State Chem. **2007**, 180, 1918–1927.

92. Frohlich, H. On the theory of dielectric breakdown in solids. Proc. R. Soc. A Math. Phys. **1947**,188, 521–532.

93. Vermeer, J. On the relation between ionic conductivity and breakdown strength of glass.Physica **1956**, 22, 1269–1278.

94. Von Hippel, A.; Maurer, R.J. Electric breakdown of glasses and crystals as a function of temperature. Phys. Rev. **1941**, 59, 820–823.

95. Gerson, R.; Marshall, T.C. Dielectric breakdown of porous ceramics. J. Appl. Phys. **1959**, 30, 1650–1653.

96. Cai, W.; Fu, C.; Lin, Z.; Deng, X. Vanadium doping effects on microstructure and dielectric properties of barium titanate ceramics. Ceram. Int. **2011**, 37, 3643–3650

97. Petnoi, N.; Bomlai, P.; Jiansirisomboon, S.; Watcharapasorn, A. Effects of Nb-doping on the micro-structure and dielectric properties of $(Bi_{0.5}Na_{0.5})TiO_3$ ceramics. Ceram. Int. **2013**, 39, S113–S117.

98. Kumar, A.; Singh, B.; Choudhary, R.; Thakur, A.K. Characterization of electrical properties of Pb-modified $BaSnO_3$ using impedance spectroscopy. Mater. Chem. Phys. **2006**, 99, 150–159.

99. Ramam, K.; Chandramouli, K. Dielectric and piezoelectric properties of combinatory effect of A-site isovalent and B-site acceptor doped PLZT ceramics. Ceram. Silik. **2009**, 53, 189.

100. Aleem, M.; Nawaz, H.; Shuaib, M.; Qaisar, S.; Akbar, M. Piezoelectric and pyroelectric properties of Sr-doped PZT (PSZT) with minor manganese additions. J. Phys. Conf. Ser. **2013**,439.

101. Koduri, R.; Lopez, M. Effect of microstructure on ferroelectric and piezoelectric properties of A-site and B-site modified PLZT. Ferroelectr. Lett. **2007**, 34, 113–123.

102. Goel, P.; Yadav, K. Substitution site effect on structural and dielectric properties of La-Bi modified PZT. J. Mater. Sci. **2007**, 42, 3928–3935.

103. Gao, F.; Dong, X.; Mao, C.; Cao, F.; Wang, G. c/a Ratio-dependent energy-storage density in (0.9 − x) $Bi_{0.5}Na_{0.5}TiO_3$-x$BaTiO_3$-0.1$K_{0.5}Na_{0.5}NbO_3$ Ceramics. J. Am. Ceram. Soc. **2011**, 94, 4162–4164.

104. Chen, X.F.; Dong, X.L.; Wang, G.S.; Cao, F.; Wang, Y.L. Doped Pb(Zr, Sn, Ti)O_3 slim-loop ferroelectric ceramics for high-power pulse capacitors application. Ferroelectrics **2008**, 363, 56–63.

105. Campbell, C.K.; Van Wyk, J.D.; Chen, R. Experimental and theoretical characterization of an antiferroelectric ceramic capacitor for power electronics. IEEE Trans. Compon. Packag. A **2002**,25, 211–216.

106. Guan, F.; Wang, J.; Yang, L.; Tseng, J.-K.; Han, K.; Wang, Q.; Zhu, L. Confinement-Induced high-field antiferroelectric-like behavior in a poly(vinylidene fluoride-co-trifluoroethylene-co-chlorotrifluoroethylene)-graft-polystyrene graft copolymer. Macromolecules **2011**, 44, 2190–2199.

107. Li, J.; Tan, S.; Ding, S.; Li, H.; Yang, L.; Zhang, Z. High-field antiferroelectric behaviour and minimized energy loss in poly(vinylidene-co-trifluoroethylene)-graft-poly(ethyl methacrylate) for energy storage application. J. Mater. Chem. **2012**, 22, 23468–23476.

108. Dong, W.D.; Carlos Valadez, J.; Gallagher, J.A.; Jo, H.R.; Sahul, R.; Hackenberger, W.; Lynch, C.S. Pressure, temperature, and electric field dependence of phase transformations in niobium modified 95/5 lead zirconate

titanate. J. Appl. Phys. **2015**, 117.

109. Jimenez, R.; Hungria, T.; Castro, A.; Jimenez-Rioboo, R. Phase transitions in $Na_{1-x}Li_xNbO_3$solid solution ceramics studied by a new pyroelectric current based method. J. Phys. D: Appl. Phys. **2008**, 41.

110. Dul'kin, E.; Mojaev, E.; Roth, M.; Jo, W.; Granzow, T. Acoustic emission study of domain wall motion and phase transition in $(1 - x - y)Bi_{0.5}Na_{0.5}TiO_3$-$xBaTiO_{3-y}K_{0.5}Na_{0.5}NbO_3$ lead-free piezoceramics. Sci. Mater. **2009**, 60, 251–253.

INDEX